DEEP TIME

Paleobiology's Perspective

A special volume commemorating the 25th anniversary of the journal *Paleobiology*

Edited by
Douglas H. Erwin and Scott L.Wing

Published by
The Paleontological Society

Douglas H. Erwin and Scott L. Wing are research paleobiologists at the National Museum of Natural History, Smithsonian Institution, Washington, D.C.

Copyeditor: Natasha Atkins
Cover Design: Mary Parrish

©2000 by The Paleontological Society
All rights reserved. Published 2000
Printed in the United States of America by Allen Press, Lawrence, Kansas
02 01 00   1 2 3

ISBN: 0-9677554-2-5 (paperback)
ISBN: 0-9677554-3-3 (hardback)
ISSN: 0094-8373

Library of Congress Cataloging-in-Publication Data
Deep Time: *Paleobiology's* Perspective / edited by Douglas H. Erwin and Scott L. Wing.

   p. cm. –(Supplement to Volume 26, number 4 of the journal *Paleobiology*)

   Includes bibliographic references.
   1. Paleobiology. 2. Paleontology. 3. History of Life.
I. Erwin, Douglas H.   II. Wing, Scott L.

Paleobiology (ISSN: 0094-8373) is published quarterly by The Paleontological Society, 810 East 10th St., Lawrence, KS 66044, USA. This book is a supplement to Volume 26, number 4.

Contents

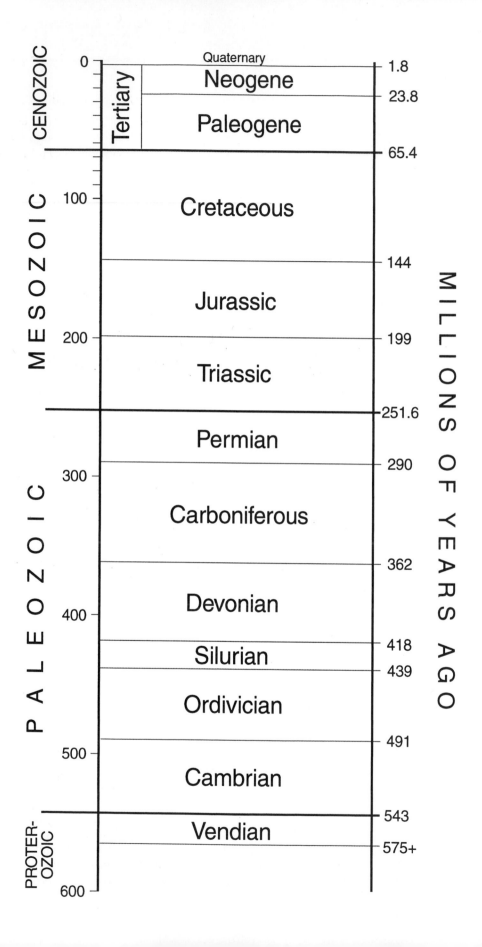

Paleobiology was founded a little more than 25 years ago to provide a forum for the greater integration of paleontology and biology. Over the last quarter century it has fulfilled that role admirably by providing an outlet for contributions on a wide variety of topics. The fifteen invited papers in this special volume represent the diversity of approaches and current activities in the field of paleobiology: the mechanical and functional properties of extinct organisms, the chemical composition of fossils, the implications of fossils for genetics and development, the factors affecting interpretation of the paleontological and stratigraphic record, the influence of environmental change on ecological systems and the evolution of lineages, the use of fossils in reconstructing phylogeny, and the implications of the fossil record for the processes that generate the largest-scale patterns in the history of life. When we solicited these papers from some of the leaders in the field, we asked the authors to provide an overview of recent advances but also to discuss the questions that remain to be answered. The papers in this volume demonstrate the accelerating strength of paleobiology as a distinct discipline as well as the breadth of its integration with both geology and biology.

The diversity of paleobiological research seems almost to defy unification, but underlying all our efforts is the conviction that the history and past behavior of living systems are key to understanding how they work. Deep time is a perspective that paleobiology alone brings to the study of living things and their interactions with one another and with the inorganic world. Like someone learning to speak a new language, biologists have often relied on the present tense, interpreting living systems solely on the basis of their current state. We believe that in the 25 years since *Paleobiology* was founded (and this may be more than a coincidence), more biologists have come to appreciate the importance of paleobiology's historical perspective. To continue the linguistic metaphor, the past tense has become more widely used and many biologists are aware that organisms and ecosystems have a past that may help explain patterns in the present. Frequently, though, these historical explanations refer only to events of the last million years or less, with the related, and curious, tendency to treat the past as a single time plane, like the present. We still need to impress on our fellow biologists that the fossil record can yield more than a snapshot of the past that may explain some aspect of the current biosphere. The next level of sophistication in thinking about the past can only be conveyed by the aptly named imperfect tense: the one we use to speak of continuing action in the past. By integrated study of fossils and the physical and chemical properties of sediments, we generate records (admittedly imperfect) of how complex biological, geochemical, and physical systems behave over periods of tens of thousands to millions of years. Each new time series reveals a particular and no doubt idiosyncratic unfolding of evolutionary and ecological processes during a specific time interval. But generalizations based on increasing numbers of these time series have the potential to reveal the processes that underlie them. This dynamic view of the fossil record is critical to understanding the processes that drive biotic change, and it is still an area where paleobiology is underappreciated. How we move from documenting patterns of change through time to understanding the processes that generated those changes remains one of the great challenges for paleobiology.

The quality and diversity of papers in this volume are a sign of the health of the field and, we think, of the journal as well. The stature of the journal reflects the excellence of the contributors and the efforts of previous editors of *Paleobiology,* whom we acknowledge here: Thomas J. M. Schopf, Ralph G. Johnson, James A. Hopson, J. John Sepkoski Jr., Peter R. Crane, Philip W. Signor, Richard Cowen, Geerat J. Vermeij, David L. Meyer, and Arnold I. Miller. On a more mundane but scarcely less important note, the ability of *Paleobiology* to produce this special twenty-fifth anniversary issue reflects the financial health of the journal, which has been nurtured by our patrons since its formation. They are listed on the first page of every issue. We thank them again here most sincerely.

Much of the energy and insight that made the journal a success during the 1980s and 1990s was provided by J. John Sepkoski Jr. of the University of Chicago. His untimely death in 1999 was a tragedy for the field of paleo-biology. His ideas, methods, and infectious enthusiasm are a lasting legacy, and the questions he posed will continue to challenge paleontologists (and, we hope, authors in *Paleobiology*) for many years to come. We dedicate this twenty-fifth anniversary issue to the memory of J. John Sepkoski Jr.

Douglas H. Erwin
Scott L. Wing
Co-editors

Directionality in the history of life: diffusion from the left wall or repeated scaling of the right?

Andrew H. Knoll and Richard K. Bambach

Abstract.—Issues of directionality in the history of life can be framed in terms of six major evolutionary steps, or megatrajectories (cf. Maynard Smith and Szathmáry 1995): (1) evolution from the origin of life to the last common ancestor of extant organisms, (2) the metabolic diversification of bacteria and archaea, (3) evolution of eukaryotic cells, (4) multicellularity, (5) the invasion of the land and (6) technological intelligence. Within each megatrajectory, overall diversification conforms to a pattern of increasing variance bounded by a right wall as well as one on the left. However, the expanding envelope of forms and physiologies also reflects—at least in part—directional evolution within clades. Each megatrajectory has introduced fundamentally new evolutionary entities that garner resources in new ways, resulting in an unambiguously directional pattern of increasing ecological complexity marked by expanding ecospace utilization. The sequential addition of megatrajectories adheres to logical rules of ecosystem function, providing a blueprint for evolution that may have been followed to varying degrees wherever life has arisen.

Andrew H. Knoll. Botanical Museum, Harvard University, 26 Oxford Street, Cambridge, Massachusetts 02138

Richard K. Bambach. Department of Geological Sciences, Virginia Polytechnic Institute and State University, Blacksburg, Virginia 24061-0420

Accepted: 15 March 2000

Introduction

That evolution has produced organisms of increasing size and complexity through time is not in doubt. A biosphere originally populated exclusively by microorganisms now supports whales, redwoods, and technologically sophisticated humans—as well as microbes of unprecedented diversity. Most biologists and paleontologists agree on the broad pattern of evolutionary history, but the interpretation of this pattern has proven contentious.

Focusing at the molecular level, Christian De Duve (1995) viewed evolutionary pattern as both determinate and directional. The vector of evolution in the structural, informational, and catalytic molecules that make up cells runs from general to specific, from inefficient to efficient, and from low specificity to high. In contrast, Stephen Jay Gould (1996) interpreted the morphological record as an unpredictable product of contingent events, its undeniable pattern principally reflecting the accumulation of variance, with directionality reduced to an inevitable and modest function of evolutionary diffusion away from life's starting point near a bounding "left wall" of minimal size and complexity. In his book *Full House,* Gould wrote, "I believe that the most knowledgeable students of life's history have always sensed the failure of the fossil record to supply the most desired ingredient of Western comfort: a clear signal of progress as measured by some form of steadily increasing complexity for life as a whole through time."

Geerat Vermeij (1999), arguably a most knowledgeable student of life's history, has positioned himself between the extremes epitomized by De Duve and Gould. Vermeij's focus is on bioenergetics, the capacity of organisms to gain nutrients and energy. From this perspective, he is accepting of contingency, diffusive evolution, and bounding walls, but nonetheless espies directionality in paleontological patterns of turnover, with high-energy species predictably replacing those with lower metabolic requirements (see also Bambach 1999).

The principal issue that confronts paleontologists who wish to evaluate such sweeping evolutionary worldviews is that the history of life is unique. How do we estimate the likelihood of events that have occurred only once? In this paper, we explore pathways toward the resolution of divergent interpretations of broad-scale macroevolutionary pattern.

0094-8373/00/2604-0001/$1.00

Framing the Issues

The views of De Duve, Gould, and Vermeij, chosen to illustrate (but not exhaust) the spectrum of opinion on evolutionary pattern, differ most obviously in their levels of biological focus—molecules vs. morphology vs. ecology. This difference alone might account, at least in part, for the divergent interpretations of these authors. For example, determinism at the molecular level could easily stand alongside morphological contingency. Indeed, Vermeij (1999) explicitly sees his and Gould's conceptions as complementary, rather than alternative, explanations of evolutionary pattern.

The viewpoints of our three exemplars differ in another way, however, that demands our attention. Gould's articulation of evolutionary diffusion is essentially a continuum model in which there is an accumulation of diversity among fundamentally similar entities bounded by a left wall, but not one on the right. It is not especially concerned with the processes that give rise to individual entities, nor with patterns of replacement among entities of a particular structural or functional type. De Duve and Vermeij, in contrast, focus explicitly on the processes of natural selection by which the functional biology of individual entities is honed. By extension, Vermeij is concerned with evolutionary turnover within guilds as well as the accumulation of guilds through time. In this view, we have two walls to worry about: the impregnable one on the left, recognized by Gould, and a second one on the right (Sterelny 1999; Sterelny and Griffiths 1999) that limits the evolutionary possibilities of any one grade of structural or biochemical entity but which can be surmounted from time to time by new entities.

Two important steps toward evaluating these worldviews are, thus, straightforward. First, is the metaphor of diffusion from a single bounding wall adequate to capture the history of evolutionary change, or can we identify right walls as well as left that limit evolutionary diversification? If we can, then simple continuum models are not sufficient to describe the history of life. Second, is observed increase in variance driven only by diffusion? If, away from bounding walls, statistical differences

between "left" and "right" trending changes can be recognized, then diffusive models must be supplemented by ones that are directionally "driven" (Stanley 1973; McShea 1994, 1996, 1999; Saunders et al. 1999).

A third, rather different question concerns the entities whose patterns of change we wish to measure. Are features of individuals, such as size or organismic complexity, the best or only available metrics for assessing directionality in the history of life? Or, should we look, as well, at the *interactions* among organisms manifest in ecosystem complexity?

The history of life as a whole may be unique, but some patterns of evolutionary change have been generated more than once. Convergence, a pattern rightly held in high esteem by Simpson (1953), helps us to assay evolutionary likelihood and, therefore, the long-term importance of contingent events (e.g., Conway Morris 1998). Multicellularity, for example, has evolved about a dozen times in distinct lineages of eukaryotic organisms (Buss 1987). Thus, given the existence of eukaryotic cell biology, the probability of evolving multicellularity appears to be high. In contrast, eukaryotic cell organization itself has arisen only once. We can rationalize why eukaryotes might have proven successful in the long term, but cannot in good faith argue that the founding events of eukaryotic biology were probable or improbable. Vermeij (1999) draws on this class of argument in his defense of life's ecological predictability: a consistent pattern of stratigraphic replacement has characterized different groups of plants and animals living in both marine and terrestrial environments.

Six Megatrajectories in the History of Life

We believe that six broad megatrajectories capture the essence of vectorial change in the history of life. The megatrajectories form a logical sequence dictated by the necessity for complexity level N to exist before N+1 can evolve. We do not, however, argue that the presence of one level makes evolution of the next level inevitable. The unique history of life makes such a claim untestable. The six megatrajectories highlighted here are similar to the major steps in evolution evaluated by Maynard Smith and Szathmáry (1995; see also

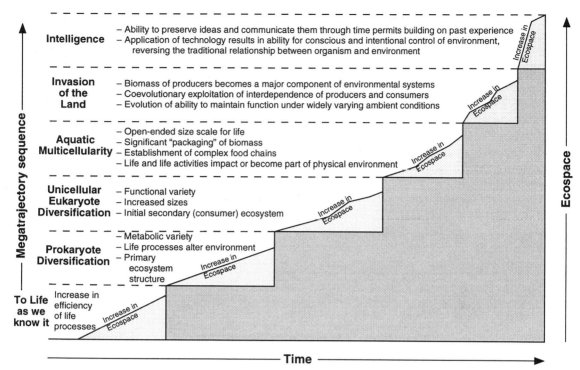

FIGURE 1. Diagram showing the increase in utilization of ecospace that accrued as each megatrajectory was added to the evolving biosphere. For each megatrajectory there is a "right wall" that limits the ecospace that can be realized by evolutionary expansion within that megatrajectory. The first organisms reaching each megatrajectory level realize only a small portion of the new potential ecospace made available to that megatrajectory. Over time both directional and diffusive evolution operate to expand the range of ecospace utilized by each megatrajectory, but eventually the bounding limits are reached. Note also that the increase in utilized ecospace is not uniform but fluctuates in rate and direction. Bounding walls are overcome only by evolutionary breakthroughs that make new dimensions or levels of ecospace available. The styles of the dimensions of ecospace new to each level are noted on the figure adjacent to the title of each megatrajectory.

Stebbins 1969). However, whereas Maynard Smith and Szathmáry stressed genomic complexity, we focus on ecosystem complexity and the multidimensional hypervolume of Hutchinsonian niche space (Hutchinson 1965), which we term *ecospace*. Relevant metrics include the number of interactions in food webs, the pervasiveness of mutualisms, and the number of organisms through which biologically important elements (carbon, nitrogen, phosphorus, and sulfur) pass before being returned to an inorganic state.

In the view offered here, each megatrajectory adds new and qualitatively distinct dimensions to the way life utilizes ecospace (Fig. 1). This complicates the evaluation of complexity in phenotypic evolution (McShea 1996), but underscores the importance of *ecological* complexity in the history of life. The megatrajectories, themselves, are directional

insofar as the descendants of organisms that break through to the next level do not re-evolve the salient features of a previous state. (No eukaryote has produced prokaryotic descendants, and with the possible exception of myxozoans, no multicellular organism has spawned a unicellular lineage.) In this sense, these additive megatrajectories pass McShea's (1999) test of "increase vs. decrease" for directionality.

Megatrajectory 1: From the Origin of Life to the Last Common Ancestor of Extant Life.—The last common ancestor (LCA) of all extant life was a sophisticated microorganism characterized (among other attributes) by DNA, RNA, ribosomes, multiple enzymes to direct transcription and translation as well as metabolism, membranes with embedded proteins to control ionic and molecular transport, and ATP. We remain in substantial ignorance about the origin(s) of living systems, but whatever route or routes led to the emergence of life, the earliest organisms must have been biochemically much simpler than this LCA. We don't know whether the first functional molecules were nucleic acids, polypeptides, or neither of these now-ubiquitous components of cell biology. Nonetheless, there is widespread agreement that early enzymes had broader and less efficient functionality, as well as lower substrate specificity, than their modern counterparts (Lazcano 1994; De Duve 1995). It also seems clear that although the earliest organisms may have contained nucleic acids, they did not synthesize DNA for the efficient storage of molecular information (Joyce 1989).

No record of the primordial biosphere has survived to the present day. There is, however, every reason to view the first evolutionary megatrajectory as truly directional—not simply, and certainly not purely, as an increase in variance. Gene duplication and subsequent specialization (Clarke 1983) as well as lateral transfer of functional gene cassettes (Lawrence 1999) led to increasingly complicated, efficient, and specialized organisms that replaced their simpler forbears. This megatrajectory from the initiation of life processes to life as we know it was driven by selection that favored increasing biochemical efficiency (De Duve 1995). It began with the stabilization of Earth's crust and hydrosphere and may have been completed by the time the earliest known microfossils were deposited 3.46 billion years ago (Schopf and Packer 1987).

Megatrajectory 2: The Metabolic Diversification of Bacteria and Archaea.—At first consideration, the diversification of prokaryotic organisms seems to provide strong evidence for Gould's view of evolution as increase in variance. Anaerobic metabolisms preceded aerobic types, but the advent of aerobiosis did not eliminate earlier physiologies. Rather, diversity *accumulated,* with ecosystem function remaining dependent on anaerobic microorganisms able to cycle carbon, nitrogen, and sulfur through anoxic environments. If we accept the Universal Tree as determined by SSU rRNA gene sequences (Woese 1987; Pace 1997), it further appears that anaerobic chemosynthetic micro-

organisms preceded photosynthetic groups. Yet, again, photosynthesis did not eliminate early chemoautotrophic metabolisms. Rather, photosynthetic organisms enabled highly productive ecosystems to expand beyond the limits of hydrothermal ridges and other habitats where strong redox gradients can be exploited (and where early-branching chemoautotrophs still reside).

The logic of microbial diversification is inextricably tied to ecosystem function, with aerobic and anaerobic, heterotrophic and autotrophic, oxidizing and reducing metabolisms closely linked in biogeochemical cycles. The metabolic products of one organism provide substrates for another, with the consequence that metabolic innovation can fuel diversification throughout a community. The redox complementarity of microbial metabolisms further ensures that materials will be cycled among environments as well as among organisms. For example, the nitrogen and sulfur cycles cannot be completed in the absence of anaerobic organisms living in anoxic habitats.

The close association of microbial metabolism and habitat means that if Earth surface environments have changed directionally, so, too, have microbial diversity and ecosystem complexity. Bacterial and archaeal diversity accumulated as environmental opportunities expanded, especially with the oxidation of the atmosphere and surface ocean (which confined but did not eliminate anoxic habitats) and, later, with the evolution of metabolically exploitable *biological* resources provided by multicellular organisms (Knoll 1994).

Although the metabolic diversification of prokaryotes and its concomitant ecological effects might appear to reflect diffusive increase in variance, a component of truly directional evolution is hidden in this picture. An impressive body of experimental research (e.g., Clarke 1983; Lenski and Travisiano 1994) indicates that bacteria introduced to novel substrates initially metabolize those substrates with extremely low efficiency. Through time, accumulating mutations result in ever more efficient substrate utilization until populations reach a point where the probability of further improvement with continued mutation effectively declines to zero (Lenski and Travisiano 1994). Thus, the earliest organisms to use, say, sulfate in respiration undoubtedly did so with less efficiency than modern sulfate reducers and were eliminated over time. Very likely, such early intervals of directional change were brief relative to subsequent periods of bacterial stasis.

In effect, then, the accumulation of prokaryotic diversity is underpinned by a pattern of directional evolution that shaped each persistent variant. Bacterial and archaeal evolution is a mixture of directional and cumulative evolution, observable at different scales.

A word about size also is appropriate, as it was the focus of Gould's arguments. The "left wall" of prokaryotic cell size is determined by the volume required to contain the minimal complement of genes, proteins, ribosomes, and membranes required for cell function (National Research Council 1999). We don't know whether the earliest microorganisms approximated this minimal size or, like some extant, early-branching archaeans, were (relatively) large (Karl Stetter *in* National Research Council 1999). Among living bacteria, however, small size is predominantly an adaptation for life in oligotrophic and other specific environments and is found most commonly within relatively late branching clades (National Research Council 1999).

Variance in prokaryotic size may have increased in the course of evolution, but not without limit. Bacterial growth depends on the diffusive importation of substrates, oxidants, and reductants into cell interiors, and because of this, cell size is sharply constrained by chemical environment. Most living bacteria have a biovolume equivalent to a sphere 300 nm to a few microns in diameter (Niklas 1992; National Research Council 1999), and only a few prokaryotes, generally chemo- and photoautotrophic bacteria that contain complex, multilayered membranes, exceed 10 microns in diameter. Indeed, at the absolute limit for bacterial size—spherical sulfur-oxidizing cells up to 750 microns in diameter found in episodically anoxic sediments off the coast of southern Africa (Schulz et al. 1999)—the microbes cheat; they are hollow. The key point about bacterial and archaeal cell size is that it

is limited by a right as well as a left wall. In four billion years, bacteria have not been notably successful in scaling either one.

Megatrajectory 3: Evolution of the Eukaryotic Cell.—The statement that no prokaryote has ever scaled the right wall that bounds cell size is, of course, not strictly true. One lineage or, perhaps more likely, one consortium managed the trick, but did so only by evolving a completely new type of biological organization—the eukaryotic cell. The origin of eukaryotic cell organization has long been a subject for debate. Consensus on this issue remains elusive, but an increasing body of data and opinion favors the hypothesis that eukaryotes are fundamentally chimeric. Notably, Martin and Müller (1998; see also Moreira and Lopez-Garcia 1998) have proposed that the domain originated via an ur-symbiosis between a facultatively anaerobic proteobacterium that fermented organic molecules to hydrogen and carbon dioxide and an archaean (possibly methanogenic) able to metabolize these waste products. No matter what triggered the origin of eukaryotic cell organization, it is clear that the principal energy-yielding metabolisms of eukaryotes were acquired by symbiosis from originally free-living bacteria (Whatley et al. 1979; Margulis 1981). Through time, several originally discrete organisms became integrated both functionally and genetically in individual eukaryotic cells. In consequence, eukaryotes cannot be viewed simply as large and complicated bacteria or associations of bacterial cells. Genetically, ultrastructurally, and functionally, they are distinct evolutionary entities (Maynard Smith and Szathmáry 1995).

In general, eukaryotes respire aerobically, and a subset of eukaryotic clades are photosynthetic. Most eukaryotes also have a limited capacity for anaerobic heterotrophy, but only a few (for example, yeast) succeed predominantly as fermenters. Insofar as the palette of eukaryotic metabolism is a limited subset of the metabolic diversity found among prokaryotic microorganisms, it is fair to ask how eukaryotes gained a foothold in prokaryotic ecosystems. The brief answer is that eukaryotic cells can do one thing prokaryotes cannot: they can engulf particles. This attribute, itself the consequence of a flexible membrane system and a dynamic cytoskeleton for organizing the cytoplasm, not only enabled eukaryotes to acquire metabolic capabilities such as photosynthesis but permitted them to add a new dimension to ecosystems via grazing and predation. Whereas bacterial diversity expanded along an axis defined by metabolism, eukaryotic cells diversified along a functional axis based on ultrastructure, life-cycle variation, and morphology. The result was not only a reinvention of the "primary" ecosystem of producers and (with fungi) decomposers but the evolution of a *new* aspect of ecosystem structure, the "secondary" or consumer ecosystem. There is some paleontological evidence for the ecological replacement of specific cyanobacteria by eukaryotic algae (e.g., Sergeev et al. 1995), but in general, protistan evolution led to the renewed accumulation of biological diversity and the exploitation of more dimensions of ecospace, increasing ecosystem complexity.

In size, as in function, eukaryotes broke through the right wall bounding bacterial diversity. Eukaryotic cells are rarely much smaller than one micron in diameter and are commonly tens to hundreds of microns in maximum dimension (e.g., Niklas 1992). The largest truly single-celled eukaryotes (excluding coenocytic organisms and large eggs produced by multicellular organisms) are a few centimeters long—think of nummulitid and fusulinid foraminiferans. Like prokaryotes, however, eukaryotic cells have both a left wall and a right wall bounding size and functional complexity (Niklas 2000). Both are displaced from those of prokaryotes, helping eukaryotic cells to invent and exploit new aspects of ecospace.

Megatrajectory 4: Multicellularity.—Multicellular organisms diversified in a third distinct way: via the functional integration of cells mediated by development. The evolutionary consequence was organisms endowed with both new functional capabilities and an unprecedented ability for "multi-tasking"—division of labor among specified cell groups within entities.

The basic molecular mechanisms of signal transduction and regulation are rooted deep in the history of life (Shapiro and Losick 1997). Among other things, the cell cycles of many single-celled prokaryotes and eukaryotes include the differentiation of distinctive resting and/or reproductive stages in response to environmental cues. With the addition of cell adhesion molecules and a more sophisticated battery of cell-surface receptors, however, several groups of eukaryotes were able to expand on this basic theme. Gene expression became regulated by molecular signals transduced from a cellular rather than physical environment to produce multicellular organisms characterized by ontogenetic cell differentiation. In so doing, these lineages opened up realms of ecospace utilization that are completely foreign to prokaryotic and unicellular eukaryotic life. Just as the eukaryotic cell is not simply a large bacterium, tissue-grade multicellular organisms are not just colonies of eukaryotic unicells. Rather, they constitute a new class of evolutionary entity in which many cells share in a unitary fate as organisms (e.g., Bonner 1998).

Tissue-grade multicellularity provides clear instances of convergence. For example, multicellular algae can tether themselves to substrates where currents are active and extend upward into the water column, enhancing their ability to garner light and nutrients. Not surprisingly, therefore, complex multicellularity arose independently in the red algae, in photosynthetic stramenopiles (brown and xanthophyte algae), and at least twice in the green algae (Graham and Wilcox 2000). Animals have fashioned multicellularity into an exceptional range of bodyplans (Ruppert and Barnes 1994), and again, some prominent features have evolved convergently. For example, muscularized appendages that facilitate rapid locomotion and complex manipulation evolved in the arthropods, cephalopods, and tetrapods—all groups that also evolved complex behavior. Some cnidarians (siphonophores, for example) gained functional complexity by the structural and functional differentiation of individuals within integrated colonies. Bilaterians, in contrast, differentiated organ systems within individuals. (Of course, some bilaterians boast differentiation of both organs *and* individuals, for example colonial bryozoans and behaviorally complex social insects.)

Some multicellular organisms remain small, overlapping in size with protists, but, with multicellularity, maximum size became limited only by mechanical strength or the functional requirements of organ systems. The multicellular megatrajectory also led to new modes of feeding, ranging from grazing and macrophagous predation to an essentially autotrophic reliance on symbiotic photo- or chemosynthetic microorganisms. At the same time, the impact of the active life habits of larger metazoans on physical conditions at the sediment-water interface and the new scale of packaging of organic nutrients (both in the mass of living tissues and in fecal pellets) changed the environment for many prokaryotes and unicellular eukaryotes, creating opportunities for parasites, pathogens, and specialized decomposers (Logan et al. 1995).

Like the other megatrajectories considered in this paper, the expansion of multicellular life in the oceans involved a large number of separate clades. At one level, this diversification provides evidence for evolution by increase in variance; animals and algae presumably started small and (relatively) simple, and small and simple species persist today beside those that are large and complex. Within each clade, however, bodyplans evolved via directional change, at least in part because of the winnowing effect of natural selection.

The directional effects ascribed to natural selection may be most apparent in establishing functional systems in new modes of life during the early evolution of a group. For instance, obligate palp-probiscide deposit feeding characterizes the nuculoid bivalves. The earliest nuculoids occur in Ordovician rocks, and the full efficiency of that particular mode of feeding and its associated feeding strategies was achieved by the Silurian–Silurian nuculoids stratified feeding depths in the same way as their modern descendants (Levinton and Bambach 1975). Likewise, the early acquisition of the full range of disparity in crinoid form, long before the peak of crinoid diversity (Foote 1994, 1999), supports the idea that clades commonly achieve their maximal functional range early in their history, after which they function within their evolved morphological constraints. Sustained directional change can occur, however, when the environment, itself, changes vectorially through time—particularly likely when selection is mediated by the biological components of an organism's effective environment, as in evolutionary "arms races" among predators and prey (Vermeij 1987).

In size, morphological complexity, genomic complexity, and energy metabolism, marine algae and animals thus show evidence of the same broad evolutionary pattern displayed by prokaryotes and protists: directional evolution within clades that leads to an increase in variance among clades. Whether observed increases in variance are diffusive or actively driven remains unexplored for most clades, but at least some groups show clear evidence of "driven" trajectories (e.g., Saunders et al. 1999).

One aspect of multicellular biology that definitely changed directionally through time is ecosystem complexity as measured by guild diversity or the number and connectivity of trophic levels in food webs (what Sterelny [1999] would call the "vertical complexity" of a community) (Bambach 1983, 1986; Vermeij 1987). Simple community structures may persist in restricted environments, but within shelf habitats of normal salinity, simple systems have given way through time to successively more complex communities.

Macroscopic animals first diversified near the end of the Proterozoic Eon, following which crown group members of bilaterian phyla radiated during the Cambrian (Conway Morris 1998; Valentine et al. 1999; Knoll and Carroll 1999). Renewed Ordovician diversification added new functional guilds, including various groups that subdivided benthic habitat space more completely by tiering both above the bottom and into the infauna (Ausich and Bottjer 1982; Bottjer and Ausich 1986; Sepkoski et al. 1991). Although total diversity of the marine biosphere appears not to have changed appreciably, the Devonian marked another time of change in ecosystem complexity, especially evident in increases of both metabolic rates and proportional diversity among predators (Signor and Brett 1984; Bambach 1999).

Further reorganization of marine ecosystems, incorporating the new modes of life and the expansion of life into previously little-exploited aspects of ecospace, occurred at the time of the Permian mass extinction (Erwin 1993; Knoll et al. 1996; Bambach and Knoll 1997, in press) and again at the end of the Mesozoic Era (Vermeij 1977; Bambach and Knoll 1997; Bambach 1999). Although it is certain that mass extinction and not just the honing of adaptations contributes to this stratigraphic pattern, directional trajectories extend across the Phanerozoic record. As Vermeij (1987, 1999) has emphasized, these focus not so much on morphology as they do the ecological attributes of animals: through time animals with high energy requirements in general, and top predators with more sophisticated means of obtaining prey in particular, replace their less metabolically needy and less specialized antecedents. Bambach (1983, 1986, 1993, 1999) has also emphasized that utilization of new or added ecospace has accompanied each major increase in diversity during the Phanerozoic.

Megatrajectory 5: Invasion of the Land.—The preceding megatrajectories reflect the origins of new evolutionary entities characterized by increasing ultrastructural, genomic, and morphological complexity. In contrast, the conquest of land was an ecological expansion made possible by *physiological* innovations. On land, photoautotrophic organisms must harness energy from the sun and nutrients and water from their substrate, and do it in a physiologically challenging setting. Cyanobacteria and some green algae colonized the land surface, presumably during the Precambrian, but it was the Paleozoic radiation of embryophytic land plants that truly changed the face of the land, making possible the subsequent invasion by several animal phyla.

It is common to view the initial colonization of terrestrial environments as an event pushed by resource limitation in the coastal ocean or pulled by the vast and little-exploited ecospace on land. The actual first steps in this megatrajectory were probably more prosaic. An aquatic charophyte algal population was likely stranded on land by a drop in water level, and because of mutations, some individuals did not die. The fact that these individuals survived stranding can be attributed in part to the accumulated biochemical features of their parental algal clade (Graham 1993) and in part to the mutations that modified these traits. Principal among these mutations must have been one involving a simple change in the timing of gene expression for sporopollenin biosynthesis from just after zygote formation to after meiosis and spore formation (Knoll et al. 1986). Such a mutation, arguably lethal in water, would have provided protection against desiccation and harmful radiation at a critically vulnerable phase of the life cycle. Subsequently, the larger suites of characters that collectively define the embryophytes, and within them the tracheophytes, were assembled. Fossils document aspects of this assembly, revealing extinct forms that document a truly directional component of early land plant evolution characterized by increases in size, organographic complexity, and biophysical performance (Knoll et al. 1984; Kenrick and Crane 1997). The succeeding 400-m.-y. history of vascular plant evolution has been characterized by the convergent evolution of novel meristems that give rise to roots, leaves, and wood, by the emergence of new clades able to exploit different physically and biologically defined habitats (increase in guild number), and by long-term gametophyte reduction (altering the plant life cycle). In these features and others, vascular plants echo the combination of patterns seen in the preceding megatrajectory: increasing ecological complexity, increasing variance in morphology among clades, and directional evolution within clades. The greening of the land surface also highlights another point: on land, the ecospace to be exploited is largely biogenic—the product of plant evolution.

Numerous animal groups have made the transition from marine to fresh water and from there to moist terrestrial habitats, but only five metazoan clades have achieved high diversity and full independence from moist settings on land: three arthropod groups (the arachnids, myriapods, and insects [Labandeira and Beall 1990]), the tetrapod chordates, and pulmonate gastropods. Possible myriapod burrows are reported in Late Ordovician rocks (Retallack and Feakes 1987) and fragmentary, arthropod-like fossils occur in Early Silurian beds (Gray and Boucot 1990). Fossils identifiable as arachnids and myriapods are reported from Late Silurian rocks in Great Britain (Jeram et al. 1990). Tetrapods first appear in the latest Devonian (Ahlberg 1995; Ahlberg et al. 1996; Janvier 1996) and the earliest pulmonate gastropods are fresh-water forms from the Early Carboniferous (Benton 1993).

The sequence of evolution of new clades of tetrapods with cumulative adaptive features that provide greater independence from the limiting problems for life on land is unambiguously directional (Kemp 1982). Late Devonian stem tetrapods (Clack 1988; Coates and Clack 1990, 1991; Ahlberg and Milner 1994) retain many features of their piscine ancestors, but by the mid-Carboniferous, amphibian-grade vertebrates had evolved more efficient limb articulation for walking on land and were clearly capable of living and reproducing with reduced dependence on water (Carroll 1988). Late in the Carboniferous Period, amniotic eggs evolved, freeing the Amniota completely from dependence on moist habitats (Sumida and Martin 1997). The coordinated evolution of increased dietary specialization and efficiency in processing food (as seen in modification and specialization of dentition and jaw mechanics), more active and rapid locomotion (as seen in alteration of limb articulation and action), and homeostasis in metabolism, especially endothermy, gave new clades greater and greater opportunity to capitalize on terrestrial ecospace. Notably, these modifications evolved convergently in synapsids, leading to the placental mammals, and in diapsids, leading to the birds.

The end result of the land radiations has been the evolution of new and complex ecosystems that include modes of life unknown in the oceans, despite the limited higher taxonomic diversity of terrestrial life. An especially important flywheel of terrestrial diversity has been the coevolution of producers and consumers, especially the interdependence of angiosperms and insects (Farrell 1998; Danforth and Ascher 1999). Like animals and algae in the marine realm, terrestrial plants and animals diversified along a functional/morphological axis; however, we emphasize again that this diversification was made possible by distinct *physiological* innovations that enabled multicellular organisms to tolerate and, subsequently, thrive in subaerial environments.

Megatrajectory 6: Intelligence and Technology.—The basic characteristics of human intelligence are present to some degree in nonhumans, but the combination of language, symbolic thinking and expression, and technology is unique to *Homo sapiens* and opened the way to the explosive exploitation of the global ecosystem. The accumulation of experience permitted by recording ideas and information and the development of systematic education have broken the constraint of time barriers that limit learning and cultural development in all other organisms to immediate direct experience. Division of labor among individuals, based on training (and socio-political systems) rather than developmental differentiation, dwarfs that displayed by any other species. The result is a technological system in which humans adapt environments to their needs, rather than the reverse. Invention has accelerated to the point where technological change now outpaces rates of biological evolution, even directed evolution, save perhaps for those of some microbial and viral pathogens.

The marriage of intelligence and technology may be limited to a single species, but its effects extend to the whole biosphere. Human intelligence and its application give one species the ability to invade every environment, including those hostile to most life, and to manipulate environments as well as most other life forms by technological processes. Unlike preceding megatrajectories, therefore, this one has been associated with a selective decline in ecosystem complexity imposed by agriculture and an overall decay of biological diversity caused by habitat alteration, global environmental change, and deliberate destruction. The pattern of increased ecospace utilization by organisms constituting a new megatrajectory continues, but only in a single species. Note that we imply no sense of wisdom or correctness to this control, only that the development of human technological civilization over the past 10,000 years has given *Homo sapiens* a unique potential (and actual) influence on the surface of the planet. Humans currently control 40% of terrestrial productivity (Vitousek et al. 1997), and our population is projected to grow by at least 50% in the next half century.

Humans form but one of myriad branches on the Tree of Life, no farther from its root than any other living organisms. Thus, the claim of special status for human intelligence is fundamentally ecological, not genealogical. We may not be "the end of evolution" (Ward 1995), but for good or ill, human societies have the power and vision to dictate future conditions, a conscious choice unique in the history of life.

Discussion

The six megatrajectories outlined in the preceding paragraphs capture much of the overall shape of life's history. We believe that, collectively, they support four general conclusions about directionality in the history of life:

1. At the broadest level of life as whole, evolutionary pattern can be depicted in terms of increasing variance bounded by a fixed wall on the left (Gould 1996) and a right wall that migrates through time (Sterelny 1999).

2. Despite this, evolutionary diversification cannot be appreciated fully in terms of a continuum model in which all organisms are seen as comparable, save in size or some other easily quantifiable trait. Rather, each megatrajectory has introduced fundamentally new evolutionary entities that garner resources in new ways, resulting in a pattern of expanding ecospace utilization subject to structural and functional constraints (Fig. 1). Viewed at this level, right walls don't move; they are fixed but can be scaled by new evolutionary entities. Variation accumulates through time, but not entirely by passive diffusion—right walls are surmounted mainly from one direction.

3. Within each megatrajectory, overall diversification conforms to a pattern of increasing variance about the starting point. If not initially constrained by a bounding wall, both the left and right walls may be "discovered" though time. The expanding envelope of form and physiology results from directional evolution within clades as well as simple diffusion from the starting point. Depending on the scale of observation, then, evolution may appear to be cumulative (and, possibly, diffusive) or directional.

4. The addition of new entities is governed by logical rules of ecosystem function. Evolution of the complementary metabolic pathways that underpin biogeochemical cycles is a microbial phenomenon. The eukaryotic innovation of endocytosis, which allowed nucleated cells to engulf particles, did little to alter this fundamental ecological circuitry but rather permitted diversifying eukaryotes to add new complexity to ecosystems. Much the same can be said about multicellularity. In light of this, the "wall" metaphor might better be turned on end: each megatrajectory adds a new story to an ecological edifice whose foundation remains bacterial. This provides what may be the most important directional pattern of all, a long-term increase in the complexity of ecosystems. It is not obvious why ecological measures of life's directionality should be considered any less appropriate or informative than compilations of shape and size. To quote Sterelny (1999: p. 467), "I suspect that vertical complexity is at least as salient to [discussions of complexity and evolution] as horizontal complexity."

In the view articulated here, there *is* evolutionary pattern best characterized by increase in variance, and in some cases it may genuinely reflect diffusion away from a bounding wall (Gould 1996). But this increase is not without limit, and only with the accumulation of a novel set of traits that permits the scaling of the right wall does diversity resume its cumulative course. There are also directional trajectories guided by natural selection, as espoused by Vermeij (1999) and many others, although whether specific morphological or molecular features are determinate, as proposed by De Duve (1995), is hard to know. This combination of breakthrough and directional change, coupled with diffusion within groups, is captured well by one of McShea's models, the one he calls "Driven (at the large scale)—

TABLE 1. Pathways of astrobiological exploration, with the potential search fields, types of observable evidence, and megatrajectory levels that each is capable of detecting.

Direct exploration:
- Our solar system
- Morphological, biochemical, and biogeochemical evidence
- Megatrajectories I and II*

Remote imaging of planetary atmospheres
- Nearby solar systems
- Biogeochemical evidence
- Megatrajectory II and (?) higher†

SETI
- The broad universe
- Intelligently transmitted signals
- Megatrajectory VI

* In principle, direct exploration could document evidence of extant or extinct life at any megatrajectory level. In practice, however, astrobiological exploration of our own solar system will focus on microbial life.
† Megatrajectory II is most critical for establishing biogeochemical signatures of life detectable in planetary atmospheres.

no boundary, strong bias (but invoked only occasionally)" (McShea 1996: Fig. 7e).

Mass extinction is generally considered to frustrate selection-mediated evolutionary trajectories, but it may also contribute to them via selectivity, the breakdown of ecological incumbency, or both (Bambach and Knoll 1997). Thus, Vermeij's (1999) trajectory of successive increase in the energetic requirements of animals is emphasized, not truncated, at the Permo/Triassic and, among tetrapods, at least, Cretaceous/Tertiary boundaries (e.g., Thayer 1983; Knoll et al. 1996; Bambach and Knoll in press).

If evolutionary pattern is a complicated and chimeric mixture of direction and accumulation, is this pattern in any way predictable? Sequenced megatrajectories, diffusion away from a wall, and directional evolution may individually or collectively be either predictable on the basis of evolutionary theory or not.

Natural selection certainly predicts a directional trajectory within clades, although there is no expectation that functional efficacy should increase without limit. Vermeij's (1999) pattern of broad, guild-level increase through time in the abundance of organisms with high energetic requirements may also be predicted by economically driven evolutionary theory. This does not equate with a statement that any specific morphology or physiology is inherently likely, but represents only the broad expectation that new organisms able to gather and utilize increasing amounts of resources will thrive in addition to or at the expense of preexisting groups.

Even (and perhaps especially) Gould's (1996) view of evolutionary pattern as an increase in variance driven by morphological diffusion away from a bounding wall carries reasonable predictions for the trajectory of the diffusional edge of biological diversity, if not for the specific characteristics of that edge. As Bonner (1988: p. 58) put it, "There is always room at the top." At the level of megatrajectories, the successive innovations that characterize these major expansions were certainly not predictable in timing, nor did the spread of each step, with its influence on the whole biosphere, take immediate effect when the new advance first evolved. Playing life's tape again (Gould 1989) might not result in dinosaurs, trilobites, or any other specific accumulation of morphological and physiological characters. But the overall trajectory of evolution, the cumulative succession of megatrajectories displayed in Earth history, may define a general pathway followed for varying distances by life wherever it has arisen.

Bridge to an Emerging Discipline

Clearly, we lack the one set of observations that might distill speculation into theory: a second biological planet. But NASA and its international partners seek to remedy this situation, and astrobiological exploration will be a major theme of planetary science during the next few decades. For this reason, questions of evolutionary predictability ("How do we extrapolate from a sample of one?") are being asked by the broad community of scientists interested in Mars, Europa, and points beyond.

What do we as paleontologists tell our colleagues? How do our interpretations of evolutionary pattern in Earth history translate into astrobiological advice? Should we find another biological planet, it is safe to assume that the aspect of its biology most likely to resemble Earth's will be that of the earliest megatrajectories. Complex chimeras comparable to eukaryotes probably have a lower probability of showing up elsewhere, and the prospects for multicellular and, in turn, intelligent entities are likely lower yet.

Our ability to determine such evolutionary odds is inevitably limited by the nature of astrobiological exploration (Table 1). Simple microscopic organisms (Megatrajectories 1 and 2) are the life forms most likely to be widespread in the universe, but we can search for them directly only within our own solar system. In a decade or so, space telescope arrays may provide the resolution needed to search for biological signatures in the atmospheres of planets in nearby solar systems, providing an expanded search space for extraterrestrial life. But for the immense broader universe, the prospects for identifying the putatively most common forms of life are grim; we can only search for the least likely form of biology, the sixth megatrajectory of intelligent life.

Until the day when unambiguous evidence of extraterrestrial life becomes available, we can seek evolutionary generality only in our own planet's history. That search is the paleontologist's task, and also the paleontologist's privilege—understanding fossils, phylogeny, and environmental history sufficiently well to discern how molecular and morphological constraints, innovations, directional evolution, increase in variance, and chance events have interacted to produce the full complexity of modern ecosystems.

Acknowledgments

This paper had its origins in a presentation prepared by R. K. B. for the Paleontological Society symposium *Evolutionary Paleoecology* organized by D. Bottjer and W. Allmon in 1997. The ideas have been sharpened by discussions with D. Erwin, M. Buzas, and D. McShea. D. McShea, W. DiMichele, and S. Porter provided helpful criticisms of the manuscript. A. H. K.'s research is supported in part by the National Aeronautic and Space Administration Astrobiology Institute.

Literature Cited

Ahlberg, P. E. 1995. *Elginerpeton pancheni* and the earliest tetrapod clade. Nature 373:420–425.

Ahlberg, P. E., and A. R. Milner. 1994. The origin and early diversification of tetrapods. Nature 368:507–514.

Ahlberg, P. E., J. A. Clack, and E. Luksevics. 1996. Rapid braincase evolution between *Panderichthys* and the earliest tetrapods. Nature 381:61–64.

Ausich, W. I., and D. J. Bottjer. 1982. Tiering in suspension feeding communities on soft substrata throughout the Phanerozoic. Science 216:173–174.

Bambach, R. K. 1983. Ecospace utilization and guilds in marine communities through the Phanerozoic. Pp. 719–746 *in* M. Tevesz and P. McCall, eds. Biotic interactions in Recent and fossil benthic communities. Plenum, New York.

———. 1986. Classes and adaptive variety: the ecology of diversification in marine faunas through the Phanerozoic. Pp. 191–253 *in* J. W. Valentine, ed. Phanerozoic diversity patterns: profiles in macroevolution. Princeton University Press, Princeton, N.J.

———. 1993. Seafood through time: changes in biomass, energetics and productivity in the marine ecosystem. Paleobiology 19:372–397.

———. 1999. Energetics in the global marine fauna: a connection between terrestrial diversification and change in the marine biosphere. Geobios 32:131–144.

Bambach, R. K., and A. H. Knoll. 1997. Fundamental physiological control on patterns of diversification in the marine biosphere. Geological Society of America Abstracts with Programs 29:7:A31.

———. In press. Physiological selectivity during the end-Permian mass extinction. Paleobiology.

Benton, M. J., ed. 1993. The fossil record 2. Chapman and Hall, London.

Bonner, J. T. 1988. The evolution of complexity. Princeton University Press, Princeton, N.J.

———. 1998. The origins of multicellularity. Integrative Biology 1:28–36.

Bottjer, D. J., and W. I. Ausich. 1986. Phanerozoic development of tiering in soft substrata suspension-feeding communities. Paleobiology 12:400–420.

Buss, L. W. 1987. The evolution of individuality. Princeton University Press, Princeton, N.J.

Carroll, R. L. 1988. Vertebrate paleontology and evolution. W. H. Freeman, New York.

Clack, J. A. 1988. New material of the early tetrapod *Acanthostega* from the Upper Devonian of Greenland. Palaeontology 31:699–724.

Clarke, P. H. 1983. Experimental evolution. Pp. 235–252 *in* D. S. Bendall, ed. Evolution from molecules to men. Cambridge University Press, Cambridge.

Coates, M. I., and J. A. Clack. 1990. Polydactyly in the earliest known tetrapod limbs. Nature 347:66–69.

———. 1991. Fish-like gills and breathing in the earliest known tetrapod. Nature 352:234–236.

Conway Morris, S. 1998. Crucible of creation: the Burgess Shale and the rise of animals. Oxford University Press, Oxford.

Danforth, B. N., and J. Ascher (with response from B. D. Farrell). 1999. Flowers and insect evolution. Science 283:143a. [Not a real publication. Only available on internet at www.sciencemag.org for as long as the AAAS keeps it there.]

De Duve, C. 1995. Vital dust: life as a cosmic imperative. Basic Books, New York.

Erwin, D. H. 1993. The great Paleozoic crisis. Columbia University Press, New York.

Farrell, B. D. 1998. "Inordinate fondness" explained: why are there so many beetles? Science 281:555–559.

Foote, M. 1994. Morphological disparity in Ordovician–Devonian crinoids and the early saturation of morphological space. Paleobiology 20:320–344.

———. 1999. Morphological diversity in the evolutionary radiation of Paleozoic and post-Paleozoic crinoids. Paleobiology Memoirs No. 1. Paleobiology 25(Suppl. to No. 2).

Gould, S. J. 1989. Wonderful life: the Burgess Shale and the nature of history. Norton, New York.

———. 1996. Full house: the spread of excellence from Plato to Darwin. Harmony Books, New York.

Graham, L. E. 1993. The origin of land plants. Wiley, New York.

Graham, L. E., and L. W. Wilcox. 2000. Algae. Prentice-Hall, Englewood Cliffs, N.J.

Gray, J., and A. J. Boucot. 1990. Early Silurian nonmarine animal remains and the nature of the early continental ecosystem. Acta Palaeontological Polonica 38:303–328.

Hutchinson, G. Evelyn. 1965. The ecological theater and the evolutionary play. Yale University Press, New Haven, Conn.

Janvier, P. 1996. Early vertebrates. Oxford Monographs on Geology and Geophysics No. 33. Oxford Science Publications, Clarendon, Oxford.

Jeram, A. J., P. A. Selden, and D. Edwards. 1990. Land animals in the Silurian: arachnids and myriapods from Shropshire, England. Science 250:658–661.

Joyce, G. F. 1989. The rise and fall of the RNA world. New Biologist 3:399–407.

Kemp, T. S. 1982. Mammal-like reptiles and the origin of mammals. Academic Press, London.

Kenrick, P., and P. R. Crane. 1997. The origin and early diversification of land plants: a cladistic study. Smithsonian Institution Press, Washington, D.C.

Knoll, A. H. 1994. Life's expanding realm. Natural History 1994(6):14–20.

Knoll, A. H., and S. B. Carroll. 1999. Early animal evolution: emerging perspectives from comparative biology and geology. Science 2129–2137.

Knoll, A. H., K. J. Niklas, P. G. Gensel, and B. H. Tiffney. 1984. Character diversification and patterns of evolution in early vascular plants. Paleobiology 10:34 47.

Knoll, A. H., S. W. F. Grant, and J. W. Tsao. 1986. The early evolution of land plants. Studies in Geology 15:45–63. Department of Geological Sciences, University of Tennessee, Knoxville.

Knoll, A. H., R. Bambach, D. Canfield, and J. P. Grotzinger. 1996. Comparative Earth history and late Permian mass extinction. Science 273:452–457.

Labandeira, C. C., and B. S. Beall. 1990. Arthropod terrestriality. In G. Mikulic, convener. Arthropod paleobiology. Short Courses in Paleontology 3:214–256. Paleontological Society, Knoxville, Tenn.

Lawrence, J. G. 1999. Gene transfer and minimal cell size. Pp. 32–38 in National Research Council (Space Studies Board) 1999. Size limits of very small organisms. NAS Press, Washington, D.C.

Lazcano, A. 1994. The transition from nonliving to living. Pp. 60–69 in S. Bengtson, ed. Early life on Earth (Nobel Symposium No. 84). Columbia University Press, New York.

Lenski, R., and M. Travisano. 1994. Dynamics of adaptation and diversification: a 10,000 generation experiment with bacterial populations. Proceedings of the National Academy of Sciences USA 91:6808–6814.

Levinton, J. S., and R. K. Bambach. 1975. A comparative study of Silurian and Recent deposit-feeding bivalve communities. Paleobiology 1:97–124.

Logan, G. A., J. M. Hayes, G. B. Hieshima, and R. E. Summons. 1995. Terminal Proterozoic reorganization of biogeochemical cycles. Nature 376:53–56.

Margulis, L. 1981. Symbiosis in cell evolution. W. H. Freeman, San Francisco.

Martin, W., and M. Müller. 1998. The hydrogen hypothesis and the first eukaryote. Nature 392:37–41.

Maynard Smith, J., and E. Szathmáry. 1995. The major transitions in evolution. W. H. Freeman Spektrum, Oxford.

McShea, D. W. 1994. Mechanisms of large-scale evolutionary trends. Evolution 48:1747–1763.

———. 1996. Perspective. Metazoan complexity and evolution: is there a trend? Evolution 50:477–492.

———. 1999. Hierarchical complexity of organisms: dynamics of a well-known trend. Geological Society of America Abstracts with Programs 31:7:A171.

Moreira, D., and P. Lopez-Garcia. 1998. Symbiosis between methanogenic archaea and δ−proteobacteria as the origin of eukaryotes: the syntrophic hypothesis. Journal of Molecular Evolution 47:517–530.

National Research Council (Space Studies Board). 1999. Size limits of very small organisms. NAS Press, Washington, D.C.

Niklas, K. J. 1992. Plant allometry. University of Chicago Press, Chicago.

———. 2000. The evolution of plant body plans: a biomechanical perspective. Annals of Botany. 85:411–438.

Pace, N. R. 1997. A molecular view of microbial diversity and the biosphere. Science 276:555–557.

Retallack, G. J., and C. R. Feakes. 1987. Trace fossil evidence for Late Ordovician animals on land. Science 23:561–563.

Ruppert, E. E., and Barnes, R. D. 1994. Invertebrate zoology, 6th ed. Saunders College Publishing, Fort Worth.

Saunders, W. B., D. M. Work, and S. V. Nikolaeva. 1999. Evolution of complexity in Paleozoic ammonoid sutures. Science 286:760–763.

Schopf, J. W., and B. Packer. 1987. Early Archean (3.3-billion to 3.5-billion-year-old) microfossils from Warrawoona Group, Australia. Science 237:70–73.

Schulz, H. N., T. Brinkhoff, T. G. Ferdelman, A. Teske, and B. B. Joergensen. 1999. Dense populations of a giant sulfur bacterium in Namibian shelf sediments. Science 284:493–495.

Sepkoski, J. J., Jr., R. K. Bambach, and M. L. Droser. 1991. Secular changes in Phanerozoic event bedding and the biological overprint. Pp. 298–312 in G. Einsele, W. Ricken, and A. Seilacher, eds. Cycles and events in stratigraphy. Springer, Berlin.

Sergeev, V. N., A. H. Knoll, and J. P. Grotzinger. 1995. Paleobiology of the Mesoproterozoic Billyakh Group, Anabar Uplift, northern Siberia. Paleontological Society Memoir 39.

Shapiro, L., and R. Losick. 1997. Protein localization and cell fate in bacteria. Science 276:712–718.

Simpson, G. G. 1953. The major features of evolution. Simon and Schuster, New York.

Signor, P. W., III, and C. E. Brett. 1984. The mid-Paleozoic precursor to the Mesozoic marine revolution. Paleobiology 10:229–245.

Sogin, M. L. 1994. The origin of eukaryotes and evolution into major kingdoms. Pp. 181–192 in S. Bengtson, ed. Early life on Earth (Nobel Symposium No. 84). Columbia University Press, New York.

Stanley, S. M. 1973. An explanation for Cope's Rule. Evolution 27:1–26.

Stebbins, G. L. 1969. The basis of progressive evolution. University of North Carolina Press, Chapel Hill.

Sterelny, K. 1999. Bacteria at the high table. Biology and Philosophy 14:459–470.

Sterelny, K., and P. E. Griffiths. 1999. Sex and death: an introduction to philosophy of biology. University of Chicago Press, Chicago.

Sumida, S. S., and K. L. M. Martin, eds. 1997. Amniote origins. Academic Press, San Diego.

Thayer, C. 1983. Sediment-mediated biological disturbance and the evolution of marine benthos. Pp. 479–625 *in* M. Tevesz and P. McCall, eds. Biotic interactions in Recent and fossil benthic communities. Plenum, New York.

Valentine, J. W., D. Jablonski, and D. H. Erwin. 1999. Fossils, molecules, and embryos: new perspectives on the Cambrian explosion. Development 126:851–859.

Vermeij, G. J. 1977. The Mesozoic marine revolution: evidence from snails, predators, and grazers. Paleobiology 3:245–258.

———. 1987. Evolution and escalation. Princeton University Press, Princeton, N.J.

———. 1999. Inequality and the directionality of history. American Naturalist 153:243–253.

Vitousek, P. M., H. A. Mooney, J. Lubchenko, and J. M. Mellilo. 1997. Human domination of Earth's ecosystems. Science 277:494–499.

Ward, P. D. 1995. End of evolution: a journey in search of clues to the third mass extinction facing planet earth. Bantam Books, New York.

Whatley, J. M., P. John, and F. R. Whatley. 1979. From extracellular to intracellular: the establishment of mitochondria and chloroplasts. Proceedings of the Royal Society of London B 204:165–187.

Woese, C. R. 1987. Bacterial evolution. Microbiological Reviews 51:221–271.

Micro- and macroevolution: scale and hierarchy in evolutionary biology and paleobiology

David Jablonski

Abstract.—The study of evolution has increasingly incorporated considerations of history, scale, and hierarchy, in terms of both the origin of variation and the sorting of that variation. Although the macroevolutionary exploration of developmental genetics has just begun, considerable progress has been made in understanding the origin of evolutionary novelty in terms of the potential for coordinated morphological change and the potential for imposing uneven probabilities on different evolutionary directions. Global or whole-organism heterochrony, local heterochrony (affecting single structures, regions, or organ systems) and heterotopies (changes in the location of developmental events), and epigenetic mechanisms (which help to integrate the developing parts of an organism into a functional whole) together contribute to profound nonlinearities between genetic and morphologic change, by permitting the generation and accommodation of evolutionary novelties without pervasive, coordinated genetic changes; the limits of these developmental processes are poorly understood, however. The discordance across hierarchical levels in the production of evolutionary novelties through time, and among latitudes and environments, is an intriguing paleontological pattern whose explanation is controversial, in part because separating effects of genetics and ecology has proven difficult. At finer scales, species in the fossil record tend to be static over geologic time, although this stasis—to which there are gradualistic exceptions—generally appears to be underlain by extensive, nondirectional change rather than absolute invariance. Only a few studies have met the necessary protocols for the analysis of evolutionary tempo and mode at the species level, and so the distribution of evolutionary patterns among clades, environments, and modes of life remains poorly understood. Sorting among taxa is widely accepted in principle as an evolutionary mechanism, but detailed analyses are scarce; if geographic range or population density can be treated as traits above the organismic level, then the paleontological and macroecological literature abounds in potential raw material for such analyses. Even if taxon sorting operates on traits that are not emergent at the species level, the differential speciation and extinction rates can shape large-scale evolutionary patterns in ways that are not simple extrapolations from short-term evolution at the organismal level. Changes in origination and extinction rates can evidently be mediated by interactions with other clades, although such interactions need to be studied in a geographically explicit fashion before the relative roles of biotic and physical factors can be assessed. Incumbency effects are important at many scales, with the most dramatic manifestation being the postextinction diversifications that follow the removal of incumbents. However, mass extinctions are evolutionarily important not only for the removal of dominant taxa, which can occur according to rules that differ from those operating during times of lower extinction intensity, but also for the dramatic diversifications that follow upon the removal or depletion of incumbents. Mass extinctions do not entirely reset the evolutionary clock, so survivors can exhibit unbroken evolutionary continuity, trends that suffer setbacks but then resume, or failure to participate in the recovery.

David Jablonski. Department of Geophysical Sciences, University of Chicago, 5734 South Ellis Avenue, Chicago, Illinois 60637. E-mail: djablons@midway.uchicago.edu

Accepted: 18 July 2000

Introduction

The landscape of evolutionary biology has changed significantly over the past quarter-century. Evolutionary and ecological studies now regularly incorporate serious considerations of history, scale, and hierarchy. This expansion of the working toolkit of the discipline—the near-routine application of ideas that were once barely developed, highly abstract, or in some circles outright anathema— heralds a more profound integration of evolutionary biology, paleobiology, systematics, ecology, and developmental biology. Whatever the intellectual forebears of these ideas (and some of their roots are very deep), the growing number and diversity of studies that directly address such factors as intrinsic constraints, phylogenetic effects, differential origination and extinction rates, and local vs. regional effects represent a true operational expansion of evolutionary theory. As a window

0094-8373/00/2604-0002/$1.00

onto a wide range of spatial and temporal scales, including extreme events not accessible to neontological study, paleontology is playing a vital role in this expansion and will certainly continue that role in the coming decades.

This is by no means to declare the demise of the highly successful microevolutionary paradigm, which is alive and well, but rather to say that the study of evolution continues to evolve and expand conceptually, and increasingly incorporates approaches that explicitly emphasize scale and hierarchy. These approaches, even those that can be traced back to Darwin, were barely visible when Tom Schopf and Ralph Johnson were mustering support for a new journal to be titled *Paleobiology*. Paleontology was one of the fields that provided a phenomenology, a conceptual base, and a battery of new quantitative methods, that fostered the expansion of evolutionary theory in the first quarter-century of this journal.

One way to survey the infusion of hierarchy and scale into evolutionary biology is through the classic darwinian two-step process: the origin and the sorting of variation. Other schemes are possible, of course, from the intrinsic/extrinsic dichotomy of causal mechanisms (e.g., Gould 1977a; Jablonski 2000a) to a nested set of temporal and spatial scales (e.g., Gould 1985; Bennett 1997), and those elements will inevitably appear here as well.

Origin of Variation

The origin of heritable variation, the raw material of the evolutionary process, is by definition a matter of genetics. One major issue in bridging scales and hierarchical levels, however, has been the correspondence between genotypic changes and the magnitude and direction of phenotypic transformation. A reasonable assumption underlying most models of evolution has been that the probability of a phenotypic change is inversely related to the complexity or magnitude of the genetic change required to generate it (among other factors, of course). Geneticists have long recognized that mutations of equal magnitude can have strikingly different phenotypic consequences depending on context, but this awareness has been somewhat lost in the empirical evidence from quantitative genetics that many traits are underlain by a large number of genes of small and mainly additive effect. However, intensive study of developmental processes in multicellular organisms has led to a new appreciation of how modest genetic changes can be amplified and channeled developmentally to yield significant variations in the magnitude and direction of phenotypic change. We are only beginning to understand how and when to apply this expanding set of approaches to the origin of novel morphologies (and the absence or rarity of certain forms), but few would deny the potential to both illuminate and be illuminated by the fossil record of morphological evolution—not least because, as discussed below, that record exhibits nonrandom patterns in space and time.

Genetic Control Pathways and Networks

At one extreme in the nonlinearities between genotypic and phenotypic change are the high-level regulatory genes now under intense study by molecular developmental biologists. Some of these genes, such as the Hox clusters that help to pattern the body axis and appendages across the entire breadth of metazoan diversity, specify positional information and thereby regulate the transcription of a large number of downstream genes. Others sit near the top of a regulatory cascade, or perhaps more commonly within a regulatory network, that determines a specific tissue or structure. For example, perhaps 2500 genes are involved in building and maintaining the *Drosophila* eye (Halder et al. 1995; 18% of the fly's genome by Adams et al.'s [2000] count, although of course many are also used elsewhere), but experimental manipulation of a single gene—most famously *eyeless*, but also *eyes absent*, *dachshund*, and others—can yield well-formed eyes in the middle of wings and other improbable sites (see reviews by Gehring and Ikeo 1999 and Hodin 2000; and see Chow et al. 1999 for similar results in the frog *Xenopus*). Such high-level regulatory genes and even entire signaling pathways have been recruited as modular units ("cassettes") in the service of novel morphologies. Thus, the *hedgehog* pathway that underlies the generation of limbs also participates in the production of eyespot patterns on butterfly wings (Keys et al. 1999); the *Toll* pathway has been recruited to the development of chick limbs, the operation of the vertebrate immune system, and the development of ventral structure in *Drosophila* (Gonzalez-Crespo and Levine 1994; Ghosh et al. 1998), and even the "master control gene" for eyes, *Pax-6*, plays multiple roles (e.g., Quinn et al. 1996; Duboule and Wilkins 1998; Hodin 2000; and see Heanue et al. 1999 and Relaix and Buckingham 1999 for an analogous situation involving *Pax-3*).

Many of the major molecular components of metazoan development appear to have arisen very early and been retained across taxa with very different bodyplans of varying complexity (reviews in Finnerty and Martindale 1998; Erwin 1999; Valentine et al. 1999; Holland 1999; Knoll and Carroll 1999, among others); although data are sketchier, the same appears to hold for plants (Purugganan 1998; Lawton-Rauh et al. 2000; Theissen et al. 2000). The extraordinary conservation (albeit in a highly dynamic fashion involving gene loss, gain, and duplication) and the multiple expression events and sites of these genes strongly suggests that the evolutionary action at this level is in the enhancer regions that govern the timing and intensity of gene expression (see Arnone and Davidson 1997 for an excellent review). The evolutionary impact of changes in the major regulatory genes will therefore be more complex than the extreme binary effects suggested by the homeotic mutants that abruptly transform segment identities. Instead, that impact should probably be visualized in terms of finer modulations in the expression of major regulatory genes, and therefore should include evolutionary changes that do not depend on the viability of extreme saltations and lone individuals (Gellon and McGinnis 1998; Akam 1998; Duboule and Wilkins 1998; Purugganan 1998, 2000; Gibson 1999; Li and McGinnis 1999; Ludwig et al. 2000). Possible examples of striking morphological changes that may have involved relatively subtle shifts in Hox expression patterns include the evolutionary differentiation of arthropod appendages (Averof and Patel 1997;

Shubin et al. 1997; Weatherbee et al. 1999), morphological transitions of vertebrae along the spinal column (Burke et al. 1995; Belting et al. 1998), the origin and diversification of tetrapod limbs (Shubin et al. 1997; Coates and Cohn 1998), and the suppression of those limbs and the homogenization of vertebral morphology along the body axis of snakes (Cohn and Tickle 1999; Greene and Cundall 2000). Vertebral identity and digit size and number appear to be dose-dependent functions of Hox gene products (Zakany et al. 1997), providing a mechanism for selection on high-level regulatory factors within populations, and thus a pathway for coordinated change effected by major genes but in a polymorphic population rather than a strictly typological context. Similar situations, again underlain by polymorphisms in regulatory genes, have been recognized by plants as well (e.g., Purugganan 1998, 2000; Lawton-Rauh et al. 2000).

These are very early days in the macroevolutionary exploration of developmental genetics. Careful study is needed in the inference of causal, evolutionary relationships between variations in Hox expression and morphological differences among taxa (e.g., Rogers et al. 1997). And the next step, taking insights derived from laboratory populations into natural evolutionary processes, has just begun. Gibson et al. (1999) found that crossing mutants of the Hox genes *Ubx* and *Antp* into natural populations of *Drosophila* could evoke a wide range of extreme phenotypes whose expression depended on the overall genetic context of the mutation, again showing how large-effect genes might impinge on morphological variation generated in the wild. More speculatively, DeSalle and Carew (1992) attribute some of the morphological extremes seen in Hawaiian drosophilid species, such as grotesquely reshaped heads and bizarre mouthpart appendages, to mutations in Hox genes because of their resemblance to *Antp* mutants in laboratory *D. melanogaster*—an attractive hypothesis that invites direct molecular study. Nevo et al. (1992) reported apparent Hox gene polymorphisms that significantly correlated with a wide array of morphological variables in the mole rat *Spalax ehrenbergi*, an early, provocative result that needs further work. On the other hand, Ahn and Gibson (1999a,b) found intraspecific variation in expression domains of Hox genes along the body axis of the three-spined stickleback, but this variation did not correlate with phenotypic variation. Finally, considerable within-species variation has been recorded for a number of major developmental genes in natural and domesticated plant populations, and several dramatic intraspecific variants have proven to be underlain by high-level developmental genes (for overview see Lawton-Rauh et al. 2000).

Evolution via changes in Hox-gene expression is hardly a prerequisite for a strong non-linearity between genetic and phenotypic change, of course, and both theory and evidence are accumulating that genes of large effect—many of them presumably involved in regulating development—do play a role in the origin of adaptations and the divergence of species (e.g., Gottlieb 1984; Orr and Coyne 1992; Palopoli and Patel 1996; Orr 1998). Much empirical work is still needed, of course, but an absolutely micromutational view of evolutionary change in natural populations seems increasingly untenable even for quantitative traits that are often seen as the bastion of the "many genes of equal and infinitesimal effect" approach championed by R. A. Fisher. One recent surprise, for example, has been the number of studies reporting quantitative trait loci (QTLs) of large phenotypic effect in both laboratory and natural populations (Orr 1998: p. 936; see also Paterson et al. 1997; Voss and Shaffer 1997; Doebley and Wang 1997; Bradshaw et al. 1998; Schemske and Bradshaw 1999; Goffinet and Gerber 2000). QTL analyses have their limitations, of course: they are only feasible among species or morphs that can be hybridized, they may not permit an absolute determination of the numbers of genes involved in important traits because minor factors may not be detected, and they can even be biased toward artificially inflating the effects of single QTLs or underestimating the number of genes involved in complex traits (Routman and Cheverud 1997; Lynch and Walsh 1998: p. 474–476; Via and Hawthorne 1998). On the other hand, laboratory selection experiments can be biased toward document-ing polygenic changes (e.g., McKenzie and Batterham 1994). The growing weight of evidence suggests that genetic differences even among closely related taxa often involve genes of large effect that in the few cases of functional analysis are involved in regulating development (e.g., Doebley and Lukens 1998); corroboration for this view has been found in a number of "candidate loci" identified through mutations that are then mapped to QTLs accounting for natural variation in populations (e.g., Long et al. 1996; Mackay 1996; Nuzhdin et al. 1999). This calls for a more complex treatment of the raw material for evolutionary change, and for new models of how this raw material can influence the direction or pace of evolution, in concert with more traditional parameters like population size or structure—models that can be applied, tested, and refined paleontologically in many instances. One potential avenue might be to use neontological data on a well-fossilized group to pinpoint some of the phenotypic changes most likely to be underlain by large-effect genes, and then to test whether those changes can predict the evolutionary trajectories of clades in the fossil record. The interplay of evolutionary factors mentioned above might be assessed in a comparative framework in which several related clades are targeted that differ in such features as genetic population structure (see below for paleontological approaches to this parameter).

Heterochrony and Heterotopy

Heterochrony, a change in the rate or timing of developmental events, has received much attention as a potential avenue to dramatic morphological evolution. As Klingenberg (1998) points out, the pervasiveness of heterochrony hinges on its definition. Developmental biologists restrict heterochrony to changes in the relative timing of developmental events, emphasizing the dissociation between developmental units (e.g., Raff and Wray 1989; Raff 1996; Hall 1999), while evolutionary biologists and paleontologists have used a broader definition that also includes changes in the relative rates of developmental processes even when the order of events is unchanged (e.g., Gould 1977b; Alberch et al. 1979; McKinney and McNamara 1991). Further difficulties are imposed by the hierarchy of biological organization: changes in timing at the molecular or cellular level need not produce heterochronic patterns at the whole-organism level, and the heterochrony at the organismal level need not involve changes in the timing of molecular events (e.g., some salamanders owe their paedomorphosis simply to the disabling of the production or reception of the hormone thyroxin, which is closer to a binary switch than a change in timing).

Global or systemic heterochrony—in which the entire phenotype is shifted relative to the ancestral ontogeny—has the greatest evolutionary potential, and is most readily detected, in species that undergo substantial phenotypic changes during ontogeny. Both continuous but nonlinear changes such as ontogenetic allometry and saltational changes such as metamorphosis (as in the classic example of neotenic salamanders that mature in the aquatic larval state) provide ample raw material for dramatic evolutionary events. This implies a little-explored predictive approach to differences in morphological diversification among clades, based on the nature of ancestral ontogenies. This might involve, for example, testing how well the ontogeny of individual organisms can predict the extent of morphological diversification of their descendant clades, a macroevolutionary equivalent of Wayne's (1986) classic conclusion that the great array of morphologically extreme dog breeds relative to cats is a consequence of their contrasting ontogenetic allometries (see also Wayne and Ostrander 1999).

The preceding example represents just one way in which the evolution of development at the individual level might indirectly shape large-scale patterns. Selection on life-history traits over ecological timescales, which as Gould (1977b) notes can often result in heterochrony, can also have far-reaching, indirect macroevolutionary effects, for example via changes in genetic population structures. For example, populations of obligate paedomorphic ambystomatid salamanders are genetically more distinct from one another than are metamorphosing populations (Shaffer 1984). Presumably the obligate paedomorphs tend to

stay in their aquatic neighborhoods, yielding lower rates of gene flow, and thus potentially higher speciation rates compared with metamorphosing relatives free to travel among ponds. Similarly, selection for small body size per se, energy economy, or short generation time in benthic invertebrates can evidently give rise to paedomorphs that are so small that they must evolve a low-dispersal, non-planktotrophic mode of development due to fecundity constraints (see Jablonski and Lutz 1983; Lindberg 1988); such a change in developmental mode would in turn be likely to result in high speciation and high extinction rates at the clade level, as discussed below (and see experiments showing that selection on egg size can in turn alter other aspects of larval biology, such as those of Sinervo and McEdward [1988] and Emlet and Hoegh-Guldberg [1997]). Again, this intriguing intersection of micro- and macroevolution, where life-history theory meets heterochrony (an important but neglected insight in Gould 1977b; see also McKinney and McNamara 1991; McKinney and Gittleman 1995), would produce dramatic shifts not only in morphology but in evolutionary dynamics.

Although global or systemic heterochrony has enjoyed the most attention, local or specific heterochrony has been much more common (e.g., McKinney and McNamara 1991). Such changes in the rate or timing of development of particular structures within an organism can break allometric relationships and generate new and coordinated morphologies, again by drawing on established developmental interactions but this time within a localized region or developmental field. For example, the most successful crinoids in modern oceans, the stemless, mobile comatulids, were derived from the stemmed crinoid order Isocrinida; intermediates such as *Eocomatula* and *Paracomatula* suggest progressively earlier offset of stem formation, with the substratum-gripping cirri, which adorn the long stem in the isocrinids, finally arising from a single centro-dorsal ossicle on the base (Simms 1988a,b, 1994; Hagdorn and Campbell 1993). In mammals, the ossification of facial bones is accelerated relative to the central nervous system in marsupials compared with both monotremes and placentals, correlated with the particular demands of neonate survival, including head movements during migration to the pouch, attachment to the teat, and suckling (Smith 1996, 1997; Nunn and Smith 1998). Such localized developmental changes, from the elongation of pterosaur digits to support a wing to the elongation of echinoid plates to create a protruding rostrum, represent coordinated changes in suites of complex tissues. Such dissociation and reintegration among local growth fields is still poorly understood but permits an enormous range of morphologies to be tapped by alterations of local growth fields. This does not mean that phenotypes are infinitely malleable, but that the scope for evolution via heterochronies of existing morphologies is far wider than implied by global heterochrony alone.

Calibration of ontogenetic trajectories against better metrics than the convenient but unreliable size criterion will be important for rigorous analysis of local and global heterochrony, particularly if life-history parameters are potential targets or by-products of selection. Fortunately, this can be achieved in paleontological material where accretionary growth leaves internal growth lines and a stable isotope record that can be calibrated at the level of monthly, seasonal, and annual periodicities (e.g., Jones 1988, 1998). In some taxa these can be tied to onset and perhaps frequency of reproduction, because reproductive growth interruptions can often be distinguished from disturbance checks (Kennish 1980; Harrington 1987; Sato 1995, 1999). Such skeletal chronometers have been used extensively in ecological studies of bivalves, but will be extremely valuable in evolutionary applications (for an exemplary study, see Jones and Gould 1999).

A final mechanism for coordinated morphological change is heterotopy, or alteration in the location of a developmental event. Heterotopy has received much less attention than heterochrony, but some authors have argued that spatial changes in development may prove to have greater evolutionary impact than temporal ones (e.g., Raff 1996; Zelditch and Fink 1996; Hall 1999). Although this perceived frequency is in part a consequence of

an expanded definition that may include virtually every developmental change that is not narrow-definition heterochrony (see Klingenberg 1998: p. 83), spatial changes in developmental events can certainly give rise to novel morphologies, as in the shifting of the scapula from outside to *inside* the ribcage in turtles (Burke 1989, 1991). Hall (1999: p. 388) goes so far as to say, "heterochrony tinkers, but heterotopy creates" and to anticipate that "heterotopy may be about to come into its own as heterochrony wanes and our knowledge of developmental mechanisms increases." This may be selling heterochrony a bit short (Hall goes on to note the intimate connections between heterochrony and heterotopy), but the recognition that the evolution of development involves more than just heterochrony—also being driven home by molecular work on gene regulation—is welcome. And, continuing one of the themes of this section, heterotopy is an effective evolutionary mechanism because, like heterochrony, it draws on preexisting developmental pathways and components: the spatial reordering of the turtle skeleton is striking by any measure, but the turtle tucks its scapula, identifiable as such during morphogenesis, under its ribcage rather than evolving a novel structure.

Epigenetics

Local heterochronies and heterotopies are effective evolutionary agents because of another aspect of developmental systems that contributes to the nonlinearities between genetic change and morphological effect: epigenetics in the classical sense, i.e., the local cell and tissue interactions that help to integrate the developing parts of an organism into a functional whole (for example the induction of vertebral cartilage formation by contact with the spinal cord [Hall 1983] or the growth response of both embryonic and postnatal bone to mechanical loading [Carter et al. 1998]). By drawing on a set of preprogrammed responses to local signals, such interactions allow the developing embryo to accommodate evolutionary changes in particular morphological elements without a host of independent but mutually beneficial mutations. We have little detailed knowledge of mechanisms, but striking epigenetic responses to experimental alterations in morphology have been described by many workers, and their evolutionary implications abundantly discussed (reviews include Rachootin and Thomson 1981; Raff and Kaufman 1983; Thomson 1988; Atchley and Hall 1991; McKinney and McNamara 1991; Hall 1999). Those experiments show that evolutionary changes in, for example, eye size need not be accompanied by independent mutations in genes governing bones, nerves, muscles, or blood vessels in the skull (see Twitty's famous transplant experiments reviewed by Müller 1990), and changes in vertebral posture need not involve independent mutations in genes that collectively control thoracic cross-sectional shape or the sites of muscle insertions on limb bones (see Slijper's bipedal goat, reviewed by Rachootin and Thomson 1981), because epigenetic interactions yield accommodations to those morphological changes.

The epigenetic interactions that help generate complex forms do have limits to the changes they can accommodate, as attested by many "failed" embryological manipulations or the more bizarre gene regulation experiments that haunt the pages of *Nature, Cell,* and similar journals. The macroevolutionarily important questions revolve around where those limits lie for particular clades or particular kinds of changes, and how evenly the potential directions of permissible change are distributed in morphospace. An approach along those lines will help to illuminate how two of the seemingly conflicting themes of development, modularity, and integration—at several hierarchical levels within the organism from molecular pathways to tissue inductions—conspire to produce evolutionary novelty.

Wanted: An Integrated Genetics

All of these aspects of genetics strongly underscore the nonrandom and nonlinear nature of variation that is the raw material for evolution. Any probability distribution of potential changes around a genotype (phenotype) will inevitably be inhomogeneous, reflecting evolutionary lines of least resistance that are conditioned by the underlying structure of developmental pathways at the molecular, cellular, and tissue levels. The challenge is to take these new insights from the organismal level to the population, species, and clade level to forge a better understanding of the links and discontinuities among those levels.

A synthesis of developmental and population genetics will not be easy, as R. A. Fisher himself recognized ("I can no longer calculate it," he said when confronted with early evidence for nonadditive genetic variation [Mayr 1992]). A general mathematical theory may not much resemble our textbook versions of microevolutionary population genetics or quantitative genetics but will contain elements of both. Various, rather disparate, pioneering attempts suggest some potential components of such a program (e.g., Arthur 1988, 1997; Atchley and Hall 1991; Atchley et al. 1994; Schluter 1996; Nijhout and Paulsen 1997; Wagner et al. 1997; Orr 1998; Rice 1998). Given that our knowledge of the genotype–phenotype relation is still heavily biased toward laboratory strains of a few model organisms, far more extensive analysis of natural populations will be a critical step whose early phases were noted in the section on genetic hierarchies and networks, above.

A discouraging possibility is that so much taxon-specific information will be required that quantitative models, or even qualitative predictions, at the macroevolutionary scale will have little power. However, the strong commonalities among phyla in basic developmental genetics hold out some hope, and certain differences among clades suggest some simple hypotheses that might be explored. For example, protostomes have evolved the genetic machinery for generating complex morphologies by keeping genomes relatively small and increasing the number of times a gene is used during development, whereas deuterostomes have enlarged the genome by two rounds of duplication so that multiple copies of a given gene are available to diverge functionally (Akam 1998; Holland 1999; Valentine 2000; among many others). Do these differences in genetic architecture impinge on phenotypic evolution to produce differences in evolutionary style between the two main metazoan branches (e.g., in dissociability of structures—that is, the relative degree of developmental independence of structures or organs within bodies, styles of heterochrony or heterotopy, etc.)? The two architectures may prove to be functionally equivalent from a macroevolutionary perspective, but demonstrating that they generate qualitatively as well as quantitatively similar variation would be equally interesting. Whether phenotypic evolution is gradual or discontinuous when regulatory changes give rise to major morphological novelties is still utterly uncertain, and probably depends on the types of characters involved. Reasonable arguments are available for either evolutionary dynamic and so data rather than theory will be required to determine relative frequencies.

Bridging the empirical and conceptual gap between developmental biology and macroevolution remains a challenge. Living and fossil organisms are all the products of developmental sequences that themselves had to evolve to yield the diverse forms of past and present taxa, but hypotheses on the nature and evolutionary impact of those developmental changes are difficult to test, and inferences on developmental changes in extinct clades can rarely be precise. Using our growing knowledge of gene-expression events within, for example, a *Drosophila* or crustacean embryo as a basis for understanding the morphological diversification of the arthropod clade (and vice-versa) represents a truly daunting shift in spatial and temporal scale. But here as elsewhere, paleontology has much to gain and much to offer. For example, paleontological comparisons across clades, habitats, or time intervals of the density distributions and sequences of novel morphologies around their ancestral starting points can make substantial contributions to understanding the large-scale consequences of the organization of developmental systems. This work will of course be most powerful if done in concert with actual developmental information from the clade in question, but paleontology's spectacular array of morphologies viewed in historical perspective can lead the way for a host of new questions and help to target organisms for developmental analyses likely to yield macroevolutionary insights.

Temporal and Spatial Variation in the Origin of Novelties

Given the many ways in which developmental changes can generate evolutionary novelties, the macroevolutionary expectation might be that novelties arise stochastically according to clade-specific pressures and opportunities. The highly nonrandom first appearances of novelties in the fossil record thus present an intriguing challenge. Although debate continues on the mechanisms behind these patterns, and more empirical work is sorely needed, these large-scale patterns in time and space hint at the potential paleontological contributions to a theory of evolutionary novelty.

Temporal Patterns.—The most striking burst of evolutionary creativity in the animal fossil record comes early in the Phanerozoic, with the Cambrian explosion of metazoan bodyplans. This extraordinary interval, which saw the first appearance of all but one of the present-day skeletonized phyla (along with an array of less familiar forms) in an interval of less than 15 m.y., has received considerable attention recently from both geological and developmental perspectives. The standing of the two major explanatory models, one involving intrinsic, developmental or genomic controls and the other involving extrinsic, environmental or ecological controls, has varied over the past decade, but resolution has been difficult (see Erwin et al. 1987; Jablonski and Bottjer 1990a; Valentine 1995; Conway Morris 1998; Erwin 1999, 2000a; Knoll and Carroll 1999; Valentine et al. 1999; Jablonski 2000a). The rival hypotheses need not be mutually exclusive, of course, for as Erwin (2000a) notes, "successful innovations require ecological opportunity, developmental possibility and an appropriate environmental setting." The problem therefore becomes a matter of assessing which of Erwin's triad of requirements— for example, open ecospace as set by the biota, intrinsic morphogenetic potential, or by favorable oxygen levels or other physical limiting factors—was the immediate trigger of the explosion. An equally intriguing and no less elusive problem is to determine which of them damped the production of bodyplan afterwards, without actually damping diversification at lower levels—including further developmental modifications—so that the production of phyla ceased but lower-level taxonomic diversity soared to new heights during the Ordovician, and again in the post-Paleozoic. And of course the trigger might have been different from the damper.

Secondary pulses of evolutionary innovation occur in the wake of mass extinctions, adding further temporal structure to the origin of novelties. These repeated pulses, measured either as the first appearances of high-ranking taxa or increases in morphological disparity, strongly suggest a role for ecological opportunity in the origin of novelties (see also below). But is this sufficient to account for the Cambrian explosion, or were genomic factors also involved? Erwin et al. (1987) reasoned that taxonomic diversity but not genomic flexibility should have approached Cambrian levels after the Permo-Triassic extinction, allowing a test of the genomic and ecological hypotheses. However, they found that even the end-Permian debacle, although cutting deep into taxonomic diversity, failed to remove major functional groups from the marine biosphere and thus did not sufficiently mimic the ecological character of Cambrian seas for a definitive test.

Another approach is to sidestep taxonomic rank and examine the evolution of morphology more directly. The rapid filling of a quantitatively defined morphospace has been documented for a number of clades originating in the early Paleozoic (see Foote's 1997 review). More daunting is the task of quantifying larger and more heterogeneous groups such as the deuterostomes or the bilaterians for the purpose of gaining an overall picture of the deployment of morphological diversity through time—but see Thomas et al.'s (2000) analysis showing that 80% of the theoretical skeleton designs available to living and extinct metazoans were occupied by the Middle Cambrian.

Comparative approaches to morphospace occupation can be used to address the rival hypotheses for temporal trends in the origin of novelties. For example, Foote (1999) found that the post-Paleozoic rediversification of crinoids, although representing a rapid increase in morphological disparity, yielded a narrower set of architectural novelties than those established in their initial, early Paleozoic radiation. He took this as evidence for less constrained developmental inputs to the Cambrian explosion than in later times. Wagner (1995) also argued for the lability of different kinds of traits during the initial and later phases of Paleozoic gastropod diversification. These are intriguing results that appear to support a genomic component to the Cambrian explosion, but they are just a first step, in part because they lack an explicitly generative or developmental component. For example, a developmentally based partitioning of traits to compare the long-term behavior of those that might have been more subject to initial freedom and later limitation, relative to those that might have been subject to earlier entrenchment (as attempted by Jacobs 1990) would be valuable. Pinpointing appropriate characters for comparative analysis may not be straightforward, however. For example, Hughes et al.'s (1999) finding that the variability of thoracic segment number in trilobites depends on the number of segments in a given species rather than its phylogenetic position or geologic age undermines a particular argument (McNamara 1983) but cannot fully address the kinds of regulatory networks that were involved in the radiation of arthropod and related clades in the Early Cambrian.

Instead of focusing on particular characters, the role of genomic changes in the damping of the Cambrian explosion might be tested by tracking levels of morphological integration through the early Paleozoic (Erwin 1994)—that is, the morphometric correlation patterns within an organism (e.g., Olson and Miller 1958; Cheverud 1996). This may require a better understanding of developmental mechanisms than presently available, because phenotypic integration can be stable even when genotypic covariances are not (e.g., Turelli 1988; Shaw et al. 1995; Schlichting and Pigliucci 1998). However, Nemeschkal's (1999) finding that avian morphometric correlation patterns correspond to the expression domains of Hox and other developmental control genes is encouraging in its implication that morphometric matrices can reflect developmental architecture (see also Leamy et al. 1999). Operational questions aside, the evolutionary role of morphological integration needs further theoretical and empirical study: maximal evolutionary lability may come at an intermediate level of integration, where the body is composed of locally integrated units that can behave as modules, as discussed above (e.g., G. P. Wagner 1996; Kirschner and Gerhart 1997). G. P. Wagner (1996) argues that the origin of multicellularity led to an decrease in integration as regional specialization of morphology led to differential gene expression, but developmental "burden" is held to increase through time as developmental interdependencies accumulate (e.g., Riedl 1978; Donoghue 1989). The two perspectives are of course potentially compatible, as they may represent very different scales (multicellularity vs. the developmental genetics of a single feature), but they do suggest that this issue would be worth exploring in greater depth.

Although direct comparisons are difficult, a major burst of novel plant architectures is associated with the invasion of land (e.g., Niklas 1997; Bateman et al. 1998). We need to develop criteria to determine whether the establishment of the major land plant designs was as profound an evolutionary event as the Cambrian explosion, as sometimes suggested. If so, we will probably need to confront the intrinsic/extrinsic debate here as well: were plant developmental systems more labile as late as the Devonian (in contrast to animal systems), or is the ecological opportunity afforded by the assembly of traits permitting terrestrial existence a sufficient explanation? Recent work suggests that the MADS-box genetic architecture that orchestrates plant development in ways reminiscent of Hox and other high-level regulatory genes in animals were in place early in land plant evolution (Theissen et al. 2000).

Evolutionary novelties at lower levels exhibit very different temporal patterns from major novelties, and may depend on different variables. As made clear by Valentine's (1969, 1973, 1980, 1990) seminal work on evolutionary and ecological hierarchies in the fossil record, this is immediately evident from the dis-

cordance between the diversity dynamics of marine taxa ranked as phyla and classes relative to the very different dynamics of those ranked as families and genera (see also Erwin et al. 1987). Such discordances are not simply an artifact of the greater inclusiveness of higher taxa (as suggested by Smith [1994], among others), because similar patterns emerge from taxon-free analyses of multivariate morphologic data (Foote 1993, 1996a, 1997, 1999; and see Lupia 1999 on an early burst of disparity followed by stability in angiosperm pollen, discordant with species-level diversity). Raup (1983) essentially recognized this as well by noting that the requirements of tree topology alone could not account for morphological divergences like the Cambrian explosion (that is, the topology of the tree cannot account for the autapomorphies on each branch).

The dynamics of novelty production at lower levels can be quite unexpected when examined in detail. For example, Jablonski et al. (1997) found the production of morphological novelties within the bryozoan orders Cyclostomata and Cheilostomata to be opposite in timing to that expected from the ecological-opportunity hypothesis that is the chief contender for explaining high-level originations. Cyclostomes generated novelties in a steady trickle despite their occurrence in the relatively low-diversity early Mesozoic world, whereas the cheilostomes produced novelties in a burst despite being embedded in the presumably more crowded, predator- and competitor-rich mid-Cretaceous environment. In this comparative study, ecological context mattered less than the relative speciation rates (and competitive abilities?) of the respective clades.

Spatial Patterns.—Evolutionary novelties also show spatial patterns in their first appearance in the fossil record. Like the major temporal patterns, onshore–offshore patterns show discordances between the first appearances of higher taxa and patterns of origination, extinction, and diversity accumulation at lower taxonomic levels: benthic marine orders tend to originate in onshore, disturbed habitats, regardless of the diversity dynamics of their constituent species, genera, and families (Jablonski and Bottjer 1990a,b, 1991; Droser et al. 1993; Jablonski et al. 1997). Although much more work is needed, this pattern can also be seen in terms of derived characters (Jablonski and Bottjer 1990b is only a crude beginning) and in the morphological divergence of the founding species for two echinoid orders relative to the disparity among species in the ancestral group, showing how the onshore initiators of these orders "broke away from the bounds of [their ancestor's] morphospace from the very start" (Eble 2000: p. 68). In apparent agreement with these patterns, major land plant originations also appear to be concentrated in disturbed habitats, in both Paleozoic and Mesozoic settings (e.g., DiMichele and Aronson 1992 and Wing and Boucher 1998, respectively), although the botanical data have not been analyzed for the marine discordance across hierarchical levels.

The geography of first occurrences of major groups is particularly subject to sampling bias, but an attempt to take these into account found disproportionate appearances of marine invertebrate orders in tropical latitudes (Jablonski 1993). Similarly, phylogenetic analysis suggests that major plant lineages tend to originate in the Tropics and spread poleward, in that primitive members of clades tend to be tropical and derived taxa tend to be restricted to or best developed in the temperate zones (e.g., Judd et al. 1994, and in a very different tradition, Meyen 1992), although of course different dynamics could underlie this pattern (but see also Askin and Spicer 1995 on paleontological data). More work is needed, however, to test whether the latitudinal trend in higher-level originations significantly exceeds the probabilistic, per-taxon expectation, given that the latitudinal trend in species richness appears to have been present over most of the Phanerozoic, albeit with varying slope and subject to considerable sampling and preservation bias (e.g., Stehli et al. 1969; Humphreville and Bambach 1979; Kelley et al. 1990; Crame 1996; Walsh 1996).

Species and genera show less striking latitudinal origination patterns, and many clearly started in high latitudes (e.g., macroinvertebrates [Feldmann et al. 1993; Crame 1997], Neogene Foraminifera [Buzas and Culver 1986; Wei and Kennett 1986; Spencer-Cervato et al. 1994], terrestrial plants [Wen 1999]). The relative fraction of high-latitude origination at low taxonomic levels has not been sufficiently quantified to compare with the preferential low-latitude appearance of the much smaller number of higher taxa, however. Plotting the mean or median geologic age, or alternatively the estimated net diversification rates, of extant lower taxa against latitude has given apparently conflicting results (e.g., for marine taxa [Wei and Kennett 1986; Flessa and Jablonski 1996 and references therein; Crame and Clarke 1997], terrestrial birds [Gaston and Blackburn 1996], birds and butterflies [Cardillo 1999]). Violation of assumed time-homogeneous dynamics may be one source of the conflict, but this entire area deserves more extensive study.

Potential mechanisms for spatial biases in evolutionary innovation are plentiful but difficult to test, and as with the temporal bias the jury is still out on whether this is a genetic or an environmental problem, or even whether the pattern is underlain by differential production or differential persistence of evolutionary novelties (or both). Environmental possibilities include high rates of local extinction and invasion and thus local opportunity for innovation, more variable and/or intense selection pressures, or the potential extinction-resistance of novelty-bearing species, in more disturbed habitats relative to more stable ones (e.g., Jablonski and Bottjer 1983, 1990b; Hoffmann and Parsons 1997). Latitudinal patterns might simply represent diversity-dependent novelty production at this spatial scale, or more complex combinations of novelty production and survival to produce the net effect recorded paleontologically (Jablonski 1993).

Genetic explanations for these patterns include the greater potential of small peripheral isolates—which would arguably be more frequently formed in disturbed, heterogeneous habitats—to have lower developmental stability or to break developmental canalization (Levin 1970; Jablonski and Bottjer 1983; Clarke 1993; Hoffmann and Parsons 1997). Most recently, Rice's (1998) model suggests that canalization might most readily be broken by strong truncation selection, when significantly less than 50% of a population is selected to produce the next generation, a situation that might obtain most often with the enormous mortality of propagules in shallow-marine benthos (although the selective basis of that mortality and its overall impact relative to mortality in other populations is poorly understood [e.g., Rumrill 1990; Pechenik 1999]). The potential role of highly variable environments in fostering evolutionary innovation (e.g., Parsons 1993, 1994; Hoffmann and Parsons 1997; Hoffmann and Hercus 2000) has gained a new developmental wrinkle with the finding that some regulatory molecules that suppress phenotypic variation can be disabled not only by mutation but by environmental extremes such as high temperatures (Rutherford and Lindquist 1998; Wagner et al. 1999).

The connection between any of these mechanisms and the empirical macroevolutionary patterns remains highly speculative. However, they do suggest testable ways of making mechanistic sense of evolutionary patterns that are not smooth extrapolations up the taxonomic hierarchy. Comparative analyses of morphological divergences within and among clades, structured along environmental gradients, would be a valuable—and tractable—step in this area, particularly because the first appearances of most orders are not as divergent from likely ancestors as are the phyla of the Cambrian explosion. Complementary approaches that examine the first appearance of major groups in terms of derived characters or multivariate morphometrics relative to patterns within clades and across environments (along the lines of Eble 2000) would be valuable.

The discordance between high and low taxonomic levels in temporal and spatial patterns of origination, and between morphological diversification at different levels within nested clades and subclades, thus provides an intriguing set of patterns that require a hierarchical approach. The spatial component also demonstrates that large-scale evolutionary processes cannot be analyzed exclusively at the global scale, because unexpected—or at least previously undetected—structure resides at the regional scale and across environmental gradients (see also Miller 1998; Jablon-

ski 2000a). These aspects of the evolutionary process can also be seen in the sorting of variation.

Sorting of Variation

Species-Level Stasis and Change

Stasis and Its Causes.—As many authors have pointed out, microevolution can occur as rapidly as needed to account for virtually any speed observed in the fossil record (e.g., Charlesworth et al. 1982; Gingerich 1993; Kirkpatrick 1996; Hendry and Kinnison 1999; and many others). This has been abundantly demonstrated not only in the laboratory but in natural populations from Galapagos finches (Grant 1986; Grant and Grant 1995) to Bahamian anoles (Losos et al. 1997) to mosquitoes in the London Underground (Byrne and Nichols 1999), although as discussed above simple extrapolations of these changes may not provide the best model for all of the inhomogeneities in the origins of major novelties. The more challenging question then becomes, why are evolutionary rates generally so slow in the fossil record? This question pertains both to the species level, which is the domain of punctuated equilibrium and its alternatives (Gould 1982; Gould and Eldredge 1993), and to the clade level, where large-scale evolutionary trends often unfold with excruciating slowness when viewed on microevolutionary timescales (e.g., Stanley 1979; McShea 1994). The relation between potential mechanisms at the different levels has been discussed mainly in broad generalities, but few workers have attempted to address whether the factors that cause, for example, species-level stasis seen in many members of the horse lineage (e.g., Prothero and Shubin 1989) are also responsible for the slow rate of body size increase in the clade. Averaged over the duration of the entire clade, this size increase was so slow as to be virtually indistinguishable from drift (see Lande 1976, and Stanley's [1979, 1982] punctuational reinterpretation, seconded by Stebbins 1982; and also Lieberman et al. 1994, who found rates so slow in a Devonian brachiopod lineage that they would have involved only three selective deaths per 10 million individuals if treated as a continuous trend).

Over the past quarter-century, evolutionary stasis has proven to be a pervasive morphological pattern in the fossil record (reviewed in Erwin and Anstey 1995a; Gould and Eldredge 1993; Hallam 1998; Jackson and Cheetham 1999). However, few of the hypotheses on the forces that maintain this stasis at the species and higher levels have been conclusively tested and again, different mechanisms may obtain in different clades. The research questions have shifted to testing for among-clade and among-habitat differences in frequencies of evolutionary tempo (abrupt vs. gradual change) and mode (anagenetic vs. cladogenetic, sometimes termed phyletic vs. branching), the roles of intrinsic and extrinsic factors that might govern those differences, and whether the direction of phenotypic change during sustained anagenesis or cladogenesis is related to the morphologic behavior of the species or its constituent populations during preceding intervals.

Given pervasive stasis, the stunning diversity and subtlety of biological adaptations must often arise episodically, in the punctuations between stable species, either in single punctuational episodes—which, of course may encompass tens or hundreds of thousands of years (e.g., Jackson and Cheetham 1999: Table 1)—or in cumulative series. This process need not rely entirely on isolation itself as the trigger for adaptive change, but may also draw on geographic variation within established species. Consider, for example, Futuyma's (1987) very attractive but still untested suggestion that speciation events cordon off local adaptations into discrete gene pools, thereby packaging ordinarily ephemeral characters into more stable evolutionary units (see also Eldredge 1989, 1995). The apparently episodic nature of this process, at least in terms of the morphologies accessible in the fossil record, underscores the need to understand stasis.

At the species level, stasis over geologic timescales has been attributed to variation in both rate and direction of change. Variation in the rate of change involves truly slow evolutionary rates between the punctuations, with temporal stability generally attributed to constant selection for intermediate phenotypes,

interrupted by rapid anagenetic or cladogenetic shifts (maximum observed rates of change may also be artificially reduced by the size of the time bin encompassing both stasis and directional change). Less often considered is the possibility that directional selection fluctuates so rapidly that populations cannot respond, with the net effect of stasis at the mean phenotype; another alternative would be time-averaging of samples rather than selection pressures, detectable if not geologically then perhaps by exceptional apparent population variances (see Kidwell and Aigner 1985).

Sustained stabilizing selection must be the force behind habitat tracking as a mechanism for stasis (Eldredge's 1985, 1989, 1995 hypothesis), in which species remain morphologically static as they move with a favorable environment during climatic and other changes. The tracking process seems well supported in the Pleistocene (e.g., Valentine and Jablonski 1993; Coope 1995; Clark 1998; see also discussions in Price et al. 1997 and Jackson and Overpeck this volume). Intrinsic differences among taxa in their ability to keep up with shifting environments have not been explored as an explanation for differences in evolutionary tempo and mode; this may be unimportant, however, if we can generalize from rates of movement in Pleistocene plants and animals (e.g., Clark 1998). For widespread species, a more realistic model might be cline translocation (coined by Koch 1986), in which a set of populations that vary along an environmental gradient shift in and out of a sampling area to give the appearance of oscillatory or even directional change as the species overall maintains a constant morphology (see for example Stanley and Yang's 1987 extensive study of late Cenozoic bivalves, in which the total range of multivariate oscillations through the history of each lineage was very similar to its present-day geographic variation).

Most species-level lineages appear to lack directionality rather than evolutionary lability; that is, they show high total rates of evolution while accumulating little net change. The most frequently invoked model is that of oscillating directional selection (e.g., Gingerich 1993; Sheldon 1996; Hendry and Kinnison 1999), a process well documented in some modern populations (e.g., Grant and Grant 1995; Via and Shaw 1996; among many others). Such evolutionary dynamics can be modeled and in principle tested against drift and other forces, although paleontological applications are still being developed and in some instances seem highly model-dependent (see Bookstein 1987, 1988; and Roopnarine et al. 1999; but see Cheetham and Jackson's [1995] overview of their superb multidisciplinary analysis of Neogene bryozoan evolution). A less-explored, explicitly hierarchical alternative involves the spatial structure within species: gene flow among highly dynamic local populations within a species might allow little net overall change (e.g., Eldredge 1985, 1989; Lieberman et al. 1995). Although molecular and other data suggest that few species lack some internal spatial structure (e.g., Hanski 1999; Avise 2000), it is not clear whether the particular metapopulation dynamics required by this model for stasis are truly pervasive in nature (e.g., S. Harrison 1998; Maurer and Nott 1998).

Also controversial is whether the apparent bounds on oscillatory stasis represent intrinsic limits of the organisms or reversals in selection pressure. I am not going to venture into the dense and tangled literature of evolutionary constraints, but the widespread existence of evolutionary trade-offs (as, for example, between age and size at first reproduction, when selection favors both early reproduction *and* large size [e.g., Stearns 1992]) seems to be a strong endorsement for some form of intrinsic constraint, at least in the short run (for morphological examples, see Nijhout and Emlen 1998). The detection of such trade-offs, however, generally carries little information on mechanisms underlying constraints, and, as with genotypic and phenotypic variance–covariance matrices (e.g., Shaw et al. 1995; Arnold and Phillips 1999), we do not know how stable they are over evolutionary time. Some must be nearly absolute, others may be quite transient and context-dependent. Plant and animal breeders hit limits all the time, and the failure to break the egg-a-day barrier in chickens (Lerner 1953), or to increase thoroughbred

racing speeds significantly over the past 70 years (Gaffney and Cunningham 1988), is not for lack of intense directional selection or high heritability of relevant traits.

Experimental work on host specificity in phytophagous insects suggests that intrinsic factors may be important in wild populations as well. Many insects exploit a restricted diet, presumably owing to plants' defensive compounds, but experiments in some groups have detected no significant relation (or, less often, a *positive* relation) between insects' performance on their host plants and their performance on other species, undermining a trade-offs hypothesis; a lack of genetic variation may actually be a limiting evolutionary factor in this instance (Futuyma et al. 1995; but see Keese 1998). This is not a trivial issue, given Farrell's (1998) contention that the overwhelming species richness of beetles is related to the macroevolutionary consequences of host shifts in phytophagous clades. The general relation between trade-offs, genotypic covariances, and other apparent limitations to evolutionary responsiveness on the one hand, and patterns of morphologic change in species over geologic timescales on the other, is clearly an attractive target for combined paleontological/neontological analysis of particular clades. To cut through the terminological morass, all of these features can be put under the rubric of *developmental constraints*, which might be defined as the resistance, owing to the configuration of developmental networks and pathways, of the phenotype to selection in certain directions. In principle this can be distinguished from *canalization*, which might be defined as the resistance, owing to the buffering or redundancy of developmental processes, of the phenotype to mutation or to environmental variation (and see Gibson and Wagner 2000 for a valuable overview).

Distribution of Stasis and Change.—The distribution of evolutionary tempos and modes at the species level remains poorly known, not least because rigorous research in this area is such a daunting task. Few studies have fully addressed all of the issues, but, drawing on the discussions of Gould and Eldredge (1977), Fortey (1985), Clarkson (1988), Erwin and Anstey (1995a), and Jackson and Cheetham (1999), an appropriate protocol would include

1. Large samples in a closely spaced time-series

2. Objective delimitation of species as operational units

3. Stratigraphic control independent of the target clade

4. Independent evidence on sedimentation and preservation rates that might vary to create artificial punctuations or protracted transitions

5. An assessment of within-species geographic variation

6. A phylogenetic hypothesis

The characterization of morphospecies has become increasingly rigorous with the availability of multivariate morphometric methods. An encouraging development has been the generally good correspondence between biological units and the morphospecies of the shelly macroinvertebrates used in most analyses of evolutionary tempo and mode at these scales. This is not the place for an extensive discussion of species concepts, but from an evolutionary perspective species-level units are most useful if they are essentially independent lineages (e.g., Simpson 1961; Wiley 1981; Mayden 1997; de Queiroz 1998, 1999, and references therein). For the outcrossing biparental species that provide most of the animal and protistan fossil record and a sizeable but unknown fraction of the plant record, that independence often involves reproductive isolation or genetic cohesion, and so coincides with any broadly defined biological species concept that can accommodate isolation, recognition, cohesion and related viewpoints (e.g., Templeton 1989, 1998; Ghiselin 1997; Coyne and Orr 1998; R. G. Harrison 1998; de Queiroz 1998). However, that evolutionary independence need not be compromised even if those barriers are not absolute (to give just two examples, the fossil record shows that cottonwoods and balsam poplars have been generating hybrids in western North America since the Miocene but have remained distinct entities [Eckenwalder 1984] and that two lineages of Neogene cyprinid fishes hybridized for 2 m.y. without subverting the evolutionary identities of the parent lineages [Smith 1992]),

and other processes besides the traditional isolating barriers may impose or contribute to evolutionary independence as well (e.g., Van Valen 1976; Hull 1997; among many others).

Analyses within paleontologically important phyla where morphometrically defined species correspond closely to biologically, usually genetically defined ones include: the cheilostome bryozoans *Stylopoma*, *Steginoporella*, and *Parasmittina* (Jackson and Cheetham 1990, 1994, who also used an extraordinary set of breeding experiments, and see also Hageman et al. 1999); the benthic foraminifer *Glabratella* (Kitazato et al. 2000, also based on breeding experiments; but see below for the uncertain situation with planktic Foraminifera); the gastropods *Amalda* (Michaux 1987, 1989, who also found congruent phylogenies using both data sets), *Nucella* (Collins et al. 1996, albeit with considerable intraspecific shell variation), *Littorina* (Rugh 1997, who compared shell morphology with such biological species indicators as genital and egg-capsule features), and *Lacuna* (Langan-Cranford and Pearse 1995, again using breeding experiments); the corals *Porites* (Potts et al. 1993; Budd et al. 1994, again with congruent phylogenies) and *Montastraea* (Weil and Knowlton 1994; Knowlton et al. 1997); the decapod *Synalpheus* (Duffy 1996, using allozymes); the articulate brachiopod *Terebratulina* (Cohen et al. 1991, using both allozymes and mtDNA); and even the notoriously nondescript inarticulate brachiopod *Glottidia* (Kowalewski et al. 1997, who lacked genetic data and relied on previous biospecies definitions). One can only hope for a steady stream of such studies, including a new round on vertebrates such as Steppan's work (1998) on the rodent *Phyllotis*, using mtDNA versus skeletal morphometrics.

These accumulating results suggest that the paleontologist need be at no greater remove from biological units than any other systematist lacking a full molecular treatment of the taxonomic units under study. And by providing concrete support for the biological reality of the morphological differences between related fossil species, they imply that the morphological punctuations in fossil lineages—an empirical pattern open to multiple interpretations—do tend to correspond to speciation events. Anagenetically evolving lineages lacking speciation-scale punctuations can be more problematic, of course, and when broken into arbitrary taxonomic segments may imply an artificially punctuational pattern (see Sheldon 1993). However, this pitfall will be avoided as long as phenotypic change is the final arbiter on questions of evolutionary tempo and mode at the species level, as seen in most recent studies including the examples cited here and by Jackson and Cheetham (1999).

On the other hand, sibling or cryptic species—that is, biological species that are virtually undetectable morphologically—are common in many taxa, both terrestrial and marine (e.g., Knowlton 1993; Avise 2000). To some authors, this imperfect correspondence between morphospecies and biologically discrete species dictates the collapse of the entire enterprise (e.g., Levinton 1988; Hoffman 1989), but this simply is not true, so long as the questions are posed appropriately. For example, a lineage is punctuational if most morphological change occurs in the context of speciation when viewed over geologic timescales. But this does not require that the converse be true, that all speciation events are accompanied by morphological change. More problematic is the generation of temporally and spatially persistent, discrete morphotypes that can arise abruptly but are not reproductively isolated, that is, are not evolutionarily independent entities (e.g., Palmer 1985; Trussell and Smith 2000). The examples cited above suggest that paleontologists are becoming adept at partitioning their morphological units in ways that are genealogically significant, but the ranking of discrete morphologies remains a potential problem and needs more attention. The same is true for character-based neontological species concepts, of course, particularly those based on "smallest diagnosably distinct units" (Cracraft 1989, 1997; Nixon and Wheeler 1990; Davis and Nixon 1992; Luckow 1995), where the taxonomic ranking and evolutionary roles of those units also can be controversial (e.g., Theriot 1992; Hull 1997; Knowlton and Weigt 1997; R. G. Harrison 1998).

The most robust analyses will be those that compare rates and patterns of morphospecies production among clades (particularly within the same geological arena, so that many potential taphonomic biases are held constant), rather than depending on absolute values. Significant differences detected in comparative analyses will be misleading only if the frequency of sibling species has a strong inverse relation to the frequency of morphospecies origination. Little or no evidence of such a relation exists, although a formal analysis would be valuable. The sparse literature on important components of the fossil record, such as marine invertebrates and terrestrial vertebrates, conveys the general impression that the numbers of morphospecies and sibling species are, if anything, *positively* correlated among clades. If this is true, or if the relation is random so that no systematic bias is introduced, then cryptic species will not be a serious problem for comparative studies of evolutionary tempo and mode, at least in large data sets.

Geographic variation has been the Achilles' heel of many paleontological studies of evolution at the species level. Analyses centered on one or a few closely spaced stratigraphic sections or cores risk confounding the lateral movements of trends in intraspecific variation with evolutionary change, the methodological pitfall created by the cline translocations mentioned above (and the potential for local populations to exhibit independent morphological trajectories without net species-level directionality adds another hierarchical level to be considered [see Lieberman et al. 1995; Bralower and Parrow 1996]). This problem was recognized over 40 years ago (Newell 1956; Arnold 1966), but its remedy is generally so labor intensive that only a few studies have risen to the challenge (but see, gratefully, Stanley and Yang 1987; Cheetham and Jackson 1995, 1996). A formidable obstacle is the inverse relation between acuity of stratigraphic resolution and geographic distance, particularly along environmental gradients or among disjunct regions: the temporal acuity often achieved by closely spaced samples in a single section declines significantly when correlating among sections (see Behrensmeyer and Hook 1992 and Behrensmeyer et al. 2000 [this volume] on analytical time-averaging). This Paleontological Uncertainty Principle—the trade-off between temporal resolution and geographic coverage—seems to be little appreciated outside the field but has implications for virtually every kind of paleobiological analysis. Quantitative stratigraphic methods, significant refinements in radiometric dating techniques, and tuning of correlations to Milankovitch cycles (e.g., Shackleton et al. 1999) will yield increasingly fine correlations, but resolution will tend to approach a limit on the order of thousands of years, if only because natural time-averaging operates at about this scale for most micro- and macrofossil records (see Kidwell and Flessa 1995; Martin 1999).

All of the end-member combinations of evolutionary tempo and mode have now been observed in fossil species transitions, and so the challenge is to assess the frequencies of the different patterns, and to test for the influence of biological traits, environmental factors, and other potential controlling variables. This effort is complicated by the strong imbalance in the evidence required to demonstrate gradualism versus stasis (see Fortey 1985, 1988; Clarkson 1988; Sheldon 1993, 1996; Pearson 1995; Wagner and Erwin 1995). Stasis can often be convincingly documented by samples from a succession of discrete sedimentary packages, even when the packages are separated by depositional hiatuses or unfavorable environments. Further, quantifying geographic and other intraspecific variation is less critical if even the local pattern is one of temporal stability. Stasis is unlikely to be artificially generated or removed by time-averaging, where successive populations are homogenized within a single sedimentary bed. Short-term directional changes can be collapsed into a single artificially variable assemblage, but trends extending over more than 10,000 years (depending on depositional environments, of course) and thus significant relative to the average duration of morphospecies, will generally be retained, and situations that would obliterate them can be recognized by independent evidence (e.g., Kidwell and Aigner 1985; Bell et al. 1987; Kidwell and Flessa 1995).

Ironically, then, gradualism is more difficult to demonstrate conclusively in the fossil record than the alternatives, even though it was long taken to be the dominant style of evolutionary change! That said, distinguishing between true punctuated equilibrium, i.e., punctuated cladogenesis, and punctuated anagenesis, in which morphological change occurs episodically but without lineage branching, is not always straightforward either. This distinction cuts to the heart of the question of speciation's role in evolutionary change: the anagenetic mode can accommodate a broad range of intraspecific evolutionary processes (e.g., Gould 1982; Wright 1982a,b; Lande 1986; and a host of others since then). As noted above, however, establishing the coexistence of ancestor and descendent species, or of multiple sister species, requires a detailed phylogeny and well-resolved stratigraphic range endpoints. Like all paleontological analyses it also hinges on the ranking of morphologically defined units: taxonomic lumpers will tend to reduce the number of branching events, while splitters are more likely to convert anagenetic patterns to cladogenetic ones by increasing the number of contemporaneous taxonomic units. The growing inventory of studies linking morphology to genetically defined species suggests that the splitters have been closer to the biological reality (with past excesses and missteps, of course). Although some cladists have rejected the possibility of identifying ancestral species on theoretical grounds (e.g., Englemann and Wiley 1977; Frost and Kluge 1994; Norell 1996), increasingly rigorous protocols have become available for the recognition of potential ancestors for both fossil and living organisms (e.g., Paul 1992; Theriot 1992; Fisher 1994; Smith 1994; Marshall 1995; P. J. Wagner 1995, 1996a; Foote 1996b; Omland 1997). The data are still sparse but suggest that ancestral species can be detected and that temporal overlap with descendants, as expected for punctuated cladogenesis, is not uncommon. The challenge now is to refine and apply methods that will permit a quantitative assessment of when, where, and how often the different evolutionary patterns obtain in nature. A vast and nearly uncharted territory is open for modeling the interplay of sampling and paleobiological pattern (see Holland and Patzkowsky 1999), but most urgently needed

is a new battery of carefully designed and selected empirical studies.

Attempts to assess the relative frequency of
evolutionary tempo and mode are premature,
but some possibilities and problems can be
defined. As already noted, stasis and punctuation appear to be the pervasive phenotypic
patterns in marine macrofossils, although the
relative proportions of anagenesis and cladogenesis remain unclear (e.g., Hallam 1998;
Jackson and Cheetham 1999). Although more
rigorous quantification would be valuable,
there is little reason to doubt Fortey's (1985)
report that gradualism occurs in fewer than
10% of the 88 trilobite species that have a
meaningful stratigraphic range in the Ordovician Valhallfonna Formation, Spitsbergen,
or Johnson's (1985) assessment, backed up by
his data-rich monograph (Johnson 1984), that
only one of the 34 scallop lineages in the
northern European Jurassic shows possible
gradual change in morphology. On the other
hand, the famous Jurassic oyster *Gryphaea*
shows a more complex mixture of stasis and
gradualism (Johnson and Lennon 1990; Johnson 1993, 1994), and whether this complexity—and contrast with other contemporaneous
bivalves—reflects the intensity of research
prompted by *Gryphaea*'s notoriety as a classic
evolutionary exemplar, difficulties of phylogenetic analysis in a morphologically difficult
and heterochrony-prone group, or a true biological difference, remains uncertain.

Sheldon (1993, 1996) made the intuitively
appealing suggestion that benthic species in
more stable offshore environments might be
more subject to gradual change, but empirical
evidence is slim: Sheldon's trilobite study involves parallel changes in a single character in
a set of lineages from a single restricted area
in which the environment is changing upsec-
tion, albeit subtly (see Sheldon 1987, 1988).
Better documented is the long-standing observation that pelagic species are more likely
to show gradual change than benthic ones
(Johnson 1982; Fortey 1985; Clarkson 1988;
Jackson and Cheetham 1999). Fortey (1985)
contrasts the evolution of a pelagic trilobite
with that of co-occurring benthic species, but
the richest data for gradualistic change come
from microfossils. Three caveats obtain here:

first, geographic coverage remains a weakness
of many analyses of pelagic organisms, although this is becoming less true; second, stasis and punctuations do occur in many microfossil lineages (see tabulations in Erwin and
Anstey 1995a and Jackson and Cheetham
1999), even when hiatuses are taken into account (see MacLeod 1991); and third, so little
is known about the population genetics, or
even how individuals are packaged into species, in these unicellular groups that interpretation of paleontological patterns is doubly
difficult (e.g., Tabachnick and Bookstein 1990;
Norris et al. 1996; Huber et al. 1997; Darling
et al. 1999, 2000; de Vargas et al. 1999; but see
Kitazato et al. 2000 for encouraging results on
a genus of benthic forams, and recall that
some genetic analyses are finding that detailed morphometry of, for example, test porosity may help to capture genetic units [e.g.,
Huber et al. 1997; de Vargas et al. 1999]).

The record for land vertebrates is difficult to
interpret because many studies lack one or
more of the elements enumerated above (for
cautionary notes see, for example, Schankler
1981; Heaton 1993). Some mammal lineages
do appear to present robust examples of gradualistic change at the species level, however
(see reviews by Barnosky [1987], Martin
[1993], Chaline et al. [1993], and Carroll
[1997]). For example, Chaline and Laurin
(1986) found gradualism in a Plio-Pleistocene
vole lineage over a broad geographic area,
with quantitative data on cheek-tooth morphology in a series of time planes extending
over an area from Spain and Britain to northern Italy, Poland, and the Czech Republic,
with additional qualitative data from localities as far east as Moldova and western Siberia. But as with microfossils, mammals are not
purely gradualistic in evolutionary tempo; indeed analyses of entire faunas or assemblages
of clades suggests that stasis and punctuation
is pervasive and perhaps prevalent (e.g., Barnosky 1987; Flynn et al. 1995; Prothero and
Heaton 1996). Again the key issue is relative
frequency and the factors that impose different frequencies among clades.

Attempts to assess the frequency of different types of speciation based exclusively on
modern species have their own pitfalls. As

Wagner and Erwin (1995) note, phylogenetic tree topology alone cannot reliably distinguish evolutionary tempo and mode. Inferences based on molecular data as a source of temporal estimates show considerable promise but remain model-dependent, not only in terms of molecular-rate constancy but in assumptions about the pattern of morphological change between nodes (e.g., Garland et al. 1999). Finally, estimates of the relative frequency of allopatric and other types of speciation based on the present-day deployment of modern species (e.g., Lynch 1989; Barraclough and Vogler 2000) are undermined by the geographic volatility of species in the recent geologic past and by extinction. Only species that have split since the last glaciation, say in the last 10,000 years, are likely to capture the relative spatial distributions of sister species at the time of speciation. Species that split, say, 2 m.y. ago have been subject to perhaps 20 episodes of geographic shuffling with the waxing and waning of Pleistocene glaciation (e.g., Valentine and Jablonski 1993; Roy et al. 1996; Jackson and Overpeck this volume), so that the relative frequencies of geographic range overlap today probably say more about competitive interactions between close relatives than about speciation events (see also Chesser and Zink 1994). Taxa separated by major geographic barriers like the Rocky Mountains or the Isthmus of Panama are reasonable candidates for allopatric speciation, of course, but these more ancient splits are subject to the problems of extinct species more closely related to one or the other living ones—i.e., of intervening speciation events that represent the true spatial and temporal pattern of lineage splitting (e.g., Schneider 1995; Jackson and Budd 1996).

With all of these caveats, and in light of the sparse and uneven nature of the data, it is unsurprising that no clear taxonomic or environmental pattern has emerged for the distribution of evolutionary tempo and mode at the species level. Perhaps, in an obvious if unsatisfyingly context-specific hypothesis, species histories depend on their geographic extent and genetic population structure—i.e., on scale and hierarchy. If gradual anagenesis is simply the expected paleontological outcome

of homogeneous directional selection, in other words, true Fisherian mass selection, then this sets some requirements on the spatial scale of gene flow relative to that of environmental variation and thus makes predictions on the distribution and genetic structure of gradualistic taxa. On the other hand, for those species that maintain genetic cohesion over different environments, or among regions with disparate selective pressures through time, the interplay of local adaptation and gene flow—intermittent or regular—will tend to impose fluctuations around a mean rather than directionality (an argument raised by Eldredge 1985, 1989; and also consistent with Futuyma 1987). Such a return to the textbook basics could explain why lineages on islands (e.g., Lister 1989) and in isolated basins (e.g., Geary 1995; and Povel 1993 in part) exhibit gradualism while related taxa in more extensive or scattered habitats often show stasis and punctuation. It also provides an approach to the presence of contrasting evolutionary patterns in co-occurring lineages, which would be unexpected if the physical environment alone (e.g., habitat stability [Sheldon 1993, 1996]) determined tempo and mode. In our present state of ignorance it may even explain the gradualistic evolution of many planktic microfossils, which may often evolve as enormous populations that occupy different depth zones in one or more otherwise relatively homogeneous oceanic water masses (e.g., Lazarus et al. 1995; but see Norris et al. 1996 and other foraminiferal references cited above). The Plio-Pleistocene vole data are, however, an apparent counterexample: Chaline and Laurin (1986) note with surprise the gradualistic trajectory of their lineage despite its likely subdivision into semi-isolated populations. This may be the exception that proves the rule, however, if the particular phenotypic changes they measured, involving increasing hypsodonty and elaboration of enamel patterning on the tooth crown, can be attributed to selection imposed by long-term vegetation changes throughout the study area.

As already noted, the relative frequency of anagenesis and cladogenesis has yet to be established. Intuitively, even excluding "pseudoextinction" (i.e., anagenetic transformation obscured by taxonomy), species extinction rates seem to be sufficiently high that frequent branching is required for lineages to persist over geologic timescales. A number of paleontological analyses of tempo and mode that consider clades of sufficient size and phylogenetic resolution for analysis do show significant numbers of species arising cladogenetically, with stratigraphic range overlaps between putative ancestors and descendants, or between sister species (see Erwin and Anstey 1995a; Jackson and Cheetham 1999; also Stanley et al. 1988; Wagner 1998). Nonetheless, all of these references, and many more besides, also contain examples of punctuated anagenesis, so that the apparent prevalence of stasis in many situations may or may not be matched by the prevalence of cladogenesis, as required by the punctuated equilibrium model. Clearly, analyses modeled on the Cheetham and Jackson (1995) studies and focused on other groups well represented in the fossils record, say bivalves or gastropods, would be valuable. Especially useful in light of the potential role of gene flow and its relation to the spatial scale of environmental variation would be to track lineages with contrasting evolutionary tempo and mode through the Neogene fossil record to their present-day populations.

Taxon Sorting

The prevalence of intraspecific oscillatory evolution and of evolutionary stasis means that the direction of speciation is difficult to predict from within-species evolutionary trajectories. Further, wherever punctuated cladogenesis is prevalent, long-term evolutionary trends will not be simple extrapolations of intraspecific evolution but instead must involve some form of sorting among species (stepwise, punctuated anagenesis patterns are less clear-cut and might also involve sorting among populations or even highly episodic, species-wide changes propelled entirely at the organismic level). That such differential speciation and extinction rates among clades might in principle shape large-scale evolutionary patterns appears to be generally accepted (e.g., Sober 1984; Maynard Smith 1989; Williams 1992). Equally important, as Slatkin

(1981) noted, differential rates can drive taxon sorting even in gradualistic systems depending on the extent of variation generated by cladogenesis and anagenesis. As in so many macroevolutionary questions the issues are the frequency of this sorting among species, the circumstances under which it occurs, and the nature of dynamics across hierarchical levels, i.e., identification of focal levels and upward and downward causation (e.g., Vrba and Gould 1986; and see Grantham 1995 for an especially thoughtful and clear review).

Species Selection and Related Processes

The term "species selection" has been used in both broad and narrow senses, sometimes by a single author. One approach, drawing on the insights of Lewontin (1970) and Hull (1980) and advocated by Vrba and Gould (1986) among others, is to maintain the neutral term "species sorting" for any pattern shaped by differential origination and extinction. Others would apply the term "species selection" here instead because fitness, i.e., differential birth and death, is being expressed at the species level, as the "emergent fitness" of species—speciation and extinction rates—within clades (e.g., Lloyd and Gould 1993; Stidd and Wade 1995; Gould and Lloyd 1999). Alternatively, species sorting can be divided into two categories depending on the hierarchical level of the characters that influence speciation and extinction rates. Then, in "effect macroevolution" differential rates are governed by organismal-level traits such as body size or habitat preferences, while in species selection the differential rates are governed by emergent, heritable properties at the species level (see Vrba 1984, 1989; Jablonski 1987; Grantham 1995).

Emergence and Heritability.—The concept of emergence in evolutionary biology has been difficult, but a simple operational approach is to recognize a feature as emergent at a given level if its evolutionary consequences do not depend on how the feature is generated at lower levels. (This approach is similar to Brandon's 1982, 1988 application of Salmon's 1971 statistical concept of "screening-off," and to a parallel view, "multiple realizabililty," that recently has been criticized as insufficiently precise in some circumstances; for discussion see Sober 1999; Sterelny and Griffiths 1999.) A classic example at the organismal level involves selection experiments in *Drosophila* where Robertson (1959) concluded that equivalent changes in wing size could be achieved either by changes in cell size or by changes in cell number, with variance in wild populations usually owing mainly to cell number, and in his experimental groups mainly to cell size (see also Stevenson et al. 1995). As the organism was the focal level of the experiment, the large-winged phenotype was the emergent property under selection, and not the cellular or genetic levels underpinning the evolutionary changes. Outside the lab, evolution of the emergent organismal property of DDT resistance is underlain by many alternative responses at the cellular level, from changes in cell walls that exclude the DDT molecule, to changes in cell metabolism that neutralize DDT when it penetrates the cell, to changes in cell physiology that sequester DDT before it can be effective (e.g., McKenzie and Batterham 1994; Feyereisen 1995).

By the same token, geographic range is an emergent property at the species level, not simply because most geographic ranges are determined by the overall distribution of conspecifics rather than by the movements of single bodies, but also because the evolutionary consequences of broad or narrow geographic ranges tend to be similar regardless of how those ranges are mediated at the organismal level (at least within broad groups, such as benthic marine invertebrates). For example, widespread species of marine gastropods are geologically longer-lived than restricted species, and the establishment and maintenance of these different ranges are statistically related to modes of larval development—an organismal trait—that differ in dispersal capabilities (Hansen 1978, 1982; Jablonski and Lutz 1983; Jablonski 1986a, 1987, 1995; Scheltema 1989, 1992; Gili and Martinell 1994; Kohn and Perron 1994). Jablonski (1987) found geographic range to be heritable at the species level (that is, closely related species showed significant correlations in the magnitudes in their geographic ranges), completing the requirements for evolution by selection at any level: the existence of heritable variation in a feature that, by interaction with the environment, imparts differential success. Cheetham and Jackson (1996) also found widespread species of bryozoans to be geologically long-lived relative to restricted species; in fact their widespread species, taken as occupying >4 regions, have a median duration of about 7.5 m.y. while the narrowly distributed species a median duration of about 2 m.y., each remarkably close to the high- and low-dispersal molluscan species, respectively, as cited above. But here the differences in geographic ranges presumably derive from the rafting of adults (e.g., Watts et al. 1998). Thus, differential taxonomic survival depends on the emergent, species-level property, i.e., the scale of the species' range and not on the underlying organismal traits.

Genetic population structures, again not a property of single organisms, can be viewed in the same way. Jablonski (1986a, 1995) attributed high per-taxon speciation rates seen in gastropod lineages having low larval dispersal ability, as inferred from their larval shells, to their genetically fragmented populations (an argument broadly supported by genetic analyses of benthic marine invertebrates [see Pechenik 1999; Bohonak 1999]). Similarly, Wilson et al. (1975) suggested that mammals with complex social structures should have genetically more fragmented populations and thus higher speciation rates than those with more open breeding systems. And more recently, Belliure et al. (2000) found that natal dispersal ability in birds is inversely related to population differentiation and therefore, they argued, to speciation propensity. If these very different routes to highly subdivided populations yield similar macroevolutionary dynamics, this again would argue for genetic population structure as an emergent property at the clade level. The consistent relationship between dispersal ability and genetic population structure in plants (Govindaraju 1988) and animals (Bohonak 1999, in his valuable meta-analysis of 333 species across all animal groups and environments) suggests that this will be a profitable avenue for macroevolutionary research. Perhaps this general mechanism underlies the decrease in speciation rate observed by Dodd et al. (1999) when angiosperm lineages switch from animal pollination to wind pollination, for example.

Recent molecular work has shown that even widely dispersing marine species can sometimes, perhaps usually, have subdivided rather than panmictic populations (e.g., Palumbi 1996; Geller 1998; Benzie 1999a; Avise 2000). This does not mean, however, that high-dispersal species are as readily subdivided as low-dispersal ones. The key issue is the stability and long-term evolutionary effects of that population structure relative to taxa with low dispersal abilities. The consistent relationships among larval type, geographic extent, and speciation/extinction rates in Cretaceous, Paleogene, and Neogene taxa (which appear to be robust to sampling [Jablonski 1988; Marshall 1991]) suggest that in at least some settings the population structures detected by mtDNA analysis may be transient or in any case do not have predictable macroevolutionary effects (see also the diversity of analyses tabulated by Bohonak 1999). An intriguing pattern that needs a more detailed evolutionary perspective is the discovery that genetic connectedness among Pacific populations of benthic invertebrates does not conform to present-day ocean circulation patterns but may be a Pleistocene holdover (Benzie and Williams 1997; Palumbi et al. 1997; Benzie 1999a,b). Spatial scale may also be important here: the vast but highly discontinuous environments of the Indo–West Pacific may impart a different evolutionary dynamic from that documented in the more linear shelves and the more continuous two-dimensional epicontinental seas that provided the paleontological data (see Valentine and Jablonski 1983).

Clearly, further analyses of evolutionary sorting of taxa would benefit greatly from a more detailed phylogenetic framework. Duda and Palumbi (1999) rightly note that the further analyses of such patterns in a phylogenetic context would be valuable. However, their emphasizing an evolutionary bias toward the production of species having low-dispersal larvae, rather than species sorting for the larval modes for Pacific *Conus*, is difficult to interpret because they lack data on extinct species and their model does not take into account empirical evidence for higher extinction and origination rates in low-dispersal lineages.

Another unresolved problem is that marine bivalves do not exhibit the same relation between larval types and species-level dynamics as the co-occurring gastropods (e.g., Jablonski and Lutz 1983; Stanley 1990). Perhaps this is because modes of larval development in bivalves are more tightly linked phylogenetically to feeding types, body sizes, and other factors that might also influence evolutionary rates. Jablonski (1986a, 1995) showed that larval modes in marine gastropods override those of adult feeding types, and if the opposite is true for bivalves then the two groups in tandem might provide a valuable system for exploring the interplay of rate-determining traits at different hierarchical levels.

The heritability of species-level traits remains a neglected area. Jablonski (1987) and Ricklefs and Latham (1992) found geographic ranges to be heritable in marine mollusks and terrestrial plants, respectively. Their comparisons of closely related species were designed as a phylogenetic analogue to the sib-sib comparisons of quantitative genetics (and see also Peterson et al. 1999, who successfully predict geographic distributions of sister species based on a model of ecological niche conservatism). Gaston and Blackburn (1997) did not find strong species-level heritability in birds using nested ANOVAs, a very different design that also has precedents in quantitative genetics but lacks a detailed phylogenetic framework, necessarily omits extinct species and Pleistocene range adjustments, and compares taxa in different geographic situations, unlike Jablonski's analysis, which is a macroevolutionary analogue of a common-garden experiment. More work is needed to assess the strengths and weaknesses of the different approaches and where they might be applied most robustly.

The Limits of Species Selection and Species Sorting.—The domain of strict-sense species selection, which depends on emergent characters, is much narrower than broad-sense species selection, which depends only on emergent fitnesses (i.e., differential origination and extinction rates regardless of the hierarchical level at which they are determined [see Vrba and Gould 1986; Lloyd and Gould 1993; Grantham 1995; Stidd and Wade 1995]). Beyond that, we simply do not know the relative frequencies of different sorting processes, overall or among clades. The theoretical literature has outstripped the empirical database, in part simply because of the scale of the databases required for rigorous analyses. However, if geographic range is arguably a species-level character, then the macroecological literature is rich in potential examples that might fit the species-selection paradigm, because so many features of living organisms can be related to geographic range and thus are candidates for hitchhiking on species sorting processes (see for example Brown 1995; Brown et al. 1996; Gaston 1998). Other components of rarity as classified by Rabinowitz (1981; Rabinowitz et al. 1986) might also be examined in this list: population sizes or densities may be emergent properties (e.g., Vrba and Eldredge 1984; and here too a large macroecological literature exists, ripe for macroevolutionary analysis, e.g., Brown 1995; Blackburn and Gaston 1997, 1999), whereas habitat specificity may reside more fully at the organismal level (e.g., Vrba 1987).

One important issue needing more attention is the stability of such species-level characteristics. Jablonski (1986b, 1987) gave evidence that marine species achieve their geographic-range magnitudes rapidly relative to their geologic durations. Tracking the magnitude, rather than the position, of geographic ranges during Pleistocene or other environmental oscillations would be interesting, as would testing for evolutionary rate differences among taxa that differ in the *amplitude* of their range-size changes over time. Population density should also be tested more fully for long-term stability (e.g., Arneberg et al. 1997). Both exciting and daunting is the loose covariation of geographic range, abundance, and body size (Brown 1995, 1999; Gaston and Blackburn 1999; Lawton 1999), and the question of how these effects spanning hierarchical levels and spatial scales interact, and become linked or decoupled on ecological and evolutionary timescales.

Given that the history of most lineages is ev-idently dominated by stasis and punctuation, other potential species-level features that might be heritable owing to factors like population sizes or genetic population structure include relative morphological inertia (and so the average duration of stasis in the phyletic mode, or the amplitude of oscillations within stasis, if these are set intrinsically) and perhaps even the size-frequency distribution of morphological divergences of daughter isolates. Sex ratios may be another example of higher-level trait, although possibly played out at an intermediate focal level if interdemic differences in sex ratios are common in some groups (e.g., Delph 1990; Graff 1999). Is the relative genetic or morphological variability of species an emergent species-level trait, or is it simply the summation of organismic properties and therefore an aggregate trait as argued by Lloyd and Gould (1993)? It depends on how that variation arises, and how sorting processes operate on that trait, and empirical work is needed here.

The potential for species sorting (= broad-sense species selection) seems extensive, given the abundant evidence for differences in intrinsic extinction and origination rates among clades (e.g., McKinney 1997; Kammer et al. 1998; Sepkoski 1998; see also "extinction risk" studies on extant organisms, e.g., Bennett and Owens 1997 and references therein). Here, too, phylogenetic hypotheses can provide a valuable framework for rigorous analysis, and methods are being developed for rigorous testing of differential origination and extinction rates in a phylogenetic context (e.g., Kirkpatrick and Slatkin 1993; Slowinski and Guyer 1993; Sanderson and Donoghue 1996; Harvey and Rambaut 1998; Paradis 1998). These methods have mostly been applied to extant taxa, where the estimation of evolutionary dynamics is made difficult by unrecorded extinction that must be ignored or assumed to be constant through time, but some also show promise for the testing of species-sorting hypotheses in the fossil record.

Species sorting, including narrow-sense species selection, will generally play a different evolutionary role from the microevolutionary sorting of organisms within populations: it will tend to determine diversity differentials among clades rather than shape adaptations. Species sorting may not construct a complex eye or a long neck, but it may determine how many species possess complex eyes or long necks over evolutionary timescales. This has two immediate implications. First, the setting of species sorting and microevolution as rival hypotheses or mechanisms is often inappropriate. And second, the mapping of species densities in morphospace need not reflect the topology of the adaptive landscape. That is, the frequency distribution of morphologies within or among clades may not be a simple indication of relative fitnesses at the organismal level.

But is sorting at the species level unequivocally ruled out as a mechanism for organismal adaptation? Rice (1995) showed that the efficacy of species sorting in character evolution depends on the speciation rate per generation, the mutation rate, and the survival rate of reproductive isolates (and the genetic complexity of the trait under consideration). Thus, species sorting would be more likely to influence character evolution directly in organisms with long generation times; for example Asian elephants fit this model, given reasonable mutation rates and selection coefficients, and a very broad taxonomic domain is possible if speciation rates are sufficiently high (Rice 1995). Perhaps a more pervasive way for species sorting to influence the evolution of complex characters is by determining the persistence and proliferation of taxa bearing characters shaped by individual selection, so that the process of assembling those characters can proceed beyond the duration of individual species (Raup 1994; Rice 1995). This means that rate differentials among clades might be important even if, as Hallam (1998) argues, many individual clades generate only a few species per unit time.

Hierarchical analyses of large-scale changes in morphology would be valuable for improving our understanding of the relative roles of processes at different levels and the potentially complex interactions between them. The infaunalization of bivalves, the evolution of tree-like growth habits with the initial assembly of terrestrial forests, and the escalation of defensive morphologies in marine benthos during the increase in predation pressure described by Vermeij (1987) as the Mesozoic Marine Revolution (MMR) would all be possible subjects. To use the last of these biotic transitions as an example (and see also Signor and Brett 1984), long-lived bivalve or gastropod genera that appear to persist through much of the MMR, such as *Aphrodina*, *Cyprimeria*, or *Solariella*, might be targeted to test whether any given species shows anagenetic change in antipredatory adaptations independent of local environmental changes. (Kelley 1989, 1991, did find intraspecific trends, but within a much later, Neogene, interval.) If species tend to be static, several alternative dynamics should be tested. A mechanism below the species level might involve directional, stepwise anagenesis, shaped by widely spaced episodes of directional selection or within-species shifting-balance processes. Next up the hierarchy would be the sorting of species to convert clades to better-defended morphologies. Operating at an even higher level would be clade sorting, in which preferential extinction of poorly defended clades or preferential origin of well-defended ones might shape the overall biota. As many authors have noted (e.g., Stanley 1973; McShea 1994; P. J. Wagner 1996b), such analyses cannot focus only on mean morphologies or the appearance of derived character states; an important component of the sorting dynamics may involve the retention of less heavily defended morphologies. Finally, spatial scale will be important in understanding mechanisms, in terms of both the establishment of well-defended species—which Roy (1996) found to vary regionally in timing and which may have been promoted by environmental perturbations (see also Miller 1998)—and the fates of less derived forms (which Vermeij [1987] argues have shifted geographically or environmentally through time).

Such large-scale evolutionary interactions of predators and prey raise the broader issue of clade interactions and how they might drive taxon sorting. As Sepkoski (1996) made clear, negative clade interactions need not produce straightforward reciprocal diversity patterns (the "double wedge pattern" of Benton 1987), but instead can be manifest in complex coupled logistic patterns, and perhaps most importantly depressed but still positive diversification rates (see also Miller and Sepkoski 1988). Still open to investigation is exactly how local ecological interactions—as many of these clade interactions surely must entail—actually scale up to determine origination and extinction rates. Simple intuitive arguments are easy: competition decreases population sizes and/or intrinsic growth rates, thereby making species more vulnerable to stochastic extinction; origination rates might be depressed in the same way, if biotic interactions reduce the population sizes of isolates and thus decrease their probability of surviving to achieve speciation. On the other hand, clade interactions via predation could arguably operate in the same fashion, although Vermeij (1994) explicitly rejects the role of escalation in extinction on mainly theoretical grounds. How negative ecological interactions actually determine per-taxon origination and extinction rates seems an important area for research.

Such analyses in the fossil record are complicated by the highly diffuse nature of many of the interactions under consideration: escalation in molluscan defenses is a response to the collective diversification of durophagous arthropods, mollusks, vertebrates, and other groups, so that simple responses to specific enemies will be unusual, difficult to detect, and probably fleeting in geologic terms. Targeting competitive interactions for combined micro/macroevolutionary analysis will also be complicated by the well-established observation that taxonomic distance is not a reliable proxy for interaction intensity (e.g., Brown and Davidson 1977; Brown et al. 1979a,b on seed-eating ants vs. rodents; Kodric-Brown and Brown 1979 on hummingbirds vs. insects; Schluter 1986 on finches vs. bees; Jackson and Hughes 1985 on spatial competition in marine encrusting communities among several phyla). Paleontologists have been aware of this— Bambach's (1983) "megaguilds" are decidedly polyphyletic, for example—but the study of the large-scale evolutionary effects of biotic interactions remains a difficult task. One powerful approach deserving more extensive application is the projection back into the fossil record of interactions that have been experimentally dissected in modern communities (e.g., Aronson 1992, 1994; Lidgard et al. 1993; Sepkoski et al. 2000).

Incumbency.—Macroevolutionary incumbency effects, such as the damping of diversification among mammals for the first two-thirds of their history (presumably by incumbent archosaurs), are perhaps the most compelling evidence of the macroevolutionary effects of competitive interactions (Jablonski and Sepkoski 1996; Jablonski 2000a). Such effects represent very different dynamics from those underlying the reciprocal diversity patterns discussed above, of course: in double-wedge or coupled logistic models, taxon A progressively excludes taxon B by virtue of A's competitive superiority; in incumbency interactions, taxon A excludes B by virtue of A's ecological preemption of resources, which need not reflect competitive superiority on a level playing field (as emphasized by Rosenzweig and McCord [1991]). The double-wedge pattern is not exclusively the hallmark of the rise of a competitive dominant, however: piecemeal extinction of the dominant incumbent may create cumulative opportunities for surviving taxa, for example. Other kinds of progressive replacements might also be underlain by incumbency effects, as Valentine (1990) suggested for the Phanerozoic decline in evolutionary rates in the marine biota: over the long run, extinction-resistant clades will tend to preempt high-volatility clades (see Sepkoski 1998 for a somewhat different view).

The acceleration in evolutionary rates, taxonomically and morphologically, among previously established clades following an extinction event (e.g., Miller and Sepkoski 1988; Patzkowsky 1995; Foote 1997; McKinney 1998; Sepkoski 1998) is the primary macroevolutionary measure of incumbency effects. Considerable evidence supports a spatial analogue, in which asymmetries in biotic interchanges appear to reflect regional differences in extinction intensities, both in the fossil record (e.g., Vermeij 1991) and in present-day biotas (e.g., Case 1996). However, not all invasions are mediated by prior extinction (see for example Williamson 1996; Lonsdale 1999), and at least for the end-Cretaceous extinction, invasion intensities need not correlate to local extinction intensities (Jablonski 1998). More work is needed that tackles the difficult task of separating the effects of diversity loss per se from regional differences in environmental change; in other words, is differential diversity loss the promoter of asymmetrical biotic interchanges, or another symptom of the effects of differential environmental perturbation? Incumbency effects may represent one of the strongest bridges between ecology and macroevolution, and cross-scale and spatially explicit analyses of how those effects are maintained, broken, or circumvented would be valuable, not only for theoretical reasons, but in order to address the pressing issues of present-day biotic interchange.

Extinction Events As Filters and Promoters

So much has been written lately on the evolutionary role of extinction events that I will touch on only a few points here (see Raup 1994; Jablonski 1995, 2000a,b; Erwin 1998, 2000b for an entry into this literature). First and foremost, we still have much to learn about the role of extinction events in evolution. Perturbations occur at all intensities and spatial scales, and as discussed above they promote biotic change in important ways, but effects are difficult to predict from magnitude alone (see Miller 1998). The differential response of large-scale biotic units like clades and regional biotas to seemingly similar perturbations at different times—whether asteroid impacts or climate changes—reflects the fundamental nonlinearities that typify most complex systems. Thresholds, and especially the importance of antecedent events, are probably an essential component to a system where four sea-level oscillations cause little change in marine community composition but a fifth brings significant turnover; asteroid impacts that form craters of 50 km have no perceptible effect on the global biota but a crater of 300 km corresponds to a Cretaceous/Tertiary (K/T) scale event; and even apparently similar mass extinction intensities can have differing effects on biosphere structure and function (Droser et al. 1997).

On the other hand, as Levin (1999: pp. 180–184) notes, there is little evidence that large-scale biotic systems, let alone global biotas, evolve to the perpetual edge of collapse like the canonical sandpile, as suggested by Bak and colleagues (e.g., Sneppan et al. 1995; Solé et al. 1999). The resilience of these biotic systems to many perturbations, and their relatively loose organization as manifested in the relatively fluid composition of Pleistocene and many earlier communities, argues against a state of self-organized criticality, and more generally argues that the major turnover events in the history of life were externally driven (rather than the product of internal dynamics). Such avalanche models may apply on local scales and the short term, but hierarchy and scale defeat their universality, because the necessary biotic interconnectedness falls away rapidly at larger scales. Spatially heterogeneous dynamics can be seen, for example, in the geographic complexities of macroevolutionary and macroecological patterns like the Ordovician radiations, the marine Mesozoic revolution, the recovery from the end-Cretaceous mass extinction, and the demise of the Pleistocene megafauna (Roy 1996; Miller 1997a,b, 1998; Jablonski 1998; Martin and Steadman 1999).

The major mass extinctions probably account for less than 5% of the species turnover in the geologic past, and their disproportionate evolutionary effects probably derive from their removal of not just minor constituents of the biota, but also established incumbents. Analyses are still sparse, but contrasts between survivorship patterns during mass extinctions and those prevalent during times of low extinction intensity have been recorded for each of the major extinction events (Jablonski 1995, 2000a,b). This does not mean that certain features favored during times of "background" extinction could never also be advantageous during mass extinction events, but even partial discordances in survivorship can have profound and lasting effects, given the intensities involved. The monotonic rather than bimodal frequency distribution of extinction intensities in the geologic past (e.g., Raup 1991), and some of the similarities in survivorship among mass extinctions despite apparently different triggers, suggests that the scale of the perturbation and the operation of threshold effects were important factors in the observed changes in selectivity.

Patterns of selectivity are difficult to assess, however, because traits can be lost as a by-product of selection on other features (or, of course, purely stochastic survivorship [Raup 1994]). For example, what was it about the end-Ordovician extinction that selected against broad selenizones in snails (P. J. Wagner 1996b), or about the end-Cretaceous extinction that selected against schizodont hinges in bivalves, elongate rostra in echinoids, or complex sutures in cephalopods? All of these losses or declines are more likely to represent correlations rather than direct causation, but they had long-term morphological effects, and additional examples are plentiful.

Finally, the long-term effects of mass extinctions are set not only by the victims and survivors of an event, but also by the dynamics of the recovery process. Here, of course, is where the incumbency effects are most vividly illustrated, by the diversifications that unfold with the removal of previously dominant taxa. As discussed by Jablonski (2000b), however, clade dynamics across extinction events can exhibit several different patterns, including (1) unbroken continuity, as in the escalation of antipredatory defenses in marine bivalves across the K/T boundary (Hansen et al. 1999); (2) continuity with setbacks, as in the increase in cheilostome bryozoan abundance relative to cyclostomes across the K/T boundary (McKinney et al. 1998) or the increase in suture complexity in Paleozoic ammonoids (Saunders et al. 1999); (3) failure to rebound and eventual extinction, as in the demise of the prolecanitid ammonoids in the Early Triassic after surviving the end-Permian debacle (Page 1996); or (4) unbridled diversification, as seen most famously in the radiation of mammals after the end-Cretaceous demise of the dinosaurs and other reptilian groups. The apportionment of survivors among these different trajectories needs much more analysis before we can understand the evolutionary roles of extinction events.

That said, mass extinctions have lasting effects across many scales. As Erwin (1998) argues, the Permo-Triassic extinction permanently restructured marine and terrestrial ecosystems, so that the raw material for microevolution and the web of potential biotic interactions was profoundly shaped by the taxonomic losses of that time (although Sepkoski [1996, 1998] held that those changes were inevitable given differential turnover among groups, and were merely hastened by the mass extinction). Similarly, the end-Cretaceous extinction removed the rudists and nearly exterminated the trigonioid bivalves, but its lasting influence can also be seen in the age distribution of living marine bivalve genera, which shows a secondary peak corresponding to the early Cenozoic. This 60-m.y. evolutionary echo is less dramatic than the mammalian rise to dominance that began at the same time, but it reflects a similar evolutionary process (Flessa and Jablonski 1996).

Conclusion

The relation between micro- and macroevolution is complementary and not mutually exclusive, with effects cascading both upwards and downwards over long timescales. The conceptual expansion of evolutionary biology with the advances in our understanding of the origin and sorting of variation has benefited many disciplines and is promoting a fuller integration across those scales and hierarchical levels. I have touched only indirectly on many of the important methodological advances discussed elsewhere in this issue: in phylogeny estimation, in dissection of developmental processes, in calculation of confidence limits on diversification rates and patterns, and in quantifying the relation of paleontological patterns to the fabric of the stratigraphic record. These developing methodologies, and the host of new questions that are emerging at the interface of biology and geology, will provide rich research opportunities for the next 25 years of *Paleobiology*.

Acknowledgments

This paper benefited from discussions with many colleagues, most recently A. K. Behrensmeyer, B. Chernoff, D. H. Erwin, M. Foote, T. A. Grantham, S. M. Kidwell, K. Roy, A. B. Smith, J. W. Valentine, P. J. Wagner, S. L. Wing, and, in the final throes, the NCEAS Working Group on Ecological Processes and Evolutionary Rates (P. M. Brakefield, N. Eldredge, S. Gavrilets, J. B. C. Jackson, R. E. Lenski, B. S. Lieberman, M. A. McPeek, W. Miller, and J. N. Thompson); few if any of these people agree fully with the views expressed here. D. J. Futuyma, S. M. Kidwell, D. W. McShea, and J. W. Valentine provided valuable reviews. I thank the editors of Paleobiology for inviting me to write this paper, and for their patience during its elephantine gestation period; Natasha Atkins significantly improved the clarity and accuracy of the text. Supported by the National Science Foundation and a John Simon Guggenheim Memorial Fellowship.

Literature Cited

Adams, M. D., et al. 2000. The genome sequence of *Drosophila melanogaster*. Science 287:2185–2195.

Ahn, D.-G., and G. Gibson. 1999a. Axial variation in the threespine stickleback: relationship to *Hox* gene expression. Development, Genes and Evolution 209:473–481.

———. 1999b. Expression patterns of the threespine stickleback *Hox* genes and insights into the evolution of the vertebrate body axis. Development, Genes and Evolution 209:482–494.

Akam, M. 1998. Hox genes, homeosis and the evolution of segment identity: no need for hopeless monsters. International Journal of Developmental Biology 42:445–451.

Alberch, P., S. J. Gould, G. F. Oster, and D. B. Wake. 1979. Size and shape in ontogeny and phylogeny. Paleobiology 5:296–317.

Arneberg, P., A. Skorping, and A. F. Read. 1997. Is population density a species character? Comparative analyses of the nematode parasites of mammals. Oikos 80:289–300.

Arnold, H. 1966. Grundsätzliche Schwierigkeiten bei der biostratigraphischen Deutung phyletischer Reihen. Senckenbergiana Lethaea 47:537–547.

Arnold, S. J., and P. C. Phillips. 1999. Hierarchical comparison of genetic variance–covariance matrices. II. Coastal-inland divergence in the garter snake, *Thamnophis elegans*. Evolution 53:1516–1527.

Arnone, M. I., and E. H. Davidson. 1997. The hardwiring of development: organization and function of genomic regulatory systems. Development 124:1851–1864.

Aronson, R. B. 1992. Biology of a scale-independent predator–prey interaction. Marine Ecology Progress Series 80:1–13.

———. 1994. Scale-independent biological processes in the marine environment. Oceanography and Marine Biology, an Annual Review 32:435–460.

Arthur, W. 1988. A theory of the evolution of development. Wiley, Chichester, England.

———. 1997. The origin of animal body plans. Cambridge University Press, Cambridge.

Askin, R. A., and R. A. Spicer. 1995. The Late Cretaceous and Cenozoic history of vegetation and climate at northern and southern high latitudes: a comparison. Pp. 156–173 *in* National Research Council Board on Earth Sciences and Resources, Effects of past global change on life. National Academy Press, Washington, D.C.

Atchley, W. R., and B. K. Hall. 1991. A model for development and evolution of complex morphological structures. Biological Reviews 66:101–157.

Atchley, W. R., S. Z. Xu, and C. Vogl. 1994. Developmental quan-

titative genetic models of evolutionary change. Developmental Genetics 15:92–103.

Averof, M., and N. H. Patel. 1997. Crustacean appendage evolution associated with changes in Hox gene expression. Nature 388:682–686.

Avise, J. 2000. Phylogeography. Harvard University Press, Cambridge.

Bambach, R. K. 1983. Ecospace utilization and guilds in marine communities through the Phanerozoic. Pp. 719–746 *in* M. J. S. Tevesz and P. L. McCall, eds. Biotic interactions in recent and fossil benthic communities. Plenum, New York.

Barnosky, A. D. 1987. Punctuated equilibria and phyletic gradualism: some facts from the Quaternary fossil record. Current Mammalogy 1:109–147.

Barraclough, T. G., and A. P. Vogler. 2000. Detecting geographical pattern of speciation from species-level phylogenies. American Naturalist 155:419–434.

Bateman, R. M., P. R. Crane, W. A. DiMichele, P. R. Kenrick, N. P. Rowe, T. Speck, and W. E. Stein. 1998. Early evolution of land plants: phylogeny, physiology, and ecology of the primary terrestrial radiation. Annual Review of Ecology and Systematics 29:263–292.

Behrensmeyer, A. K., and R. W. Hook. 1992. Paleoenvironmental contexts and taphonomic modes. Pp. 15–136 *in* A. K. Behrensmeyer et al., eds. Terrestrial ecosystems through time. University of Chicago Press, Chicago.

Behrensmeyer, A. K., S. M. Kidwell, and R. A. Gastaldo. 2000. Taphonomy and paleobiology. Paleobiology 26:103–147.

Bell, M. A., M. S. Sadagursky, and J. V. Baumgartner. 1987. Utility of lacustrine deposits for the study of variation within fossil samples. Palaios 2:455–466.

Belliure, J., G. Sorci, A. P. Møller, and J. Clobert. 2000. Dispersal distances predict subspecies richness in birds. Journal of Evolutionary Biology 13:480–487.

Belting, H. G., C. S. Shashikant, and F. H. Ruddle. 1998. Modification of expression and cis-regulation of Hoxc8 in the evolution of diverged axial morphology. Proceedings of the National Academy of Sciences USA 95:2355–2360.

Bennett, K. D. 1997. Evolution and ecology: the pace of life. Cambridge University Press, Cambridge.

Bennett, P. M., and I. P. F. Owens. 1997. Variation in extinction risk among birds: chance or evolutionary predisposition? Proceedings of the Royal Society of London B 264:401–408.

Benton, M. J. 1987. Progress and competition in macroevolution. Biological Reviews 62:305–338.

Benzie, J. A. H. 1999a. Genetic structure of coral reef organisms: ghosts of dispersal past. American Zoologist 39:131–145.

———. 1999b. Major genetic differences between crown-of-thorns starfish (*Acanthaster planci*) populations in the Indian and Pacific Oceans. Evolution 53:1782–1795.

Benzie, J. A. H., and S. T. Williams. 1997. Genetic structure of giant clam (*Tridacna maxima*) populations in the west Pacific is not consistent with dispersal by present-day ocean currents. Evolution 51:768–783.

Blackburn, T. M., and K. J. Gaston. 1997. A critical assessment of the form of the interspecific relationship between abundance and body size in animals. Journal of Animal Ecology 66:233–249.

———. 1999. The relationship between animal abundance and body size: a review of the mechanisms. Advances in Ecological Research 28:181–210.

Bohonak, A. J. 1999. Dispersal, gene flow, and population structure. Quarterly Review of Biology 74:21–45.

Bookstein, F. L. 1987. Random walk and the existence of evolutionary rates. Paleobiology 11:258–271.

———. 1988. Random walk and the biometrics of morphological characters. Evolutionary Biology 23:369–398.

Bradshaw, H. D., K. G. Otto, B. E. Frewen, J. K. McKay, and D.

W. Schemske. 1998. Quantitative trait loci affecting differences in floral morphology between two species of monkeyflower (*Mimulus*). Genetics 149:367–382.

Bralower, T. J., and M. Parrow. 1996. Morphometrics of the Paleocene coccolith genera *Cruciplacolithus, Chiasmolithus,* and *Sullivania*: a complex evolutionary history. Paleobiology 22: 352–385.

Brandon, R. N. 1982. The levels of selection. Pp. 315–323 *in* P. Asquith and T. Nichols, eds. PSA 1982, Vol. 1. Philosophy of Science Association, East Lansing, Mich.

———. 1988. The levels of selection: a hierarchy of interactors. Pp. 51–71 *in* H. Plotkin, ed. The role of behavior in evolution. MIT Press, Cambridge.

Brown, J. H. 1995. Macroecology. University of Chicago Press, Chicago.

———. 1999. Macroecology: progress and prospect. Oikos 87: 3–14.

Brown, J. H., and D. W. Davidson. 1977. Competition between seed-eating rodents and ants in desert ecosystems. Science 196:880–882.

Brown, J. H., D. W. Davidson, and O. J. Reichman. 1979a. An experimental study of competition between seed-eating desert rodents and ants. American Zoologist 19:1129–1143.

Brown, J. H., O. J. Reichman, and D. W. Davidson. 1979b. Granivory in desert ecosystems. Annual Review of Ecology and Systematics 10:201–227.

Brown, J. H., G. C. Stevens, and D. M. Kaufman. 1996. The geographic range: size, shape, boundaries, and internal structure. Annual Review of Ecology and Systematics 27:597–623.

Budd, A. F., K. G. Johnson, and D. C. Potts. 1994. Recognizing morphospecies in colonial reef corals. 1. Landmark-based methods. Paleobiology 20:484–505

Burke, A. C. 1989. Development of the turtle carapace: implications for the evolution of a novel Bauplan. Journal of Morphology 199:363–378.

———. 1991. The development and evolution of the turtle body plan: inferring intrinsic aspects of the evolutionary process from experimental embryology. American Zoologist 31:616–627.

Burke, A. C., C. E. Nelson, B. A. Morgan, and C. Tabin. 1995. *Hox* genes and the evolution of vertebrate axial morphology. Development 121:333–346.

Buzas, M. A., and S. J. Culver. 1986. Geographic origin of benthic foraminiferal species. Science 232:775–776.

Byrne, K., and R. A. Nichols. 1999. *Culex pipiens* in London Underground tunnels: differentiation between surface and subterranean populations. Heredity 82:7–15.

Cardillo, M. 1999. Latitude and rates of diversification in birds and butterflies. Proceedings of the Royal Society of London B 266:1221–1225.

Carroll, R. L. 1997. Patterns and processes of vertebrate evolution. Cambridge University Press, Cambridge.

Carter, D. R., B. Miki, and K. Padian. 1998. Epigenetic mechanical factors in the evolution of long bone epiphyses. Zoological Journal of the Linnean Society 123:163–178.

Case, T. J. 1996. Global patterns in the establishment and distribution of exotic birds. Biological Conservation 78:69–96.

Chaline, J., and B. Laurin. 1986. Phyletic gradualism in a European Plio-Pleistocene *Mimomys* lineage (Arvicolidae, Rodentia). Paleobiology 12:203–216.

Chaline, J., P. Brunet-Lecomte, S. Montuire, L. Viriot, and F. Courant. 1993. Morphological trends and rates of evolution in arvicolids (Arvicolidae, Rodentia): towards a punctuated equilibria/disequilibria model. Quaternary International 19: 27–39.

Charlesworth, B., R. Lande, and M. Slatkin. 1982. A neo-Darwinian commentary on macroevolution. Evolution 36:474–498.

Cheetham, A. H., and J. B. C. Jackson. 1995. Process from pattern: tests for selection versus random change in punctuated bryozoan speciation. Pp. 184–207 *in* Erwin and Anstey 1995b.

———. 1996. Speciation, extinction, and the decline of arborescent growth in Neogene and Quaternary cheilostome Bryozoa of tropical America. Pp. 205–233 *in* J. B. C. Jackson, A. F. Budd, and A. G. Coates, eds. Evolution and environment in tropical America. University of Chicago Press, Chicago.

Chesser, R. T., and R. M. Zink. 1994. Modes of speciation in birds: a test of Lynch's method. Evolution 48:490–497.

Cheverud, J. M. 1996. Developmental integration and the evolution of pleiotropy. American Zoologist 36:44–50.

Chow, R. L., C. R. Altmann, R. A. Lang, and A. Hemmati-Brivanlou. 1999. *Pax6* induces ectopic eyes in a vertebrate. Development 126:4213–4222.

Claridge, M. F., H. A. Dawah, and M. R. Wilson, eds. Species: the units of biodiversity. Chapman and Hall, London.

Clark, J. S. 1998. Why trees migrate so fast: Confronting theory with dispersal biology and the paleorecord. American Naturalist 152:204–224.

Clarke, G. M. 1993. The genetic basis of developmental stability. 1. Relationships between stability, heterozygosity and genomic coadaptation. Genetica 9:15–23.

Clarkson, E. N. K. 1988. The origin of marine invertebrate species: a critical review of microevolutionary transformations. Proceedings of the Geologists' Association 99:153–171.

Coates, M. I., and M. J. Cohn. 1998. Fins, limbs, and tails: outgrowths and axial patterning in vertebrate evolution. BioEssays 20:371–381.

Cohen, B. L., P. Balfe, M. Cohen, and G. B. Curry. 1991. Molecular evolution and morphological speciation in North Atlantic brachiopods (*Terebratulina* spp.). Canadian Journal of Zoology 69:2903–2911.

Cohn, M. J., and C. Tickle. 1999. Developmental basis of limblessness and axial patterning in snakes. Nature 399:474–479.

Collins, T. M., K. Frazer, A. R. Palmer, G. J. Vermeij, and W. M. Brown. 1996. Evolutionary history of northern hemisphere *Nucella* (Gastropoda, Muricidae): molecular, morphological, ecological, and paleontological evidence. Evolution 50:2287–2304.

Conway Morris, S. 1998. The evolution of diversity in ancient ecosystems: a review. Philosophical Transactions of the Royal Society of London B 353:327–345.

Coope, G. R. 1995. Insect faunas in Ice Age environments: why so little extinction? Pp. 55–74 *in* J. H. Lawton and R. M. May, eds. Extinction rates. Oxford University Press, Oxford.

Coyne, J. A., and H. A. Orr. 1998. The evolutionary genetics of speciation. Philosophical Transactions of the Royal Society of London B 353:287–305.

Cracraft, J. 1989. Speciation and its ontology: The empirical consequences of alternative species concepts for understanding patterns and processes of differentiation. Pp. 28–59 *in* Otte and Endler 1989.

———. 1997. Species concepts in systematics and conservation biology—an ornithological viewpoint. Pp. 325–339 *in* Claridge et al. 1997.

Crame, J. A. 1996. Antarctica and the evolution of taxonomic diversity gradients in the marine realm. Terra Antarctica 3:121–134.

———. 1997. An evolutionary framework for the polar regions. Journal of Biogeography 24:1–9.

Crame, J. A., and A. Clarke. 1997. The historical component of marine taxonomic diversity gradients. Pp. 258–273 *in* R. F. G. Ormond, J. D. Gage, and M. V. Angel, eds. Marine biodiversity: patterns and processes. Cambridge University Press, Cambridge.

Darling, K. F., C. M. Wade, D. Kroon, A. J. L. Brown, and J. Bijma. 1999. The diversity and distribution of modern planktic foraminiferal small subunit ribosomal RNA genotypes and their potential as tracers of present and past ocean circulations. Paleoceanography 14:3–12.

Darling, K. F., C. M. Wade, I. A. Stewart, D. Kroon, R. Dingle, and A. J. L. Brown. 2000. Molecular evidence for genetic mixing of Arctic and Antarctic subpolar populations of planktonic foraminifers. Nature 405:43–47.

Davis, J. I., and K. C. Nixon. 1992. Populations, genetic variation, and the delimitation of phylogenetic species. Systematic Biology 41:421–435.

Delph, L. F. 1990. Sex-ratio variation in the gynodioecious shrub *Hebe strictissima* (Scrophulariaceae). Evolution 44:134–142.

de Queiroz, K. 1998. The general lineage concept of species, species criteria, and the process of speciation. Pp. 57–75 *in* Howard and Berlocher 1998.

———. 1999. The general lineage concept of species and the defining properties of the species category. Pp. 49–89 *in* R. A. Wilson, ed. Species: new interdisciplinary essays. MIT Press, Cambridge.

DeSalle, R., and E. Carew. 1992. Phyletic phenocopy and the role of developmental genes in morphological evolution in the Drosophilidae. Journal of Evolutionary Biology 5:363–374.

de Vargas, C., R. Norris, L. Zaninetti, S. W. Gibb, and J. Pawlowski. 1999. Molecular evidence of cryptic speciation in planktonic foraminifers and their relation to oceanic provinces. Proceedings of the National Academy of Sciences USA 96: 2864–2868.

DiMichele, W. A., and R. B. Aronson. 1992. The Pennsylvanian-Permian vegetational transition: a terrestrial analogue of the onshore–offshore hypothesis. Evolution 46:807–824.

Dodd, M. E., J. Silvertown, and M. W. Chase. 1999. Phylogenetic analysis of trait evolution and species diversity variation among angiosperm families. Evolution 53:732–744.

Doebley, J., and L. Lukens. 1998. Transcriptional regulators and the evolution of plant form. Plant Cell 10:1075–1082.

Doebley, J., and R.-L. Wang. 1997. Genetics and the evolution of plant form: an example from maize. Cold Spring Harbor Symposia in Quantitative Biology 62:361–367.

Donoghue, M. J. 1989. Phylogenies and the analysis of evolutionary sequences, with examples from the seed plants. Evolution 43:1137–1156.

Droser, M. L., G. Hampt, and S. J. Clements. 1993. Environmental patterns in the origin and diversification of rugose and deep-water scleractinian corals. Courier Forschungsinstitut Senckenberg 164:47–54.

Droser, M. L., D. J. Bottjer, and P. M. Sheehan. 1997. Evaluating the ecological architecture of major events in the Phanerozoic history of marine invertebrate life. Geology 25:167–170.

Duboule, D., and A. S. Wilkins. 1998. The evolution of 'bricolage.' Trends in Genetics 14:54–59.

Duda, T. F., and S. R. Palumbi. 1999. Developmental shifts and species selection in gastropods. Proceedings of the National Academy of Sciences USA 96:10272–10277.

Duffy, J. E. 1996. Species boundaries, specialization, and the radiation of sponge-dwelling alpheid shrimp. Biological Journal of the Linnean Society 58:307–324.

Eble, G. J. 2000. Contrasting evolutionary flexibility in sister groups: disparity and diversity in Mesozoic atelostomate echinoids. Paleobiology 26:56–79.

Eckenwalder, J. E. 1984. Natural intersectional hybridization between North American species of *Populus* (Salicaceae) in sections *Aigeiros* and *Tacamahaca*. III. Paleobotany and evolution. Canadian Journal of Botany 62:336–342.

Eldredge, N. 1985. Unfinished synthesis. Oxford University Press, New York.

———. 1989. Macroevolutionary dynamics. McGraw-Hill, New York.

———. 1995. Species, speciation, and the context of adaptive change in evolution. Pp. 39–63 *in* Erwin and Anstey 1995b.

Emlet, R. B., and O. Hoegh-Guldberg. 1997. Effects of egg size on postlarval performance: experimental evidence from a sea urchin. Evolution 51:141–152.

Englemann, G. F., and E. O. Wiley. 1977. The place of ancestor-descendant relationships in phylogeny reconstruction. Systematic Zoology 26:1–11.

Erwin, D. H. 1994. Early evolution of major morphological innovations. Acta Palaeontologica Polonica 38:281–294.

———. 1998. The end and the beginning: recoveries from mass extinctions. Trends in Ecology and Evolution 13:344–349.

———. 1999. The origin of bodyplans. American Zoologist 39: 617–629.

———. 2000a. Macroevolution is more than repeated rounds of microevolution. Evolution and Development 2:78–84.

———. 2000b. Lessons from the past: biotic recoveries from mass extinctions. Proceedings of the National Academy of Sciences USA (in press).

Erwin, D. H., and R. L. Anstey. 1995a. Speciation in the fossil record. Pp. 11–38 *in* Erwin and Anstey 1995b.

———. 1995b. New approaches to speciation in the fossil record. Columbia University Press, New York.

Erwin, D. H., J. W. Valentine, and J. J. Sepkoski Jr. 1987. A comparative study of diversification events: the early Paleozoic versus the Mesozoic. Evolution 41:1177–1186.

Farrell, B. D. 1998. "Inordinate fondness" explained: why are there so many beetles? Science 281:555–559.

Feldmann, R. M., D. M. Tshudy, and M. R. A. Thomson. 1993. Late Cretaceous and Paleocene decapod crustaceans from James Ross Basin, Antarctic Peninsula. Paleontological Society Memoir 28. Journal of Paleontology 67(Suppl. to No. 1).

Feyereisen, R. 1995. Molecular biology of insecticide resistance. Toxicology Letters 82–3:83–90.

Finnerty, J. R., and M. Q. Martindale. 1998. The evolution of the Hox cluster: insights from outgroups. Current Opinion in Genetics and Development 8:681–687.

Fisher, D. C. 1994. Stratocladistics: Morphological and temporal patterns and their relation to phylogenetic process. Pp. 133–171 *in* L. Grande and O. Rieppel, eds. Interpreting the hierarchy of nature. Academic Press, Orlando, Fla.

Flessa, K. W., and D. Jablonski. 1996. The geography of evolutionary turnover: a global analysis of extant bivalves. Pp. 376–397 *in* D. Jablonski, D. H. Erwin, and J. H. Lipps, eds. Evolutionary Paleobiology. University of Chicago Press, Chicago.

Flynn, L. J., J. C. Barry, M. E. Morgan, D. Pilbeam, L. L. Jacobs, and E. H. Lindsay. 1995. Neogene Siwalik mammalian lineages: species longevities, rates of change, and modes of speciation. Palaeogeography, Palaeoclimatology, Palaeoecology 115:249–264.

Foote, M. 1993. Discordance and concordance between morphological and taxonomic diversity. Paleobiology 19:185–204.

———. 1996a. Perspective: evolutionary patterns in the fossil record. Evolution 50:1–11.

———. 1996b. On the probability of ancestors in the fossil record. Paleobiology 22:141–151.

———. 1997. The evolution of morphological diversity. Annual Review of Ecology and Systematics 28:129–152.

———. 1999. Morphological diversity in the evolutionary radiation of Paleozoic and post-Paleozoic crinoids. Paleobiology Memoirs No. 1. Paleobiology 25(Suppl. to No. 2).

Fortey, R. A. 1985. Gradualism and punctuated equilibria as competing and complementary theories. Special Papers in Palaeontology 33:17–28.

———. 1988. Seeing is believing: gradualism and punctuated equilibria in the fossil record. Science Progress 72:1–19.

Frost, D. R., and A. G. Kluge. 1994. A consideration of epistemology in systematic biology, with special reference to species. Cladistics 10:259–294.

Futuyma, D. J. 1987. On the role of species in anagenesis. American Naturalist 130:465–475.

Futuyma, D. J., M. C. Keese, and D. J. Funk. 1995. Genetic constraints on macroevolution: the evolution of host affiliation in the leaf beetle genus *Ophraella*. Evolution 49:797–809.

Gaffney, B., and E. P. Cunningham. 1988. Estimation of genetic trend in racing performance of thoroughbred horses. Nature 332:722–724.

Garland, T., P. E. Midford, and A. R. Ives. 1999. An introduction to phylogenetically based statistical methods, with a new method for confidence intervals on ancestral values. American Zoologist 39:374–388.

Gaston, K. J. 1998. Species-range size distributions: products of speciation, extinction and transformation. Philosophical Transactions of the Royal Society of London B 353:219–230.

Gaston, K. J., and T. M. Blackburn. 1996. The tropics as a museum of biological diversity: an analysis of the New World avifauna. Proceedings of the Royal Society of London B 263: 63–68.

———. 1997. Age, area and avian diversification. Biological Journal of the Linnean Society 62:239–253.

———. 1999. A critique for macroecology. Oikos 84:353–368.

Geary, D. H. 1995. The importance of gradual change in species-level transitions. Pp. 67–86 *in* Erwin and Anstey 1995b.

Gehring, W. J., and K. Ikeo. 1999. *Pax-6*: mastering eye morphogenesis and eye evolution. Trends in Genetics 15:371–377.

Geller, J. B. 1998. Molecular studies of marine invertebrate biodiversity: Status and prospects. Pp. 359–376 *in* K. E. Cooksey, ed. Molecular approaches to the study of the ocean. Chapman and Hall, London.

Gellon, G., and W. McGinnis. 1998. Shaping animal body plans in development and evolution by modulation of *Hox* expression patterns. BioEssays 20:116–125.

Ghiselin, M. T. 1997. Metaphysics and the origin of species. State University of New York Press, Albany.

Ghosh, S., M. J. May, and E. B. Kopp. 1998. NF-kappa B and rel proteins: evolutionarily conserved mediators of immune responses. Annual Review of Immunology 16:225–260.

Gibson, G. 1999. Insect evolution: Redesigning the fruitfly. Current Biology 9:R86–R89.

Gibson, G., and G. Wagner. 2000. Canalization in evolutionary genetics: a stabilizing theory? BioEssays 22:372–380.

Gibson, G., M. Wemple, and S. van Helden. 1999. Potential variance affecting homeotic Ultrabithorax and Antennipedia phenotypes in *Drosophila melanogaster*. Genetics 151:1081–1091.

Gili, C., and J. Martinell. 1994. Relationship between species longevity and larval ecology in nassariid gastropods. Lethaia 27:291–299.

Gingerich, P. D. 1993. Quantification and comparison of evolutionary rates. American Journal of Science 293-A:453–478.

Goffinet, B., and S. Gerber. 2000. Quantitative Trait Loci: a meta-analysis. Genetics 155:463–473.

Gonzalez-Crespo, S., and M. Levine. 1994. Related target enhancers for dorsal and NF-kappa-B signalling pathways. Science 264:255–258.

Gottlieb, L. D. 1984. Genetics and morphological evolution in plants. American Naturalist 123:681–709.

Gould, S. J. 1977a. Eternal metaphors of paleontology. Pp. 1–26 *in* A. Hallam, ed. Patterns of evolution. Elsevier, Amsterdam.

———. 1977b. Ontogeny and phylogeny. Harvard University Press, Cambridge.

———. 1982. The meaning of punctuated equilibrium and its role in validating a hierarchical approach to macroevolution. Pp. 83–104 *in* R. Milkman, ed. Perspectives on evolution. Sinauer, Sunderland, Mass.

———. 1985. The paradox of the first tier: an agenda for paleobiology. Paleobiology 11:2–12.

Gould, S. J., and N. Eldredge. 1977. Punctuated equilibria: the tempo and mode of evolution reconsidered. Paleobiology 3: 115–151.

———. 1993. Punctuated equilibrium comes of age. Nature 366: 223–227.

Gould, S. J., and E. A. Lloyd. 1999. Individuality and adaptation across levels of selection: how shall we name and generalize the unit of Darwinism? Proceedings of the National Academy of Sciences USA 96:11904–11909.

Govindaraju, D. R. 1988. Relationship between dispersal ability and levels of gene flow in plants. Oikos 52:31–35.

Graff, A. 1999. Population sex structure and reproductive fitness in gynodioecious *Sidalcea malviflora malviflora* (Malvaceae). Evolution 53:1714–1722.

Grant, P. R. 1986. Ecology and evolution of Darwin's finches. Princeton University Press, Princeton, N.J.

Grant, P. R., and B. R. Grant. 1995. Predicting microevolutionary responses to directional selection on heritable variation. Evolution 49:241–251.

Grantham, T. A. 1995. Hierarchical approaches to macroevolution: recent work on species selection and the "effect hypothesis." Annual Review of Ecology and Systematics 26:301–322.

Greene, H. W., and D. Cundall. 2000. Limbless tetrapods and snakes with legs. Science 287:1939–1941.

Hagdorn, H., and H. J. Campbell. 1993. *Paracomatula triadica* sp. nov.—an early comatulid crinoid from the Otapirian (Late Triassic) of New Caledonia. Alcheringa 17:1–17.

Hageman, S. J., M. M. Bayer, and C. D. Todd. 1999. Partitioning phenotypic variation: genotypic, environmental and residual components from bryozoan skeletal morphology. Journal of Natural History 33:1713–1735.

Halder, G., P. Callaerts, and W. J. Gehring. 1995. Induction of ectopic eyes by targeted expression of the *eyeless* gene in *Drosophila*. Science 267:1788–1792.

Hall, B. K. 1983. Epigenetic control in development and evolution. Pp. 353–379 *in* B. C. Goodwin, N. Holder, and C. C. Wylie, eds. Development and evolution. Cambridge University Press, Cambridge.

———. 1999. Evolutionary developmental biology, 2d ed. Kluwer Academic, Dordrecht, Netherlands.

Hallam, A. 1998. Speciation patterns and trends in the fossil record. Geobios 7:921–930.

Hansen, T. A. 1978. Larval dispersal and species longevity in Lower Tertiary gastropods. Science 199:885–887.

———. 1982. Modes of larval development in early Tertiary neogastropods. Paleobiology 8:367–372.

Hansen, T. A., P. H. Kelley, V. D. Melland, and S. E. Graham. 1999. Effect of climate-related mass extinctions on escalation in molluscs. Geology 27:1139–1142.

Hanski, I. 1999. Metapopulation ecology. Oxford University Press, Oxford.

Harrington, R. J. 1987. Skeletal growth histories of *Protothaca staminea* (Conrad) and *Protothaca grata* (Say) throughout their geographic ranges, northeastern Pacific. Veliger 30:148–158.

Harrison, R. G. 1998. Linking evolutionary pattern and process: the relevance of species concepts for the study of speciation. Pp. 19–31 *in* Howard and Berlocher 1998.

Harrison, S. 1998. Do taxa persist as metapopulations in evolutionary time? Pp. 19–30 *in* McKinney and Drake 1998.

Harvey, P. H., and A. Rambaut. 1998. Phylogenetic extinction rates and comparative methodology. Proceedings of the Royal Society of London B 265:1691–1696.

Heanue, T. A., R. Rashef, R. J. Davis, G. Mardon, G. Oliver, S. Tomarev, A. B. Lassar, and C. J. Tabin. 1999. Synergistic regulation of vertebrate development by *Dach2*, *Eya2*, and *Six1*, homologs of genes required for *Drosophila* eye formation. Genes and Development 13:3231–3243.

Heaton, T. H. 1993. The Oligocene rodent *Ischyromys* of the Great Plains: replacement mistaken for anagenesis. Journal of Paleontology 67:297–308.

Hendry, A. P., and M. T. Kinnison. 1999. Perspective: the pace of modern life: measuring rates of contemporary microevolution. Evolution 53:1637–1653.

Hodin, J. 2000. Plasticity and constraints in development and evolution. Journal of Experimental Zoology (Molecular and Developmental Evolution) 288:1–20.

Hoffman, A. 1989. Arguments on evolution. Oxford University Press, Oxford.

Hoffmann, A. A., and M. J. Hercus. 2000. Environmental stress as an evolutionary force. BioScience 50:217–226.

Hoffmann, A. A., and P. A. Parsons. 1997. Extreme environmental change and evolution. Cambridge University Press, Cambridge.

Holland, P. W. H. 1999. The future of evolutionary developmental biology. Nature 402(Suppl.):C41–C44.

Holland, S. M., and M. E. Patzkowsky. 1999. Models for simulating the fossil record. Geology 27:491–494.

Howard, D. J., and S. H. Berlocher, eds. Endless forms: species and speciation. Oxford University Press, New York.

Huber, B. T., J. Bijma, and K. Darling. 1997. Cryptic speciation in the living planktonic foraminifer *Globigerinella siphonifera* (d'Orbigny). Paleobiology 23:33–62.

Hughes, N. C., R. E. Chapman, and J. M. Adrain. 1999. The stability of thoracic segmentation in trilobites: a case study in developmental and ecological constraints. Evolution and Development 1:24–35.

Hull, D. L. 1980. Individuality and selection. Annual Review of Ecology and Systematics 11:311–332.

———. 1997. The ideal species concept—and why we can't get it. Pp. 357–380 *in* Claridge et al. 1997.

Humphreville, R., and R. K. Bambach. 1979. Influence of geography, climate and ocean circulation on the pattern of generic diversity of brachiopods in the Permian. Geological Society of America Abstracts with Programs 11:447.

Jablonski, D. 1986a. Larval ecology and macroevolution of marine invertebrates. Bulletin of Marine Science 39:565–587.

———. 1986b. Background and mass extinctions: the alternation of macroevolutionary regimes. Science 231:129–133.

———. 1987. Heritability at the species level: analysis of geographic ranges of Cretaceous mollusks. Science 238:360–363.

———. 1988. Response [to Russell and Lindberg]. Science 240: 969.

———. 1993. The tropics as a source of evolutionary novelty: the post-Palaeozoic fossil record of marine invertebrates. Nature 364:142–144.

———. 1995. Extinction in the fossil record. Pp. 25–44 *in* J. H. Lawton and R. M. May, eds. Extinction rates. Oxford University Press, Oxford.

———. 1998. Geographic variation in the molluscan recovery from the end-Cretaceous extinction. Science 279:1327–1330.

———. 2000a. The interplay between physical and biotic factors in evolution. *In* A. Lister and L. Rothschild, eds. Evolution on planet Earth: the impact of the physical environment. Linnean Society and Academic Press, London (in press).

———. 2000b. Lessons from the past: Evolutionary impacts of mass extinctions. Proceedings of the National Academy of Sciences USA (in press).

Jablonski, D., and D. J. Bottjer. 1983. Soft-substratum epifaunal suspension-feeding assemblages in the Late Cretaceous: Implications for the evolution of benthic paleocommunities. Pp. 747–812 *in* M. J. Tevesz and P. L. McCall, eds. Biotic interactions in Recent and fossil benthic communities. Plenum, New York.

———. 1990a. The ecology of evolutionary innovations: the fossil record. Pp. 253–288 *in* M. H. Nitecki, ed. Evolutionary innovations. University of Chicago Press, Chicago.

———. 1990b. The origin and diversification of major groups: environmental patterns and macroevolutionary lags. Pp. 17–57 *in* P. D. Taylor and G. P. Larwood, eds. Major evolutionary radiations. Clarendon, Oxford.

———. 1991. Environmental patterns in the origins of higher taxa: the post-Paleozoic fossil record. Science 252:1831–1833.

Jablonski, D., and R. A. Lutz. 1983. Larval ecology of marine benthic invertebrates: paleobiological implications. Biological Reviews 58:21–89.

Jablonski, D., and J. J. Sepkoski Jr. 1996. Paleobiology, community ecology, and scales of ecological pattern. Ecology 77: 1367–1378.

Jablonski, D., S. Lidgard, and P. D. Taylor. 1997. Comparative ecology of bryozoan radiations: origin of novelties in cyclostomes and cheilostomes. Palaios 12:505–523.

Jackson, J. B. C., and A. F. Budd. 1996. Evolution and environment: Introduction and overview. Pp. 1–20 *in* J. B. C. Jackson, A. F. Budd, and A. G. Coates, eds. Evolution and environment in tropical America. University of Chicago Press, Chicago.

Jackson, J. B. C., and A. H. Cheetham. 1990. Evolutionary significance of morphospecies: a test with cheilostome Bryozoa. Science 248:579–583.

———. 1994. Phylogeny reconstruction and the tempo of speciation in cheilostome Bryozoa. Paleobiology 20:407–423.

———. 1999. Tempo and mode of speciation in the sea. Trends in Ecology and Evolution 14:72–77.

Jackson, J. B. C., and T. P. Hughes. 1985. Adaptive strategies of coral-reef invertebrates. American Scientist 73:265–274.

Jackson, S. T., and J. T. Overpeck. 2000. Responses of plant populations and communities to environmental changes of the Late Quaternary. Paleobiology 26:194–220.

Jacobs, D. K. 1990. Selector genes and the Cambrian radiation of Bilateria. Proceedings of the National Academy of Sciences USA 87:4406–4410.

Johnson, A. L. A. 1984. The paleobiology of the bivalve families Pectinidae and Propeamussidae in the Jurassic of Europe. Zitteliana 11:235 p.

———. 1985. The rate of evolutionary change in European Jurassic scallops. Special Papers in Palaeontology 33:91–102.

———. 1993. Punctuated equilibria vs. phyletic gradualism in European Jurassic *Gryphaea* evolution. Proceedings of the Geologists' Association 104:209–222.

———. 1994. Evolution of European Lower Jurassic *Gryphaea* (*Gryphaea*) and contemporaneous bivalves. Historical Biology 7:167–186.

Johnson, A. L. A., and C. D. Lennon. 1990. Evolution of gryphaeate oysters in the mid-Jurassic of western Europe. Palaeontology 33:453–485.

Johnson, J. G. 1982. Occurrence of phyletic gradualism and punctuated equilibria through time. Journal of Paleontology 56:1329–1331.

Jones, D. S. 1988. Sclerochronology and the size versus age problem. Pp. 93–108 *in* M. L. McKinney, ed. Heterochrony in evolution. Plenum, New York.

———. 1998. Isotopic determination of growth and longevity in fossil and modern invertebrates. *In* R. D. Norris and R. M. Corfield, eds. Isotope paleobiology and paleoecology. Paleontological Society Papers 4:37–67. Paleontological Society, Knoxville, Tenn.

Jones, D. S., and S. J. Gould. 1999. Direct measurement of age in fossil *Gryphaea*: the solution to a classic problem in heterochrony. Paleobiology 25:158–187.

Judd, W. S., R. W. Sanders, and M. J. Donoghue. 1994. Angiosperm family pairs: preliminary phylogenetic analysis. Harvard Papers in Botany 5:1–51.

Kammer, T. W., T. K. Baumiller, and W. I. Ausich. 1998. Evolutionary significance of differential species longevity in Osagean–Meramecian (Mississippian) crinoid clades. Paleobiology 24:155–176.

Keese, M. C. 1998. Performance of two monophagous leaf feeding beetles (Coleoptera: Chrysomelidae) on each other's host plant: do intrinsic factors determine host plant specialization? Journal of Evolutionary Biology 11:403–419.

Kelley, P. H. 1989. Evolutionary trends within bivalve prey of Chesapeake Group naticid gastropods. Historical Biology 2:139–156.

———. 1991. The effect of predation intensity on rate of evolution of five Miocene bivalves. Historical Biology 5:65–78.

Kelley, P. H., A. Raymond, and C. B. Lutken. 1990. Carboniferous brachiopod migration and latitudinal diversity; a new palaeoclimatic method. Pp. 325–332 in W. S. McKerrow and C. R. Scotese, eds. Palaeozoic palaeogeography and biogeography. Geological Society of London Memoir 12.

Kennish, M. J. 1980. Shell microgrowth analysis: Mercenaria mercenaria as a type example for research in population dynamics. Pp. 255–294 in D. C. Rhoads and R. A. Lutz, eds. Skeletal growth of aquatic organisms. Plenum, New York.

Keys, D. N., D. L. Lewis, J. E. Selegue, B. J. Pearson, L. V. Goodrich, R. J. Johnson, J. Gates, M. P. Scott, and S. B. Carroll. 1999. Recruitment of a hedgehog regulatory circuit in butterfly eyespot evolution. Science 283:532–534.

Kidwell, S. M., and T. A. Aigner. 1985. Sedimentary dynamics of complex shell beds: implications for ecologic and evolutionary patterns. Pp. 383–395 in U. Bayer and A. Seilacher, eds. Sedimentary and evolutionary cycles. Springer, Berlin.

Kidwell, S. M., and K. W. Flessa. 1995. The quality of the fossil record: populations, species, and communities. Annual Review of Ecology and Systematics 26:269–299.

Kirkpatrick, M. 1996. Genes and adaptation: a pocket guide to the theory. Pp. 125–146 in M. R. Rose and G. V. Lauder, eds. Adaptation. Academic Press, San Diego.

Kirkpatrick, M., and M. Slatkin. 1993. Searching for evolutionary patterns in the shape of a phylogenetic tree. Evolution 47:1171–1181.

Kirschner, M., and J. Gerhart. 1997. Cells, embryos, and evolution. Blackwell Science, Malden, Mass.

Kitazato, H., M. Tsuchiya, and K. Takahara. 2000. Recognition of breeding populations in foraminifera: an example using the genus Glabratella. Paleontological Research 4:1–15.

Klingenberg, C. P. 1998. Heterochrony and allometry: the analysis of evolutionary change in ontogeny. Biological Reviews 73:79–123.

Knoll, A. H., and S. B. Carroll. 1999. Early animal evolution: emerging views from comparative biology and geology. Science 284:2129–2137.

Knowlton, N. 1993. Sibling species in the sea. Annual Review of Ecology and Systematics 24:189–216.

Knowlton, N., and L. A. Weigt. 1997. Species of marine invertebrates: a comparison of the biological and phylogenetic species concepts. Pp. 199–219 in Claridge et al. 1997.

Knowlton, N., J. Maté, H. M. Guzmán, R. Rowan, and J. Jara. 1997. Direct evidence for reproductive isolation among the three species of the Montastraea annularis complex in Central America (Panamá and Honduras). Marine Biology 127:705–711.

Koch, P. L. 1986. Clinal geographic variation in mammals: implications for the study of chronoclines. Paleobiology 12:269–281.

Kodric-Brown, A., and J. H. Brown. 1979. Competition between distantly related taxa and the co-evolution of plants and pollinators. American Zoologist 19:1115–1127.

Kohn, A. J., and F. E. Perron. 1994. Life history and biogeography: patterns in Conus. Clarendon, Oxford.

Kowalewski, M., E. Dyreson, J. D. Marcot, J. A. Vargas, K. W. Flessa, and D. P. Hallman. 1997. Phenetic discrimination of biometric simpletons: paleobiological implications of morphospecies in the lingulide brachiopod Glottidia. Paleobiology 23:444–469.

Lande, R. 1976. Natural selection and random genetic drift in phenotypic evolution. Evolution 30:314–334.

———. 1986. The dynamics of peak shifts and the pattern of morphological evolution. Paleobiology 12:343–354.

Langan-Cranford, K. M., and J. S. Pearse. 1995. Breeding experiments confirm species status of two morphologically similar gastropods (Lacuna spp.) in central California. Journal of Experimental Marine Biology and Ecology 186:17–31.

Lawton, J. H. 1999. Are there general laws in ecology? Oikos 84:177–192.

Lawton-Rauh, A. L., E. R. Alvzarez-Buylla, and M. D. Purugganan. 2000. Molecular evolution of flower development. Trends in Ecology and Evolution 15:144–149.

Lazarus, D., H. Hilbrecht, C. Spencer-Cervas, and H. Thierstein. 1995. Sympatric speciation and phyletic change in Globorotalia truncatulinoides. Paleobiology 21:28–51.

Leamy, L. J., E. J. Routman, and J. M. Cheverud. 1999. Quantitative trait loci for early- and late-developing skull characters in mice: a test of the genetic independence model of morphological integration. American Naturalist 153:201–214.

Lerner, I. M. 1953. Genetic homeostasis. Wiley, New York.

Levin, D. A. 1970. Developmental instability and evolution in peripheral populations. American Naturalist 104:343–353.

Levin, S. A. 1999. Fragile dominion: complexity and the commons. Perseus Books, Reading, Mass.

Levinton, J. 1988. Genetics, paleontology, and macroevolution. Cambridge University Press, New York.

Lewontin, R. C. 1970. The units of selection. Annual Review of Ecology and Systematics 1:1–18.

Li, X., and W. McGinnis. 1999. Activity regulation of Hox proteins, a mechanism for altering functional specificity in development and evolution. Proceedings of the National Academy of Sciences USA 96:6802–6807.

Lidgard, S., F. K. McKinney, and P. D. Taylor. 1993. Competition, clade replacement, and a history of cyclostome and cheilostome diversity. Paleobiology 19:352–371.

Lieberman, B. S., C. E. Brett, and N. Eldredge. 1994. Patterns and processes of stasis in two species lineages of brachiopods from the Middle Devonian of New York State. American Museum Novitates 3114·1–23.

———. 1995. A study of stasis and change in two species lineages of brachiopods from the Middle Devonian of New York State. Paleobiology 21:15–27.

Lindberg, D. R. 1988. Heterochrony in gastropods: a neontological view. Pp. 197–216 in M. L. McKinney, ed. Heterochrony in evolution. Plenum, New York.

Lister, A. M. 1989. Rapid dwarfing of red deer on Jersey in the last interglacial. Nature 342:539–542.

Lloyd, E. A., and S. J. Gould. 1993. Species selection on variability. Proceedings of the National Academy of Sciences USA 90:595–599.

Long, A. D., S. L. Mullaney, T. F. C. Mackay, and C. H. Langley. 1996. Genetic interactions between naturally occurring alleles at quantitative trait loci and mutant alleles at candidate loci affecting bristle number in Drosophila melanogaster. Genetics 144:1497–1510.

Lonsdale, W. M. 1999. Global patterns of plant invasions and the concept of invasibility. Ecology 80:1522–1536.

Losos, J. B., K. I. Warheit, and T. W. Schoener. 1997. Adaptive differentiation following experimental island colonization in Anolis lizards. Nature 387:70–73.

Luckow, M. 1995. Species concepts: assumptions, methods, and applications. Systematic Botany 20:589–605.

Ludwig, M. Z., C. Bergman, N. H. Patel, and M. Kreitman. 2000. Evidence for stabilizing selection in a eukaryotic enhancer element. Nature 403:564–567.

Lupia, R. 1999. Discordant morphological disparity and taxonomic diversity during the Cretaceous angiosperm radiation: North American pollen record. Paleobiology 25:1–28.

Lynch, J. B. 1989. The gauge of speciation: on the frequencies of modes of speciation. Pp. 527–553 in Otte and Endler 1989.

Lynch, M., and J. B. Walsh. 1998. Genetics and analysis of quantitative traits. Sinauer, Sunderland, Mass.

Mackay, T. F. C. 1996. The nature of quantitative genetic variation revisited: lessons from Drosophila bristles. BioEssays 18: 113–121.

MacLeod, N. 1991. Punctuated anagenesis and the importance of stratigraphy to paleobiology. Paleobiology 17:167–188.

Marshall, C. R. 1991. Estimation of taxonomic ranges from the fossil record. In N. L. Gilinsky and P. W. Signor, eds. Analytical paleobiology. Short Courses in Paleontology 4:19–38. Paleontological Society, Knoxville, Tenn.

———. 1995. Stratigraphy, the true order of species originations and extinctions, and testing ancestor-descendant-hypotheses among Caribbean Neogene bryozoans. Pp. 208–235 in Erwin and Anstey 1995b.

Martin, P. S., and D. W. Steadman. 1999. Prehistoric extinctions on islands and continents. Pp. 17–55 in R. D. E. MacPhee, ed. Extinctions in near time. Kluwer Academic/Plenum, New York.

Martin, R. A. 1993. Patterns of variation and speciation in Quaternary rodents. Pp. 226–280 in R. A. Martin and A. D. Barnosky, eds. Morphological change in Quaternary mammals of North America. Cambridge University Press, Cambridge.

Martin, R. E. 1999. Taphonomy: a process approach. Cambridge University Press, New York.

Maurer, B. A., and M. P. Nott. 1998. Geographic range fragmentation and the evolution of biological diversity. Pp. 31–50 in McKinney and Drake 1998.

Mayden, R. L. 1997. A hierarchy of species concepts: the denouement in the saga of the species problem. Pp. 381–424 in Claridge et al. 1997.

Maynard Smith, J. 1989. The causes of extinction. Philosophical Transactions of the Royal Society of London B 325:241–252.

Mayr, E. 1992. Controversies in retrospect. Oxford Surveys in Evolutionary Biology 8:1–34.

McKenzie, J. A., and P. Batterham. 1994. The genetic, molecular and phenotypic consequences of selection for insecticide resistance. Trends in Ecology and Evolution 9:166–169.

McKinney, F. K., S. Lidgard, J. J. Sepkoski Jr., and P. D. Taylor. 1998. Decoupled temporal patterns of evolution and ecology in two post-Paleozoic clades. Science 281:809–809.

McKinney, M. L. 1997. Extinction vulnerability and selectivity: combining ecological and paleontological views. Annual Review of Ecology and Systematics 28:495–516.

———. 1998. Biodiversity dynamics: Niche preemption and saturation in diversity equilibria. Pp. 1–16 in McKinney and Drake 1998.

McKinney, M. L., and J. A. Drake, eds. 1998. Biodiversity dynamics. Columbia University Press, New York.

McKinney, M. L., and J. L. Gittleman. 1995. Ontogeny and phylogeny: tinkering with covariation in life history, morphology and behaviour. Pp. 21–47 in K. J. McNamara, ed. Evolutionary change and heterochrony. Wiley, Chichester, England.

McKinney, M. L., and K. J. McNamara. 1991. Heterochrony: the evolution of ontogeny. Plenum, New York.

McNamara, K. J. 1983. Progenesis in trilobites. Special Papers in Palaeontology 30:59–68.

McShea, D. W. 1994. Mechanisms of large-scale evolutionary trends. Evolution 48:1747–1763.

Meyen, S. V. 1992. Geography of macroevolution in higher plants. Soviet Scientific Reviews G (Geology) 1:39–70.

Michaux, B. 1987. An analysis of allozymic characters of four species of New Zealand Amalda (Gastropoda: Olividae: Ancillinae). New Zealand Journal of Zoology 14:359–366.

———. 1989. Morphological variation of species through time. Biological Journal of the Linnean Society 38:239–255.

Miller, A. I. 1997a. Comparative diversification dynamics among palaeocontinents during the Ordovician radiation. Geobios Mémoire Spécial 20:397–406.

———. 1997b. Dissecting global diversity patterns: examples from the Ordovician Radiation. Annual Review of Ecology and Systematics 28:85–104.

———. 1998. Biotic transitions in global marine diversity. Science 281:1157–1160.

Miller, A. I., and J. J. Sepkoski Jr. 1988. Modeling bivalve diversification: the effect of interaction on a macroevolutionary system. Paleobiology 14:364–369.

Müller, G. B. 1990. Developmental mechanisms at the origin of morphological novelty: a side-effect hypothesis. Pp. 99–130 in M. H. Nitecki, ed. Evolutionary innovations. University of Chicago Press, Chicago.

Nemeschkal, H. L. 1999. Morphometric correlation patterns of adult birds (Fringillidae: Passeriformes and Columbiformes) mirror the expression of developmental control genes. Evolution 53:899–918.

Nevo, E., R. Ben-Shlomo, A. Beiles, C. P. Hart, and F. H. Ruddle. 1992. Homeobox DNA polymorphisms (RFLPs) in subterranean mammals of the Spalax ehrenbergi superspecies in Israel: patterns, correlates, and evolutionary significance. Journal of Experimental Zoology 263:430–441.

Newell, N. D. 1956. Fossil populations. In A. C. Sylvester-Bradley, ed. The species concept in palaeontology. Systematics Association Publication 2:63–82.

Nijhout, H. F., and D. J. Emlen. 1998. Competition among body parts in the development and evolution of insect morphology. Proceedings of the National Academy of Sciences USA 95: 3685–3689.

Nijhout, H. F., and S. M. Paulsen. 1997. Developmental models and polygenic characters. American Naturalist 149:394–405.

Niklas, K. J. 1997. The evolutionary biology of plants. University of Chicago Press, Chicago.

Nixon, K. C., and Q. D. Wheeler. 1990. An amplification of the phylogenetic species concept. Cladistics 6:211–223.

Norell, M. A. 1996. Ghost taxa, ancestors, and assumptions: a comment on Wagner. Paleobiology 22:453–455.

Norris, R. D., R. M. Corfield, and J. E. Cartlidge. 1996. What is gradualism? Cryptic speciation in globorotaliid foraminifera. Paleobiology 22:386–405.

Nunn, C. L., and K. K. Smith. 1998. Statistical analyses of developmental sequences: the craniofacial region in marsupial and placental mammals. American Naturalist 152:82–101.

Nuzhdin, S. V., C. L. Dilda, and T. F. C. Mackay. 1999. The genetic architecture of selection response: inferences from fine-scale mapping of bristle number quantitative trait loci in Drosophila melanogaster. Genetics 153:1317–1331.

Olson, E. C., and R. L. Miller. 1958. Morphological integration. University of Chicago Press, Chicago (reprinted 1999).

Omland, K. E. 1997. Examining two standard assumptions of ancestral reconstructions: repeated loss of dichromatism in dabbling ducks (Anatini). Evolution 51:1636–1646.

Orr, H. A. 1998. The population genetics of adaptation: the distribution of factors fixed during adaptive evolution. Evolution 52:935–949.

Orr, H. A., and J. A. Coyne. 1992. The genetics of adaptation revisited. American Naturalist 140:725–742.

Otte, D., and J. A. Endler. 1989. Speciation and its consequences. Sinauer, Sunderland, Mass.

Page, K. N. 1996. Mesozoic ammonoids in time and space. Pp. 755–794 in N. H. Landman, K. Tanabe, and R. A. Davis, eds. Ammonoid paleobiology. Plenum, New York.

Palmer, A. R. 1985. Quantum changes in gastropod shell morphology need not reflect speciation. Evolution 39:699–705.

Palopoli, M. F., and N. H. Patel. 1996. Neo-Darwinian developmental evolution: Can we bridge the gap between pattern and process? Current Opinion in Genetics and Development 6:502–508.

Palumbi, S. R. 1996. Macrospatial genetic structure and speciation in marine taxa with high dispersal abilities. Pp. 101–117 in J. Ferraris and S. R. Palumbi, eds. Molecular zoology. Wiley, New York.

Palumbi, S. R., G. Grabowsky, T. Duda, L. Geyer, and N. Tachino. 1997. Speciation and population genetic structure in tropical Pacific sea urchins. Evolution 51:1506–1517.

Paradis, E. 1998. Detecting shifts in diversification rates without fossils. American Naturalist 152:176–187.

Parsons, P. A. 1993. Stress, extinctions and evolutionary change: from living organisms to fossils. Biological Reviews 68:313–333.

———. 1994. Habitats, stress, and evolutionary rates. Journal of Evolutionary Biology 7:387–397.

Paterson, A. H., Y.-R. Lin, Z. Li, K. F. Schertz, J. F. Doebley, S. R. M. Pinson, S.-C. Liu, J. W. Stansel, and J. E. Irvine. 1997. Convergent domestication of cereal crops by independent mutation at corresponding genetic loci. Science 269:1714–1718.

Patzkowsky, M. E. 1995. A hierarchical branching model of evolutionary radiations. Paleobiology 21:440–460.

Paul, C. R. C. 1992. The recognition of ancestors. Historical Biology 6:239–250.

Pearson, P. N. 1995. Investigating age-dependency of species extinction rates using dynamic survivorship analysis. Historical Biology 10:119–136.

Pechenik, J. A. 1999. On the advantages and disadvantages of larval stages in benthic marine invertebrate life cycles. Marine Ecology Progress Series 177:269–297.

Peterson, A. T., J. Soberon, and V. Sanchez-Cordero. 1999. Conservatism of ecological niches in evolutionary time. Science 285:1265–1267.

Potts, D. C., A. F. Budd, and R. L. Garthwaite. 1993. Soft tissue vs. skeletal approaches to species recognition and phylogeny reconstruction in corals. Courier Forschungsinstitut Senckenberg 16:221–231.

Povel, G. D. E. 1993. The main branch of Miocene Gyraulus (Gastropoda; Planorbindae) of Steinheim (southern Germany): a reconsideration of Mensink's data set. Scripta Geologica Special Issue 3:371–386.

Price, T. D., A. J. Helbig, and A. D. Richman. 1997. Evolution of breeding distributions in the Old World leaf warblers (genus Phylloscopus). Evolution 51:552–561.

Prothero, D. R., and T. H. Heaton. 1996. Faunal stability during the Early Oligocene climatic crash. Palaeogeography, Palaeoclimatology, Palaeoecology 127:257–283.

Prothero, D. R., and N. Shubin. 1989. The evolution of Oligocene horses. Pp. 142–175 in D. R. Prothero and R. M. Schoch, eds. The evolution of perissodactyls. Oxford University Press, New York.

Purugganan, M. D. 1998. The molecular evolution of development. BioEssays 20:700–711.

———. 2000. The molecular population genetics of regulatory genes. Molecular Ecology (in press).

Quinn, J. C., J. D. West, and R. E. Hill. 1996. Multiple functions for Pax6 in mouse eye and nasal development. Genes and Development 10:435–446.

Rabinowitz, D. 1981. Seven forms of rarity. Pp. 205–217 in H. Synge, ed. The biological aspects of rare plant conservation. Wiley, New York.

Rabinowitz, D., S. Cairns, and T. Dillon. 1986. Seven forms of rarity and their frequency in the flora of the British Isles. Pp. 182–204 in M. E. Soulé, ed. Conservation biology. Sinauer, Sunderland, Mass.

Rachootin, S. P., and K. S. Thomson. 1981. Epigenetics, paleontology, and evolution. Pp. 181–193 in G. G. E. Scudder and J. L. Reveal, eds. Evolution today. Proceedings of the second international congress of systematic and evolutionary biology. Hunt Institute for Botanical Documentation, Carnegie-Mellon University, Pittsburgh, Penn.

Raff, R. A. 1996. The shape of life: genes, development, and the evolution of animal form. University of Chicago Press, Chicago.

Raff, R. A., and T. C. Kaufman. 1983. Embryos, genes, and evolution. Macmillan, New York (reprinted 1991, Indiana University Press, Bloomington).

Raff, R. A., and G. A. Wray. 1989. Heterochrony: developmental mechanisms and evolutionary results. Journal of Evolutionary Biology 2:409–434.

Raup, D. M. 1983. On the early origins of major biologic groups. Paleobiology 9:107–115.

———. 1991. A kill curve for Phanerozoic marine species. Paleobiology 17:37–48.

———. 1994. The role of extinction in evolution. Proceedings of the National Academy of Sciences USA 91:6758–6763.

Relaix, F., and M. Buckingham. 1999. From insect eye to vertebrate muscle: redeployment of a regulatory network. Genes and Development 13:3171–3178.

Rice, S. H. 1995. A genetical theory of species selection. Journal of Theoretical Biology 177:237–245.

———. 1998. The evolution of canalization and the breaking of von Baer's laws: modeling the evolution of development with epistasis. Evolution 52:647–656.

Ricklefs, R. E., and R. E. Latham. 1992. Intercontinental correlation of geographic ranges suggests stasis in ecological traits of relict genera of temperate perennial herbs. American Naturalist 139:1305–1321.

Riedl, R. 1978. Order in living organisms. Wiley, Chichester, England.

Robertson, F. W. 1959. Studies in quantitative inheritance. XIII. Interrelations between genetic behavior and development in the cellular constitution in the Drosophila wing. Genetics 44:1113–1130.

Rogers, B. T., M. D. Peterson, and T. C. Kaufman. 1997. Evolution of the insect body plan as revealed by the Sex combs reduced expression pattern. Development 124:149–157.

Roopnarine, P. D., G. Byars, and P. Fitzgerald. 1999. Anagenetic evolution, stratophenetic patterns, and random walk models. Paleobiology 25:41–57.

Rosenzweig, M. L., and R. D. McCord. 1991. Incumbent replacement: evidence for long-term evolutionary progress. Paleobiology 17:202–213.

Routman, E. J., and J. M. Cheverud. 1997. Gene effects on a quantitative trait: two-locus epistatic effects measured at microsatellite markers and at estimated QTL. Evolution 51:1654–1662.

Roy, K. 1996. The roles of mass extinction and biotic interaction in large-scale replacements: a reexamination using the fossil record of stromboidean gastropods. Paleobiology 22:436–452.

Roy, K., J. W. Valentine, D. Jablonski, and S. M. Kidwell. 1996. Scales of climatic variability and time averaging in Pleistocene biotas: implications for ecology and evolution. Trends in Ecology and Evolution 11:458–463.

Rugh, N. S. 1997. Differences in shell morphology between the sibling species Littorina scutulata and Littorina plena (Gastropoda: Prosobranchia). Veliger 40:350–357.

Rumrill, S. S. 1990. Natural mortality of marine invertebrate larvae. Ophelia 32:163–198.

Rutherford, S. L., and S. Lindquist. 1998. Hsp 90 as a capacitor for morphological evolution. Nature 396:336–342.

Salmon, W. C. 1971. Statistical explanation and statistical relevance. University of Pittsburgh Press, Pittsburgh, Penn.

Sanderson, M. J., and M. J. Donoghue. 1996. Reconstructing shifts in diversification rates on phylogenetic trees. Trends in Ecology and Evolution 11:15–20.

Sato, S. 1995. Spawning periodicity and shell microgrowth patterns of the venerid bivalve *Phacosoma japonicum* (Reeve, 1850). Veliger 38:61–72.

———. 1999. Temporal change of life-history traits in fossil bivalves: an example of *Phacosoma japonicum* from the Pleistocene of Japan. Palaeogeography, Palaeoclimatology, Palaeoecology 154:313–323.

Saunders, W. B., D. M. Work, and S. V. Nikolaeva. 1999. Evolution of complexity in Paleozoic ammonoid sutures. Science 286:760–763.

Schankler, D. M. 1981. Local extinction and ecological re-entry of early Eocene mammals. Nature 293:135–138.

Scheltema, R. S. 1989. Planktonic and non-planktonic development among prosobranch gastropods and its relationship to the geographic range of species. Pp. 183–188 *in* J. S. Ryland and P. A. Tyler, eds. Reproduction, genetics and distribution of marine organisms. Olsen and Olsen, Fredensborg, Denmark.

———. 1992. Passive dispersal of planktonic larvae and the biogeography of tropical sublittoral invertebrate species. Pp. 195–202 *in* G. Colombo et al., eds. Marine eutrophication and population dynamics. Olsen and Olsen, Fredensborg, Denmark.

Schemske, D. W., and H. D. Bradshaw. 1999. Pollinator preference and the evolution of floral traits in monkeyflowers (*Mimulus*). Proceedings of the National Academy of Sciences USA 96:11910–11915.

Schlichting, C. D., and M. Pigliucci. 1998. Phenotypic evolution: a reaction norm perspective. Sinauer, Sunderland, Mass.

Schluter, D. 1986. Character displacement between distantly related taxa? Finches and bees in the Galapagos. American Naturalist 127:95–102.

———. 1996. Adaptive radiation along genetic lines of least resistance. Evolution 50:1766–1774.

Schneider, J. A. 1995. Phylogenetic relationships of transisthmian Cardiidae (Bivalvia) and the usage of fossils in reinterpreting the geminate species concept. Geological Society of America Abstracts with Programs 27(6):A-52.

Sepkoski, J. J., Jr. 1996. Competition in macroevolution: the double wedge revisited. Pp. 211–255 *in* D. Jablonski, D. H. Erwin, and J. H. Lipps, eds. Evolutionary paleobiology. University of Chicago Press, Chicago.

———. 1998. Rates of speciation in the fossil record. Philosophical Transactions of the Royal Society of London B 353:315–326.

Sepkoski, J. J., Jr., F. K. McKinney, and S. Lidgard. 2000. Competitive displacement among post-Paleozoic cyclostome and cheilostome bryozoans. Paleobiology 26:7–18.

Shackleton, N. J., I. N. McCave, and G. P. Weedon, eds. 1999. Astronomical (Milankovitch) calibration of the geological timescale. Philosophical Transactions of the Royal Society of London A 357:1733–2007.

Shaffer, H. B. 1984. Evolution in a paedomorphic lineage. I. An electrophoretic analysis of the Mexican ambystomatid salamanders. Evolution 38:1194–1206.

Shaw, F. H., R. G. Shaw, G. S. Wilkinson, and M. Turelli. 1995. Changes in genetic variances and covariances: G whiz! Evolution 49:1260–1267.

Sheldon, P. R. 1987. Parallel gradualistic evolution in Ordovician trilobites. Nature 330:561–563.

———. 1988. Trilobite size-frequency distributions, recognition of instars, and phyletic size change. Lethaia 21:293–306.

———. 1993. Making sense of microevolutionary patterns. Pp. 19–31 *in* D. R. Lees and D. Edwards, eds. Evolutionary patterns and processes. Academic Press, London.

———. 1996. Plus ça change—a model for stasis and evolution in different environments. Palaeogeography, Palaeoclimatology, Palaeoecology 127:209–227.

Shubin, N., C. Tabin, and S. Carroll. 1997. Fossil, genes and the evolution of animal limbs. Nature 388:639–648.

Signor, P. W., III, and C. E. Brett. 1984. The mid-Paleozoic precursor to the Mesozoic marine revolution. Paleobiology 10:229–245.

Simms, M. J. 1988a. The role of heterochrony in the evolution of post-Paleozoic crinoids. Pp. 97–102 *in* R. D. Burke, P. V. Mladenov, P. Lambert, and R. L. Parsley, eds. Echinoderm biology. Balkema, Rotterdam.

———. 1988b. The phylogeny of post-Palaeozoic crinoids. Pp. 269–284 *in* C. R. C. Paul and A. B. Smith, eds. Echinoderm phylogeny and evolution. Clarendon, Oxford.

———. 1994. Crinoids from the Chambara Formation, Pucara Group, central Peru. Palaeontographica, Abteilung A 233:169–175.

Simpson, G. G. 1961. Principles of animal taxonomy. Columbia University Press, New York.

Sinervo, B., and L. R. McEdward. 1988. Developmental consequences of an evolutionary change in egg size: an experimental test. Evolution 42:885–899.

Slatkin, M. 1981. A diffusion model of species selection. Paleobiology 7:421–425.

Slowinski, J. B., and C. Guyer. 1993. Testing whether certain traits have caused amplified diversification: an improved method based on a model of random speciation and extinction. American Naturalist 142:1019–1024.

Smith, A. B. 1994. Systematics and the fossil record. Blackwell Scientific, Oxford.

Smith, G. R. 1992. Introgression in fishes: significance for paleontology, cladistics, and evolutionary rates. Systematic Biology 41:41–57.

Smith, K. K. 1996. Integration of craniofacial structures during development in mammals. American Zoologist 36:70–79.

———. 1997. Comparative patterns of craniofacial development in eutherian and metatherian mammals. Evolution 51:1663–1678.

Sneppan, K., P. Bak, H. Flybjerg, and M. H. Jensen. 1995. Evolution as a self-organized critical phenomenon. Proceedings of the National Academy of Sciences USA 92:5209–5213.

Sober, E. 1984. The nature of selection. MIT Press, Cambridge.

———. 1999. The multiple realizability argument against reductionism. Philosophy of Science 66:542–564.

Solé, R. V., S. C. Manrubia, M. Benton, S. Kauffman, and P. Bak. 1999. Criticality and scaling in evolutionary ecology. Trends in Ecology and Evolution 14:156–160.

Spencer-Cervato, C., H. R. Thierstein, D. B. Lazarus, and J.-P. Beckmann. 1994. How synchronous are Neogene marine plankton events? Paleoceanography 9:739–763.

Stanley, S. M. 1973. An explanation for Cope's Rule. Evolution 27:1–26.

———. 1979. Macroevolution. W. H. Freeman, San Francisco.

———. 1982. Macroevolution and the fossil record. Evolution 36:460–473.

———. 1990. The general correlation between rate of speciation and rate of extinction: fortuitous causal linkages. Pp. 103–127 *in* R. M. Ross and W. D. Allmon, eds. Causes of evolution. University of Chicago Press, Chicago.

Stanley, S. M., and X. Yang. 1987. Approximate evolutionary stasis for bivalve morphology over millions of years: a multivariate, multilineage study. Paleobiology 13:113–139.

Stanley, S. M., K. L. Wetmore, and J. P. Kennett. 1988. Macroevolutionary differences between two major clades of Neogene planktonic Foraminifera. Paleobiology 14:235–249.

Stearns, S. C. 1992. The evolution of life histories. Oxford University Press, Oxford.

Stebbins, G. L. 1982. Perspectives in evolutionary theory. Evolution 36:1109–1119.

Stehli, F. G., R. G. Douglas, and N. D. Newell. 1969. Generation and maintenance of gradients in taxonomic diversity. Science 164:947–949.

Steppan, S. J. 1998. Phylogenetic relationships and species limits within *Phyllotis* (Rodentia: Sigmodontinae): concordance between mtDNA sequence and morphology. Journal of Mammalogy 79:573–593.

Sterelny, K., and P. E. Griffiths. 1999. Sex and death: an introduction to philosophy of biology. University of Chicago Press, Chicago.

Stevenson, R. D., M. F. Hill, and P. J. Bryant. 1995. Organ and cell allometry in Hawaiian *Drosophila*: how to make a big fly. Proceedings of the Royal Society of London B 259:105–110.

Stidd, B. M., and D. L. Wade. 1995. Is species selection dependent upon emergent characters? Biology and Philosophy 10: 55–76.

Tabachnick, R. E., and F. L. Bookstein. 1990. The structure of individual variation in Miocene *Globorotalia*. Paleobiology 44: 416–434.

Templeton, A. R. 1989. The meaning of species and speciation: a genetic perspective. Pp. 3–27 *in* Otte and Endler 1989.

———. 1998. Species and speciation: Geography, population structure, ecology, and gene trees. Pp. 32–43 *in* Howard and Berlocher 1998.

Theissen, G., A. Becker, A. Di Rosa, A. Kanno, J. T. Kim, T. Münster, K.-U. Winter, and H. Saedler. 2000. A short history of MADS-box genes in plants. Plant Molecular Biology 42:115–149.

Theriot, E. 1992. Clusters, species concepts, and morphological evolution of diatoms. Systematic Biology 41:141–157.

Thomas, R. D. K., R. M. Shearman, and G. W. Stewart. 2000. Evolutionary exploitation of design options by the first animals with hard skeletons. Science 288:1239–1242.

Thomson, K. S. 1988. Morphogenesis and evolution. Oxford University Press, New York.

Trussell, G. C., and L. D. Smith. 2000. Induced defenses in response to an invading crab predator: an explanation of historical and geographic phenotypic change. Proceedings of the National Academy of Sciences USA 97:2123–2127.

Turelli, M. 1988. Phenotypic evolution, constant covariances, and the maintenance of additive variance. Evolution 42:1342–1347.

Valentine, J. W. 1969. Taxonomic and ecological structure of the shelf benthos during Phanerozoic time. Palaeontology 12: 684–709.

———. 1973. Evolutionary paleoecology of the marine biosphere. Prentice-Hall, Englewood Cliffs, N.J.

———. 1980. Determinants of diversity in higher taxonomic categories. Paleobiology 6:444–450.

———. 1990. The fossil record: a sampler of life's diversity. Philosophical Transactions of the Royal Society of London B 330: 251–268.

———. 1995. Why no new phyla after the Cambrian? Genome and ecospace hypotheses revisited. Palaios 10:190–194.

———. 2000. Two paths to complexity in metazoan evolution. Paleobiology 26:513–519.

Valentine, J. W., and D. Jablonski. 1983. Speciation in the shallow sea: General patterns and biogeographic controls. *In* R. W. Sims, J. H. Price, and P. E. S. Whalley, eds. Evolution, time and space. Systematics Association Special Volume 23:201–226. Academic Press, London.

———. 1993. Fossil communities: compositional variation at many time scales. Pp. 341–349 *in* R. E. Ricklefs and D. Schluter, eds. Species diversity in ecological communities: historical and geographical perspectives. University of Chicago Press, Chicago.

Valentine, J. W., D. Jablonski, and D. H. Erwin. 1999. Fossils, molecules and embryos: new perspectives on the Cambrian explosion. Development 126:851–859.

Van Valen, L. 1976. Ecological species, multispecies, and oaks. Taxon 25:233–239.

Vermeij, G. J. 1987. Evolution and escalation. Princeton University Press, Princeton, N.J.

———. 1991. When biotas meet: understanding biotic interchange. Science 253:1099–1104.

———. 1994. The evolutionary interaction among species: selection, escalation, and coevolution. Annual Review of Ecology and Systematics 25:219–236.

Via, S., and D. J. Hawthorne. 1998. The genetics of speciation: promises and prospects of Quantitative Trait Locus mapping. Pp. 352–354 *in* Howard and Berlocher 1998.

Via, S., and A. J. Shaw. 1996. Short-term evolution in size and shape of pea aphids. Evolution 50:163–173.

Voss, S. R., and H. B. Shaffer. 1997. Adaptive evolution via a major gene effect: paedomorphosis in the Mexican axolotl. Proceedings of the National Academy of Sciences USA 94:14185–14189.

Vrba, E. S. 1984. What is species selection? Systematic Zoology 33:318–328.

———. 1987. Ecology in relation to speciation rates: some case histories of Miocene-Recent mammal clades. Evolutionary Ecology 1:283–300.

———. 1989. Levels of selection and sorting with special reference to the species level. Oxford Surveys in Evolutionary Biology 6:111–168.

Vrba, E. S., and N. Eldredge. 1984. Individuals, hierarchies and processes: towards a more complete evolutionary theory. Paleobiology 10:146–171.

Vrba, E. S., and S. J. Gould. 1986. The hierarchical expansion of sorting and selection: sorting and selection cannot be equated. Paleobiology 12:217–228.

Wagner, G. P. 1996. Homologues, natural kinds and the evolution of modularity. American Zoologist 36:36–43.

Wagner, G. P., G. Booth, and H. Bagheri-Chaichian. 1997. A population genetic theory of canalization. Evolution 51:329–347.

Wagner, G. P., C. H. Chiu, and T. H. Hansen. 1999. Is Hsp 90 a regulator of evolvability? Journal of Experimental Zoology (Molecular and Developmental Evolution) 285:116–118.

Wagner, P. J. 1995. Testing evolutionary constraint hypotheses: examples with early Paleozoic gastropods. Paleobiology 21: 248–272.

———. 1996a. Ghost taxa, ancestors, and assumptions: A reply to Norell. Paleobiology 22:456–460.

———. 1996b. Contrasting the underlying patterns of active trends in morphologic evolution. Evolution 50:990–1007.

———. 1998. A likelihood approach for evaluating estimates of phylogenetic relationships among fossil taxa. Paleobiology 24:430–449.

Wagner, P. J., and D. H. Erwin. 1995. Phylogenetic patterns as tests of speciation models. Pp. 87–122 *in* Erwin and Anstey 1995b.

Walsh, J. A. 1996. No second chances? New perspectives on biotic interactions in post-Paleozoic brachiopod history. Pp. 281–288 *in* P. Copper and J. Jin, eds. Brachiopods. Balkema, Rotterdam.

Watts, P. C., J. P. Thorpe, and P. D. Taylor. 1998. Natural and anthropogenic dispersal mechanisms in the marine environment: a study using cheilostome Bryozoa. Philosophical Transactions of the Royal Society of London B 353:453–464.

Wayne, R. K. 1986. Cranial morphology of domestic and wild canids: the influence of development on morphological change. Evolution 40:243–261.

Wayne, R. K., and E. A. Ostrander. 1999. Origin, genetic diversity, and genome structure of the domestic dog. BioEssays 21: 247–257.

Weatherbee, S. D., H. F. Nijhout, L. W. Grunert, G. Halder, R. Galant, J. Selegue, and S. Carroll. 1999. *Ultrabithorax* function in butterfly wings and the evolution of insect wing patterns. Current Biology 9:109–115.

Wei, K.-Y., and J. P. Kennett. 1986. Taxonomic evolution of Neogene planktonic Foraminifera and paleoceanographic relations. Paleoceanography 1:67–84.

Weil, E., and N. Knowlton. 1994. A multi-character analysis of the Caribbean coral *Montastraea annularis* (Ellis and Solander, 1786) and its two sibling species, *M. faveolata* (Ellis and Solander, 1786) and *M. franksi* (Gregory, 1895). Bulletin of Marine Science 55:151–175.

Wen, J. 1999. Evolution of the eastern Asian and eastern North American disjunct distributions in flowering plants. Annual Review of Ecology and Systematics 30:421–455.

Wiley, E. O. 1981. Phylogenetics. Wiley, New York.

Williams, G. C. 1992. Natural selection: domains, levels, and challenges. Oxford University Press, New York.

Williamson, M. 1996. Biological invasions. Chapman and Hall, London.

Wilson, A. C., G. L. Bush, S. M. Case, and M.-C. King. 1975. Social structuring of mammalian populations and rate of chromosomal evolution. Proceedings of the National Academy of Sciences USA 72:5061–5065.

Wing, S. L., and L. D. Boucher. 1998. Ecological aspects of the Cretaceous flowering plant radiation. Annual Review of Earth and Planetary Sciences 26:379–421.

Wright, S. 1982a. Character change, speciation and higher taxa. Evolution 36:427–443.

———. 1982b. The shifting balance theory and macroevolution. Annual Review of Genetics 16:1–19.

Zakany, J., C. Fromental-Ramain, X. Wartot, and D. Duboule. 1997. Regulation of number and size of digits by posterior Hox genes: a dose-dependent mechanism with potential evolutionary implications. Proceedings of the National Academy of Sciences USA 94:13695–13700.

Zelditch, M. L., and W. L. Fink. 1996. Heterochrony and heterotopy: stability and innovation in the evolution of form. Paleobiology 22:241–254.

Conversations about Phanerozoic global diversity

Arnold I. Miller

Abstract.—The emergence of Phanerozoic global diversity as a central theme of investigation has resulted from a confluence of factors, including the assembly by several researchers of global taxonomic databases; the advent of computers, which permitted construction and analysis of global Phanerozoic diversity trajectories; and the recognition that Phanerozoic diversity trends are important bellwethers of the evolutionary processes that cause biotic transitions. Despite the enormous progress in the measurement and interpretation of Phanerozoic diversity over the past quarter century, much of which has been reported in *Paleobiology*, these studies have collectively generated at least as many new questions as they have answered—arguably the mark of an area of inquiry that continues to be vital. In this essay, I discuss several outstanding issues in the investigation of Phanerozoic diversity, ranging from the viability of literature-derived databases for investigating global diversity trends, to the biological significance of the myriad biotic transitions that have taken place throughout the Phanerozoic.

Arnold I. Miller. Department of Geology, Post Office Box 210013, University of Cincinnati, Cincinnati, Ohio 45221-0013. E-mail: arnold.miller@uc.edu

Accepted: 22 June 2000

"Do these patterns compel me to believe that bivalves and brachiopods have been locked hand to hand (or, foot to pedicle) in some sort of mortal combat for the last 500 million years? My answer is no, not really. . . ." (Sepkoski 1996: p. 242).

Introduction

In 1993, when I was co-editor of *Paleobiology*, I phoned Richard Bambach to let him know that I was accepting for publication his *Seafood through time* manuscript (Bambach 1993); that really was the title—or the first part of it. In that paper, Bambach suggested that major transitions among marine biotas through the Phanerozoic were related closely to an increase in the availability of food in the world's oceans associated with the rise of new sources of organic detritus, including land plants. He argued that many of the marine animals living on early Phanerozoic seafloors (e.g., brachiopods) required less energy than the comparatively fleshy animals that came to dominate later (e.g., clams and snails), and that the diversification of the later groups could only take place after sufficient food was available to meet their needs.

When we talked on the phone that day, Richard was curious to know what I thought of the paper. My simple response was, "I liked it."

This wasn't enough for Richard; he was not going to let me off the hook with a polite response. So, he asked, "Yes, but what did you *think* of it?"

I paused for a moment, and said, again, "I liked it." But this time, I added an additional comment: "I'd give you about a five or ten percent share of the truth."

At that point, I heard the expected snickering at the other end of the phone line. Not surprisingly, Richard thought I was sarcastically dismissing his paper. But, as I explained to him, I wasn't being dismissive at all. In my view, a five or ten percent share of the truth is actually quite good.

What I liked about Bambach's analysis was his linkage of transitions in terrestrial ecosystems to those in marine settings. It is reasonable and compelling to suggest, as several researchers now have, that changes through time in global inputs of food, nutrients, and other materials to marine ecosystems should affect the composition of marine biotas.

At the same time, despite their many virtues, I have become concerned about the large and still-growing number of studies that, like the *Seafood through time* hypothesis, suggest overarching explanations for the history of Phanerozoic marine global biodiversity, or major subsets thereof. It is certainly understandable that researchers would seek out such explanations: the global Phanerozoic trajectory looks simple, can be described successfully with relatively few parameters (e.g., Sepkoski 1984), and can be parsed readily into groups of one's choosing that exhibit diversity zeniths during different portions of the Phanerozoic, with one group giving way sequentially to the next. However, there is an expanding body of evidence suggesting that a good deal of complexity resides beneath the global veneer: at local and regional levels, biodiversity transitions tend to be highly episodic (cf. Ivany and Schopf 1996) and, during at least some extended intervals, there is, at best, only limited global synchronicity to these episodes (Miller 1997a).

During the past several years, I have gravitated to the view that the global Phanerozoic diversity trajectory is the product of a spectrum of mechanisms that operate at multiple scales (Miller 1998). On the one hand, this might seem so blatantly obvious as to be trivial. And yet, the alternative view, that global patterns have been mediated mainly by global-scale processes—be they biotic interactions among biotas played out over geological time (e.g., Thayer 1979; Sepkoski 1984) or, at the other end of the spectrum, the direct consequences of mass extinctions (Gould 1985)—has certainly figured prominently in the literature.

This viewpoint might seem like a surrender to the complexities of a world that is dynamic physically and biologically at a variety of spatio-temporal scales. In fact, it is nothing of the kind. Because of our rapidly improving understanding of local and regional biotic patterns and their relationships—or not—to physical attributes of the settings in which they lived, there is good reason to be optimistic about our chances of one day understanding the intricacies of local and regional patterns, and how they weave together to produce the global signals with which we are all familiar.

Despite the enormous progress of the past quarter century, or maybe because of it, nearly every facet of the study of Phanerozoic diversity trends remains contentious, and I cannot pretend to suggest definitive research programs to solve the most pressing issues. What I can hope to do, however, is to frame some of these issues in a way that might make them more accessible to the broad audience interested in global diversity. As a vehicle for doing so, my discussion below will be woven with descriptions or excerpts of various conversations that I have engaged in over the years with friends and colleagues. Snippets from these conversations are worth recounting because they have helped me to (1) crystallize my viewpoints, (2) set a personal research agenda, and (3) recognize my own foolishness at times. Maybe they can help some readers to do the same.

I will focus below on a set of intertwined issues related to the taxonomic diversity of Phanerozoic marine biotas. These include the assembly and use of taxonomic databases, the calibration of Phanerozoic diversity trajectories, the causes of major global transitions in biotic composition, the extent to which Phanerozoic diversity trends transcend geographic and ecological scale, and the extent to which biotic transitions have been mediated by physical transitions and perturbations.

There are several additional themes that I will not pursue in this presentation despite their obvious relevance to Phanerozoic diversification, including the relationship between taxonomic and morphologic diversity (e.g., Foote 1993; Wagner 1995); similarities and differences in factors that affect diversity in marine and terrestrial settings (e.g., Valentine et al. 1991; Eble 1999); and the correlation—or not—between transitions in taxonomic diversity and relative abundance (e.g., McKinney et al. 1998; Lupia et al. 1999). While these and other themes will clearly be of significance to the eventual understanding of biotic transitions, my aim here is to discuss issues that must be settled before we can get our taxonomic diversity house in order.

Eavesdropping on an Internet Conversation, July, 1999: Counting Taxa

Okay, this first one wasn't *my* conversation. Rather, it was a portion of an Internet conversation that took place during July, 1999, on the *PaleoNet Listserver*, an online discussion forum for paleontologists. But first, a bit of perspective:

Over the years, perhaps more than any other endeavor in the field of paleontology, the study of global biodiversity has been emblematic of the changes that the discipline has undergone. Paleontology has been transformed from a science that was once largely descriptive to a more synthetic enterprise in which information about fossils is assembled into databases, and the data are then analyzed to address large-scale questions that could not possibly have been evaluated exhaustively by our pre-1950 forbears, because of the lack of computers (but see below). This transition is hardly unique to our field and, undoubtedly, many other sciences have experienced the kind of resultant squabbling that has popped up intermittently on the pages of paleontological journals and at scientific meetings. Still, one would think that, by now, the primary issues had long been settled concerning the assembly of published information about fossils into databases, the tabulation of these data to produce global diversity curves, and the use of assembled databases to test hypotheses about the history of diversity. But the squabble goes on. With that as a backdrop, here are some excerpts from one participant in the *PaleoNet* discussion (references to taxa, periods, and dates have been deleted to protect the identity of the *PaleoNet* correspondent):

> Bluntly: . . . non-systematists, people with no detailed knowledge of the group at hand, shouldn't bother compiling taxonomic databases. The taxonomic literature is byzantine, wildly variable in quality, approach, basic philosophy, etc., etc. Databasing it requires subjective taxonomic decisions to be made at every turn. Even a very thorough trawl of the primary literature (as opposed to the usual Treatise / Zoo Rec productions) may generate little more than noise without the application of specialist knowledge. Ok, I better qualify that 1) I have direct experience only of [taxon deleted]; but for [taxon deleted], I can affirm that the situation is absolutely, positively that

> bad—in many groups, the white noise of bad, weird, lunatic fringe taxonomy generates distortion of over 100% in terms of raw counts, and [period and taxon deleted] taxa still have basically no philosophical / methodological connection to post-[period deleted] taxa. 2) The [taxon deleted] Treatise still mainly dates to [date deleted]. . . .the Treatise in general is very much a Good Thing, and may be an excellent source of standardized data for many animal groups for which recent revisions are available.

> Anyway, the point of the preceding rant is that "meaningful" taxonomic data aren't objective entities that anyone can gather up, and require an investment of more than just time. Sure, you can just go and gather s(tuff) up, and people obviously have.

These comments suggest that data collected by "non-specialists" not only are bound to be deeply flawed, but may well produce distorted patterns that cannot be trusted. (I will not bother to address the suggestion that "non-systematists" who assemble synoptic, biodiversity databases typically do so blindly, and mainly by mining data directly from the *Treatise* or the *Zoological Record*, except to note that the days when database assemblers *limited* their compilations to those kinds of sources, rather than the primary literature, ended more than 20 years ago.)

Several papers in recent years have analyzed directly the significance of possible shortcomings in paleontological data, focusing on the question of taxonomic validity of the taxa under investigation, as well as the extent to which improvements and additions to available data over a period of time change perceptions about paleontological patterns. These include Sepkoski's (1993) recognition that accumulated changes to his global compendium of fossil marine families over ten years yielded remarkably little change to the Phanerozoic trajectory of global marine diversity; Maxwell and Benton's (1990) demonstration that 87 yr of improvement to the data available on tetrapod families did little to alter the major features of the diversity trajectory of Phanerozoic tetrapod families; Sepkoski and Kendrick's (1993) finding that paraphyly is probably not a significant problem in the analysis of biodiversity patterns at the genus and family levels; Benton's (1995) publication of an aggregate diversity trajectory for marine families for the Phanerozoic that closely paralleled Sepkoski's earlier depictions, despite the use of a different database; and Foote's (1997) demonstration that, for several higher taxa, the addition of data from future fossil discoveries is unlikely to change the already-emergent picture of their overall morphological variability and disparity. More recently, Benton (1999) offered a thoughtful rejoinder to those who continue to suggest that paleontological databases are not up to the tasks for which they have been utilized, and I encourage interested readers to consult his paper.

Still, the comments of the *PaleoNet* correspondent point to the need for further, *direct* evaluation of the concern that data not properly vetted or otherwise updated by specialists may yield—or perhaps has already yielded—untrustworthy signals. In the "ideal" analysis, global-scale diversity patterns extracted from databases compiled by nonspecialists would be compared with those based on compendia for the same higher taxa produced and vetted by specialists. Just such an analysis was conducted for trilobites in a recent paper by Adrain and Westrop (2000), who compared lower Tremadocian through upper Wenlockian genus diversification patterns based on Sepkoski's unpublished compendium to those generated from their own state-of-the-art database. Adrain and Westrop's database, which has permitted an extensive investigation of trilobite macroevolution throughout the early Paleozoic (e.g., Adrain et al. 1998, 2000; Westrop and Adrain 1998), was compiled largely from direct, critical evaluation and vetting of previous systematic research on trilobites, as well as from extensive field and laboratory investigations. In comparison to their database, Adrain and Westrop (2000) demonstrated that Sepkoski's compendium contained a large number of systematic and stratigraphic errors. Nevertheless, the global diversity trajectory extracted from Sepkoski's compendium was nearly identical to that developed with their database (Adrain and Westrop 2000: Fig 3). Thus, the authors

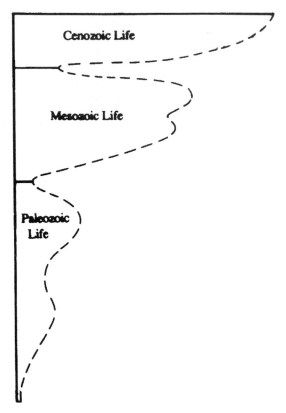

FIGURE 1. Phillips's depiction of the changing number of marine species through geological time, based on a compendium of fossil data from Great Britain (redrawn from Phillips 1860: Fig. 1).

concluded that Sepkoski's compendium produced an accurate depiction of the large-scale, global diversity trajectory.

Adrain et al. (1998, 2000; Westrop and Adrain 1998) produced significant conceptual advances in perceptions about the macroevolutionary history of trilobites that clearly could not have been achieved through reliance on Sepkoski's compendium. However, these advances resulted not so much from taxonomic revision to global compendia as they did from the thoughtful, and novel, ways in which these workers have analyzed their data and extended their databases to include geographic regions neglected in previous compilations. Just as importantly, they have moved beyond the relatively coarse global scale to rigorously address paleogeographic and paleoenvironmental aspects of diversification, both of which should now be viewed as prerequisites for the interpretation of global diversity patterns (see below).

While I have no doubt that concerns about taxonomic problems are well justified at relatively fine scales, and certainly call for continued taxonomic analyses and improvements to fossil databases, I doubt strongly that further improvements to the *taxonomic quality* of paleontological databases will affect how we perceive large-scale biodiversity patterns, at least among marine biotas. This is not to claim that the interpretation of global diversity trajectories is uncontroversial. But it is time to acknowledge, once and for all, that, from a taxonomic standpoint, the data already at hand are of sufficient quality—and have been for quite some time—to provide a faithful rendition of large-scale, global diversity trends as preserved in the marine fossil record. There are two primary reasons that this is the case. First, most of the major features of global diversity curves, including global radiations and mass extinctions, are not subtle and would not turn on the misassignment of even a moderate number of taxa. Second, many of the trends delineated in global diversity trajectories have been recognized through protracted intervals of geological time, and if bad taxonomy did produce them, it would have taken something approaching a series of nonrandom conspiracies of bad taxonomy, sometimes among workers focusing on diverse sets of higher taxa. If anything, taxonomic problems should obscure, rather than produce, the patterns observed (Benton 1999).

Washington Island, Wisconsin, August, 1996: Forging a Consensus

In recent years, it has become fashionable to cite the groundbreaking work of Phillips (1860) as the source of the first Phanerozoic diversity curve; clearly, Phillips deserves credit for this and more. His depiction of Phanerozoic species diversity (Fig. 1), which was based on a comprehensive compilation of fossils described from British strata (Morris 1854), involved more than simply tabulating the total number of species present within each of several stratigraphic intervals through the Phanerozoic. Because there were drastic differences in stratigraphic thickness (and vol-

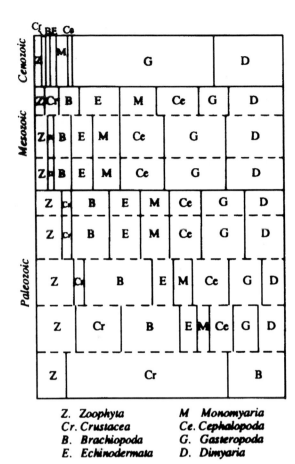

Z. *Zoophyta*   M *Monomyaria*
Cr. *Crustacea*   Ce. *Cephalopoda*
B. *Brachiopoda*   G. *Gasteropoda*
E. *Echinodermata* D. *Dimyaria*

FIGURE 2.   Phillips's depiction of the changing composition of marine invertebrate biotas through the Phanerozoic (redrawn from Phillips 1860: Fig. 6; taxonomic designations, some of which are outdated, are from Phillips).

ume) from interval to interval and, all else being equal, a thicker interval might be expected to contain more species simply because of its greater volume, Phillips sought to overcome variation in interval thickness by recalibrating diversity values for each stratigraphic interval as the number of species per unit thickness (1000 feet) (see Phillips 1860: Fig. 3 for an example from the Ordovician and Silurian). The resulting curve mitigated the overwhelming dominance of Paleozoic strata relative to younger rocks in Great Britain and suggested that, per unit thickness, Cenozoic strata were far more species-rich than those of the Paleozoic. Without Phillips's volumetric correction, species diversity on the curve would have peaked sometime during the Paleozoic.

For any number of reasons, Phillips's curve

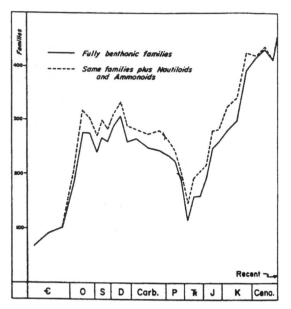

FIGURE 3. Valentine's depiction of the diversity of Phanerozoic marine families, compiled from an aggregate database for nine well-skeletonized marine phyla (reprinted from Valentine 1969: Fig. 5; published by permission of the Palaeontological Association).

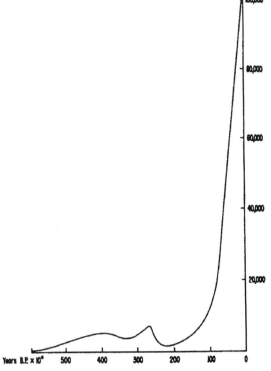

FIGURE 4. Valentine's estimated trajectory of Phanerozoic species diversity (reprinted from Valentine 1970: Fig. 2).

might not accurately depict Phanerozoic trends in global marine biodiversity. For one thing, the database on which it is based was confined to Great Britain. Moreover, no diversity values are actually recorded on the graph, although these numbers would only be of limited use in any absolute sense, given the correction for thickness. In addition, it is not clear how many separate Phanerozoic stratigraphic intervals (data points) were evaluated to produce the curve. Nevertheless, the general trends that Phillips depicted are strikingly similar to those recognized in more recent compilations that were global in scope, as are the transitions among major faunal elements that he also illustrated (Fig. 2).

The strength of this signal became apparent during the '60s and '70s, when several researchers recognized similar patterns using data collected at various taxonomic levels with disparate methodologies (see Newell 1967 and other references cited below). However, a major debate ensued about the biological veracity of the signal, starting with Valentine's (1969, 1970) compilations of Phanerozoic global diversity from an aggregate database for nine well-skeletonized marine phyla. Valentine

presented diversity curves at taxonomic levels ranging from phylum to species (e.g., Figs. 3, 4), based on data contained mainly in three sources published in the 1950s and 1960s: the *Treatise on Invertebrate Paleontology, Osnovy Paleontologii,* and *The Fossil Record.* Whereas the curves for higher taxonomic levels were compiled directly from these data, these literature sources did not contain species-level data. Thus, Valentine's species-level curve was an approximation, based primarily on an estimate of the total number of present-day species yielded by the nine phyla in question; inferences about the ratios of the number of taxa expected—all else being equal—from one taxonomic level to the next; and changes in inferred levels of faunal provinciality.

Among the Phanerozoic diversification patterns that Valentine documented with his compilations were the striking differences in trajectories exhibited at different taxonomic levels and a dramatic post-Paleozoic increase at the family level and lower. He explained these patterns in the context of ecological and geological transitions that characterized the history of life and the physical history of the earth, suggesting that diversification at higher taxonomic levels characterized the initial, early Paleozoic radiation, whereas post-Paleozoic diversification took place mainly at lower taxonomic levels as a consequence of increasing provinciality. These differences in taxonomic attributes of diversification were confirmed subsequently by Erwin et al. (1987), who compared diversification in the wake of the Late Permian mass extinction to that during the early Paleozoic radiations.

Valentine's species-level curve (Fig. 4) is particularly striking, in that it suggests a Cenozoic increase in standing diversity of approximately one order of magnitude. Moreover, Valentine clearly considered the patterns on his curves to be *real*, in the sense that he did not view them as consequences of artifacts that would compromise their biological significance.

Shortly thereafter, however, a different view was presented by Raup (1972), who showed that that there were strong similarities between diversity trajectories compiled for Phanerozoic families and genera and those depicting the changing availability of sedimentary rock through geological time. Recognizing that the quality and quantity of fossil sampling probably improved through the Phanerozoic, Raup further showed with computer simulation that differences between apparent trajectories produced at various taxonomic levels could have been direct consequences of a secular improvement in sampling (see Raup 1972: Fig. 5).

In a pair of subsequent papers, Raup (1976a,b) presented his own compilation of Phanerozoic marine species diversity with a database generated by sampling the *Zoological Record* for citations to new fossil species belonging to 19 higher taxonomic groupings. The resulting diversity trajectory exhibited a two- to threefold increase in Cenozoic diversity, which was far more modest than Valentine's estimate. Following on his earlier paper, Raup demonstrated correlations between Phanerozoic species diversity and sedimentary rock volume/area (see Fig. 5). While

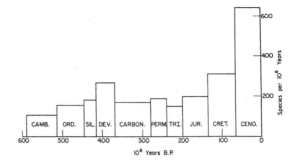

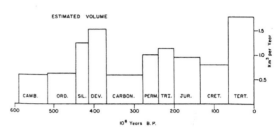

FIGURE 5. A comparison from Raup of apparent species diversity through the Phanerozoic (upper graph) versus the estimated volume of sedimentary rock for the corresponding stratigraphic intervals (lower graph). (Upper graph is reprinted from Raup 1976a: Fig. 2; lower graph is from Raup 1976b: Fig. 2.)

Raup was careful not to rule out the possibility of a biologically meaningful post-Paleozoic diversity increase, he noted that "there is no compelling evidence for a general increase in the number of invertebrate species from Paleozoic to Recent."

Against the backdrop of the ambiguities suggested by Raup, Bambach (1977) presented a new approach intended to overcome the problem of sedimentary rock volume. Bambach determined the species richness for each of 386 fossil communities delineated mainly in North America and England/Wales, and then calculated, for several Phanerozoic intervals, the *median* species richness of communities contained within each of three designated "habitat types." Recognizing that the number of lists would likely be influenced by the amount of the sedimentary rock within the intervals available for sampling, Bambach used medians, and thus his richness values were not dependent directly on the number of faunal lists contained within each stratigraphic interval. Interestingly, Bambach's period-to-period median-richness trajectory exhibited by the values from "open marine environments" closely paralleled that observed in Raup's (1976a) compilation of species diversity, suggesting that there is a robust, broad Phanerozoic diversification trajectory that can be recognized regardless of whether (1) there is a "correction" for secular changes in rock volume; (2) relative richness is based on within-community values or on sampling of the aggregate, global record; or (3) the geographical purview of the analysis is limited to a fair extent.

This theme of comparability among different databases was expanded and addressed explicitly in a landmark study published by Sepkoski, Bambach, Raup, and Valentine (1981). Period-by-period Phanerozoic diversity trajectories were juxtaposed for five different databases (Fig. 6): Bambach's (1977) median species richness for open marine environments; Raup's (1976a) global species diversity; median ichnospecies richness in neritic and flysch facies based on data provided in sources published by Seilacher (e.g., Seilacher 1974); global genus diversity based on a transformation to the data provided by Raup (1978); and global family diversity based on data from Sepkoski (see Sepkoski 1982). The remarkable similarities exhibited by the five trajectories motivated the authors to conclude first that there is a strong, underlying trajectory to Phanerozoic diversity that transcends the varying idiosyncrasies of the five data bases, and second that this trajectory is biologically meaningful, and not a consequence of preservational or other artifacts.

That this paper is known commonly by two different nicknames (the "consensus paper" and the "kiss and make up paper") is strong testimony to its influence on the field. After a decade of disagreement about the meaning of the emerging Phanerozoic diversity signal, two main protagonists in the debate, Raup and Valentine, signed on to a paper that reached conclusions at odds with positions that they championed earlier. For Valentine, the concession was relatively mild: by accepting Raup's species-level trajectory, he was conceding tacitly that there had not been an order-of-magnitude diversity increase in the Cenozoic. Raup's concession was arguably far

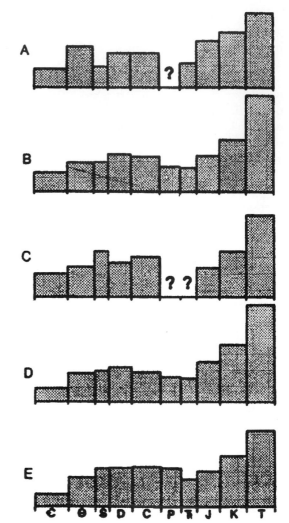

FIGURE 6. A comparison of five different compilations of taxonomic diversity through the Phanerozoic. A, Median richness of ichnospecies assemblages, using data acquired from Seilacher (1974 and other papers). B, A slightly modified version of Raup's (1976a) depiction of species diversity. C, Median species richness of communities in "open marine environments," from Bambach 1977. D, Genus diversity, using data from Raup 1978. E, Family Diversity, using data from Sepkoski. (Reprinted from Sepkoski et al. 1981: Fig. 1, by permission from Nature copyright 1981 Macmillan Magazines Ltd.)

more significant, and its effects reverberate even today: he abandoned his position that the post-Paleozoic increase might be an artifact. Sepkoski's (1981) much more highly resolved global Phanerozoic diversity trajectory (Fig. 7), in which marine families were parsed with factor analysis into the three "evolutionary faunas," was published shortly before the con-

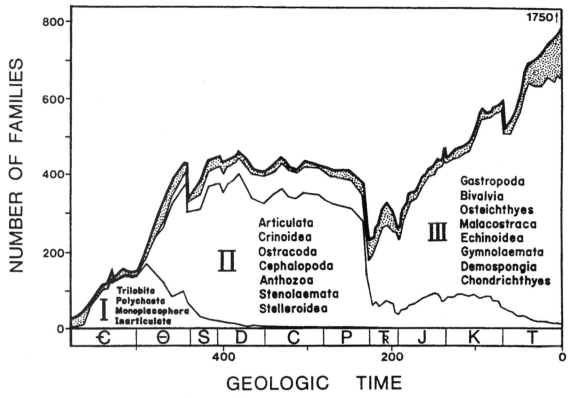

FIGURE 7. Depiction of the diversity of Phanerozoic marine families, with the three "evolutionary faunas" delineated, based on a factor analysis of a global compendium of fossil marine families. On the graph, the evolutionary faunas are depicted with roman numerals: I, Cambrian Fauna; II, Paleozoic Fauna; III, Modern Fauna. Subsequent analyses of updated databases (e.g., Sepkoski 1993) did not appreciably alter the patterns depicted in this figure. (Reprinted from Sepkoski 1981: Fig. 5.)

sensus paper, and subsequently became a kind of "industry standard," the major features of which numerous researchers have sought to explain in biological contexts. However, it can be said with some justification that this status could not have been achieved without the cover provided by the consensus paper.

To me, Raup's turnaround had always been rather puzzling: what had motivated him to change his mind? In 1997, I had an opportunity to ask him about this directly, during a visit to his home on Washington Island, Wisconsin. Given that more than a decade and a half had transpired since publication of the consensus paper, his recollection about his decision to join the paper was a bit hazy. Nevertheless, he remembered that he had been unhappy about his disagreement during the 1970s with Valentine over the magnitude of the species-level diversity increase, and he

saw the consensus paper as a kind of numerical splitting of the difference between two rather disparate views:

I was so delighted to see support for the view that a lot of the increase that Jim Valentine had observed couldn't be substantiated—it was a numerical compromise.... [A]s I read this second '76 paper [Raup 1976b]—"there is no compelling evidence for the existence of a trend"—that doesn't say that there *isn't* a trend and, to me, this consensus paper was supportive in that it reduced the increase to something almost down in the noise level of the sample. It was clearly a step in my direction.... Also, Jim Valentine had... plotted and published the factor of ten [graph]—his certainly represented the very strong conventional wisdom that everybody had—and so the consensus paper—I saw as a severe break with the conventional wisdom—not only Jim Valentine, but everybody else. . . .[A]lthough I happily agreed to be an author on the consensus paper, I don't remember agreeing with it. But it was a lot closer to my position than I thought possible.. . . [A]n important element of this is that at no point did I think I had a case for level or declining diversity. The only thing I had a case for was the *possibility* of level or declining diversity. Therefore, I didn't have an advocacy position to lose.

Nevertheless, it became clear later in our conversation that, even if his earlier work had not *proven* that Phanerozoic diversity had achieved and maintained a fairly steady level through most of the Phanerozoic, Raup continues to believe that the effects of potential sampling artifacts on the Phanerozoic diversity trajectory have not been tested adequately, and we spent much of our remaining conversation discussing how one might conduct such tests. In fact, he went so far as to note humorously, in reference to his 1976 papers: "And, damn it, this rock volume curve looks so much like these [diversity curves]."

Therefore, while most readers would perceive the consensus paper as a strong statement about the biological reality of a threefold late Mesozoic and Cenozoic increase in diversity at the family level and below, it is clear that at least one author of the consensus paper continues to harbor doubts. But, given the results highlighted in the consensus paper, why should doubts remain? Two primary reasons come quickly to mind: First, of the five graphs depicted in the consensus paper, three (Raup's curves at the genus and species levels, and Sepkoski's at the family level) are subject directly to the rock volume problem. Second, while the remaining two graphs (based on Bambach's within-habitat median species-richness values and Seilacher's trace-fossil data) should not be similarly affected by changing rock volumes, this does not mean that they are free of sampling concerns. For example, the major Cenozoic increase in Bambach's North-America-dominated data could reflect the transition from lithified to unconsolidated sediments in the North American

stratigraphic record, as exemplified in the Cenozoic record of the coastal plain. Well-preserved fossils are much more easily extracted from unconsolidated sediments, which would tend to inflate the taxonomic richness of a Cenozoic sample relative to one collected from an older, lithified stratum. In addition, there is no reason to expect a priori that within-habitat (alpha) diversity trajectories, like those based on Bambach's and Seilacher's data, should correlate directly with synoptic, global-scale trajectories (see below).

Rather than relying on sedimentary rock quantity as an indirect proxy of sample size, a more effective approach is to assess *directly* from fossil data whether the Phanerozoic trajectory is affected by variations in sampling intensity from interval to interval. This approach requires the development of databases that record the occurrences of taxa in multiple locations/strata for each sampling interval; the frequency of occurrence of a particular genus in a given database thus reflects the relative frequency of its occurrence in strata from the interval under study. These data can then be "sampling standardized" using rarefaction and other techniques that permit a researcher to ask whether the genus richness of a larger sample from one time interval exhibits an inflated taxonomic richness relative to an interval with a smaller overall sample simply because of the difference in sample size. This approach has been used by Miller and Foote (1996) to assess the series-to-series diversity trajectory of global marine genera during the Ordovician Radiation, and by Alroy (1996, 1998) to correct the genus- and lineage-level diversity trajectories of Cenozoic mammals in North America. In both instances, sampling-standardization produced notable changes to the raw, uncorrected diversity trajectories. The task of "correcting" the entire Phanerozoic marine trajectory seems daunting indeed, but just such an effort is now underway under the auspices of a working group anchored at the National Center for Ecological Analysis and Synthesis. One goal of this group project, entitled "A sampling-standardized analysis of Phanerozoic marine diversification and extinction," is to construct a database of Phanerozoic genus occurrences that will permit the application of sampling-standardization techniques to extensive Phanerozoic intervals. To date, the majority of effort has focused on the acquisition of data for two major intervals, including one in the Paleozoic (Middle Ordovician through Lower Carboniferous) and another straddling the Mesozoic/Cenozoic boundary (Upper Jurassic through Eocene). Preliminary results (J. Alroy et al. unpublished) suggest that published raw diversity trajectories may exaggerate the increase in diversity during the late Mesozoic–Cenozoic.

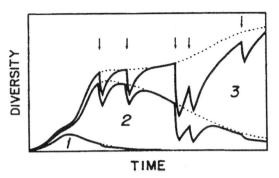

FIGURE 8. A three-phase, coupled logistic model of Phanerozoic familial diversification, with the five major Phanerozoic mass extinctions incorporated. The dotted lines illustrate the trajectories that each of the three modeled evolutionary faunas followed when mass extinctions were not included. (Reprinted from Sepkoski 1984: Fig. 8C.)

Telephone Conversation, Warwick to Cincinnati, August, 1998: Is Global Diversity Mediated by Global-Scale Processes?

The empirical pattern of marine diversity presented by Sepkoski (1981) provided immediate impetus for the investigation of several intertwined issues. In a sequence of papers, Sepkoski (1978, 1979, 1984) argued that the major features of Phanerozoic diversity could be modeled successfully and meaningfully with logistic equations. In doing so, he contended that global taxonomic diversity could be treated in much the same way that MacArthur and Wilson (1963, 1967) treated the colonization of an island over ecological time, but scaled-up spatially and temporally to model the changing taxonomic richness of the earth over geological time. However, Sepkoski focused on two features of Phanerozoic diversity that argued for something more than simple logistic growth and stabilization in taxonomic diversity: (1) the apparent failure of diversity, measured at the family level, to achieve and maintain a steady state; (2) the transition in global biotic composition from clades with high turnover rates that characterized the Cambrian and Paleozoic Evolutionary Faunas, to others with lower turnover rates representative of the Modern Evolutionary Fauna (Sepkoski 1981). To incorporate these features, Sepkoski (1984) developed three coupled (i.e., interactive) logistic equations (or "phases") that accommodated the apparent behavior in raw, global diversity trajectories of the three evolutionary faunas. The system was driven by differences in the initial rates of diversification of each of the three

modeled evolutionary faunas, as well as by differences in the rates at which their respective origination and extinction rates converged as diversity increased. The fit of the three-phase model to the actual Phanerozoic family diversity trajectory was quite impressive, especially when mass extinctions were incorporated (Fig. 8).

In Sepkoski's view, the coupled logistic model provided more than just a means to summarize Phanerozoic global marine diversity. The success of the model motivated two fundamental suggestions about the processes responsible for the observed patterns: first, that the history of biodiversity was mediated by global-scale interactions among higher taxa or groups of higher taxa (evolutionary faunas in this case) over geological time (see also Kitchell and Carr 1985), and second, that mass extinctions only perturbed the pattern temporarily, but did not alter it permanently. To these, a third implication can be added: that the biotic transitions observed in synoptic, global diversity curves have been driven by global-scale dynamics.

Sepkoski was not the only researcher to view the history of Phanerozoic diversity in the context of equilibrium dynamics. However, his vision of multiple equilibria, with biotic transitions driven by global-scale interactions, was a unique application of the equilibrium concept, with its own implications for the history of global biodiversity. By contrast, Rosenzweig (1995 and elsewhere) echoed Raup's

(1976b) suggestions that Phanerozoic diversity has been at equilibrium since the Paleozoic—except temporarily during mass extinctions—and, failing that, that global diversity increased because of the colonization of new habitats or the realization of previously unexplored niches (cf. Bambach 1985). Courtillot and Gaudemer (1996) provided yet another view, arguing that Phanerozoic could be modeled with a sequence of four simple logistic functions, the parameters of which were altered and reset by mass extinctions. Thus, they envisioned a central role for mass extinctions in mediating the history of biodiversity.

Regardless of which, if any, of these alternatives more closely describes what transpired in the history of global diversity, they all suggest some level of de facto global mediation associated with the presumably finite amount of ecospace available throughout the world. This assessment appears to be supported by a host of studies suggesting that taxonomic origination rates and probabilities declined through time, both within higher taxa (e.g., Gilinsky and Bambach 1987; Alroy 1998; Eble 1999) and in the global biota tracked through the Phanerozoic (e.g., Gilinsky 1998), an expectation of a system that is filling up with taxa. Because secular declines in origination rates within higher taxa would not be expected in a system where diversification occurs without limitation, these observations also appear to obviate Benton's (1995) claim that Phanerozoic diversity in marine and terrestrial settings has not approached equilibrium and has exhibited a long-term exponential increase. Benton has since modified this claim, at least for marine organisms (Benton 1997).

Recently, I suggested an alternative to Sepkoski's vision of a central role for biotic interactions in causing the long-term Phanerozoic transitions among constituents of evolutionary faunas (Miller 1998), observing that (1) the Phanerozoic-scale transition from taxa with higher turnover rates to others with lower rates, which can be explained with the coupled logistic model, could instead have been caused by physical perturbations at myriad scales that "weeded out" the more volatile higher turnover taxa (see Gilinsky 1994 and

others); and (2) when dissected into their local and regional components, diversity trends that appear gradual at the synoptic, global scale, appear, instead, to be highly episodic and typically linked to physical perturbations. Carrying these observations to the extreme, I argued that biotic transitions throughout the Phanerozoic might have been mediated not by global-scale interactions, but rather by physical perturbations ranging in scale from the global events that induced mass extinctions to the local events that induced regional biotic turnover within a single depositional basin (e.g., Patzkowsky and Holland 1993, 1999; Brett and Baird 1995).

Shortly after I published this view, Jack Sepkoski called me from his home in Rhode Island to talk about my paper and our apparent disagreement. While he noted that he was comfortable with the prospect raised in several of my earlier papers of a possible link between diversification and tectonic activity during the Ordovician Radiation (e.g., Miller and Mao 1995), he pointed out that changes through the entire Phanerozoic in levels of tectonic activity or in any other physical aspect of the earth were unlikely to be reflected in the biodiversity changes that we observe through the Phanerozoic. Further, he suggested that my observations about the apparent breakdown of global-scale patterns at the local level could be viewed mainly as reflecting simple patchiness at a relatively fine paleogeographic and paleoenvironmental scale of a pattern that is nevertheless apparent and governed by a broader dynamic at the global scale.

In considering this issue of scale, it is worth drawing an analogy to biotic patterns that we expect on a seafloor along a sampling transect that traverses an environmental gradient; in this analogy, the environmental gradient as a whole corresponds to the global biota, and individual locations along the gradient correspond to the biota within limited, constituent geographic regimes. Although Whittaker (1975) once likened environmental gradients to continuous spectra, the reality is that no environmental gradient is infinitely continuous. If we sample any environmental gradient with sufficiently fine lateral spacing across its entire extent, we should discover, instead, that every gradient contains some number of zones, or "steps," within which we would expect samples to vary randomly because of local patchiness, rather than in continuous lock-step with the gradient (Miller and Cummins 1990). Of course, the discovery of patchy distributions at fine lateral scales within these zones would not negate the reality of the gradient at a broader scale.

The validity of this analogy with respect to Phanerozoic global diversity hinges on the relationship between the finer-scale patches and the broader-scale pattern. In the case of the diagnosis of an environmental gradient on a seafloor, it would not be worthwhile or particularly informative to map all of the patches within a single step, because there is no reason to expect that variation on the scale of the patches *causes* the broader patterns that we observe along the gradient. By contrast, in the case of Phanerozoic diversity (1) we can do an effective job of mapping out the relevant "patches" in space and time; i.e., the regional units of relative faunal uniformity—if not co-ordinated stasis—that are bounded by intervals in which the rate of biotic change is accelerated (see Patzkowsky and Holland 1997 for an example); and (2) it is possible to test whether episodic, physically induced transitions at scales that range from global to regional combine to produce, and therefore "cause," global patterns. This issue is further considered below.

On a Train, Somewhere between Valencia and Barcelona, Spain, October, 1999: Avoiding Straw Men in the Search for Physically Mediated Scale-Independence

In a series of papers, Aronson (e.g., 1994; Aronson and Plotnick 1998) has established *scale-independence* as a valuable unifying concept with respect to biotic transitions: processes that govern and explain patterns observed on relatively fine spatio-temporal scales may also scale up to explain large-scale biotic transitions that are observed in geological time. There are numerous examples of this concept, some of which involve interactions among clades and others involving interactions among ecological groups that may have diverse taxonomic memberships. Martin

(1998) has extended the concept further by suggesting that nutrient-related phenomena affecting biotic patterns on limited spatio-temporal scales might also explain global biotic transitions associated with Phanerozoic-scale changes in energy availability (e.g., Bambach 1993; Vermeij 1995).

(I will not discuss here the emerging body of literature suggesting that biodiversity patterns exhibit a hierarchical structure that is self-similar and that their behavior through time can be explained by models that incorporate self-organized criticality [see Solé et al. 1999 for a concise review]. For the moment, I am not sure how to interpret the admittedly seductive results emerging from this body of work, given that they require a degree of global connectedness among species that remains unsupported. However, I recognize that this research is still in its infancy, and I look forward over the coming years to the continued development and discussion of this perspective.)

Benton (1987, 1991, 1996) summarized a spectrum of possible relationships between the global diversity histories of two clades through geological time in cases where one clade declines to extinction and the second undergoes diversification. These relationships ranged from a pattern of waning in one clade matched by waxing in another (the "double wedge" of some authors), to a total lack of an apparent relationship between the two clades in which their respective histories were mediated by differential responses to a mass extinction episode (Fig. 9).

As recounted by Gould and Calloway (1980) the notion has been around at least since Darwin that the demise of one taxon might be caused by the rise of another. Among marine and nonmarine tetrapod families that he investigated, however, Benton (1996) showed that instances in which one family replaced another because of competition are probably quite uncommon, when information on geographic, ecological, and temporal overlap were considered. Typically, suggestions of competitive replacements among higher taxa are based on the contention that a group exhibiting a diversity increase was, in some way, competitively superior to a group exhibiting a

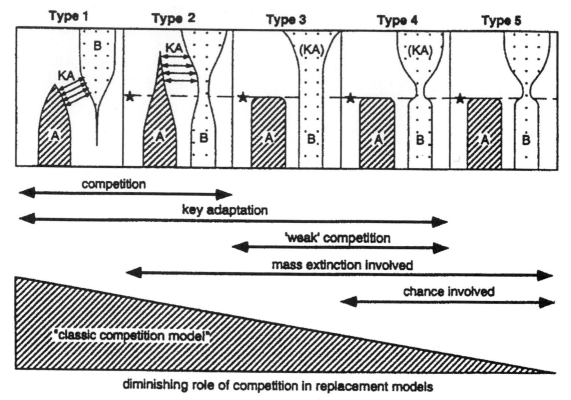

FIGURE 9. A spectrum of models depicting possible modes of biotic replacement of one clade by another, with the role of competition diminishing, and that of perturbation increasing, from left to right. The star in four of the figures denotes a mass extinction, and "KA" denotes a "key adaptation." Type 1 illustrates a classic "double wedge," and type 5 illustrates a transition contingent entirely on the occurrence of a mass extinction. Intermediate possibilities include scenarios in which a group that is competitively superior is nevertheless impeded until competitively inferior incumbents are decimated by a perturbation. (Reprinted from Benton 1996: Fig. 8.1, by permission of the University of Chicago Press; for further discussion, see Benton 1987, 1991, 1996; Rosenzweig and McCord 1991.)

decrease, and that, when scaled up to geological time, the superior group won out, as measured by complementary changes in constituent taxonomic richness. However, an array of studies have pointed, sometimes unintentionally, to the difficulty of demonstrating that such transitions have actually transpired through competitive displacement. In itself, competitive superiority is difficult to document convincingly, and even if it were possible to make such a case, the geometry of comparative clade shapes rarely, if ever, presents an instance of a definitive double wedge.

These concerns are well-illustrated by the classic case of clams versus articulate brachiopods. Owing to their superficial morphological resemblance, and the recognition that articulate brachiopods gave way to bivalves as major contributors to Phanerozoic benthic marine diversity, it was commonplace to believe that bivalves had outcompeted brachiopods through the course of geological time. Indeed, some workers proffered evidence to suggest that, on a physiological basis, bivalves were competitively superior to brachiopods (e.g., Steele-Petrovic 1979), but a series of studies by Thayer (e.g., 1985, 1986) and Rhodes and Thompson (1993) illustrated the difficulty of making this determination definitively. From the standpoint of their respective diversity histories, Gould and Calloway (1980) showed that there was not the negative correlation in the Paleozoic or post-Paleozoic diversity trajectories of brachiopods and bivalves that would be expected if bivalves had outcompeted brachiopods through the Phanerozoic. Rather, they argued that there was a significant upturn in the diversity of bivalves and a downturn in that of articulate brachiopods in association with the Late Permian mass extinction, indicating that the extinction was pivotal in governing the post-Paleozoic fates of both groups (see further discussion of mass extinctions, below). Miller and Sepkoski (1988) questioned this view by demonstrating that the per-genus rate of diversification of bivalves was unaffected by the Late Permian mass extinction and by illustrating, with a two-phase coupled logistic model, that bivalve diversification may have been dampened throughout the group's history by interactions with other taxa.

Miller and Sepkoski's (1988) analysis should not be construed as suggesting that bivalves outcompeted brachiopods, however. While one interacting phase (group) in the two-phase model was intended to represent bivalves, the other phase did not represent, from the standpoint of its diversity trajectory, any other clade, and might have appropriately been interpreted as some combination of other taxa on the Paleozoic seafloor with which bivalves might have interacted. Indeed, given this variety of potentially interacting higher taxa, one is left to suspect that the only reason bivalves and brachiopods were singled out in the first place for comparison is their superficial resemblance (Eldredge 1987).

Sepkoski (1996), however, demonstrated elegantly how one might diagnose quantitatively the long-term competitive displacement of one clade by another. Sepkoski et al. (2000) applied this approach to an analysis of the diversity trajectories of cyclostome and cheilostome bryozoans, and developed a solution with a two-phase coupled logistic model that captures most of the important attributes of their Mesozoic and Cenozoic diversity trajectories. This case is compelling because of the evidence that cheilostomes have tended to overgrow cyclostomes where they have encountered one another, both in Recent and fossil cases (McKinney 1992, 1995), suggesting that cheilostomes have historically enjoyed a competitive advantage over cyclostomes. Moreover, application of the coupled logistic model in this instance is unique compared with other studies (e.g., Sepkoski 1984; Miller and Sepkoski 1988): the "younger" cheilosto-

me fauna exhibited turnover rates that were less than those of the "older" cyclostome fauna, and the transition was modeled with these differences incorporated. By contrast, as alluded to earlier, the Phanerozoic-scale transition among evolutionary faunas involved a sequential transition that was the opposite of that exhibited by cyclostomes and cheilostomes: on average, there was a *decrease* in the turnover rates exhibited by successive evolutionary faunas. Thus, in contrast to the Phanerozoic-scale transition among evolutionary faunas, it is not possible to argue, as an alternative to the coupled logistic framework, that the earlier group, the cyclostomes, was simply "weeded out" by perturbations because of its greater evolutionary volatility (see earlier discussion). If anything, the parameters reported by Sepkoski et al. (2000) suggest that the later group, the cheilostomes, was the more volatile.

That said, this is not a classic case of the double wedge because, as recognized by Sepkoski et al. (2000) and Lidgard et al. (1993) and noted by Benton (1996), cyclostome genus diversity has persisted at a rather steady level throughout the Cenozoic. As Lidgard et al. (1993) pointed out, the global diversity trajectories of cheilostomes and cyclostomes may have been affected by a combination of processes played out at different scales.

Donovan and Gale (1990) argued that the post-Paleozoic decline of articulate brachiopods was linked to the early Mesozoic rise of asteroids (the echinoderm variety, not the extraterrestrial kind) that preyed on them. However, Donovan and Gale's own discussion points out the difficulty of demonstrating even today that asteroids like to eat brachiopods. Further, there is little direct evidence of asteroid predation on brachiopods in the fossil record, nor are there even indications that they tended to co-occur in the same assemblages (more extensive critiques of the Donovan and Gale hypothesis are offered by Blake and Guensburg [1990] and Vermeij [1990]).

As with the question of clade-versus-clade dynamics discussed earlier, the Donovan and Gale example illustrates how unlikely it is for the demise of a class-level taxon to be tied to a single kind of ecological interaction. An alternative approach—that of suggesting an ecological mediation to morphological transitions within groups—has met with much more success. This approach has been perhaps best exemplified by Vermeij's (1977) argument that a suite of morphological transitions exhibited by gastropods after the Jurassic were consequences of the diversification of predators capable of crushing the shells of gastropod prey that were less well protected. This perspective was extended to Paleozoic biotas by Signor and Brett (1994) and was later broadened considerably by Vermeij (1987) to cover a spectrum of cases in marine and non-marine settings, in which morphological arms races are thought to have taken place through geological time between taxa that were "enemies" in the sense that they interacted with one another competitively, or in predator/prey couplets. Vermeij considered these arms races to be a form of "escalation," which he viewed as a pervasive theme of macroevolution.

Vermeij's arguments apply most clearly to morphological characteristics, rather than to taxonomic diversity per se. Despite the failings of cases like that of articulate brachiopods versus asteroids, this is certainly not to say that the taxonomic richness of a given group cannot itself be affected by ecological interactions. However, given the focus on ecology in these instances, it makes much more sense to chart the diversity trajectories of groups that share ecological/trophic attributes, rather than to necessarily limit oneself to evaluating higher taxa linked phylogenetically as monophyletic groups. In many cases, morphological characters that are fundamentally meaningful as phylogenetic indicators are of little ecological significance (e.g., bivalve dentition), and using such characters to chart diversity histories may provide less insight into diversity transitions than using ecological groupings.

Thayer's (1979, 1983) "biological bulldozing" hypothesis constitutes a classic case in which taxa were grouped ecologically across taxonomic boundaries, and the global diversification of one group was shown to be at the expense of another group. In this instance, recognizing that active burrowers ("biological bulldozers") are typically segregated from sedentary suspension feeders on soft-bottom seafloors in the present day, Thayer reasoned that the documented rise of biological bulldozers was responsible for the concomitant decline of immobile suspension feeders on soft substrates (ISOSS), which had dominated Paleozoic seafloors; this transition was initiated in the mid-Paleozoic.

On the face of it, Vermeij's and Thayer's analyses, and others like them, are quite compelling. They are grounded strongly in actualistic evidence, and the actualistic data successfully predict broad transitions that can be observed in synoptic global diversity trajectories. So then why do they make me fidgety? I fidget because, in the documentation of scale-independence, we have sidestepped a scale at which the patterns do not appear to fit our expectations: the local and regional level on timescales of a few to about ten million years. If predators have induced morphological transitions in prey lineages or biological bulldozers have evicted ISOSS, why have we rarely, if ever, observed this happening in the high-resolution stratigraphic records available to us regionally during key global transition intervals (see Brett et al. 1996)? Surely, five million years is sufficient time to effect a morphological change in a lineage or an eviction of an ecological group. Instead, as described earlier, what we observe at local and regional levels are units with relatively little biotic change, bracketed by intervals in which major biotic transitions occur abruptly. Moreover, while this pattern may be accentuated artificially because of the coincidence of turnover intervals with stratigraphic sequence boundaries, the relative abruptness of at least some regional faunal transitions remains even after accommodating these factors (Patzkowsky and Holland 1997).

The biological meaning of this kind of pattern is open to interpretation (Ivany 1996), and it may simply be that rates of evolution are accelerated during these intervals to such an extent that morphological transitions and evictions actually happened, but at rates that were too rapid to be observed in the geological record. Dovetailing on this possibility is the suggestion that the perturbations manifested by the abrupt transitions were required to remove incumbents, even in cases when the replacement groups had demonstrable competitive edges over the incumbents (Rosenzweig and McCord 1991).

As alluded to earlier, however, I see a clear, and testable, alternative that likely applies in some instances: that regional perturbations induce net physical changes to regional environmental milieus, and that, following these perturbations, replacement groups become predominant and incumbents are lost not because of interactions, but because of fundamental environmental changes. The example of biological bulldozers versus ISOSS may represent a case in point: clearly, the Paleozoic environments in which ISOSS would have been expected to thrive best were rather different than the muddy substrates favored by deposit-feeding, biological bulldozers. The loss of ISOSS incumbents regionally may have been induced by transitions to muddy substrates; the global shift would thus have been "caused" by the summation of regional changes induced by environmental transitions. A close examination of Devonian strata in New York might reveal, for example, that these two biotas were tracking their preferred lithofacies, which are known to have shifted abruptly on local and regional scales, and that, ultimately, biological bulldozers "won out" in conjunction with the overwhelming progradation of terrigenous sediments associated with the Acadian Orogeny. This is not to say that the entire world became muddy during the Devonian and never went back the other way (see below). But as the history of life illustrates, broad evolutionary transitions are not easily reversed: if a substantial portion of the marine world did become muddy for some protracted interval, say a period or two (e.g., the Devonian and Carboniferous), this would have effected a decline in ISOSS from which it would have been difficult to recover.

In October, 1999, Mike Foote and I were attending a conference in Spain, and we had a chance to talk about some of these issues in detail on a train ride from Valencia to Barcelona. Foote wondered whether, in citing the biological bulldozing example, I was setting up a straw man. As he explained, he was not certain that the bulldozing hypothesis was

considered viable anymore, and thus that I was questioning the validity of a hypothesis that had already been dismissed. While I was able to convince him that this was not the case and that, at least to my reading, the bulldozing hypothesis remains as a prominent example of an ecologically mediated global diversity transition, I still had a hard time dismissing the straw man issue in a slightly different context: that I am misinterpreting the general notion of what constitutes a competitive advantage and thus have unrealistic expectations about what we should see when we dissect a global transition into its local and regional components. It might be argued that the ability to thrive in a particular kind of environment constitutes an advantage in itself, and that distinguishing this from a case in which one group displaces another amounts to hairsplitting, given that both may require perturbations to transpire. While the question of whether this matters at all with respect to evolutionary theory is debatable, I will simply assert that it is worth distinguishing between these two alternatives if we have the data at hand to do so: it is worth knowing whether long-term, global biotic transitions can actually be caused by the summation of abrupt, regional environmental transitions, and with a concerted effort, we can do a much more extensive job than we have in the past of mapping the relationship of biotic to physical transitions at a variety of scales.

In a paper published in the tenth anniversary issue of *Paleobiology*, Gould (1985) broadened the theme articulated by Gould and Calloway (1980) in suggesting that mass extinctions collectively constitute an overarching class of macroevolutionary events capable of superseding or undoing evolutionary changes that accumulated during the intervals between them (see also Bennett 1990, 1997). My viewpoint, articulated in this paper and elsewhere (Miller 1997b, 1998), does not really question whether mass extinctions were important mediating events in evolutionary history, but it does remove the firewall between mass extinctions and the so-called background intervals between them (see also Raup 1991, 1992; Martin 1998; Patzkowsky 1999). A closer look at the local and regional data from which the synoptic, global pattern is constructed reveals that biotic turnover is punctuated by perturbations at a spectrum of geographic scales and that mass extinctions are simply the largest, most globally extensive end-members (see Miller 1998 for additional discussion of this issue). In this sense, the mediation of Phanerozoic diversity is scale-independent in a manner that extends Aronson's (1994; Aronson and Plotnick 1998) interpretation to include physical forcing agents, alongside the ecological and evolutionary processes already categorized in this fashion.

Strolling up to the State Capitol Building, Salt Lake City, October, 1997: The Three (or More) Great Phanerozoic Evolutionary Lithofacies?

While in Salt Lake City for the 1997 Geological Society of America meeting, Dan McShea, Steve Holland, and I walked up the hill to the state capitol building one night to admire the spectacular view. On our walk, we discussed several of the issues recounted herein, and after I admitted to McShea that I was advocating a form of historicism (after he explained to me what "historicism" is), we carried this to its logical extreme spurred on by a question from Holland: Was I suggesting that a global categorization and quantification through time of marine paleoenvironments, if graphed in a style similar to that used for Sepkoski's global diversity curves, would yield a pattern of Phanerozoic transitions paralleling those exhibited by global biodiversity? I confess that my answer at the time was "yes," or at least "maybe." We then discussed how we might actually go about developing the database for such an analysis and convinced ourselves that it would obviously not be an easy thing to do, although it would be interesting to try.

I have come to recognize that it was hopelessly naïve to contemplate this project, not only because it would be a difficult task, but because there is no real reason to expect that the entire history of Phanerozoic diversity can be explained simply by secular changes in the availability of physical habitats. Just as Sepkoski (1996) understood that global competition would not be played out as combat among individuals in two competing groups

through geological time, I concede that the history of Phanerozoic life is not simply the outcome of a battle among myriad "competing" paleoenvironments and geochemistries.

Furthermore, the evidence is overwhelming that biologically mediated pathways have also played major roles in governing the contents of marine biotas throughout the Phanerozoic (e.g., Vermeij 1987; Conway Morris 1998). What I *do* believe, however, is that a broad attempt to exhaustively map Phanerozoic biotic transitions at regional scales in the context of lithostratigraphic and geochemical transitions would improve our wherewithal to choose among alternative mechanisms that purport to explain Phanerozoic biotic transitions.

When we limit ourselves to the global level in seeking explanations for Phanerozoic diversity transitions, we can choose any number of morphological or other features with which to divide up the marine fossil biota into two (or more) mutually exclusive groups. In many cases, regardless of the features chosen, we will end up with a picture of declining diversity for one of the groups and increasing diversity for the other, a pattern that would appear to call for an explanation. However, we can feel much more confident that the global pattern is meaningful if, when we investigate it at the more local levels of individual regions or paleocontinents, we witness the same transition, albeit more abruptly and timed differently from place to place because of differences in the timing of perturbations that permitted the transition to take place (see Rosenzweig and McCord 1991).

Along the way, we must also ascertain whether the diversification patterns exhibited in synoptic compilations transcend scale, as discussed earlier. In the Ordovician and Silurian, for example, there is considerable evidence suggesting that "within-community" (alpha) patterns are at variance with those exhibited at the "between-community" (beta) and global levels (Patzkowsky 1995; Miller and Mao 1998; Adrain et al. 2000).

So, in the end, the simple message of this paper is this: DISSECT, DISSECT, DISSECT. To do so will require the continued development of extensive databases that capture information from the literature already catalogued by generations of paleobiologists and other geo-scientists, combined with a new generation of fieldwork tailored to acquire data from previously uncharted strata or with a higher degree of stratigraphic resolution from some venues than that acquired in the past.

Afterward

A few months before his death, Jack Sepkoski agreed to coauthor this paper. We both recognized that it would be a good vehicle to discuss our different perspectives on global diversity and to identify for readers a palette of research approaches with which to choose among different explanations for Phanerozoic diversity transitions. We never got any further than that, and I cannot honestly say whether Jack would be pleased with anything I have written here. All I know is that every thought I have ever had about biodiversity bears his imprint.

Acknowledgments

I thank the friends whose discussions were highlighted in this view for patiently dealing with my queries over the years, and for serving as my teachers, formally and informally. I also thank D. Erwin, S. Holland, D. McShea, and M. Foote for reviewing an earlier draft of this paper. My research on global diversity has been supported by grants from the National Aeronautic and Space Administration's Program in Exobiology (grants NAGW-3307 and NAG5-6946).

Literature Cited

Adrain, J. A., and S. R. Westrop. 2000. An empirical assessment of taxic paleobiology. Science (in press).

Adrain, J. A., R. A. Fortey, and S. R. Westrop. 1998. Post-Cambrian trilobite diversity and evolutionary faunas. Science 280: 1922–1925.

Adrain, J. A., S. R. Westrop, B. D. E. Chatterton, and L. Ramsköld. 2000. Silurian trilobite alpha diversity and the end-Ordovician mass extinction. Paleobiology 26:625–646.

Alroy, J. 1996. Constant extinction, constrained diversification, and uncoordinated stasis in North American mammals. Palaeogeography, Palaeoclimatology, Palaeoecology 127:285–312.

———. 1998. Equilibrial diversity dynamics in North American mammals. Pp. 232–287 in McKinney and Drake 1998.

Aronson, R. B. 1994. Scale-independent biological interactions in the marine environment. Oceanography and Marine Biology, an Annual Review 32:435–460.

Aronson, R. B., and R. E. Plotnick. 1998. Scale-independent interpretations of macroevolutionary dynamics. Pp. 430–450 in McKinney and Drake 1998.

Bambach, R. K. 1977. Species richness in marine benthic habitats through the Phanerozoic. Paleobiology 3:152–167.

———. 1985. Classes and adaptive variety: the ecology of diversification in marine faunas through the Phanerozoic. Pp. 191–253 in J. W. Valentine, ed. Phanerozoic diversity patterns: profiles in macroevolution. Princeton University Press, Princeton, N.J.

———. 1993. Seafood through time: changes in biomass, energetics, and productivity in the marine ecosystem. Paleobiology 19:372–397.

Bennett, K. D. 1990. Milankovitch cycles and their effects on species in ecological and evolutionary time. Paleobiology 16:11–21.

———. 1997. Evolution and ecology: the pace of life. Cambridge University Press, Cambridge.

Benton, M. J. 1987. Progress and competition in macroevolution. Biological Reviews 62:305–338.

———. 1991. Extinction, biotic replacements, and clade interactions. Pp. 89–102 in E. C. Dudley, ed. The unity of evolutionary biology. Dioscorides, Portland, Ore.

———. 1995. Diversification and extinction in the history of life. Science 268:52–58.

———. 1996. On the nonprevalance of competitive replacement in the evolution of tetrapods. Pp. 185–210 in D. Jablonski, D. H. Erwin, and J. H. Lipps, eds. Evolutionary paleobiology. University of Chicago Press, Chicago.

———. 1997. Models for the diversification of life. Trends in Ecology and Evolution 12:490–495.

———. 1999. The history of life: large databases in palaeontology. Pp. 249–283 in D. A. T. Harper, ed. Numerical palaeobiology: computer based modelling of fossils and their distributions. Wiley, Chichester, England.

Blake, D. B., and T. E. Guensburg. 1990. Predatory asteroids and the fate of brachiopods—a comment. Lethaia 23:429–430.

Brett, C. E., and G. C. Baird. 1995. Coordinated stasis and evolutionary ecology of Silurian to Middle Devonian faunas in the Appalachian Basin. Pp. 285–315 in D. H. Erwin and R. L. Anstey, eds. New approaches to speciation in the fossil record. Columbia University Press, New York.

Brett, C. E., L. C. Ivany, and K. M. Schopf. 1996. Coordinated stasis: an overview. Palaeogeography, Palaeoclimatology, Palaeoecology 127:1–20.

Conway Morris, S. 1998. The crucible of creation: the Burgess Shale and the rise of animals. Oxford University Press, Oxford.

Courtillot, V., and Y. Gaudemer. 1996. Effects of mass extinctions on biodiversity. Nature 381:146–148.

Donovan, S. K., and A. S. Gale. 1990. Predatory asteroids and the decline of the articulate brachiopods. Lethaia 23:77–86.

Eble, G. J. 1999. Originations: land and sea compared. Geobios 32:223–234.

Eldredge, N. 1987. Life pulse: episodes from the story of the fossil record. Facts on File, New York.

Erwin, D. H., J. W. Valentine, and J. J. Sepkoski Jr. 1987. A comparative study of diversification events: the early Paleozoic versus the Mesozoic. Evolution 41:1177–1186.

Foote, M. 1993. Discordance and concordance between morphological and taxonomic diversity. Paleobiology 19:185–204.

———. 1997. Sampling, taxonomic description, and our evolving knowledge of morphological diversity. Paleobiology 23: 181–206.

Gilinsky, N. L. 1994. Volatility and the Phanerozoic decline of background extinction intensity. Paleobiology 20:445–458.

———. 1998. Evolutionary turnover and volatility in higher taxa. Pp. 162–184 in McKinney and Drake 1998.

Gilinsky, N. L., and R. K. Bambach. 1987. Asymmetrical patterns of origination and extinction in higher taxa. Paleobiology 13: 427–445.

Gould, S. J. 1985. The paradox of the first tier: an agenda for paleobiology. Paleobiology 11:2–12.

Gould, S. J., and C. B. Calloway. 1980. Clams and brachiopods—ships that pass in the night. Paleobiology 6:383–396.

Ivany, L. C. 1996. Coordinated stasis or coordinated turnover? Exploring intrinsic vs. extrinsic controls on pattern. Palaeogeography, Palaeoclimatology, Palaeoecology 127:239–256.

Ivany, L. C., and K. M. Schopf, eds. 1996. New perspectives on faunal stability in the fossil record. Palaeogeography, Palaeoclimatology, Palaeoecology 127.

Kitchell, J. A., and T. R. Carr. 1985. Nonequilibrium model of diversification: faunal turnover dynamics. Pp. 277–309 in J. W. Valentine, ed. Phanerozoic diversity patterns: profiles in macroevolution. Princeton University Press, Princeton, N.J.

Lidgard, S., F. K. McKinney, and P. D. Taylor. 1993. Competition, clade replacement, and a history of cyclostome and cheilostome bryozoan diversity. Paleobiology 19:352–371.

Lupia, R., S. Lidgard, and P. R. Crane. 1999. Comparing palynological abundance and diversity: implications for biotic replacement during the Cretaceous angiosperm radiation. Paleobiology 25:305–340.

MacArthur, R. H., and E. O. Wilson. 1963. An equilibrium theory of insular zoogeography. Evolution 17:373–387.

———. 1967. The theory of island biogeography. Princeton University Press, Princeton, N.J.

Martin, R. E. 1998. Catastrophic fluctuations in nutrient levels as an agent of mass extinction: upward scaling of ecological processes. Pp. 405–429 in McKinney and Drake 1998.

Maxwell, W. D., and M. J. Benton. 1990. Historical tests of the absolute completeness of the fossil record of tetrapods. Paleobiology 16:322–335.

McKinney, F. K. 1992. Competitive interactions between related clades: evolutionary implications of overgrowth between encrusting cyclostome and cheilostome bryozoans. Marine Biology 114:645–652.

———. 1995. One hundred million years of competitive interactions between bryozoan clades: asymmetrical but not escalating. Biological Journal of the Linnean Society 56:465–481.

McKinney, F. K., S. Lidgard, J. J. Sepkoski Jr., and P. D. Taylor. 1998. Decoupled temporal patterns of evolution and ecology in two post-Paleozoic clades. Science 281:807–809.

McKinney, M. L., and J. A. Drake, eds. 1998. Biodiversity dynamics: turnover of populations, taxa, and communities. Columbia University Press, New York.

Miller, A. I. 1997a. Comparative diversification dynamics among palaeocontinents during the Ordovician Radiation. Geobios Mémoire Spécial 20:397–406.

———. 1997b. Coordinated stasis or coincident relative stability? Paleobiology 23:155–164.

———. 1998. Biotic transitions in global marine diversity. Science 281:1157–1160.

Miller, A. I., and H. Cummins. 1990. A numerical model for the formation of fossil assemblages: estimating the amount of post-mortem transport along environmental gradients. Palaios 5:303–316.

Miller, A. I., and M. Foote. 1996. Calibrating the Ordovician radiation of marine life: implications for Phanerozoic diversity trends. Paleobiology 22:304–309.

Miller, A. I., and S. Mao. 1995. Association of orogenic activity with the Ordovician radiation of marine life. Geology 23:305–308.

Miller, A. I., and J. J. Sepkoski Jr. 1988. Modeling bivalve diversification: the effect of interaction on a macroevolutionary system. Paleobiology 14:364–369.

Morris, J. 1854. A catalogue of British fossils. The author, London.

Newell, N. D. 1967. Revolutions in the history of life. Geological Society of America Special Paper 89:63–91.

Patzkowsky, M. E. 1995. Ecological aspects of the Ordovician radiation of articulate brachiopods. Pp. 413–414 in J. D. Cooper, M. L. Droser, and S. C. Finney, eds. Ordovician odyssey: short papers for the seventh international symposium on the Ordovician System. Pacific Section of the Society for Sedimentary Geology (SEPM), Fullerton, Calif.

———. 1999. A new agenda for evolutionary paleoecology—or would you in the background please step forward. Palaios 14: 195–197.

Patzkowsky, M. E., and S. M. Holland. 1993. Biotic response to a Middle Ordovician paleoceanographic event in eastern North America. Geology 21:619–622.

———. 1997. Patterns of turnover in Middle and Upper Ordovician brachiopods of the eastern United States: a test of coordinated stasis. Paleobiology 23:420–443.

———. 1999. Biofacies replacement in a sequence stratigraphic framework: Middle and Upper Ordovician of the Nashville Dome, Tennessee, USA. Palaios 14:301–323.

Phillips, J. 1860. Life on the earth; its origin and succession. Macmillan, Cambridge, England.

Raup, D. M. 1972. Taxonomic diversity during the Phanerozoic. Science 177:1065–1071.

———. 1976a. Species diversity in the Phanerozoic: a tabulation. Paleobiology 2:279–288.

———. 1976b. Species diversity in the Phanerozoic: an interpretation. Paleobiology 2:289–297.

———. 1978. Cohort analysis of generic survivorship. Paleobiology 4:1–15.

———. 1991. A kill curve for Phanerozoic marine species. Paleobiology 17:37–48.

———. 1992. Large-body impact and extinction in the Phanerozoic. Paleobiology 18:80–88.

Rhodes, M. C., and R. J. Thompson. 1993. Comparative physiology of suspension-feeding in living brachiopods and bivalves: evolutionary implications. Paleobiology 19:322–334.

Rosenzweig, M. L. 1995. Species diversity in space and time. Cambridge University Press, Cambridge.

Rosenzweig, M. L., and R. D. McCord. 1991. Incumbent replacement: evidence for long-term evolutionary progress. Paleobiology 17:202–213.

Seilacher, A. 1974. Fossil-Vergesellschaftungen 20, flysch trace fossils: evolution of behavioural diversity in the deep sea. Neues Jarbuch für Geologie und Paläontologie, Monatshefte 4:233–245.

Sepkoski, J. J., Jr. 1978. A kinetic mode of Phanerozoic taxonomic diversity I. Analysis of marine orders. Paleobiology 4:223–251.

———. 1979. A kinetic model of Phanerozoic taxonomic diversity II. Early Phanerozoic families and multiple equilibria. Paleobiology 5:222–251.

———. 1981. A factor analytic description of the Phanerozoic marine fossil record. Paleobiology 7:36–53.

———. 1982. A compendium of fossil marine families. Milwaukee Public Museum, Milwaukee.

———. 1984. A kinetic model of Phanerozoic taxonomic diversity III. Post-Paleozoic families and multiple equilibria. Paleobiology 10:246–267.

———. 1993. Ten years in the library: new data confirm paleontological patterns. Paleobiology 19:43–51.

———. 1996. Competition in macroevolution: the double wedge revisited. Pp. 211–255 in D. Jablonski, D. H. Erwin, and J. H. Lipps, eds. Evolutionary paleobiology. University of Chicago Press, Chicago.

Sepkoski, J. J., Jr., and D. C. Kendrick. 1993. Numerical experiments with model monophyletic and paraphyletic taxa. Paleobiology 19:168–184.

Sepkoski, J. J., Jr., R. K. Bambach, D. M. Raup, and J. W. Valentine.

1981. Phanerozoic marine diversity and the fossil record. Nature 293:435–437.

Sepkoski, J. J., Jr., F. K. McKinney, and S. Lidgard. 2000. Competitive displacement between post-Paleozoic cyclostome and cheilostome bryozoans. Paleobiology 26:7–18.

Signor, P. W., and C. E. Brett. 1994. The mid-Paleozoic precursor to the Mesozoic marine revolution. Paleobiology 10:229–245.

Solé, R. V., S. C. Manrubia, M. Benton, S. Kauffman, and P. Bak. 1999. Criticality and scaling in evolutionary ecology. Trends in Ecology and Evolution 14:156–160.

Steele-Petrovic, H. M. 1979. The physiological differences between articulate brachiopods and filter-feeding bivalves as a factor in the evolution of marine level-bottom communities. Palaeontology 22:101–134.

Thayer, C. W. 1979. Biological bulldozers and the evolution of marine benthic communities. Science 203:458–461.

———. 1983. Sediment-mediated biological disturbance and the evolution of marine benthos. Pp. 479–625 in M. J. S. Tevesz and P. L. McCall, eds. Biotic interactions in Recent and fossil benthic communities. Plenum, New York.

———. 1985. Brachiopods versus mussels: competition, predation and palatability. Science 228:1527–1528.

———. 1986. Are brachiopods better than bivalves? Mechanisms of turbidity tolerance and their interaction with feeding in articulates. Paleobiology 12:161–174.

Valentine, J. W. 1969. Patterns of taxonomic and ecological structure of the shelf benthos during Phanerozoic time. Palaeontology 12:684–709.

———. 1970. How many marine invertebrate fossil species? A new approximation. Journal of Paleontology 44:410–415.

Valentine, J. W., B. H. Tiffney, and J. J. Sepkoski Jr. 1991. Evolutionary dynamics of plants and animals: a comparative approach. Palaios 6:81–88.

Vermeij, G. J. 1977. The Mesozoic marine revolution: evidence from snails, predators and grazers. Paleobiology 3:245–258.

———. 1987. Evolution and escalation: an ecological history of life. Princeton University Press, Princeton, N.J.

———. 1990. Asteroids and articulates: is there a causal link? Lethaia 23:431–432.

———. 1995. Economics, volcanoes, and Phanerozoic revolutions. Paleobiology 21:125–152.

Wagner, P. J. 1995. Testing evolutionary constraint hypotheses with early Paleozoic gastropods. Paleobiology 21:248–272.

Westrop, S. R., and J. A. Adrain. 1998. Trilobite alpha diversity and the reorganization of Ordovician benthic marine communities. Paleobiology 24:1–16.

Whittaker, R. H. 1975. Communities and ecosystems. Macmillan, New York.

Origination and extinction components of taxonomic diversity: general problems

Mike Foote

Abstract.—Mathematical modeling of cladogenesis and fossil preservation is used to explore the expected behavior of commonly used measures of taxonomic diversity and taxonomic rates with respect to interval length, quality of preservation, position of interval in a stratigraphic succession, and taxonomic rates themselves. Particular attention is focused on the independent estimation of origination and extinction rates. Modeling supports intuitive and empirical arguments that single-interval taxa, being especially sensitive to variation in preservation and interval length, produce many undesirable distortions of the fossil record. It may generally be preferable to base diversity and rate measures on estimated numbers of taxa extant at single points in time rather than to adjust conventional interval-based measures by discarding single-interval taxa.

A combination of modeling and empirical analysis of fossil genera supports two major trends in marine animal evolution. (1) The Phanerozoic decline in taxonomic rates is unlikely to be an artifact of secular improvement in the quality of the fossil record, a point that has been argued before on different grounds. (2) The post-Paleozoic rise in diversity may be exaggerated by the essentially complete knowledge of the living fauna, but this bias is not the principal cause of the pattern. The pattern may partly reflect a secular increase in preservation nevertheless.

Apparent temporal variation in taxonomic rates can be produced artificially by variation in preservation rate. Some empirical arguments suggest, however, that much of the short-term variation in taxonomic rates observed in the fossil record is real. (1) For marine animals as a whole, the quality of the fossil record of a higher taxon is not a good predictor of its apparent variability in taxonomic rates. (2) For a sample data set covering a cross-section of higher taxa in the Ordovician, most of the apparent variation in origination and extinction rates is not statistically attributable to independently measured variation in preservation rates. (3) Previous work has shown that standardized sampling to remove effects of variable preservation and sampling yields abundant temporal variation in estimated taxonomic rates. While modeling suggests which rate measures are likely to be most accurate in principle, the question of how best to capture true variation in taxonomic rates remains open.

Mike Foote. Department of the Geophysical Sciences, University of Chicago, Chicago, Illinois 60637. E-mail: mfoote@midway.uchicago.edu

Accepted: 21 June 2000

Interval Estimates of Diversity and Taxonomic Rates

Do changes in taxonomic diversity tend to be associated preferentially with changes in origination rate or with changes in extinction rate? This rather simple empirical question, which I address in a companion paper (Foote 2000), involves a number of methodological problems that I will consider here: (1) How are interval estimates of diversity and taxonomic rates affected by interval length? (2) How do true rates affect our ability to estimate rates? (3) How are diversity and rate estimates affected by incompleteness of the fossil record? (4) How are they affected by the finite stratigraphic extent of the fossil record? (5) How do the foregoing factors affect our ability to mea-

sure origination and extinction rates independently? This paper focuses on measuring diversity and rate changes over discrete stratigraphic or temporal intervals, rather than in calculating long-term average rates (cf. Foote in press a).

Using mathematical modeling of cladogenesis and fossil preservation, I will explore the behavior of several interval measures of diversity and taxonomic rates. This exercise points to a number of undesirable properties of single-interval taxa. The difficulties discussed here stand in addition to those that arise from differences in species abundance and, consequently, in preservation probability (Buzas et al. 1982). Taxa confined to single intervals are problematic even if preservation is complete or if all taxa and time intervals are

0094-8373/00/2604-0004/$1.00

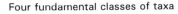

Four fundamental classes of taxa

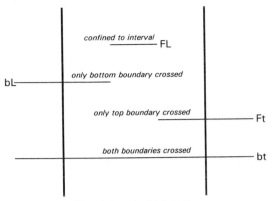

Time interval of interest

FIGURE 1. Illustration of four fundamental classes of taxa present during a stratigraphic interval. N_{FL} is the number of taxa confined to the interval, N_{bL} is the number that cross the bottom boundary only, N_{Ft} is the number that cross the top boundary only, and N_{bt} is the number that cross both boundaries.

characterized by the same quality of preservation. It is therefore advantageous to base measures on taxa that cross between stratigraphic or temporal intervals. Simply modifying conventional measures by discarding single-interval taxa may not be the best approach, however. Instead, it is preferable under a wide range of circumstances to use measures for which single-interval are simply irrelevant by the very nature of the measures.

Four Fundamental Classes of Taxa

Any taxon known or inferred from a stratigraphic interval can be classified into one of four mutually exclusive categories (Fig. 1) (see Barry et al. 1995 for a similar classification): (1) taxa confined to the interval, i.e., taxa whose first and last appearance are both within the interval; (2) taxa that cross the bottom boundary and make their last appearance during the interval; (3) taxa that make their first appearance during the interval and cross the top boundary; and (4) taxa that range through the entire interval, crossing both the top and bottom boundaries. Using b and t to refer to crossing the bottom and top boundaries of an interval and using F and L to refer to first and last appearance within the interval, I will denote the numbers of taxa in the four categories N_{FL}, N_{bL}, N_{Ft}, and N_{bt}. The term

singleton is commonly used to refer to species represented by a single specimen (Buzas and Culver 1994, 1998). I will denote as a *singleton* any taxon that is confined to a single stratigraphic interval at the given level of resolution (any member of the FL category). Although I will refer to boundaries, these need not be times of major biotic turnover; any recognizable temporal or stratigraphic division can serve as a point of reference. As I will discuss below, there are many useful combinations of these categories of taxa. Two that are especially important are the total number of taxa crossing the bottom boundary, N_b (= $N_{bL}+N_{bt}$), and the total number crossing the top boundary, N_t (= $N_{Ft}+N_{bt}$).

All measures of diversity and taxonomic rates for an interval are derived from combinations of the numbers of taxa in the four fundamental classes. The numbers denoted by N with subscripts are the true numbers of taxa with the given properties. Although these are useful for modeling, in the fossil record we always deal with observed numbers, which I will denote by X with the corresponding subscripts (Foote 1999: Appendix 7). The relationships between true and observed numbers are central to what will be developed below (see Appendix for details). Of course, taxa can shift categories when filtered through incomplete preservation. For example, a taxon that truly belongs to the bt category can contribute to X_{FL}, X_{bL}, X_{Ft}, or X_{bt}, depending on how much of its range is truncated. It should also be kept in mind that observed taxa (X) in the first three categories must necessarily be found within the interval, while the last category includes taxa that are found before and after, but not necessarily during, the interval.

Measures of Diversity and Rate

Numerous measures of diversity and rate have been applied to paleontological data, and their intuitive advantages and disadvantages have been discussed at length (Gingerich 1987; Gilinsky 1991; Foote 1994; Harper 1996; Sepkoski and Koch 1996). Rather than review all of them, I will explore the properties of some commonly used measures that have been proposed to solve particular problems or to take advantage of particular aspects

TABLE 1. Definitions of taxonomic diversity measures and taxonomic rate metrics for intervals of length Δt. Measures are expressed in terms of numbers belonging to the four fundamental classes of taxa, N_{FL}, N_{bL}, N_{Ft}, and N_{bt} (Fig. 1), or combinations derived from them (Appendix).

Diversity measures	
Measure	Definition
Total diversity, N_{tot}	$N_{FL} + N_{bL} + N_{Ft} + N_{bt}$
Total diversity minus singletons	$N_{bL} + N_{Ft} + N_{bt}$
Bottom-boundary crossers, N_b	$N_{bL} + N_{bt}$
Top-boundary crossers, N_t	$N_{Ft} + N_{bt}$
Number of originations, N_o	$N_{FL} + N_{Ft}$
Number of extinctions, N_e	$N_{FL} + N_{bL}$
Estimated mean standing diversity	$(N_b + N_t) / 2$
	$= (N_{bL} + N_{Ft} + 2N_{bt}) / 2$
	$= (N_{tot} - N_o/2 - N_e/2)$

Rate measures	
Measure	Definition
Per-taxon rate	Origination: $(N_{FL} + N_{Ft})/(N_{tot})/\Delta t$
	Extinction: $(N_{FL} + N_{bL})/(N_{tot})/\Delta t$
Van Valen metric	Origination: $(N_{FL} + N_{Ft})/[(N_b + N_t)/2]/\Delta t$
	Extinction: $(N_{FL} + N_{bL})/[(N_b + N_t)/2]/\Delta t$
Van Valen metric without singletons	Origination: $(N_{Ft})/[(N_b + N_t)/2]/\Delta t$
	Extinction: $(N_{bL})/[(N_b + N_t)/2]/\Delta t$
Estimated per-capita rate, $\hat{p}$ and $\hat{q}$	$\hat{p}$: $-\ln(N_{bt}/N_t)/\Delta t$
	$\hat{q}$: $-\ln(N_{bt}/N_b)/\Delta t$

of data. I will not discuss the well-known problems associated with poor constraints on interval length (Gilinsky 1991; McGhee 1996; Sepkoski and Koch 1996). Long-term averages, appropriately calculated, are not very sensitive to errors in interval length (Foote 1999: Appendix 7). It is often short-term variation that is of interest, however, and the fundamental distortion is that an overestimate of interval length yields an underestimate of taxonomic rates, and conversely for an underestimate of interval length. Thus, while dating error may produce spurious extremes in taxonomic rates, it is unlikely to produce long-term secular patterns. The diversity measures differ in whether they seek an estimate of standing diversity at a point in time or of the total number of taxa that exist during any part of an interval. Most rate metrics start with a tabulation of the number of events within an interval and normalize this number by some measure of diversity and by the length of the interval. The goal of these normalizations is to obtain an estimate of the instantaneous, per-capita rates of origination and extinction, p and q, per lineage-million-years (Lmy) (Raup 1985). The per-capita rate estimates advocated here, $\hat{p}$ and $\hat{q}$ (Table 1), are derived directly

from branching theory rather than as a normalization of the number of events observed within an interval.

Table 1 gives definitions of diversity and rate measures in terms of the four fundamental classes of taxa and the interval length, Δt. The expected values of these measures under various conditions of completeness are derived in the Appendix. Expressing these expectations in terms of fundamental parameters such as origination and extinction rates, rather than counts of number of events, helps to make sense of their behavior. Since the Appendix contains exact expressions for a number of quantities that have sometimes been derived less directly with Monte Carlo methods, it is hoped that it will be of some use to the paleontologist. Table 2 provides a "road map" to the Appendix.

The first two sections that follow will discuss measures of diversity and rate in terms of true numbers of taxa (N) in order to emphasize problems that would exist even if the fossil record were complete. The discussion will then switch to observed numbers (X) in order to emphasize problems related to paleontological incompleteness.

Effects of Interval Length

For the problems discussed here, it is natural and convenient to measure time in multiples of $1/q$, the average taxon duration. This yields in effect a dimensionless expression of time. Similarly, expressing origination and preservation rates relative to q yields dimensionless rate measures. Thus, an increase in interval length with taxonomic rates held constant has the same effect as an increase in taxonomic rates with interval length held constant. The number of top- or bottom-boundary crossers may be larger or smaller depending on where we choose our time lines, and if origination is concentrated shortly after boundaries and extinction shortly before them, average standing diversity may be underestimated (Raup 1991; Alroy 1992; Foote 1994). Nevertheless, the number of boundary crossers is not systematically affected by interval length (Fig. 2) (Bambach 1999). This is just one potential advantage of measuring diversity using boundary crossers (see below).

The per-taxon rates have been used as a way to normalize the number of origination or extinction events by total diversity and interval length. As interval length increases, a progressively larger proportion of total diversity consists of singletons (Fig. 2). Because these taxa first appear and last appear within the same interval, proportional origination and extinction asymptotically approach unity as Δt increases, and the per-taxon rates consequently decline as interval length increases (Fig. 3; Appendix, section 2) (Gingerich 1987; Foote 1994).

Van Valen (1984) used a rate metric designed to normalize the number of originations and extinctions by estimated average standing diversity, $(N_b + N_t)/2$, since this number rather than total diversity better expresses the number of taxa susceptible to origination or extinction at an instant in time (Harper 1975; Van Valen 1984). This normalization implicitly assumes a linear change in standing diversity, a change that is expected to be exponential if rates are constant within an interval. As Δt increases, the linear approximation becomes progressively worse, and $(N_b + N_t)/2$ overestimates mean standing

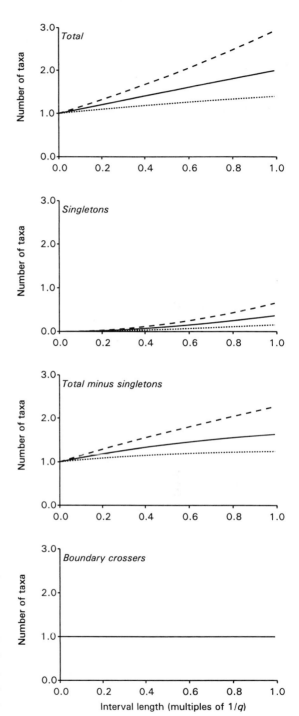

FIGURE 2. Effect of interval length on number of taxa within an interval. Rates are time-homogeneous and fossil record is complete. Interval length is expressed as multiples of $1/q$. Solid lines, $p = q$; dashed lines, $p = 1.5q$; dotted lines, $p = 0.5q$. In this and subsequent figures, diversity at the start of the interval is unity. Only the number of boundary crossers is independent of interval length.

TABLE 2. Guide to equations in Appendix.

Quantity of interest	Preservation	Edge effect?	Equations*
1A. Total diversity	Complete	—	9
1B. Total diversity	Constant	No	24c,25c,26–29,34,35e
1C. Total diversity	Constant	Yes	24b,25b,26–29,34,35e
1D. Total diversity	Variable	Yes	24a,25a,26–29,34,35e
2A. Taxa crossing bottom boundary only	Complete	—	4
2B. Taxa crossing bottom boundary only	Constant	No	24c,25c,27,31b,32a,34b,37d
2C. Taxa crossing bottom boundary only	Constant	Yes	24b,25b,27,31b,32a,34b
2D. Taxa crossing bottom boundary only	Variable	Yes	24a,25a,27,31b,32a,34b
3A. Taxa crossing top boundary only	Complete	—	5
3B. Taxa crossing top boundary only	Constant	No	24c,25c,28,31c,33a,34c,37e
3C. Taxa crossing top boundary only	Constant	Yes	24b,25b,28,31c,33a,34c
3D. Taxa crossing top boundary only	Variable	Yes	24a,25a,28,31c,33a,34c
4A. Total taxa crossing bottom boundary	Complete	—	1
4B. Total taxa crossing bottom boundary	Constant	No	24c,25c,30c,35a
4C. Total taxa crossing bottom boundary	Constant	Yes	24b,25b,30b,35a
4D. Total taxa crossing bottom boundary	Variable	Yes	24a,25a,30a,35a
5A. Total taxa crossing top boundary	Complete	—	2
5B. Total taxa crossing top boundary	Constant	No	24c,25c,30c,35b
5C. Total taxa crossing top boundary	Constant	Yes	24b,25b,30b,35b
5D. Total taxa crossing top boundary	Variable	Yes	24a,25a,30a,35b
6A. Taxa crossing both boundaries	Complete	—	3
6B. Taxa crossing both boundaries	Constant	No	24c,25c,30c,31a,34a,37c
6C. Taxa crossing both boundaries	Constant	Yes	24b,25b,30b,31a,34a,36c
6D. Taxa crossing both boundaries	Variable	Yes	24a,25a,30a,31a,34a
7A. Singletons	Complete	—	6
7B. Singletons	Constant	No	24c,25c,29,31d,32b,33b,34d,38b
7C. Singletons	Constant	Yes	24b,25b,29,31d,32b,33b,34d
7D. Singletons	Variable	Yes	24a,25a,29,31d,32b,33b,34d
8A. Number of originations or extinctions	Complete	—	7,8
8B. Number of originations or extinctions	Constant	No	35c; lines 2B,3B,7B
8C. Number of originations or extinctions	Constant	Yes	35c; lines 2C,3C,7C
8D. Number of originations or extinctions	Variable	Yes	35c; lines 2D,3D,7D
9A. Proportional origination or extinction	Complete	—	10,11
9B. Proportional origination or extinction	Constant	No	10,11; line 8B
9C. Proportional origination or extinction	Constant	Yes	10,11; line 8C
9E. Proportional origination or extinction	Variable	Yes	10,11; line 8D
10A. Per taxon rate metric	Complete	—	12,13
10B. Per taxon rate metric	Constant	No	12,13; line 8B
10C. Per taxon rate metric	Constant	Yes	12,13; line 8C
10D. Per taxon rate metric	Variable	Yes	12,13; line 8D
11A. Van Valen rate metric	Complete	—	14,15,16,17
11B. Van Valen rate metric	Constant	No	39
11C. Van Valen rate metric	Constant	Yes	14a; lines 4C,5C,8C
11D. Van Valen rate metric	Variable	Yes	14a; lines 4D,5D,8D

Table 2. Continued.

Quantity of interest	Preservation	Edge effect?	Equations*
12A. Van Valen rate metric (minus singletons)	Complete	—	18,19,20,21
12B. Van Valen rate metric (minus singletons)	Constant	No	18a; lines 2B,3B,4B,5B
12C. Van Valen rate metric (minus singletons)	Constant	Yes	18a; lines 2C,3C,4C,5C
12D. Van Valen rate metric (minus singletons)	Variable	Yes	18a; lines 2D,3D,4D,5D
13A. Estimated per-capita rates	Complete	—	22,23
13B. Estimated per-capita rates	Constant	No	22,23; lines 4B,5B,6B
13C. Estimated per-capita rates	Constant	Yes	22,23; lines 4C,5C,6C
13D. Estimated per-capita rates	Variable	Yes	22,23; lines 4D,5D,6D
14A. Probability of preservation before boundary	Constant	No	24c
14B. Probability of preservation before boundary	Constant	Yes	24b
14C. Probability of preservation before boundary	Variable	Yes	24a
15A. Probability of preservation after boundary	Constant	No	25c
15B. Probability of preservation after boundary	Constant	Yes	25b
15C. Probability of preservation after boundary	Variable	Yes	25a
16A. Probability of preservation during interval	Constant during interval	—	26–29
16B. Probability of being observed to cross a boundary	Constant	No	30c
16C. Probability of being observed to cross a boundary	Constant	Yes	30b
17A. Probability of being observed to cross a boundary	Variable	Yes	30a
18. Probability that singleton is observed	Constant within interval	—	29
19A. Probability that bL taxon is observed as such	Constant	No	24c,27,32a
19B. Probability that bL taxon is observed as such	Constant	Yes	24b,27,32a
19C. Probability that bL taxon is observed as such	Variable	Yes	24a,27,32a
20A. Probability that bL taxon is observed as singleton	Constant	No	24c,27,32b
20B. Probability that bL taxon is observed as singleton	Constant	Yes	24b,27,32b
20C. Probability that bL taxon is observed as singleton	Variable	Yes	24a,27,32b
21A. Probability that Ft taxon is observed as such	Constant	No	25c,28,33a
21B. Probability that Ft taxon is observed as such	Constant	Yes	25b,28,33a
21C. Probability that Ft taxon is observed as such	Variable	Yes	25a,28,33a
22A. Probability that Ft taxon is observed as singleton	Constant	No	25c,28,33b
22B. Probability that Ft taxon is observed as singleton	Constant	Yes	25b,28,33b
22C. Probability that Ft taxon is observed as singleton	Variable	Yes	25a,28,33b
23A. Probability that bt taxon is observed as such	Constant	No	24c,25c,31a
23B. Probability that bt taxon is observed as such	Constant	Yes	24b,25b,31a
23C. Probability that bt taxon is observed as such	Variable	Yes	24a,25a,31a
24A. Probability that bt taxon is observed as bL taxon	Constant	No	24c,25c,26,31b
24B. Probability that bt taxon is observed as bL taxon	Constant	Yes	24b,25b,26,31b
24C. Probability that bt taxon is observed as bL taxon	Variable	Yes	24a,25a,26,31b
25A. Probability that bt taxon is observed as Ft taxon	Constant	No	24c,25c,26,31c
25B. Probability that bt taxon is observed as Ft taxon	Constant	Yes	24b,25b,26,31c
25C. Probability that bt taxon is observed as Ft taxon	Variable	Yes	24a,25a,26,31c
26A. Probability that bt taxon is observed as singleton	Constant	No	24c,25c,26,31d
26B. Probability that bt taxon is observed as singleton	Constant	Yes	24b,25b,26,31d
27C. Probability that bt taxon is observed as singleton	Variable	Yes	24a,25a,26,31d

* Lines in lists of equations refer to numbered lines in this table.

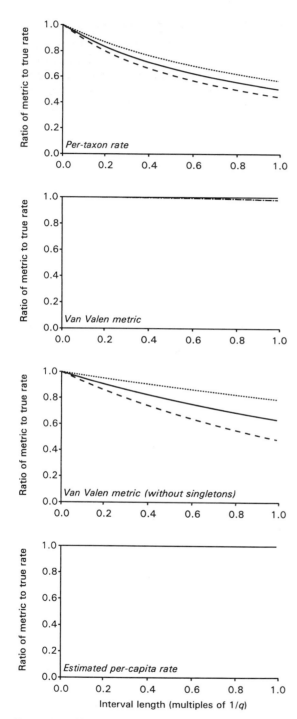

FIGURE 3. Effect of interval length on extinction metrics with time-homogeneous rates and complete record. Similar effects result for origination metrics. See Figure 2 for explanation. Dotted and dashed lines coincide for the Van Valen metric. The per-capita rate estimated from numbers of boundary crossers is independent of interval length. This is also the only metric for which origination rate does not affect the estimation of extinction rate and vice versa.

diversity when $p \neq q$. The Van Valen metric therefore underestimates true rates more as Δt increases (Fig. 3), although the error is relatively small if p and q are not very different. Recognizing potential problems with singletons, Harper (1996) suggested a variation on Van Valen's metric in which singletons are omitted. As interval length increases, more of the extinctions and originations are singletons. Discarding them completely therefore leads to an underestimate of taxonomic rates that becomes worse as the interval length increases (Fig. 3).

The estimates of per-capita rates advocated here, $\hat{p}$ and $\hat{q}$, are unaffected by interval length. N_{bt}/N_b gives the proportion of lineages extant at the start of the interval that survive to the end, and N_{bt}/N_t gives the proportion of lineages extant at the end that were already extant at the start. These ratios decay exponentially with time if rates are constant within the interval; thus the logarithm of each ratio declines linearly with time. The magnitude of the slope of this decline is exactly equal to the extinction rate in the first case and the origination rate in the second case. In fact, even if rates are not constant, $\hat{p}$ and $\hat{q}$ provide unbiased estimates of the mean rate within an interval (Appendix, section 2).

Alroy (1996b) independently proposed rate metrics similar to Van Valen's and Harper's, normalizing by N_b rather than $(N_b + N_t)/2$. When $p = q$, Alroy's metrics are expected to be identical to those of Van Valen and Harper (since $N_b = N_t$ in this case). Moreover, if $p = q$, if $p\Delta t$ and $q\Delta t$ are relatively low, and if singletons are ignored, Alroy's metric and Van Valen's metric are approximately equal to $\hat{p}$ and $\hat{q}$ (see Appendix, section 2, and Alroy et al. 2000).

In some cases, for example those involving deep-sea microfossils or Neogene macroinvertebrates, temporal resolution may be sufficiently fine that equal-length intervals can be established relatively easily (Wei and Kennett 1983, 1986; Pearson 1992, 1996; Budd et al. 1994; Budd and Johnson 1999). Alroy (1992, 1996b, 1998, 1999) has developed an alternative approach to circumventing the problem of unequal interval lengths. Using an exhaustive compilation of faunal lists, he has ordinated first and last appearance data of North American Cenozoic mammals and has used this ordination to interpolate between well-constrained absolute ages. This has allowed an arbitrarily fine, equal-interval temporal grid to be superimposed on the data. Although Alroy's approach is labor-intensive, it holds great promise for avoiding problems associated with uneven and uncertain interval lengths.

In general, the only measures of diversity and taxonomic rates that are expected to be independent of interval length are those based exclusively on boundary crossers. This suggests that, if interval length varies substantially, it is a good idea to measure diversity as N_b or N_t and to measure taxonomic rates as $\hat{p}$ and $\hat{q}$. It is nevertheless often the case empirically that there is ample true variation in diversity and rates, with the result that most of the apparent variation in rates is not attributable to variation in interval length (Raup 1986; Gingerich 1987; Collins 1989; Foote 1994; Patzkowsky and Holland 1997).

Effects of Rates on Rate Estimation

For certain rate metrics, true rates of taxonomic evolution affect our ability to estimate rates in two principal ways, concerning the magnitude of rates and the difference between origination and extinction rates.

Because an increase in taxonomic rates has the same effect as an increase in interval length, the per-taxon rate and the Van Valen metric, with or without singletons, become less accurate as taxonomic rates increase. The only exception to this is that the Van Valen metric is insensitive to interval length and taxonomic rates if $p = q$ and if singletons are included.

The greater the difference between p and q, the greater the discrepancy between the true taxonomic rates and the estimates given by the per-taxon rate and the Van Valen metric, with or without singletons (Fig. 3). This is largely because of the increasing discrepancy between an exponential diversity change within an interval and the linear change implicit in the normalization by $(N_b + N_t)/2$. The per-taxon rate and the singleton-free Van Valen metric exhibit a more serious problem. Suppose that extinction rate is held fixed. As origina-

tion rate increases (from the dotted to the solid to the dashed lines in Fig. 3), the estimate of extinction rate decreases when either of these rate metrics is used. (Likewise, if origination rate is constant and extinction rate changes, the estimate of origination rate changes in the opposite direction.) The reason the per-taxon rate behaves this way can be seen by inspecting equations (8a) and (9a) in the Appendix, which give the numerator and denominator of the rate metric. The normalization makes sense only if $p \approx q$. The reason for the bias in the singleton-free Van Valen metric is different. Equation (21) shows that this metric approximates the extinction rate only when $p \approx q$ and when both rates are low. With these two rate metrics, true variation in rates will therefore contribute to a spurious negative correlation between origination and extinction. This will complicate the independent measurement of origination and extinction rates.

In contrast to the other rate metrics, $\hat{p}$ is unaffected by extinction rate, while $\hat{q}$ is unaffected by origination rate. This makes $\hat{p}$ and $\hat{q}$ especially useful if one desires independent estimates of origination and extinction rates.

Incomplete Preservation

Preservation can be modeled in a number of realistic ways that include variation in time and space (Shaw 1964; Koch and Morgan 1988; Marshall 1994; Holland 1995; Holland and Patzkowsky 1999; Weiss and Marshall 1999). As a heuristic tool for understanding the behavior of diversity and rate measures, it is convenient to focus on the temporal aspect and to start by assuming time-homogeneous fossil preservation at a constant per-capita rate r per Lmy (Paul 1982, 1988; Pease 1985; Strauss and Sadler 1989; Marshall 1990; Foote and Raup 1996; Solow and Smith 1997; Foote 1997). This simple assumption will be relaxed below. In the time-homogeneous case, the proportion of lineages preserved is equal to $r/(q + r)$ if $p = q$ and if the fossil record is of effectively infinite length (Pease 1985; Solow and Smith 1997) (see Edge Effects, below). It is therefore natural for many problems to express preservation rate as a multiple of q. Throughout this discussion I will assume tax-

onomic homogeneity of taxonomic and preservational rates. For modeling, this assumption can easily be relaxed by performing calculations for an arbitrary number of rate classes and combining the results (see Buzas et al.
1982, Koch and Morgan 1988, Holland 1995,
Holland and Patzkowsky 1999, and Weiss and
Marshall 1999 for explicit treatments of taxonomic heterogeneity of preservation).

As preservation rate decreases, there is a
regular decrease in observed numbers in nearly all categories of taxa within an interval. The
sole exception is singletons, whose behavior is
especially problematic. This is shown in Figure 4, which portrays the observed number of
taxa relative to the true number. As preservation rate decreases, singletons constitute an
ever greater proportion of observed taxa. For
a certain range of values of preservation rate
and interval length, the observed number of
singletons is not only relatively high but also
absolutely greater than the true number. In
general, the observed excess of singletons
causes the per-taxon rate and the Van Valen
metric to increase as preservation rate decreases (Fig. 5). Because all categories of
boundary crossers are diminished in the same
proportion by incomplete, time-homogenous
preservation, however, $\hat{p}$, $\hat{q}$, and the Van Valen
metric without singletons are unaffected by
incomplete preservation, as are other metrics
based only on boundary crossers (Alroy
1996b). This result is related to the fact that the
observed age distribution of taxa, exclusive of
singletons, is expected to be identical to the
true age distribution (Foote and Raup 1996;
Foote 1997; Solow and Smith 1997). Without
independent estimates of preservation rate, it
may be difficult to distinguish a truly high
number of singletons (reflecting high taxonomic rates) from a preservational artifact.

I have been unable to develop a measure of
absolute diversity that is insensitive to incomplete preservation. It is nevertheless possible
to estimate changes in diversity accurately if
we accept the point that a proportional or log-
arithmic scale is a natural one with which to
measure diversity (Sepkoski 1991). The quantity $\ln(N_t/N_b)$ gives the proportional change
in diversity through an interval, i.e., the fundamental growth rate $(p - q)$ times the inter-

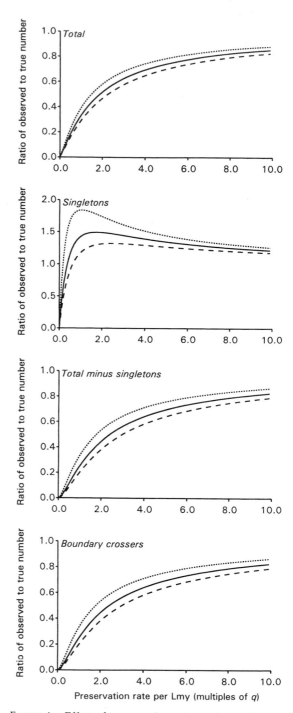

FIGURE 4.   Effect of preservation rate on observed numbers of taxa within an interval. Preservation is time-homogeneous and there are no edge effects. Interval length
is fixed at $0.5/q$, and preservation rate r is expressed as
multiples of extinction rate q. Solid, dashed, and dotted
lines are as in Figure 2. Most categories of observed taxa
increase monotonically with preservation rate, but the
number of observed singletons is disproportionately
large when preservation rate is low.

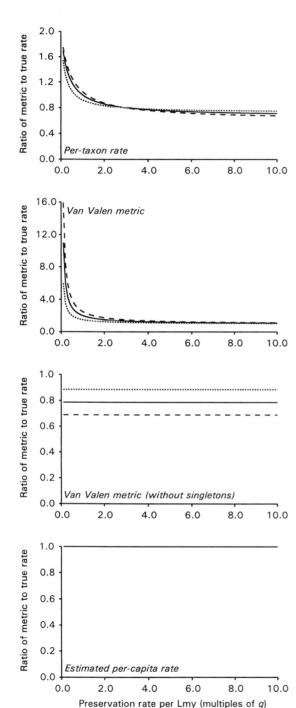

FIGURE 5. Effects of preservation rate on extinction metrics. Similar effects obtain for origination metrics. See Figure 4 for explanation. Because of the inclusion of singletons, the per-taxon rate and the Van Valen metric are strongly influenced by preservation. If singletons are excluded, the Van Valen metric is independent of preservation rate, but it is sensitive to the difference between origination and extinction rates.

val length Δt. In the time-homogeneous case, this quantity is unaffected by incomplete preservation.

If one has estimates of preservation rate and is interested in true levels of diversity, then it is a relatively straightforward matter to adjust observed numbers of boundary crossers in the time-homogeneous case: the true number is equal to the observed number times $[(p + r)(q + r)]/r^2$ (see Appendix, section 5). The case of variable rates is more realistic, however. How to adjust diversity measures in the face of variable preservation remains an important problem. Rarefaction and other approaches involving standardized resampling and subsampling of data have been used extensively (Hessler and Sanders 1967; Sanders 1968; Jackson et al. 1993; Rex et al. 1993, 1997; Raymond and Metz 1995; Alroy 1996a,b, 1998, 1999; Miller and Foote 1996; Markwick 1998; Marshall et al. 1999), as have methods that rely on an estimated phylogeny to identify and fill gaps (Norell 1992; Benton 1994; Johnson 1998). The relative performance of these various approaches under different conditions still needs further exploration.

In summary, many effects of incompleteness can be overcome in the case of time-homogeneous preservation if taxonomic rates and changes in diversity are measured on the basis of relative numbers of boundary crossers.

Edge Effects

The fossil record as a whole, or any part of it we investigate, has a discrete beginning and end. During any interval of measurement, whether coarse or fine, the presence of a taxon can be inferred because the taxon is actually preserved during the interval or because it is preserved before and after the interval; this is the standard range-through approach. As the interval in question falls toward either edge of the record, our ability to infer the presence of taxa by the range-through method diminishes. This creates a series of related edge effects (Figs. 6, 7). Apparent diversity declines. The number of first appearances is high toward the lower edge (as taxa that truly extend below the beginning of the record make their first appearance) and the number of last appearances is high toward the top. The number of

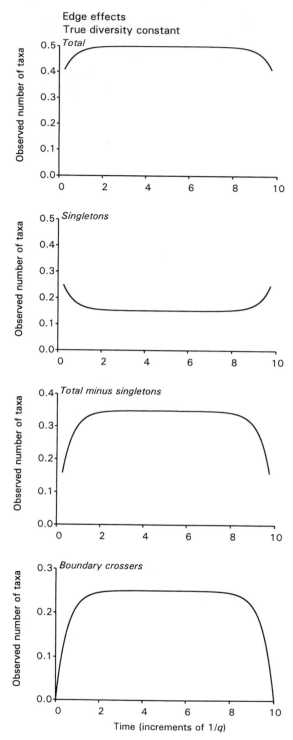

singletons increases, as taxa that would have been observed to cross a boundary are now less likely to be found before the interval (near the lower edge) or after the interval (near the upper edge). The singletons contribute heavily to counts of origination and extinction. As a result of these edge effects, the per-taxon rate and the Van Valen metric with singletons are inflated toward the edges. Note that estimates of both origination and extinction rate for these metrics are inflated near both edges.

The increase in first and last appearances relative to total numbers also means that $\hat{p}$ and the Van Valen origination metric without singletons are inflated toward the beginning of the record, while the corresponding extinction metrics are inflated toward the end. Note that the total number of last appearances is low near the beginning of the record and that the apparent change in standing diversity (the difference between X_t and X_b) can be substantial within a single interval near an edge. Because the Van Valen metric without singletons compares a reduced number of events to an estimate of average standing diversity that is changing substantially near the edges, the extinction metric is also depressed near the beginning of the record while the origination metric is also depressed near the end. In contrast to the three other metrics, $\hat{p}$ and $\hat{q}$ are affected only toward the beginning or end of the record, respectively.

Edge effects are significant only to the extent that a taxon extant at some point in time is likely to intersect either edge. At a distance t from the bottom edge, the edge is no longer felt when e^{-qt} is acceptably small; near the top edge, the relevant quantity is e^{-pt}. For example, suppose $p = q$ and the interval is placed near the bottom edge. If the interval is separated from the edge by 2.3 times the mean taxon duration ($2.3/q$), then e^{-qt} is about 0.10, and $\hat{p}$, $\hat{q}$, X_{FL}, X_{bL}, X_{Ft}, X_b, X_t, and X_{bt} are within about 1% of the values they would have in the absence of edge effects. Practically speaking, then, an edge is no longer felt within about two or so average taxon lengths.

I have presented edge effects that result from incomplete preservation within the window of observation as if they were created only by the termination of the fossil record at

FIGURE 6. Edge effects on taxonomic diversity. Figure shows number of observed taxa in relation to distance of the interval from beginning or end of fossil record. In all cases $p = q$, $r = q$, and interval length is equal to $0.5/q$. The height of curves depends on the value of r. The effect decays exponentially and the distance from the edge at which the curves level off is a function of the taxonomic rates (see text for further explanation).

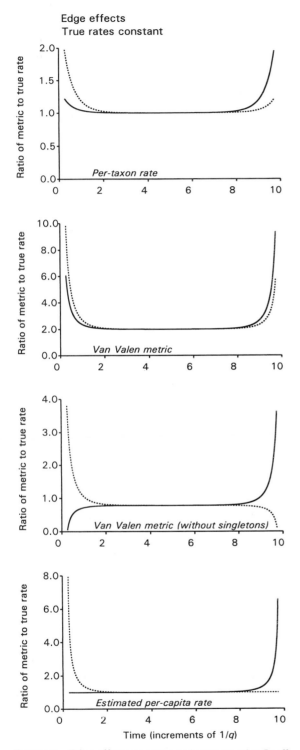

Edge effects
True rates constant

Per-taxon rate

Van Valen metric

Van Valen metric (without singletons)

Estimated per-capita rate

Time (increments of $1/q$)

FIGURE 7. Edge effects on taxonomic rate metrics. In all cases $p = q$, $r = q$, and interval length is equal to $0.5/q$. Solid line, extinction metric; dotted line, origination metric. All rate metrics except estimated per-capita rate are affected at both edges.

either end. Qualitatively different but mathematically identical edge effects are also created by sudden drops in preservation rate (Holland 1995) and by major changes in evolutionary rates such as mass extinctions and evolutionary radiations (Signor and Lipps 1982). Gilinsky and Bambach (1987) discuss an edge effect that results from the definition of certain rate metrics even if preservation is complete; proportional origination and extinction (number of events divided by total diversity) must be unity in the first and last intervals, respectively. Similarly, $\hat{p}$ and $\hat{q}$ are undefined in the first and last intervals.

A different kind of edge effect is created by a singular increase in preservation rate. This is obviously relevant to the effects of Recent taxa on patterns of diversity and taxonomic rates (Raup 1972, 1979; Pease 1988a,b, 1992). Because the Recent fauna of skeletonized marine animals is very well known, taxa that lack a late Cenozoic fossil record can have their ranges pulled forward, with the result that apparent diversity is likely to be inflated toward the Recent and apparent extinction rate is likely to decline (Figs. 8, 9). Whether apparent origination rate increases, decreases, or is unaffected depends on which rate metric is used (Fig. 9).

For fossil marine animal genera, a substantial number of taxa extend from the Recent back into the early to mid Tertiary. It is thus conceivable that the rise in diversity and decline in background rates seen during the Phanerozoic (Raup and Sepkoski 1982; Van Valen 1984; Sepkoski 1996, 1997, 1998) are partly artificial (Raup 1972; Pease 1985, 1988a,b, 1992). As Sepkoski (1997) showed, however, the increase in genus diversity, measured as total non-singleton genera in an interval, persists even when Recent genera are included only if they are known to have a Plio-Pleistocene fossil record. Figure 10 shows the same result for boundary-crossing diversity. (For the sake of completeness, Figures 10 and 11 also show the effect of removing all genera that extend to the Recent, a culling that is unreasonably extreme since many of these genera have a fossil record near the Recent.) The decline in extinction rate is seen in both the Paleozoic and post-Paleozoic even when Re-

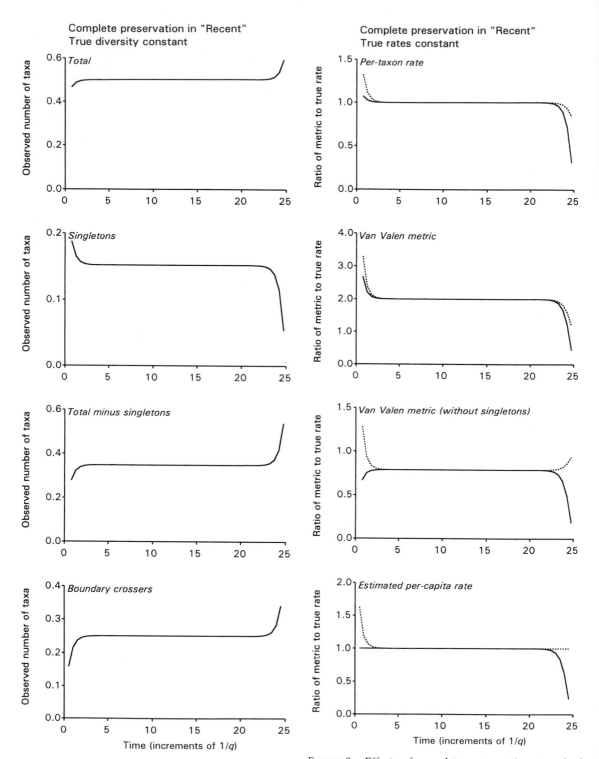

FIGURE 8. Effects of complete preservation at end of fossil record on observed numbers of taxa. Interval length is equal to $0.5/q$, $p = q$, and $r = q$. As with other edge effects, the edge is no longer relevant beyond about two taxon lengths.

FIGURE 9. Effects of complete preservation at end of fossil record on taxonomic rate metrics. See Figure 8 for explanation. Solid line, extinction metric; dotted line, origination metric.

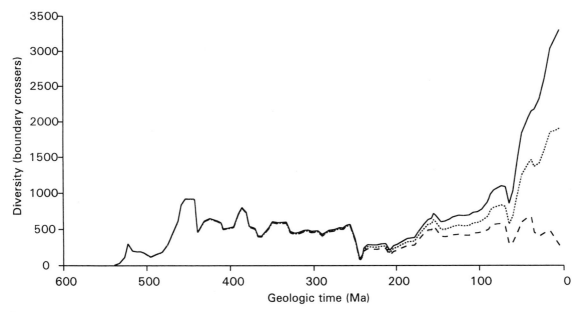

FIGURE 10. Genus diversity through the Phanerozoic. Data are from Sepkoski's unpublished compendium (Sepkoski 1996). Diversity curves show number of boundary crossers based on 25,049 fossil genera whose first and last appearances are fully resolved to one of 107 stratigraphic intervals of about 5.1 m.y. average duration (Foote and Sepkoski 1999; ages based mainly on Harland et al. 1990, Tucker and McKerrow 1995, and Bowring and Erwin 1998). Solid curve, all fossil genera included. Dashed curve, 3759 fossil genera still extant today are excluded. This is an overly extreme culling, since many extant genera do in fact have a fossil record near the Recent. Dotted curve, 1630 extant genera that are known to have a Plio-Pleistocene fossil record are included (Sepkoski 1997). Dotted curve is an absolute minimum, since it is unknown to what extent Sepkoski had documented Plio-Pleistocene occurrences. Curves have similar shapes except for the Tertiary portions. The Cenozoic rise in diversity may be exaggerated by nearly complete knowledge of the Recent fauna, but it is unlikely to be a complete artifact of this bias. Similar results hold for total diversity (Sepkoski 1997).

cent genera lacking a Plio-Pleistocene record are omitted (Fig. 11). Moreover, the decline in origination rate is seen, at least in the Paleozoic, with $\hat{p}$, a metric that is not expected to feel the upper edge of the fossil record (Fig. 11), and with the singleton-free Van Valen metric (data not presented), which is expected to increase as a result of essentially complete preservation in the Recent. The early Paleozoic decline in origination rate may be exaggerated by the left-hand edge effect, but the decline continues far beyond the point where this edge has a substantial influence. These results suggest that the Cenozoic increase in diversity and the Phanerozoic decline in taxonomic rates seen in marine animals are not artifacts of our relatively complete knowledge of the Recent fauna.

Temporal Variation in Taxonomic Rates

Consider the effect of an increase in extinction rate that lasts for one stratigraphic interval (Fig. 12). (An increase in origination rate yields converse results.) Because of incomplete preservation, the last appearances are smeared back in time (Signor and Lipps 1982; Raup 1989; Meldahl 1990; Koch 1991; Stanley and Yang 1994; Rampino and Adler 1998). The estimate of extinction rate in the interval is therefore lower than it should be, while it is higher than it should be in earlier intervals. As discussed above, changes in one rate affect estimates of both rates with the per-taxon metric, the Van Valen metric, and the singleton-free Van Valen metric. Incompleteness and the correlation of rate estimates combine to produce complicated signals in apparent origination and extinction rates even if only one rate varies. With $\hat{p}$ and $\hat{q}$, however, a change in one rate leaves the estimate of the other unaffected, even in the case of incomplete but homogeneous preservation. This property, lacking in other rate metrics, should lead us to favor $\hat{p}$ and $\hat{q}$, especially in cases where inde-

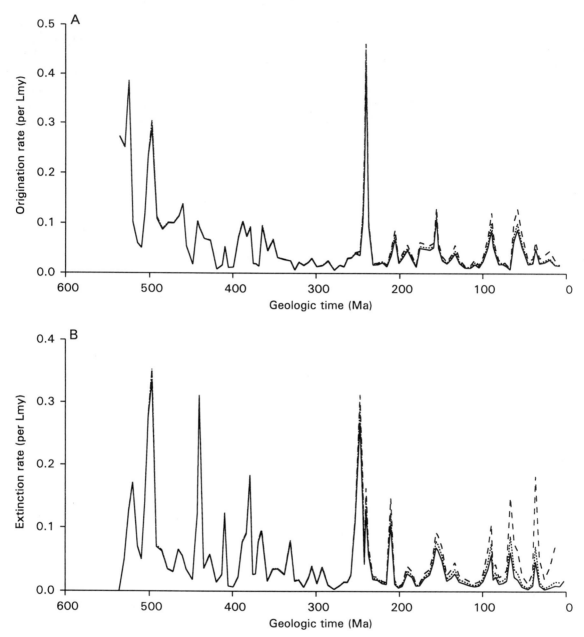

FIGURE 11.   Per-capita origination rate (A) and extinction rate (B) for genus data portrayed in Figure 10. Solid, dashed, and dotted curves as in Figure 10. Complete culling of extant taxa yields extinction rates that rise toward the Recent (exhibiting the edge effect of Figure 7), but this culling is unreasonably extreme (see Fig. 10). The similarity of solid and dotted curves suggests that the Phanerozoic decline in rates is not a consequence of nearly complete knowledge of the living fauna.

pendent estimates of origination and extinction are desired.

Temporal Variation in Rate of Preservation

If preservation rate increases or decreases gradually over time while taxonomic rates re-

main constant, the number of singletons will gradually change in the opposite direction. Thus, the per-taxon rate and the Van Valen metric will show spurious secular changes (Fig. 13). Partly for this reason, Pease (1988a,b, 1992) argued that the Phanerozoic decline in

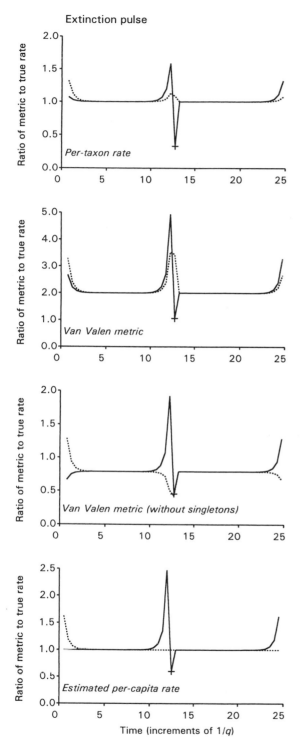

Extinction pulse

taxonomic rates may be an artifact of increasing completeness of the fossil record. If the change in preservation rate is smooth, however, then top- and bottom-boundary crossers will not experience significantly different preservational histories. Thus, rate metrics based on boundary crossers should barely be affected by long-term secular changes in preservation rate. The fact that the Phanerozoic decline in rates is seen in the singleton-free Van Valen metric and in $\hat{p}$ and $\hat{q}$ (Fig. 11) suggests that this decline is not an artifact of a secular increase in the quality of the fossil record. Holman (1985) also argued that the decline in rates is real, since the frequency of gaps, estimated from the stratigraphic ranges of lower-level taxa within the ranges of the higher taxa containing them, does not show an obvious decrease through the Phanerozoic.

Figures 14 and 15 illustrate the more complicated effects of a sudden increase in preservation rate. (A sudden decrease has converse effects.) All categories of taxa of course increase. Top- and bottom-boundary crossers are affected equally only if origination and extinction rates are equal. When diversity is truly increasing, the magnitude of the increase is exaggerated, and conversely when diversity is decreasing (Appendix, section 5). Taxa that would have been singletons in adjacent intervals had preservation been homogeneous now extend into the interval in question; the number of singletons in adjacent intervals therefore declines. The number of first and last appearances increases with the increase in preservation rate, so that both the origination and extinction rates appear to increase. In preceding intervals, the number of taxa that would have made their last appearance is reduced because they now appear last in the interval with better preservation. The same is true of

$\leftarrow$

in the interval of the extinction spike, and the extra extinctions are spread backward in time. With all metrics except estimated per-capita rate, the estimate of origination rate is also affected (see Fig. 3). As with related edge effects (Fig. 7), the effect of a transient rate pulse decays exponentially until it is no longer detectable after about two taxon lengths. A transient increase in origination rate has converse effects, which propagate in the opposite direction in time.

FIGURE 12. Effect of short-lived increase in extinction rate on taxonomic rate metrics. Solid line, extinction metric; dotted line, origination metric. Extinction rate q is constant except for a fivefold increase that lasts for one interval (indicated by the cross). Interval length is equal to $0.5/q$, and $r = p = q$. Because the record is incomplete, extinction rate appears lower than it should

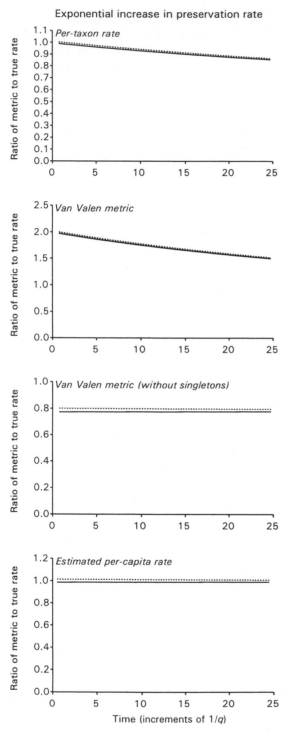

FIGURE 13. Effect of secular trend in preservation rate on taxonomic rate metrics. Interval length is equal to $0.5/q$, and $p = q$. Dotted line, origination; solid line, extinction. Preservation rate increases exponentially from q to $2q$. There are no edge effects. Results for linear change in preservation rate are similar. Metrics that disregard singletons are barely affected by a smooth change in preservation rate.

first appearances in succeeding intervals. As a result, all metrics underestimate extinction rate before the interval of increased preservation and underestimate origination rate afterwards. In addition, the per-taxon rate and the Van Valen metric underestimate origination beforehand, while they underestimate extinction afterwards. This is because they include singletons in the origination and extinction counts. The singleton-free Van Valen metric overestimates origination before the pulse in preservation and overestimates extinction afterwards. A strength of the proposed estimates of per-capita rates is that $\hat{p}$ is affected only during and after the interval of unusually high (or low) preservation, and $\hat{q}$ is affected only during and before this interval.

It is often observed that certain intervals of time appear to have unusually high rates of both origination and extinction (Allmon et al. 1993). Because apparent rates increase when preservation increases, abrupt variation in preservation rates can induce a spurious positive correlation between origination and extinction metrics even if the two rates are in fact independent. This effect is especially strong for the per-taxon rate and the Van Valen metric, since both rates also apparently decrease before and after an interval with better preservation. The effect is quite evident empirically as well. Mark and Flessa (1977) and Alroy (1996b, 1998) showed that apparent origination and extinction rates are less strongly correlated if singletons are removed. For Phanerozoic marine animal genera, the per-taxon rate and the Van Valen metric tend to show higher correlations than do the singleton-free Van Valen metric and $\hat{p}$ and $\hat{q}$ (Table 3). The correlations tend to be lowest for the singleton-free Van Valen metric, but this may partly reflect the fact that true, independent variation in origination and extinction rates tends to yield a spurious negative correlation between the apparent rates with this metric (Fig. 3).

More generally, variation in preservation rate can artificially create patterns that resemble true temporal variation in origination and extinction rates. This is a fundamental problem that has long plagued paleontology. The extreme case of Lagerstätten is relatively easy

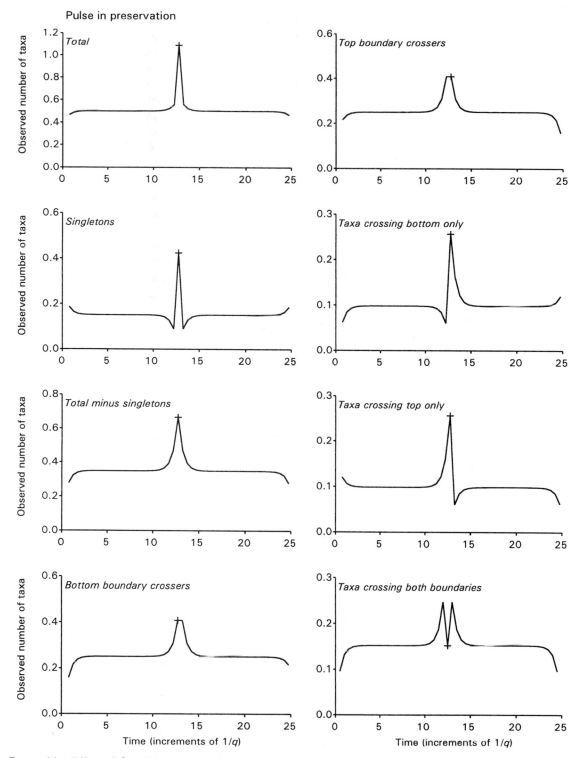

FIGURE 14. Effect of short-lived increase in preservation rate on number of observed taxa. Preservation rate r is constant except for a fivefold increase that lasts for one interval (indicated by the cross). Interval length is equal to $0.5/q$, and $r = p = q$. Transient decrease in preservation rate has opposite effects.

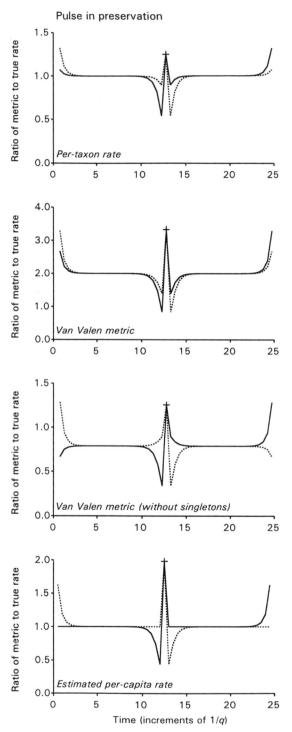

to identify, but more subtle variation in preservation rate is likely to be less obvious.

Distinguishing Variation in Taxonomic Rates from Variation in Rates of Preservation

To what extent is apparent variation in origination and extinction rates real, and to what extent is it an artifact of variation in preservation rate? This question must be addressed case by case (Koch and Morgan 1988; Koch 1991; Alroy 1996b). I will offer two arguments, suggestive but certainly not conclusive, that there is a real signal of variation in taxonomic rates for marine animal genera as a whole. First, if the apparent variation in taxonomic rates were an artifact of variable preservation, groups with lower preservation potential should have more apparent variation in taxonomic rates. This is because the effect of changes in preservation rate is greater when preservation rate on average is lower (Appendix, section 3). There are diminishing returns, such that an increment in preservation rate has a substantial effect if the record is poor but a negligible effect if the record is good. Figure 16 compares the estimated probability of preservation per genus per stratigraphic interval, measured as the *FreqRat* (Foote and Raup 1996; Wagner 1997; Cheetham and Jackson 1998; Foote and Sepkoski 1999), with the variability in taxonomic rates, measured as the median absolute difference in log rate between adjacent stratigraphic intervals, for a number of higher taxa of animals. The data are those used in Figures 10 and 11. Contrary to the expectations of the artifact hypothesis, the correlations between preservability and rate variability are in fact positive (though not significantly so). Clearly, overall quality of preservation is not a good predictor of apparent variation in taxonomic rates.

FIGURE 15. Effect of short-lived increase in preservation rate on taxonomic rate metrics. Solid line, extinction metric; dotted line, origination metric. Preservation rate r is constant except for a fivefold increase that lasts for one interval (indicated by the cross). Interval length is equal to $0.5/q$ and $r = p = q$. All metrics increase artificially as a result of spike in preservation. Because singletons are depressed on both sides of the preservation-

$\leftarrow$

al spike (Fig. 14), rate metrics that include singletons are depressed on both sides of the spike. Note that with the estimated per-capita rate, apparent origination is affected only during and after the preservational spike, while extinction is affected only before and during the spike. With other metrics, both rates are affected in both directions. Transient decrease in preservation rate has opposite effects.

TABLE 3. Correlation between change in origination metric and change in extinction metric for Phanerozoic marine animal genera (unpublished data from Sepkoski; see Sepkoski 1996, 1997, and Figs. 10 and 11 of this paper). First differences are used in order to detrend the data. Correlation coefficient is Kendall's τ.

Metric	Correlation
Per-taxon rate	0.599
Van Valen metric	0.669
Van Valen metric (without singletons)	0.256
Estimated per-capita rate ($\hat{p}$ and $\hat{q}$)	0.381

The second argument concerns the proportion of variation in apparent taxonomic rates that is potentially attributable to variation in preservation rates. This last quantity is most directly estimated not with first and last appearance data alone, but rather with occurrences within stratigraphic ranges. I have used data on occurrences of Ordovician brachiopods, mollusks, and trilobites kindly supplied by Arnold I. Miller of the University of Cincinnati. The data are part of an ongoing effort to analyze temporal, geographic, and environmental patterns of diversity, origination, and extinction through the Ordovician (Miller and Mao 1995, 1998; Miller and Foote 1996; Miller 1997a,b, 1998). At the time of writing, the data mainly cover Laurentia, China, Baltica, East Avalonia, Bohemia, Australia, and South America (including the Precordillera). I analyzed occurrences and ranges at the subseries level of resolution and ignored occurrences that could not be adequately resolved. All told, I included 1075 genera and 7461 occurrences.

I first used gap analysis (Paul 1982, 1998) to estimate the preservation probability, R_i, for each Ordovician subseries. I tallied the number of genera, X_{bt}, known both before and after the interval in question and the number of these genera actually sampled during the interval, $X_{bt,samp}$, and estimated R_i as $X_{bt,samp}/X_{bt}$. (Note that first and last occurrences of genera are necessarily discarded by this approach. To include them, as Paul did, would bias the estimate of R_i upward [see Holman 1985; Maas et al. 1995; Foote and Raup 1996; Markwick 1998; Foote in press b].) As is usually the case, R_i is the joint probability that a taxon is preserved, collected, published, and entered into

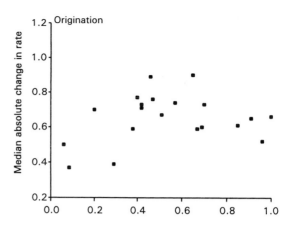

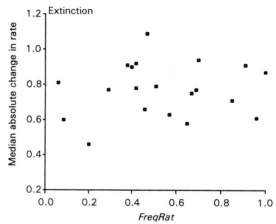

FIGURE 16. Quality of preservation versus apparent variability in per-capita origination and extinction rates for genera within higher taxa. The *FreqRat* (Foote and Raup 1996) is used to estimate probability of genus preservation within stratigraphic intervals of about 5.1 m.y. duration (Foote and Sepkoski 1999). Data are from Sepkoski (see Figs. 10, 11). To measure abrupt rather than long-term variation, variability is measured as the median absolute difference in log taxonomic rates between adjacent stratigraphic intervals. Quality of preservation is not a good predictor of apparent variability in rates, suggesting that this variability may be at least partly real. Taxa analyzed are Ammonoidea, Anthozoa, Asterozoa, Bivalvia, Blastozoa, Brachiopoda, Bryozoa, Cephalopoda, Chondrichthyes, Conodonta, Crinoidea, Echinoidea, Gastropoda, Graptolithina, Malacostraca, Nautiloidea, Osteichthyes, Ostracoda, Polychaeta, Porifera, and Trilobita. Note that Cephalopoda contains Ammonoidea and Nautiloidea; thus not all points are independent.

the database. Since R_i is expected to be equal to $1 - \exp(-r_i \Delta t)$, the preservation rate r_i is estimated as $-\ln(1 - R_i)/\Delta t$ (Appendix: eq. 26). Because Miller's data cover the Ordovician only, it would be impossible to estimate preservation probability in this way for the first and last intervals of the Ordovician. I therefore also used Sepkoski's genus data to tabulate Cambrian first occurrences and post-Ordovician last occurrences of genera present in Miller's data. Even though I have used data outside the Ordovician, it is still likely that there are edge effects. Of 1187 total genera in Miller's data, 108 (9.1%) are not found at all in Sepkoski's data.

If variation in apparent taxonomic rates were dominated by variation in preservation rates, then correlations between r and $\hat{p}$ and between r and $\hat{q}$ should both be large and positive. In fact, the first is negative and the second is positive, and neither is statistically significant (product-moment correlation coefficients: $r_{r,\hat{p}} = -0.10$; $r_{r,\hat{q}} = 0.36$). (Results are similar if the Kendall's τ is used, except that $\tau_{r,\hat{p}}$ is small and positive rather than small and negative.) It is possible that an effect of preservation rates is obscured because all three rates show temporal trends (Fig. 17). If we take first differences, there is a weak negative effect of change in preservation rate on change in apparent origination rate, while the effect of change in preservation rate on change in apparent extinction rate is significant and positive ($r_{\Delta r,\Delta\hat{p}} = -0.16$; $r_{\Delta r,\Delta\hat{q}} = 0.66$). Even though this last correlation is statistically significant ($p < 0.05$), the proportion of variation in apparent extinction rate that can be explained by variation in preservation rate is less than 40%. These results certainly argue against taking all variation in taxonomic rates at face value, but at the same time they suggest that there is substantial variation in estimated origination and extinction rates that is not an artifact of variation in preservation rate.

Discussion

Using boundary crossers to estimate origination and extinction rates is relatively insensitive to secular trends in the quality of preservation. Except very near the Recent, this approach is also affected but little by the nearly complete knowledge of the living fauna. Why is it then that Pease (1988a,b, 1992) interpreted the Phanerozoic decline in taxonomic rates as an artifact of improving preservation and of the Pull of the Recent? There seem to be at least two reasons. First, he used origination and extinction metrics, such as the per-taxon

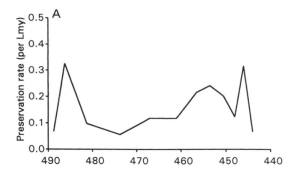

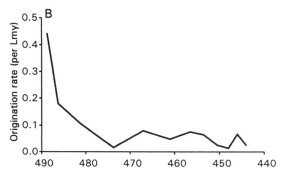

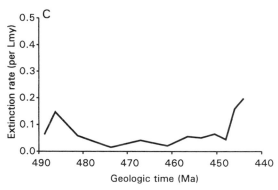

FIGURE 17. Estimated preservation rate (A), per-capita origination rate (B), and per-capita extinction rate (C) during the Ordovician. Data are from A. I. Miller, supplemented from Sepkoski for pre- and post-Ordovician occurrences of genera in Miller's data set. Probability of preservation R_i for each interval is estimated as inverse of ratio of number of genera known both before and after the interval to the number of these genera actually occurring within the interval. R_i is converted to a rate per Lmy: $r_i = -\ln(1 - R_i)/\Delta t$. See text for discussion.

rate, that are sensitive to secular changes in preservability. Second, some of his arguments were based on analysis of bivalve families. Because families tend to be long-lived taxa in general, and bivalve taxa are especially long-lived, the edge effect of the Recent extends far back in time in this case.

Because variation in preservation can mimic

variation in taxonomic rates, it is important to consider how much true variation in taxonomic rates is contained in stratigraphic range data and how this variation may be distorted (Koch and Morgan 1988). For the Phanerozoic as a whole, the magnitude of apparent variation in taxonomic rates is not correlated with a group's preservation potential. This suggests that there is substantial, short-term variation in taxonomic rates that is potentially measurable. The same conclusion is indicated by the correlations among preservation rate and apparent rates of origination and extinction in Ordovician invertebrates. If the results for marine animals are typical, then apparent variation in origination and extinction may be largely real. But knowing that the variation is largely real in a statistical sense is not the same as knowing the exact pattern of this variation. There are at least two obvious approaches to uncovering this pattern: (1) We can exploit occurrence data to adjust first and last appearances in a way that takes variable preservation and sampling into consideration. For example, by basing origination and extinction metrics on sampling-standardized first and last appearances, Alroy (1996b, 1998) has estimated the pattern of origination and extinction in North American Cenozoic mammals that one would likely have observed if preservation and sampling had been uniform through time. This approach has shown considerable variation in taxonomic rates that is not easily attributable to variation in the quality of the record or our knowledge of it. Extensive data of the sort that Miller has collected for the Ordovician and Alroy for the Cenozoic have not yet been compiled for most of geologic time, so we are far from being able to follow the approach of after-the-fact, standardized resampling of the entire fossil record. Such a comprehensive compilation has been started for Phanerozoic marine animals, however (Marshall et al. 1999), and preliminary results suggest that there are some similarities between genus origination and extinction curves based on raw data and those based on sampling-standardized data (J. Alroy et al. unpublished). (2) We can seek to develop methods that enable preservation rates to be estimated and first and last appearances

to be correspondingly adjusted given only the more readily available data on range endpoints, with no information on occurrences within ranges. I hope to report on one such method in a future contribution to this journal.

Conclusion

Incomplete preservation and variation in interval length cause most diversity measures and origination and extinction metrics to be inaccurate. Moreover, some metrics of taxonomic rate by their very nature preclude the independent estimation of origination and extinction rates even under the ideal assumption of a complete record. Modeling of cladogenesis and preservation supports previous intuitive and empirical arguments that diversity and rates are best estimated if single-interval taxa are disregarded (Sepkoski 1990, 1993; Buzas and Culver 1994, 1998; Raymond and Metz 1995; Alroy 1996b, 1998, 1999; Harper 1996; Sepkoski and Koch 1996; Bambach 1999; see also Pease 1985). Using measures for which singletons are simply irrelevant is preferable to adapting conventional measures by discarding singletons. Thus it appears advantageous in principle to measure relative changes in diversity using the proportional difference between the number of taxa crossing into an interval and the number crossing out, and to measure taxonomic rates using the number of taxa that range completely through the interval relative to the total number that cross into or out of the interval. Although the fossil record is incomplete to an extent that varies substantially over time, past and current developments aimed at coping with this variability suggest that it may be possible to extract true signals of origination rate, extinction rate, and taxonomic diversity through time, provided that one avoids the dual pitfalls of taking the record at face value and assuming that it is so distorted as to be uninformative. The fact that methods for uncovering these signals are still being developed attests to the vibrancy of paleontology today.

Acknowledgments

I am especially grateful to R. H. Foote for years of discussing this work and for suggesting substantial improvements in analysis and presentation. This work also benefited from discussions with J. Alroy, D. Jablonski, A. I. Miller, and the late J. J. Sepkoski Jr. A. I. Miller and J. J. Sepkoski Jr. generously provided data. J. Alroy, R. H. Foote, A. I. Miller, and M. Newman read and improved the manuscript. This research was supported by the National Science Foundation (grant EAR 95-06568).

Literature Cited

Allmon, W. D., G. Rosenberg, R. W. Portell, and K. S. Schindler. 1993. Diversity of Atlantic Coastal Plain mollusks since the Pliocene. Science 260:1626–1629.

Alroy, J. 1992. Conjunction among taxonomic distributions and the Miocene mammalian biochronology of the Great Plains. Paleobiology 18:326–343.

———. 1996a. Four methods of correcting diversity curves for sampling effects: which is best? Geological Society of America Abstracts with Programs 28:A107.

———. 1996b. Constant extinction, constrained diversification, and uncoordinated stasis in North American mammals. Palaeogeography, Palaeoclimatology, Palaeoecology 127:285–311.

———. 1998. Equilibrial diversity dynamics in North American mammals. Pp. 233–287 in M. L. McKinney and J. A. Drake, eds. Biodiversity dynamics: turnover of populations, taxa and communities. Columbia University Press, New York.

———. 1999. Putting North America's end-Pleistocene megafaunal extinction in context: large scale analyses of spatial patterns, extinction rates, and size distributions. Pp. 105–143 in R. D. E. MacPhee, ed. Extinctions in near time: causes, contexts, and consequences. Plenum, New York.

Alroy, J., P. L. Koch, and J. C. Zachos. 2000. In D. H. Erwin and S. L. Wing, eds. Deep time: Paleobiology's perspective. Paleobiology 26(Suppl. to No. 4):259–288.

Bambach, R. K. 1999. Energetics in the global marine fauna: a connection between terrestrial diversification and change in the marine biosphere. Geobios 32:131–144.

Barry, J. C., M. E. Morgan, L. J. Flynn, D. Pilbeam, L. L. Jacobs, E. H. Lindsay, S. M. Raza, and N. Solounias. 1995. Patterns of faunal turnover and diversity in the Neogene Siwaliks of northern Pakistan. Palaeogeography, Palaeoclimatology, Palaeoecology 115:209–226.

Benton, M. J. 1994. Palaeontological data and identifying mass extinctions. Trends in Ecology and Evolution 9:181–185.

Bowring, S. A., and D. H. Erwin. 1998. A new look at evolutionary rates in deep time: uniting paleontology and high-precision geochronology. GSA Today 8(9):1–8.

Budd, A. F., and K. G. Johnson. 1999. Origination preceding extinction during late Cenozoic turnover of Caribbean reefs. Paleobiology 25:188–200.

Budd, A. F., T. Stemann, and K. G. Johnson. 1994. Stratigraphic distributions of genera and species of Neogene to Recent Caribbean reef corals. Journal of Paleontology 68:951–977.

Buzas, M. A., and S. J. Culver. 1994. Species pool and dynamics of marine paleocommunities. Science 264:1439–1441.

———. 1998. Assembly, disassembly, and balance in marine communities. Palaios 13:263–275.

Buzas, M. A., C. F. Koch, S. J. Culver, and N. F. Sohl. 1982. On the distribution of species occurrence. Paleobiology 8:143–150.

Cheetham, A. H., and J. B. C. Jackson. 1998. The fossil record of cheilostome Bryozoa in the Neogene and Quaternary of trop-

ical America. Pp. 227–242 in S. K. Donovan and C. R. C. Paul, eds. The adequacy of the fossil record. Wiley, Chichester, England.

Collins, L. S. 1989. Evolutionary rates of a rapid radiation: the Paleogene planktic foraminifera. Palaios 4:251–263.

Foote, M. 1994. Temporal variation in extinction risk and temporal scaling of extinction metrics. Paleobiology 20:424–444.

———. 1997. Estimating taxonomic durations and preservation probability. Paleobiology 23:278–300.

———. 1999. Morphological diversity in the evolutionary radiation of Paleozoic and post-Paleozoic crinoids. Paleobiology Memoirs No. 1. Paleobiology 25(Suppl. to No. 2).

———. 2000. Origination and extinction components of taxonomic diversity: Paleozoic and post-Paleozoic dynamics. Paleobiology 26:578–605.

———. In press a. Evolutionary rates and the age distributions of living and extinct taxa. In J. B. C. Jackson, F. K. McKinney, and S. Lidgard, eds. Evolutionary patterns: growth, form, and tempo in the fossil record. University of Chicago Press, Chicago.

———. In press b. Estimating completeness of the fossil record. In D. E. G. Briggs and P. R. Crowther, eds. Palaeobiology II. Blackwell Scientific, Oxford.

Foote, M., and D. M. Raup. 1996. Fossil preservation and the stratigraphic ranges of taxa. Paleobiology 22:121–140.

Foote, M., and J. J. Sepkoski Jr. 1999. Absolute measures of the completeness of the fossil record. Nature 398:415–417.

Gilinsky, N. L. 1991. The pace of taxonomic evolution. In N. L. Gilinsky and P. W. Signor, eds. Analytical paleobiology. Short Courses in Paleontology 4:157–174. Paleontological Society, Knoxville, Tenn.

Gilinsky, N. L., and R. K. Bambach. 1987. Asymmetrical patterns of origination and extinction in higher taxa. Paleobiology 13:427–445.

Gingerich, P. D. 1987. Extinction of Phanerozoic marine families. Geological Society of America Abstracts with Programs 19:677.

Harland, W. B., R. L. Armstrong, A. V. Cox, L. E. Craig, A. G. Smith, and D. G. Smith. 1990. A geologic time scale 1989. Cambridge University Press, Cambridge.

Harper, C. W., Jr. 1975. Standing diversity of fossil groups in successive intervals of geologic time: a new measure. Journal of Paleontology 49:752–757.

———. 1996. Patterns of diversity, extinction, and origination in the Ordovician-Devonian Stropheodontacea. Historical Biology 11:267–288.

Hessler, R. R., and H. L. Sanders. 1967. Faunal diversity in the deep sea. Deep-Sea Research 14:65–78.

Holland, S. M. 1995. The stratigraphic distribution of fossils. Paleobiology 21:92–109.

Holland, S. M., and M. E. Patzkowsky. 1999. Models for simulating the fossil record. Geology 27:491–494.

Holman, E. W. 1985. Gaps in the fossil record. Paleobiology 11:221–226.

Jackson, J. B. C., P. Jung, A. G. Coates, and L. S. Collins. 1993. Diversity and extinction of tropical American mollusks and emergence of the Isthmus of Panama. Science 260:1624–1626.

Johnson, K. G. 1998. A phylogenetic test of accelerated turnover in Neogene Caribbean brain corals (Scleractinia: Faviidae). Palaeontology 41:1247–1267.

Kendall, D. G. 1948. On the generalized "birth-and-death" process. Annals of Mathematical Statistics 19:1–15.

Koch, C. F. 1991. Species extinctions across the Cretaceous-Tertiary boundary: observed patterns versus predicted sampling effects, stepwise or otherwise? Historical Biology 5:355–361.

Koch, C. F., and J. P. Morgan. 1988. On the expected distribution of species' ranges. Paleobiology 14:126–138.

Maas, M. C., M. R. L. Anthony, P. D. Gingerich, G. F. Gunnell, and D. W. Krause. 1995. Mammalian generic diversity and turnover in the Late Paleocene and Early Eocene of the Bighorn and Crazy Mountains Basins, Wyoming and Montana (USA). Palaeogeography, Palaeoclimatology, Palaeoecology 115:181–207.

Mark, G. A., and K. W. Flessa. 1977. A test for evolutionary equilibria: Phanerozoic brachiopods and Cenozoic mammals. Paleobiology 3:17–22.

Markwick, P. J. 1998. Crocodilian diversity in space and time: the role of climate in paleoecology and its implications for understanding K/T extinctions. Paleobiology 24:470–497.

Marshall, C. R. 1990. Confidence intervals on stratigraphic ranges. Paleobiology 16:1–10.

———. 1994. Confidence intervals on stratigraphic ranges: partial relaxation of the assumption of randomly distributed fossil horizons. Paleobiology 20:459–469.

Marshall, C. R., J. Alroy, and the NCEAS Phanerozoic Diversity Working Group. 1999. Towards a sample-standardized Phanerozoic diversity curve. Geological Society of America Abstracts with Programs 31:A336.

McGhee, G. R., Jr. 1996. The Late Devonian mass extinction. Columbia University Press, New York.

Meldahl, K. H. 1990. Sampling, species abundance, and the stratigraphic signature of mass extinction: a test using Holocene tidal flat molluscs. Geology 18:890–893.

Miller, A. I. 1997a. Dissecting global diversity patterns: examples from the Ordovician Radiation. Annual Review of Ecology and Systematics 28:85–104.

———. 1997b. A new look at age and area: the geographic and environmental expansion of genera during the Ordovician Radiation. Paleobiology 23:410–419.

———. 1998. Biotic transitions in global marine diversity. Science 281:1157–1160.

Miller, A. I., and M. Foote. 1996. Calibrating the Ordovician Radiation of marine life: implications for Phanerozoic diversity trends. Paleobiology 22:304–309.

Miller, A. I., and S. G. Mao. 1995. Association of orogenic activity with the Ordovician Radiation of marine life. Geology 23: 305–308.

———. 1998. Scales of diversification and the Ordovician Radiation. Pp. 288–310 in M. L. McKinney and J. A. Drake, eds. Biodiversity dynamics: turnover of populations, taxa, and communities. Columbia University Press, New York.

Norell, M. A. 1992. Taxic origin and temporal diversity: the effect of phylogeny. Pp. 89–118 in M. A. Novacek and Q. D. Wheeler, eds. Extinction and phylogeny. Columbia University Press, New York.

Patzkowsky, M. E., and S. M. Holland. 1997. Patterns of turnover in Middle and Upper Ordovician brachiopods of the eastern United States: a test of coordinated stasis. Paleobiology 23: 420–443.

Paul, C. R. C. 1982. The adequacy of the fossil record. Pp. 75–117 in K. A. Joysey and A. E. Friday, eds. Problems of phylogenetic reconstruction. Academic Press, London.

———. 1998. Adequacy, completeness and the fossil record. Pp. 1–22 in S. K. Donovan and C. R. C. Paul, eds. The adequacy of the fossil record. Wiley, Chichester, England.

Pearson, P. N. 1992. Survivorship analysis of fossil taxa when real-time extinction rates vary: the Paleogene planktonic foraminifera. Paleobiology 18:115–131.

———. 1996. Cladogenetic, extinction, and survivorship patterns from a lineage phylogeny: the Paleogene planktonic foraminifera. Micropaleontology 42:179–188.

Pease, C. M. 1985. Biases in the durations and diversities of fossil taxa. Paleobiology 11:272–292.

———. 1988a. Biases in the total extinction rates of fossil taxa. Journal of Theoretical Biology 130:1–7.

———. 1988b. Biases in the per-taxon origination and extinction rates of fossil taxa. Journal of Theoretical Biology 130:9–30.

———. 1992. On the declining extinction and origination rates of fossil taxa. Paleobiology 18:89–92.

Rampino, M. R., and A. C. Adler. 1998. Evidence for abrupt latest Permian mass extinction of foraminifera: results of tests for the Signor-Lipps effect. Geology 26:415–418.

Raup, D. M. 1972. Taxonomic diversity during the Phanerozoic. Science 177:1065–1071.

———. 1979. Biases in the fossil record of species and genera. Bulletin of the Carnegie Museum of Natural History 13:85–91.

———. 1985. Mathematical models of cladogenesis. Paleobiology 11:42–52.

———. 1986. Biological extinction in Earth history. Science 231: 1528–1533.

———. 1989. The case for extraterrestrial causes of extinction. Philosophical Transactions of the Royal Society of London B 325:421–435.

———. 1991. A kill curve for Phanerozoic marine species. Paleobiology 17:37–48.

Raup, D. M., and J. J. Sepkoski Jr. 1982. Mass extinctions in the marine fossil record. Science 215:1501–1503.

Raymond, A., and C. Metz. 1995. Laurussian land-plant diversity during the Silurian and Devonian: mass extinction, sampling bias, or both? Paleobiology 21:74–91.

Rex, M. A., C. T. Stuart, R. R. Hessler, J. A. Allen, H. L. Sanders, and G. D. F. Wilson. 1993. Global-scale latitudinal patterns of species diversity in the deep-sea benthos. Nature 365:636–639.

Rex, M. A., R. J. Etter, and C. T. Stuart. 1997. Large-scale patterns of species diversity in the deep-sea benthos. Pp. 94–121 in R. F. G. Ormond, J. D. Gage, and M. V. Angel, eds. Marine biodiversity: patterns and processes. Cambridge University Press, Cambridge.

Sanders, H. L. 1968. Marine benthic diversity: a comparative study. American Naturalist 102:243–282.

Sepkoski, J. J., Jr. 1990. The taxonomic structure of periodic extinction. Geological Society of America Special Paper 247:33–44.

———. 1991. Population biology models in macroevolution. In N. L. Gilinsky and P. W. Signor, eds. Analytical paleobiology. Short Courses in Paleontology 4:136–156. Paleontological Society, Knoxville, Tenn.

———. 1993. Phanerozoic diversity at the genus level: problems and prospects. Geological Society of America Abstracts with Programs 25:A50.

———. 1996. Patterns of Phanerozoic extinctions: a perspective from global databases. Pp. 35–52 in O. H. Walliser, ed. Global events and event stratigraphy. Springer, Berlin.

———. 1997. Biodiversity: past, present, and future. Journal of Paleontology 71:533–539.

———. 1998. Rates of speciation in the fossil record. Philosophical Transactions of the Royal Society of London B 353:315–326.

Sepkoski, J. J., Jr., and C. F. Koch. 1996. Evaluating paleontologic data relating to bio-events. Pp. 21–34 in O. H. Walliser, ed. Global events and event stratigraphy. Springer, Berlin.

Shaw, A. B. 1964. Time in stratigraphy. McGraw-Hill, New York.

Signor, P. W., III, and J. H. Lipps. 1982. Sampling bias, gradual extinction patterns and catastrophes in the fossil record. Geological Society of America Special Paper 190:291–296.

Solow, A. R., and W. Smith. 1997. On fossil preservation and the stratigraphic ranges of taxa. Paleobiology 23:271–278.

Stanley, S. M., and X. Yang. 1994. A double mass extinction at the end of the Paleozoic Era. Science 266:1340–1344.

Strauss, D., and P. M. Sadler. 1989. Classical confidence intervals and Bayesian probability estimates for ends of local taxon ranges. Mathematical Geology 21:411–427.

Tucker, R. D., and W. S. McKerrow. 1995. Early Paleozoic chronology: a review in light of new U-Pb zircon ages from Newfoundland and Britain. Canadian Journal of Earth Sciences 32:368–379.

Van Valen, L. M. 1984. A resetting of Phanerozoic community evolution. Nature 307:50–52.

Wagner, P. J. 1997. Patterns of morphologic diversification among the Rostroconchia. Paleobiology 23:115–150.

Wei, K.-Y., and J. P. Kennett. 1983. Nonconstant extinction rates of Neogene planktonic foraminifera. Nature 305:218–220.

———. 1986. Taxonomic evolution of Neogene planktonic foraminifera and paleoceanographic relations. Paleoceanography 1:67–84.

Weiss, R. E., and C. R. Marshall. 1999. The uncertainty in the true end point of a fossil's stratigraphic range when stratigraphic sections are sampled discretely. Mathematical Geology 31:435–453.

Appendix

Equations Regarding True and Apparent Taxonomic Rates and Diversity

This appendix develops equations for observed measures of diversity and taxonomic rates given true taxonomic rates and rates of fossil preservation. The basic equations for branching theory are taken mainly from Kendall (1948), Raup (1985), and Foote (in press a), while those for preservation are based upon Foote (1997) and simplifications of Pease (1985). A header file in C, containing library functions for all relevant calculations, is available from the author.

A time window over which observations can be made extends from time $T = 0$ to $T = w$. In cases where a discrete interval of time is of interest, it has bottom and top boundaries $T = t_b$ and $T = t_t$ and duration $\Delta t = t_t - t_b$. $p(T)$, $q(T)$, and $r(T)$ are the time-specific, per-capita rates of origination, extinction, and preservation at an instant in time T. For simplicity, I will assume that rates are constant within a discrete interval, although they may vary among intervals. Interval rates will be denoted p_i, q_i, and r_i. This assumption greatly reduces computational time for the time-heterogeneous numerical integrations, but it can easily be relaxed by allowing intervals to be arbitrarily short.

1. True Numbers of Taxa in a Stratigraphic Interval

There are four fundamental, exclusive kinds of lineages that can exist during an interval (see Barry et al. 1995 for a similar classification): (1) those that have both first and last appearances within the interval, i.e., singletons; (2) those that cross into the interval and last appear within it; (3) those that first appear within the interval and extend beyond it; and (4) those that cross into the interval from below and extend beyond the top of the interval. Let the corresponding numbers be denoted N_{FL}, N_{bL}, N_{Ft}, and N_{bt}, where the subscripts refer to *first-last, bottom-last, first-top,* and *bottom-top.* These numbers can be combined to yield several composite groups: (5) all lineages that cross into the interval, $N_b = N_{bt} + N_{bL}$; (6) all lineages that cross out of the interval, $N_t = N_{bt} + N_{Ft}$; (7) lineages that become extinct during the interval, $N_e = N_{bL} + N_{FL}$; (8) lineages that originate during the interval, $N_o = N_{Ft} + N_{FL}$; and (9) all lineages, $N_{tot} = N_{bt} + N_{bL} + N_{Ft} + N_{FL}$.

Let $\rho(t)$ be the accumulated difference between origination and extinction from $T = 0$ until $T = t$:

$$\rho(t) = \int_0^t \{p(T) - q(T)\} \, dT.$$

Let $N(t)$ be the expected diversity at time t. Then

$$N(t) = N(0)e^{\rho(t)}.$$

For simplicity, I will take $N(0)$ to be equal to unity. Thus, the number of taxa extant at the start of the interval is given by

$$N_b = N(t_b) = e^{\rho(t_b)}. \tag{1a}$$

In the special case where rates are constant, this is equal to

$$N_b = N(t_b) = e^{(p-q)t_b}. \tag{1b}$$

Similarly,

$$N_t = N_b e^{(p_i - q_i)\Delta t}. \tag{2}$$

The probability that a lineage entering the interval will still be extant at the end is equal to $e^{-q_i\Delta t}$. Likewise, the probability that a lineage leaving the interval was already extant at the start of the interval is equal to $e^{-p_i\Delta t}$. Thus,

$$N_{bt} = N_b e^{-q_i\Delta t} = N_t e^{-p_i\Delta t}, \tag{3}$$

$$N_{bL} = N_b(1 - e^{-q_i\Delta t}), \quad \text{and} \tag{4}$$

$$N_{Ft} = N_t(1 - e^{-p_i\Delta t}) = N_b e^{(p_i - q_i)\Delta t}(1 - e^{-p_i\Delta t}). \tag{5}$$

The number of lineages confined to the interval is found by integrating the expected number of originations at any time during the interval (i.e., origination rate times standing diversity), multiplied by the probability of not surviving from the time of origination to the end of the interval. Thus,

$$N_{FL} = N_b \int_0^{\Delta t} p_i e^{(p_i - q_i)T}[1 - e^{-q_i(\Delta t - T)}] \, dT. \tag{6a}$$

This is equal to

$$N_{FL} = N_b(e^{-q_i\Delta t} + p_i\Delta t - 1) \qquad \text{if } p_i = q_i, \text{ and} \tag{6b}$$

$$N_{FL} = N_b \frac{q_i e^{(p_i - q_i)\Delta t} + (p_i - q_i)e^{-q_i\Delta t} - p_i}{p_i - q_i} \quad \text{if } p_i \neq q_i. \tag{6c}$$

The number of originations during the interval is the integral of the origination rate times the standing diversity. Thus

$$N_o = N_b \int_0^{\Delta t} p_i e^{(p_i - q_i)} \, dT, \tag{7a}$$

which is equal to

$$N_o = \begin{cases} N_b p_i \Delta t & \text{if } p_i = q_i \text{ and} \\ N_b \dfrac{p_i[e^{(p_i - q_i)\Delta t} - 1]}{p_i - q_i} & \text{if } p_i \neq q_i. \end{cases} \tag{7b}$$

Similarly,

$$N_e = N_b \int_0^{\Delta t} q_i e^{(p_i - q_i)} \, dT, \tag{8a}$$

which is equal to

$$N_e = \begin{cases} N_b q_i \Delta t & \text{if } p_i = q_i \text{ and} \\ N_b \dfrac{q_i[e^{(p_i - q_i)\Delta t} - 1]}{p_i - q_i} & \text{if } p_i \neq q_i. \end{cases} \tag{8b}$$

Finally, the total number of lineages within the interval (the total progeny of Kendall 1948) is equal to

$$N_{tot} = N_b\left(1 + \int_0^{\Delta t} p_i e^{(p_i - q_i)} \, dT\right), \tag{9a}$$

which is equal to

$$N_{tot} = \begin{cases} N_b(1 + p_i \Delta t) & \text{if } p_i = q_i \text{ and} \\ N_b \dfrac{p_i e^{(p_i - q_i)\Delta t} - q_i}{p_i - q_i} & \text{if } p_i \neq q_i. \end{cases} \quad (9b)$$

2. Taxonomic Rate Metrics for an Interval, Assuming Complete Preservation

Various metrics have been devised to measure taxonomic rates over some extended interval of time, usually by counting originations and extinctions and normalizing by diversity and/or interval length. One problem that has been noted (Gilinsky 1991) is that some normalizations implicitly assume constancy of rates within an interval. Inspection of the foregoing equations shows that N_o, N_e, N_{FL}, and N_{tot} are affected by variation in rates within an interval, not just by mean rates over the interval. In contrast, N_b, N_t, N_{bt}, N_{bL} ($= N_b - N_{bt}$), and N_{Ft} ($= N_t - N_{bt}$) depend only on the average rates within the interval. Therefore, measures of taxonomic rate that are based only on N_b, N_t, and N_{bt} are, at least in theory, insensitive to rate variation within an interval and capable of accurately estimating average rates.

Proportional origination is the ratio of number of originations to total diversity. Substituting into the equations for N_o and N_{tot} yields

$$P_o = \begin{cases} \dfrac{p_i \Delta t}{1 + p_i \Delta t} & \text{if } p_i = q_i \text{ and} \\ \dfrac{p_i[e^{(p_i - q_i)\Delta t} - 1]}{p_i e^{(p_i - q_i)\Delta t} - q_i} & \text{if } p_i \neq q_i. \end{cases} \quad (10)$$

Similarly, for proportional extinction

$$P_e = \begin{cases} \dfrac{q_i \Delta t}{1 + p_i \Delta t} & \text{if } p_i = q_i \text{ and} \\ \dfrac{q_i[e^{(p_i - q_i)\Delta t} - 1]}{p_i e^{(p_i - q_i)\Delta t} - q_i} & \text{if } p_i \neq q_i. \end{cases} \quad (11)$$

Normalizing these expressions by interval length yields the so-called per-taxon rates:

$$P_{o/m.y.} = \begin{cases} \dfrac{p_i}{1 + p_i \Delta t} & \text{if } p_i = q_i \text{ and} \\ \dfrac{p_i[e^{(p_i - q_i)\Delta t} - 1]}{[p_i e^{(p_i - q_i)\Delta t} - q_i]\Delta t} & \text{if } p_i \neq q_i \end{cases} \quad (12)$$

and

$$P_{e/m.y.} = \begin{cases} \dfrac{q_i}{1 + q_i \Delta t} & \text{if } p_i = q_i \text{ and} \\ \dfrac{q_i[e^{(p_i - q_i)\Delta t} - 1]}{[p_i e^{(p_i - q_i)\Delta t} - q_i]\Delta t} & \text{if } p_i \neq q_i. \end{cases} \quad (13)$$

Clearly, P_o and P_e increase nonlinearly with Δt, whereas $P_{o/m.y.}$ and $P_{e/m.y.}$ decrease nonlinearly with Δt. This nonlinear dependence on Δt, which cannot be eliminated by a simple normalization, complicates the use of proportional origination and extinction when interval length varies (Gilinsky 1991; Foote 1994).

Van Valen proposed a measure of taxonomic rates that normalizes the observed number of events by the estimated standing diversity within the interval, which is simply the mean of N_b and N_t (algebraically identical to $N_{tot} - N_o/2 - N_e/2$; [see Table 1 and Harper 1975, 1996]). The corresponding origination metric is given by

$$V_o = \frac{N_o}{(N_b + N_t)/2}, \quad (14a)$$

which is equal to

$$V_o = \begin{cases} p_i \Delta t & \text{if } p_i = q_i \text{ and} \\ \dfrac{2p_i[e^{(p_i - q_i)\Delta t} - 1]}{(p_i - q_i)[1 + e^{(p_i - q_i)\Delta t}]} & \text{if } p_i \neq q_i. \end{cases} \quad (14b)$$

The time-normalized origination metric is thus equal to

$$V_{o/m.y.} = \begin{cases} p_i & \text{if } p_i = q_i \text{ and} \\ \dfrac{2p_i[e^{(p_i - q_i)\Delta t} - 1]}{(p_i - q_i)[1 + e^{(p_i - q_i)\Delta t}]\Delta t} & \text{if } p_i \neq q_i. \end{cases} \quad (15)$$

The corresponding extinction metrics are given by

$$V_e = \begin{cases} q_i \Delta t & \text{if } p_i = q_i \text{ and} \\ \dfrac{2q_i[e^{(p_i - q_i)\Delta t} - 1]}{(p_i - q_i)[1 + e^{(p_i - q_i)\Delta t}]} & \text{if } p_i \neq q_i, \end{cases} \quad (16)$$

and

$$V_{e/m.y.} = \begin{cases} q_i & \text{if } p_i = q_i \text{ and} \\ \dfrac{2q_i[e^{(p_i - q_i)\Delta t} - 1]}{(p_i - q_i)[1 + e^{(p_i - q_i)\Delta t}]\Delta t} & \text{if } p_i \neq q_i. \end{cases} \quad (17)$$

Thus, if origination and extinction rates are equal, Van Valen's metric provides an accurate estimate of these rates that is independent of interval length. The Van Valen metric progressively underestimates origination and extinction rates as the true difference between these rates increases.

Harper (1996) used a number of measures of origination and extinction that disregard singletons. Harper's modification of the Van Valen metric is given by

$$V_o^* = \frac{N_o - N_{FL}}{(N_b + N_t)/2} = \frac{N_{Ft}}{(N_b + N_t)/2}, \quad (18a)$$

which is equal to

$$V_o^* = \begin{cases} 1 - e^{-p_i \Delta t} & \text{if } p_i = q_i \text{ and} \\ \dfrac{2[e^{(p_i - q_i)\Delta t} - e^{-p_i \Delta t}]}{1 + e^{(p_i - q_i)\Delta t}} & \text{if } p_i \neq q_i. \end{cases} \quad (18b)$$

Thus, when origination and extinction rates are equal, the singleton-free Van Valen metric before time-normalization gives the probability that a lineage present at either boundary will extend all the way through the interval. (Alroy [1996b] also used a similar approach, normalizing by N_b rather than by $(N_b + N_t)/2$. The two metrics are equivalent if $p_i = q_i$.) Normalizing by interval length yields

$$V_{o/m.y.}^* = \begin{cases} \dfrac{1 - e^{-p_i \Delta t}}{\Delta t} & \text{if } p_i = q_i \text{ and} \\ \dfrac{2[e^{(p_i - q_i)\Delta t} - e^{-p_i \Delta t}]}{[1 + e^{(p_i - q_i)\Delta t}]\Delta t} & \text{if } p_i \neq q_i. \end{cases} \quad (19)$$

The corresponding extinction metrics are given by

$$V_e^* = \begin{cases} 1 - e^{-q_i \Delta t} & \text{if } p_i = q_i \text{ and} \\ \dfrac{2(1 - e^{-q_i \Delta t})}{1 + e^{(p_i - q_i)\Delta t}} & \text{if } p_i \neq q_i, \end{cases} \quad (20)$$

and

$$V_{e/m.y.}^* = \begin{cases} \dfrac{1 - e^{-q_i \Delta t}}{\Delta t} & \text{if } p_i = q_i \text{ and} \\ \dfrac{2(1 - e^{-q_i \Delta t})}{[1 + e^{(p_i - q_i)\Delta t}]\Delta t} & \text{if } p_i \neq q_i. \end{cases} \quad (21)$$

In the general case, $V_{o/m.y.}^*$ and $V_{e/m.y.}^*$ do not provide accurate rate estimates. They decrease nonlinearly with interval length, and they deviate more from p_i and q_i as the true rates increase and as the difference between them increases. Because $1 - e^{-x}$

$\approx x$ for small x, however, $V^*_{o/m.y.} = V^*_{e/m.y.} \approx p_i = q_i$ in the special case where $p_i = q_i$ and p_i, q_i, and Δt are not too large. The same is true of Alroy's (1996b) metrics when singletons are disregarded (see Alroy et al. 2000).

The number of lineages extending through the entire interval depends only on the mean taxonomic rates for the interval. Rearranging the simple equivalence $N_{bt} = N_b e^{-q_i \Delta t} = N_t e^{-p_i \Delta t}$ yields estimates of per-capita rates:

$$\hat{p} = \frac{-\ln(N_{bt}/N_t)}{\Delta t} \quad \text{and} \tag{22}$$

$$\hat{q} = \frac{-\ln(N_{bt}/N_b)}{\Delta t}. \tag{23}$$

$\hat{p}$ and $\hat{q}$ reduce to p_i and q_i, but, in contrast to the singleton-free Van Valen metric, this is true regardless of the magnitude of the rates. If origination and extinction rates are unequal, then, in contrast to the Van Valen metrics, with or without singletons, one rate is not expected to affect the estimation of the other. Note that $\hat{p}$ and $\hat{q}$ do not rely on counting events within the interval and normalizing by total diversity or estimated standing diversity. Also, as with the singleton-free Van Valen metric, single-interval taxa play no role. This will be important when the incompleteness of the fossil record is taken into consideration.

3. Fundamental Preservation Probabilities

A lineage with time-specific preservation rate $r(T)$ per m.y. and duration t from t_1 to t_2 has a net probability of preservation equal to $1 - e^{-\int_{t_1}^{t_2} r(T) \, dT}$. In the case of constant preservation rates, this is equal to $1 - e^{-rt}$. This relationship and the probability distribution of taxon durations are used to calculate probabilities of preservation for taxa spanning all or part of relevant intervals of time.

Probability of Preservation before a Point in Time.—The duration of a lineage before any arbitrary point in time depends on prior rates of origination and on the span of time. The probability of preservation before time t, assuming that the lineage is in fact extant at time t (not assuming that we know this fact), is therefore equal to

$$P_B = \int_t^0 \left(p(T) \exp\left[-\int_t^T p(x) \, dx \right] \left\{ 1 - \exp\left[-\int_t^T r(x) \, dx \right] \right\} \right) dT. \tag{24a}$$

In the special case where p and r are constant, this reduces to

$$P_B = \frac{r}{p + r}[1 - e^{-(p+r)t}]. \tag{24b}$$

In the special case where p and r are constant and the time span t is effectively infinite (i.e., the probability that a lineage extends from $T = 0$ to $T = t$ is approximately nil), this reduces to

$$P_B = \frac{r}{p + r}. \tag{24c}$$

The corresponding probability of preservation after an arbitrary point in time, given that the lineage is extant at that point, depends on the extinction and preservation rates and the span of time after that point. Thus

$$P_A = \int_t^w \left(q(T) \exp\left[-\int_t^T q(x) \, dx \right] \left\{ 1 - \exp\left[-\int_t^T r(x) \, dx \right] \right\} \right) dT. \tag{25a}$$

where, as above, w is the upper bound of the stratigraphic window over which observations can be made. If q and r are constant, this reduces to

$$P_A = \frac{r}{q + r}[1 - e^{-(q+r)(w-t)}]. \tag{25b}$$

If q and r are constant and the time span $(w - t)$ is effectively infinite, this reduces to

$$P_A = \frac{r}{q + r}. \tag{25c}$$

Equations (24c) and (25c) are the backward and forward preservation probabilities of Pease (1985).

Probability of Preservation during an Interval.—The probability of preservation depends on whether a lineage (1) spans the entire interval, (2) crosses into it from below and terminates within it, (3) originates within it and extends beyond it, or (4) originates and becomes extinct during the interval. For lineages that span the entire interval, we have

$$P_{D|bt} = 1 - e^{r_i \Delta t}. \tag{26}$$

For lineages that originate before the interval and terminate within it, the probability of preservation depends on how far into the interval they extend, which is a function of the extinction rate. Thus

$$P_{D|bL} = \frac{\int_0^{\Delta t} q_i e^{-q_i T}(1 - e^{-r_i T}) \, dT}{1 - e^{-q_i \Delta t}}. \tag{27a}$$

The denominator in this equation is a normalization reflecting the probability of extinction during the interval if the lineage is extant at the start. This equation reduces to

$$P_{D|bL} = \left\{ \frac{[r_i + q_i e^{-(q_i+r_i)\Delta t}]}{q_i + r_i} - e^{-q_i \Delta t} \right\} \Big/ (1 - e^{-q_i \Delta t}). \tag{27b}$$

For lineages that originate during the interval and extend beyond it, the preservation probability is analogous to the foregoing, except that the origination rate is the relevant parameter:

$$P_{D|Ft} = \frac{\int_0^{\Delta t} p_i e^{-p_i T}(1 - e^{-r_i T}) \, dT}{1 - e^{-p_i \Delta t}}. \tag{28a}$$

This reduces to

$$P_{D|Ft} = \left\{ \frac{[r_i + p_i e^{-(p_i+r_i)\Delta t}]}{p_i + r_i} - e^{-p_i \Delta t} \right\} \Big/ (1 - e^{-p_i \Delta t}). \tag{28b}$$

For lineages that originate and become extinct during the same interval, we need to consider the density of origination at any point during the interval (which is uniform only if $p_i = q_i$). This is obtained by multiplying the origination rate p_i by the standing diversity at time T within the interval, $N_b e^{(p_i-q_i)T}$, and normalizing by the total number of single-interval lineages, N_{FL}. The density of origination is then weighted by the density of a given duration, which depends on the extinction rate q_i, and the probability of preservation given that duration. Thus

$$P_{D|FL} = \int_0^{\Delta t} \frac{N_b p_i e^{(p_i-q_i)T}}{N_{FL}} \left[\int_0^{\Delta t-T} q_i e^{-q_i x}(1 - e^{-r_i x}) \, dx \right] dT. \tag{29a}$$

If $p_i = q_i$ this is equal to

$$P_{D|FL} = \frac{N_b p_i}{N_{FL}} \left[\frac{r_i \Delta t}{p_i + r_i} - \frac{1 - e^{-p_i \Delta t}}{p_i} - \frac{p_i(1 - e^{-(p_i+r_i)\Delta t})}{(p_i + r_i)^2} \right]. \tag{29b}$$

If $p_i \neq q_i$ this is equal to

$$P_{D|FL} = \frac{N_b}{N_{FL}} \left\{ \frac{p_i r_i[e^{(p_i-q_i)\Delta t} - 1]}{(q_i + r_i)(p_i - q_i)} + \frac{p_i q_i e^{-(q_i+r_i)\Delta t}[e^{(p_i+r_i)\Delta t} - 1]}{(p_i + r_i)(q_i + r_i)} \right. $$
$$\left. - e^{-q_i \Delta t}(e^{p_i \Delta t} - 1) \right\}. \tag{29c}$$

In equations (29b) and (29c), N_b and N_{FL} are from equations (1b) and (6b).

4. Compound Probabilities of Preservation

Probabilities of preservation before, during, and after an interval can be combined to yield probabilities of observed ranges given true durations. Because I am interested in the relationship between true and observed durations and rates, I will not present equations expressing the probability of not being preserved at all. These are easily derived from the fundamental probabilities (see Pease 1985; Foote and Raup 1996; Foote 1997; Solow and Smith 1997). The probabilities depend on a particular span of time. To avoid ambiguities, I will use notations such as $P_B(t)$ to indicate, for example, the probability of preservation before a point in time when there is a time span t over which preservation can occur. Let $P_{\times|\times}$ denote the probability that a lineage is observed to cross a particular time line, given that the lineage truly crosses it. This probability is simply the product of the probabilities of preservation before and after:

$$P_{\times|\times} = P_B P_A. \tag{30a}$$

If the rates are constant and t_B and t_A are the time spans before and after the time line, then this is equal to

$$P_{\times|\times} = \frac{r^2}{(p+r)(q+r)}[1 - e^{-(p+r)t_B}][1 - e^{-(q+r)t_A}]. \tag{30b}$$

If the rates are constant and t_B and t_A are effectively infinite, then this reduces to

$$P_{\times|\times} = \frac{r^2}{(p+r)(q+r)}. \tag{30c}$$

In the time-homogeneous case, the foregoing expression gives the ratio of observed to true standing diversity at any point in time. Note that this is substantially smaller than the proportion of taxa preserved (see below).

Lineages that truly span an entire interval can appear a number of different ways. They can appear to span the entire interval, to cross the bottom boundary only, to cross the top boundary only, or to be confined to the interval. The corresponding probabilities are

$$P_{bt|bt} = P_B(t_b) \cdot P_A(w - t_t), \tag{31a}$$

$$P_{bL|bt} = P_B(t_b) \cdot P_{D|bt} \cdot [1 - P_A(w - t_t)], \tag{31b}$$

$$P_{Ft|bt} = [1 - P_B(t_b)] \cdot P_{D|bt} \cdot P_A(w - t_t), \quad \text{and} \tag{31c}$$

$$P_{FL|bt} = [1 - P_B(t_b)] \cdot P_{D|bt} \cdot [1 - P_A(w - t_t)]. \tag{31d}$$

Note that the relevant time span for $P_{D|bt}$ is Δt in all cases. The probabilities that a lineage will appear to cross the bottom boundary (whether or not it also crosses the top) or that it will appear to cross the top boundary (whether or not it also crosses the bottom) are given by

$$P_{b|b} = P_B(t_b) \cdot P_A(w - t_t + \Delta t) \quad \text{and} \tag{31e}$$

$$P_{t|t} = P_B(t_b + \Delta t) \cdot P_A(w - t_t). \tag{31f}$$

Lineages that truly cross only the bottom boundary of an interval can appear to cross that boundary or to be confined to the interval. The corresponding probabilities are

$$P_{bL|bL} = P_B(t_b) \cdot P_{D|bL} \quad \text{and} \tag{32a}$$

$$P_{FL|bL} = [1 - P_B(t_b)] \cdot P_{D|bL}. \tag{32b}$$

Similarly, lineages that truly cross only the top boundary can appear to cross that boundary or to be confined to the interval. The probabilities are

$$P_{Ft|Ft} = P_{D|Ft} \cdot P_A(w - t_t), \quad \text{and} \tag{33a}$$

$$P_{FL|Ft} = P_{D|Ft} \cdot [1 - P_A(w - t_t)]. \tag{33b}$$

Lineages truly confined to an interval can be preserved only in that interval. The corresponding preservation probability was given above as $P_{D|FL}$ (eq. 29).

5. Observed Numbers of Taxa in an Interval

Given the expected true numbers of taxa from section 1 above and the probabilities of preservation from section 4, it is easy to derive expressions for the number of taxa in various categories that are observed in an interval. For a taxon to appear to span the entire interval, it must have done so in reality. The number of taxa that are observed to cross both top and bottom boundaries is thus equal to

$$X_{bt} = N_{bt} P_{bt|bt}. \tag{34a}$$

A taxon that appears to cross the bottom boundary only may have crossed only that boundary in reality, or it may have spanned the entire interval. The number of taxa observed to cross the bottom boundary only is thus equal to

$$X_{bL} = N_{bt} P_{bL|bt} + N_{bL} P_{bL|bL}. \tag{34b}$$

Similarly, the number of taxa observed to cross only the top boundary is equal to

$$X_{Ft} = N_{bt} P_{Ft|bt} + N_{Ft} P_{Ft|Ft}. \tag{34c}$$

Taxa that are observed to be confined to the interval may truly have been confined to the interval, or they may in fact have crossed either or both boundaries. The number of observed single-interval taxa is thus equal to

$$X_{FL} = N_{bt} P_{FL|bt} + N_{bL} P_{FL|bL} + N_{Ft} P_{FL|Ft} + N_{FL} P_{D|FL}. \tag{34d}$$

The four fundamental classes of observed lineages can be combined in a number of ways to yield the total observed taxa crossing the bottom and top boundaries, the number of taxa first appearing (apparently originating) within the interval, the number last appearing (apparently becoming extinct) within the interval, and the total number of taxa known from the interval (including those preserved before and after but not during the interval). Thus,

$$X_b = X_{bt} + X_{bL}. \tag{35a.i}$$

Note that this is also equal to

$$X_b = N_b P_B(t_b) \cdot P_A(w - t_t + \Delta t). \tag{35a.ii}$$

Similarly,

$$X_t = X_{bt} + X_{Ft}, \tag{35b.i}$$

which is equal to

$$X_t = N_t P_B(t_b + \Delta t) \cdot P_A(w - t_t). \tag{35b.ii}$$

Finally,

$$X_o = X_{bL} + X_{FL}, \tag{35c}$$

$$X_e = X_{Ft} + X_{FL}, \quad \text{and} \tag{35d}$$

$$X_{tot} = X_{bt} + X_{bL} + X_{Ft} + X_{FL}. \tag{35e}$$

Section 2 discussed methods for estimating taxonomic rates that use only boundary-crossing lineages. The observed numbers of boundary crossers, X_b, X_t, and X_{bt}, are given by simple expressions in the case of constant rates:

$$X_b = N_b \frac{r^2}{(p+r)(q+r)}[1 - e^{-(p+r)t_b}][1 - e^{-(q+r)(w-t_t+\Delta t)}], \quad (36a)$$

$$X_t = N_t \frac{r^2}{(p+r)(q+r)}[1 - e^{-(p+r)(t_b+\Delta t)}][1 - e^{-(q+r)(w-t_t)}], \quad (36b)$$

and

$$X_{bt} = N_{bt} \frac{r^2}{(p+r)(q+r)}[1 - e^{-(p+r)t_b}][1 - e^{-(q+r)(w-t_t)}]. \quad (36c)$$

As the interval in question gets farther from the beginning or end of the window of observation, the exponential terms in these equations become less important. When the interval is far enough from the edges that t_b and $w - t_t$ are effectively infinite, the exponential terms vanish and we have

$$X_b = N_b \frac{r^2}{(p+r)(q+r)}, \quad (37a)$$

$$X_t = N_t \frac{r^2}{(p+r)(q+r)}, \quad \text{and} \quad (37b)$$

$$X_{bt} = N_{bt} \frac{r^2}{(p+r)(q+r)}. \quad (37c)$$

From these it follows that

$$X_{bL} = N_{bL} \frac{r^2}{(p+r)(q+r)}, \quad \text{and} \quad (37d)$$

$$X_{Ft} = N_{Ft} \frac{r^2}{(p+r)(q+r)}. \quad (37e)$$

If we add the further constraint that $p = q$, then we have simple relationships for X_o, X_e, and X_{FL} as well:

$$X_o = X_e = N_b p \Delta t \frac{r}{p+r} = N_b q \Delta t \frac{r}{q+r} \quad \text{and} \quad (38a)$$

$$X_{FL} = N_b \left[p \Delta t \frac{r}{p+r} + (e^{-p\Delta t} - 1) \frac{r^2}{(p+r)^2} \right]. \quad (38b)$$

Note that $[r/(p+r)]$ $(= [r/(q+r)])$ in this case is simply the proportion of lineages preserved (Solow and Smith 1997).

Equations (37a) through (37e) have important implications. The ratios of observed numbers of boundary crossers are identical to the ratios of actual numbers of boundary crossers. If preservation rate is constant and the rates are calculated for an interval that is far from the beginning or end of the window of observation, the rate estimates of equations (22) and (23) are unaffected by incompleteness of the fossil record. This result is related to fact that the observed age distribution of taxa exclusive of singletons is identical to the true age distribution in the time-homogeneous case (Foote and Raup 1996; Foote 1997). Several authors have advocated excluding single-interval taxa from measures of taxonomic rates (Pease 1985; Sepkoski 1990; Alroy 1996b, 1998, 1999; Harper 1996), and many have advocated measuring diversity as the number of taxa crossing time lines rather than the number accumulated over an interval (Raymond and Metz 1995; Alroy 1996b, 1998, 1999; Bambach 1999). Whether the exclusion of singletons is seen as an adjustment of a conventional metric (Harper 1996) or part of a less conventional approach to rate estimation, the practice has much to recommend it, at least in the time-homogeneous case. The text explores behavior of the proposed metric and other metrics when taxonomic rates and preservation rates vary and when the interval is close enough to the beginning or end of the window of observation to experience a noticeable edge effect.

The singleton-free Van Valen metrics discussed above are also insensitive to rate of preservation in the time-homogeneous case when edge effects are absent. As shown earlier, however, these metrics yield inaccurate estimates of origination and extinction rates, and the degree of inaccuracy increases as the magnitude of rates or the difference between them increases.

Because the Van Valen metric and the per-taxon rate include singletons in the count of originations and extinctions, these metrics increase as the rate of preservation decreases. Combining equations (37a), (37b), and (38a) yields a simple and striking result. When origination and extinction rates are constant and equal and when there are no edge effects, the normalized Van Valen metric is equal to

$$V_{o/m.y.} = V_{e/m.y.} = p\frac{p+r}{r} = q\frac{q+r}{r}. \quad (39)$$

Since $r/(p+r) = r/(q+r)$ is the proportion of lineages preserved, the observed Van Valen metric in this special case is simply the true taxonomic rate divided by proportion of lineages preserved. Thus, in practice, the average ratio of $\hat{p}$ to $V_{o/m.y.}$ or of $\hat{q}$ to $V_{e/m.y.}$ may provide a measure of the completeness of the fossil record. To my knowledge, this possibility has not yet been explored in detail.

Taphonomy and paleobiology

Anna K. Behrensmeyer, Susan M. Kidwell, and Robert A. Gastaldo

Abstract.—Taphonomy plays diverse roles in paleobiology. These include assessing sample quality relevant to ecologic, biogeographic, and evolutionary questions, diagnosing the roles of various taphonomic agents, processes and circumstances in generating the sedimentary and fossil records, and reconstructing the dynamics of organic recycling over time as a part of Earth history. Major advances over the past 15 years have occurred in understanding (1) the controls on preservation, especially the ecology and biogeochemistry of soft-tissue preservation, and the dominance of biological versus physical agents in the destruction of remains from all major taxonomic groups (plants, invertebrates, vertebrates); (2) scales of spatial and temporal resolution, particularly the relatively minor role of out-of-habitat transport contrasted with the major effects of time-averaging; (3) quantitative compositional fidelity; that is, the degree to which different types of assemblages reflect the species composition and abundance of source faunas and floras; and (4) large-scale variations through time in preservational regimes (megabiases), caused by the evolution of new bodyplans and behavioral capabilities, and by broad-scale changes in climate, tectonics, and geochemistry of Earth surface systems. Paleobiological questions regarding major trends in biodiversity, major extinctions and recoveries, timing of cladogenesis and rates of evolution, and the role of environmental forcing in evolution all entail issues appropriate for taphonomic analysis, and a wide range of strategies are being developed to minimize the impact of sample incompleteness and bias. These include taphonomically robust metrics of paleontologic patterns, gap analysis, equalizing samples via rarefaction, inferences about preservation probability, isotaphonomic comparisons, taphonomic control taxa, and modeling of artificial fossil assemblages based on modern analogues. All of this work is yielding a more quantitative assessment of both the positive and negative aspects of paleobiological samples. Comparisons and syntheses of patterns across major groups and over a wider range of temporal and spatial scales present a challenging and exciting agenda for taphonomy in the coming decades.

Anna K. Behrensmeyer. Department of Paleobiology, National Museum of Natural History, Smithsonian Institution, MRC 121, Washington, D.C. 20560. E-mail: Behrensmeyer.Kay@NMNH.SI.edu
Susan M. Kidwell. Department of Geophysical Sciences, University of Chicago, 5734 South Ellis Avenue, Chicago, Illinois 60637. E-mail: skidwell@midway.uchicago.edu
Robert A. Gastaldo. Department of Geology, Colby College, Waterville, Maine 04901-4799. E-mail: ragastal@colby.edu

Accepted: 30 June 2000

What is Taphonomy?

The fossil record is rich in biological and ecological information, but the quality of this information is uneven and incomplete. The same might be said for many types of neobiological information, but in such cases, sampling biases are imposed by scientists and are explicable as part of a research design. With fossils, natural processes have done the sampling and created the biases before research begins. Taphonomy seeks to understand these processes so that data from the fossil record can be evaluated correctly and applied to paleobiological and paleoecological questions.

Efremov (1940: p. 85) first defined taphonomy as "the study of the transition (in all its details) of animal remains from the biosphere into the lithosphere," naming a field that we characterize more generally as "the study of processes of preservation and how they affect information in the fossil record" (Behrensmeyer and Kidwell 1985). Since the 1950s, the analysis of postmortem bias in paleobiologic data has been one of the prime motivations of the field, but taphonomy has always been a multi-tasking science (e.g., see historical reviews in Behrensmeyer and Kidwell 1985; Cadée 1991), and this remains true today. States of preservation of biotic remains are not only (1) indicators of how faithfully biological history has been recorded (issues of paleobiologic data fidelity and resolution), but are also (2) testaments to environmental conditions (the aegis of "taphofacies"), and (3) evidence of important aspects of biological evolution (skeletal and biochemical novelties, live/dead

0094-8373/00/2604-0005/$1.00

interactions and feedbacks), because organisms not only produce potential fossils but also are highly effective recyclers of plant and animal material. Strictly speaking, the logical limits of taphonomy are defined by its focus on processes and patterns of *fossil preservation*[1], but in practice, taphonomy serves a broader role in stimulating research on all types of biases affecting paleontological information, including those introduced by collecting, publication, and curation methods on the one hand, and stratigraphic incompleteness on the other (see also Lyman 1994; Donovan and Paul 1998; Holland this volume).

Taphonomy today is focused first and foremost on a geobiological understanding of the earth, grounded on the postmortem processes that recycle biological materials and affect our ability—positively and negatively—to reconstruct past environments and biotas. The classic flowchart of taphonomic transformations (Fig. 1) is now underpinned by a much fuller and quantitative understanding of interim states and pathways of fossilization, owing to an explosion of interest in the field since the early 1980s. Some of the most notable advances have been in (1) microbial, biogeochemical, and larger-scale controls on the preservation of different tissue types; (2) processes that concentrate biological remains; (3) the spatio-temporal resolution and ecological fidelity of species assemblages; and (4) the outlines of "megabiases" (large-scale patterns in the quality of the fossil record that affect paleobiologic analysis at provincial to global levels and at timescales usually exceeding ten million years). These advances are highlighted in this review because of their impact on paleobiologic analysis and their promise as research themes in the coming decades. Such advances reflect an increasingly ecumenical approach in terms of scientific methods (field measurements, ma-

nipulative experiments, analyses of synoptic data sets, probabilistic models) and scientific disciplines (tools and expertise from biogeochemistry, geomicrobiology, isotope geochemistry, geochronology, ecology, biomechanics, archeozoology, anthropology, sedimentology, sequence stratigraphy; see recent reviews and syntheses in Wilson 1988a; Allison and Briggs 1991a; Donovan 1991; Gifford-Gonzalez 1991; Lyman 1994; Brett 1995; Briggs 1995; Haglund and Sorg 1997; Claassen 1998; Martin 1999).

Taphonomy still is strongly oriented toward modern analogues as a means of identifying and quantifying processes, but increasingly exploits the stratigraphic record for hypothesis testing. Reliance on the fossil record to "bear its own witness" is an absolute necessity for some facies and taxa, but constitutes a powerful independent method even for environments and groups that are well represented in the Recent world. Regardless of subject, however, most taphonomists remain determinedly empirical in approach, dedicated to assembling baseline information on taphonomic patterns and processes. Such work usually targets individual fossil assemblages or modern analogues for particular groups of organisms (protists to vertebrates) and types of environments (glacial to abyssal plain). This fact-gathering focus is typical of a relatively new field of study, but a theoretical component also is beginning to develop, with proposals for general models for organic preservation (e.g., Lyman 1994; Kowalewski 1997). There have been a number of forays into the realm of taphonomic theory by paleobiologists seeking to distinguish sampling biases from biological patterns. These include attempts to account for preservational biases using assumptions of random preservation and "hollow curve" models for original taxonomic abundance as well as models that test the effects of incomplete fossilization, stratigraphic incompleteness, nonrandom distributions of facies and hiatuses, and blurring of generations by time-averaging on our ability to evaluate phylogenies, rates of evolution, and tempo and mode of speciation (Marshall 1990, 1994; Gilinsky and Bennington 1994; Foote

---

[1] It often falls to Taphonomy to answer the most basic of paleontological questions, "What is a fossil?" Material definitions concerning degree of mineralization and criteria based on age considerations are problematic for Holocene to sub-Recent organic remains. Hence, we prefer a more flexible definition: "A fossil is any nonliving, biologically generated trace or material that paleontologists study as part of the record of past life."

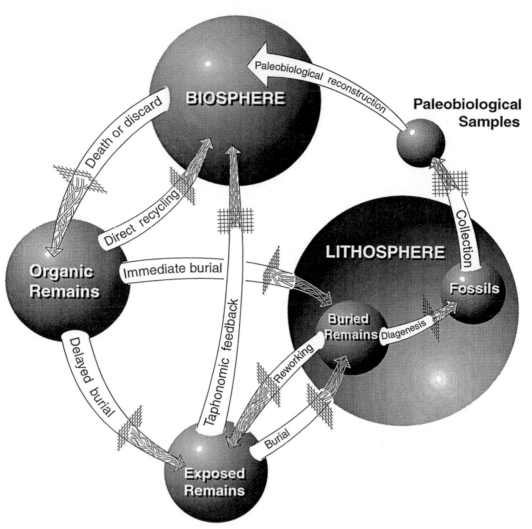

FIGURE 1. The main pathways for organic remains from death to paleobiological inference. Each path is affected by taphonomic processes and circumstances that filter the information as it passes to the next stage. Taphonomy is the study of how biological, chemical, and physical processes operating between each stage preserve or destroy organic remains and affect information in the fossil record (Behrensmeyer and Kidwell 1985).

1996; Foote and Raup 1996; Roopnarine 1999; Wagner 2000a; Simões et al. 2000a; and see papers in this volume by Wagner, Holland, Alroy et al., and Jablonski). Arguments about whether the lack of a fossil record is evidence for original absence (e.g., Vrba 1995; Foote et al. 1999; Valentine et al. 1999; Novacek et al. 2000) also draw upon underlying assumptions about how taphonomy works at a more general level.

Here we review highlights of taphonomic research from the past 15 years, since the tenth anniversary issue of *Paleobiology*, and suggest

some promising directions for the future. This review is organized by scales of processes in order to underscore two key points. One is the wide array of different qualities of the fossil record that paleobiologic analysis depends upon and that taphonomic analysis is relevant to; these qualities range from the preservation of DNA molecules to the analytic comparability of samples from disparate regions and geologic periods (Table 1). A second is the multidisciplinary nature of taphonomic analysis at all scales, illustrated by the variety of new techniques and lines of evidence that are

TABLE 1.   Research in taphonomy has demonstrated many sources of potential bias affecting qualities of the fossil record relevant to various paleobiological questions. (Adapted from Kidwell and Brenchley 1996.)

Aspects of quality in the fossil record	Sources of bias
Biochemical fidelity	Shifting of original compositions (e.g., isotopic and molecular) by diagenesis and metamorphism
Anatomical fidelity	Destruction or incomplete mineralization of soft tissues; disarticulation, fragmentation, recrystallization, and physical deformation
Spatial fidelity	Transport out of life position, rearrangement within life habitat, transport out of life habitat or biogeographic province (e.g., necroplanktonic organisms, pollen)
Temporal resolution	Mixing of noncontemporaneous remains within single sedimentary units via physical or biological processes (taphonomic time-averaging)
Compositional fidelity	Selective destruction/preservation of species, morphs, discarded body parts; bias from introduction of exotics and noncontemporaneous remains
Completeness of time series	Episodicity in deposition; taphonomic or diagenetic obliteration of fossils in surviving lithofacies (producing gaps and condensation of the record); poor preservation of some environments (deposits thin, localized, or readily eroded)
Consistency in preservation over Geologic time	Major shifts in intrinsic and extrinsic properties of organisms, including morphology and behavior in relation to other organisms—or shifts in the global environment, which can cause secular or long-term cyclic changes in preservation (megabiases)

now being brought to bear on both established and new issues in paleobiology.[2]

Highlights and Research Outlooks

Much of the progress in taphonomy has occurred via an environment-by-environment search for patterns and processes. Although many environments have not been explored fully, it is clear from available actualistic and stratigraphic studies that depositional context is extremely important in controlling the quality and nature of fossil preservation. Environmental setting determines such important factors as the likelihood of immediate burial, exhumation and reworking, the biogeochemistry of the early diagenetic environment, and the nature of the local community that generates or is capable of recycling tissues (i.e., is mortality typically attritional or catastrophic, are biominerals undersaturated or in surplus?). From such considerations the general taphonomic attributes of most major fossil-preserving facies now can be sketched as a framework for more detailed testing (e.g., ta-

bles in Gastaldo 1988; Kidwell and Bosence 1991; Speyer and Brett 1991; Behrensmeyer and Hook 1992; Martin 1999).

Detailed studies show that taphonomic systems are more complex than originally supposed, but many of these complications are shared across major environments and taxonomic groups, which is good news for data comparability and the potential for unifying theory. Contrary to the impressions given by basic paleoecology texts, some taphonomic features previously thought to be diagnostic of a particular taphonomic process or circumstance are now recognized as having a different dominant cause or resulting from multiple processes (the concept of "equifinality" [Lyman 1994]). A good example is the disarticulation and fragmentation of animal hardparts: a growing body of actualistic evidence indicates that, in both continental and marine settings, such damage is overwhelmingly biogenic (from predation, scavenging, etc.) rather than an index of physical energy (e.g., Haynes 1991; Jodry and Stanford 1992; Behrensmeyer 1993; Cadée 1994; Cate and Evans 1994; Lyman 1994; Oliver and Graham 1994; Best and Kidwell 2000a; for Cambrian exam-

---

[2] Taphonomy also provides guidelines concerning how humans can become fossilized. See Mirsky 1998 and Haglund and Sorg 1996 for user-friendly reviews.

ple see Pratt 1998). Moreover, in the absence of recycling metazoans, damage is dependent upon the state of decay of connective tissues rather than the distance of hydraulic transport (e.g., Allison 1986; Kidwell and Baumiller 1990; Greenstein 1991; Ferguson 1995). Similarly, rounding of hardparts is more likely to result from repeated reworking within a high-energy environment than from abrasion during long-distance transport, as demonstrated by comparing indigenous shells from beaches versus exotic shells in turbidites, or bones that have been trampled or chewed versus those transported in rivers (Behrensmeyer 1982, 1990; Potts 1988; Davies et al. 1989; Andrews 1990; Meldahl and Flessa 1990; Kidwell and Bosence 1991; Spicer 1991; Gastaldo 1994; Lyman 1994; Llona et al. 1999; Nebelsick 1999). Paleoanthropologists and archeologists have learned that many taphonomic agents, including humans, can cause similar patterns of bone modification, skeletal-part representation, and faunal composition; these patterns are heavily influenced by which bones and taxa are the most durable and identifiable in the face of destructive processes (Grayson 1989; Gifford-Gonzalez 1991; Lyman 1994).

A second complicating realization, derived primarily from experiments on marine macroinvertebrates, is that many taphonomic processes are inconstant in rate over time. Carcasses of regular echinoids, for example, fracture like live echinoids until microbial decay is sufficiently advanced for connective ligaments to be weaker than the calcite plates, a period of "ambiguous" behavior that lasts a few hours in tropical temperatures but days or weeks in cold water; once this decay threshold is passed, the disintegration of the test proceeds at a much faster rate than in pre-threshold specimens (Kidwell and Baumiller 1990). As a second example, the postmortem "disappearance," probably by dissolution, of aragonitic shells from early postlarval mollusks in Texas lagoons is very rapid initially but slows logarithmically, so that loss is best described as a taphonomic half-life (Cummins et al. 1986a,b). In contrast, the episodic movement of plant debris downstream, the alternating burial/exposure of shells on seafloors and the reworking of bones in channels all provide examples of non-

steady rates of postmortem modification, which are linked to chaotic aspects of the extrinsic environment and over long periods of time could appear to be linear.

Controls on the Preservation of Biological Remains

Most individual organisms never become fossils, but taphonomic research has discovered much about the circumstances that capture rich samples of past life. These samples may be quite different from those of living systems because of postmortem processes, but there is plenty to work with, whether the tissues of interest (1) are composed exclusively of volatile organics (e.g., nucleic acids, amino acids, simple sugars, starch; see Briggs this volume), (2) include refractory organics (lignin, collagen, cellulose, chitin, glycolipids, resins, sporopollenin [see Briggs 1993]), or (3) are mineralized during life (major biominerals are aragonite, calcite, apatite, various forms of silica).

Preservation depends on an array of processes and conditions operating at different scales (Fig. 1). These are

1. *the supply side of the equation*: rate of input, total volume, and composition (durability) of biological remains delivered to the environment;

2. *the nature of the pre-burial environment*: selectivity and intensity of modification by local physical, chemical and biological agents at the sediment-air or sediment-water interface. Modification may be destructive (as in the case of bioerosion, scavenging, and dissolution) or stabilizing (as in the case of bioencrustation and den/burrow formation;

3. *the rate (immediacy) and permanence of burial*, which determines how long tissues are exposed to processes operating on the sediment surface as opposed to those within the sedimentary column; and

4. *diagenetic conditions within the upper part of the sedimentary column* (highly dynamic mixed zone), where organic remains and sediments are still subject to bioturbation, meteoric effects, microbial processes, and possible physical reworking. Postburial modification may stabilize (e.g., mineral coatings, infillings, replacements) or reduce biochemical fidelity

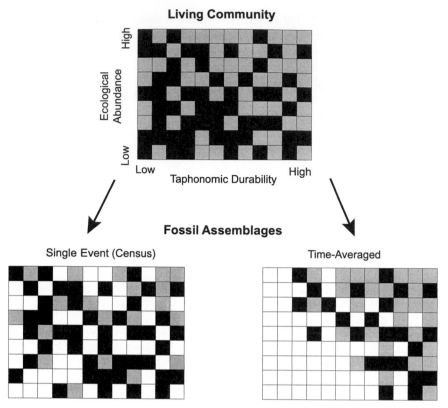

FIGURE 2. Schematic portrayal of information changes in two different types of fossil assemblages compared with a hypothetical living community at one point in time. Each cell is a species characterized by two variables, ecological abundance during life and taphonomic durability of the remains; the living community consists of equal numbers (50:50) of species that are permanent residents (black cells) and transients (gray cells, e.g., highly mobile forms or those on seasonal or longer-term population cycles). White cells in the fossil assemblages indicate species that are not preserved. The single-event assemblage (census) will capture most of the resident and some of the transient species (in this example, 80% and 37% of the cells (species) in the life assemblage, respectively) and is not strongly affected by taphonomic durability. In contrast, the attritional, time-averaged assemblage will be biased toward higher durability and more abundant species, capturing 44% of the resident and 52% of the transient species. A similar graphic model could be applied to all types of organisms or organic parts, with varying results for census versus time-averaged assemblages depending on the range of durability, type of community, environment, and length of time-averaging.

and anatomical detail (hydrolysis, continued maceration, dissolution, recrystallization).

Beyond these near-surface factors, which embed the remains in a consolidated sedimentary matrix, the long-term survival of fossilized material is determined by

5. *the fate of the larger sedimentary body.* The key factors here are strongly linked to tectonic setting, which determines rates of sediment aggradation and compaction, depth of eventual burial (and thus nature of later diagenesis and metamorphism), and structural deformation. The longest-surviving fossil-bearing sequences occur in stable cratonic margins or

interiors and in continental rift margins or aulocogens (failed branches of continental rifts) that have escaped tectonic recycling. Examples include Archaean and Proterozoic earliest life deposits (Grotzinger 1994; Walter et al. 1995) and Devonian through Carboniferous land plants and animals (Kidston and Lang 1920; Rolfe et al. 1994).

Fossil assemblages commonly are parsed according to the way remains initially accumulated in the depositional system, i.e., in terms of major types of supply-side input (Fig. 2). Attritional (time-averaged) assemblages reflect the release of discarded organic products (e.g., pollen, leaves) and input from normal mortality over periods of years to millennia. Single-event (census) assemblages reflect unusual events such as sudden anoxia, severe storms, pathogen outbreaks, droughts, and volcanic eruptions that kill large numbers of individuals at one time (minutes to months). In the former case, it may take considerable time and slow net sediment accumulation to amass a dense concentration of organic remains in a single bed in the absence of some other concentrating process; in the latter, this may happen literally overnight.

Soft-Part Preservation.—The taphonomy and diagenetic biogeochemistry of metazoan soft tissues and biomolecules have been the subject of intense field and laboratory study in the last 15 years (Allison and Briggs 1991b; Henwood 1992a; Briggs 1993; Briggs and Kear 1994a,b; Allison and Pye 1994; Westall et al. 1995; Bartley 1996; Briggs et al. 1997, 1998; Bartels et al. 1998; Davis and Briggs 1998; Duncan et al. 1998; Orr et al. 1998; Briggs this volume). Delicate molecules like DNA are extremely difficult to preserve, as would not surprise anyone who has struggled to extract good material from living organisms, and the oldest confidently identified DNA is less than 100 k.y. old (Bada et al. 1999; Wayne et al. 1999). Although cinematically fabled, amber is not favorable for DNA preservation. Resins are not airtight, and so generally only the most refractory portions of insects are fossilized (Stankiewicz et al. 1998; but see preservation of volatile structures in amber via dehydration [Henwood 1992a,b]).

Laboratory degradation of metazoan carcasses under oxic and anoxic conditions demonstrates the relative reactivities of tissue types, means of retarding decay, and absolute rates of decay. Such data are used not only to (1) rank tissue reactivities, but (2) rank rates of mineral precipitation in fossil specimens, (3) establish criteria for recognizing oxic and anoxic subenvironments (diagenetic minerals precipitate in distinctive Eh-pH fields created by different anaerobic microbial communities), and (4) provide absolute time limits on the contemporaneity of co-occurring fossils (Allison 1988; and see McGree 1984, and for plants Ferguson 1995). Apparently, only phos-

phatization proceeds rapidly enough to preserve undegraded volatile muscle and visceral tissues in three dimensions (including embryos [Xiao and Knoll 1999]), whereas calcite and pyrite are sufficient to preserve structures composed of more slowly decaying chitin, lignin, and collagen (Allison 1988; Allison and Briggs 1991b; Underwood and Bottrell 1994). Instances of successful in vitro precipitation of minerals provide further insights into the dynamics of fossilization of soft tissues: "phosphatization" may consist of (1) fine-grained 0.3-μ apatite that precipitates in the tissues themselves (subcell features preserved), (2) 1-μ apatite that replaces invasive bacteria and creates a fully 3-D pseudomorph of cells or tissues, and (3) comparable replacement but of *noninvasive* bacterial coats, which replicate only the outlines of cells or tissues (Wilby and Briggs 1997; and see Franzen 1985; Martill 1990; Xiao and Knoll 1999; also Taylor 1990 and Evans and Todd 1997 for replication of soft tissues by biological overgrowth).

Laboratory and field studies on animal soft-tissue preservation and konservat-lagerstätten have greatly elaborated and deepened our understanding of the multiple advantages of anaerobic decomposition. Anaerobic decomposition is in fact slower (Kristensen et al. 1995) and far less effective (efficient) in decomposing refractory material than aerobic decomposition, thus prolonging the window for preservation. If it is linked to low oxygen in overlying water, this excludes predators and scavengers and keeps them from destroying tissues before these can be encased by microbial mats (fostering local anoxic conditions) or become buried in sediment below the redox discontinuity level (Seilacher 1984; Seilacher et al. 1985; Wilby et al. 1996; Palaios 1999; and see Janzen 1977 on microbial strategies against metazoan scavenging). Even though anaerobic decomposition proceeds almost as fast as aerobic decomposition on volatile material, only anaerobic microbial processes liberate the appropriate cations and levels of alkalinity to precipitate early diagenetic phosphate, calcite, pyrite, siderite, and other minerals (Allison and Briggs 1991b). In terms of soft-tissue preservation, a little decay of organic matter is thus good (Allison 1988; Chafetz and Buczynski 1992) because, by depleting oxygen, the local chemical environment is driven to anaerobic conditions that favor mineral precipitation in and around the organics, which is essential to their long-term preservation.

A general model for superb soft-part preservation thus has the following requisites: (a) a carcass (microbe, catkin, worm, wombat) in good condition at time of death (death without significant morbidity or other damage to body parts); (b) postmortem isolation of the carcass from scavengers and physical disruption; (c) decomposition retarded until mineralization is accomplished; and (d) avoidance of later reworking (advantages of entombment within tree stump, incised valley fill, karst depressions, structural graben, or aulocogen [e.g., Lyell and Dawson 1853; Dawson 1882; Archer et al. 1991; Cunningham et al. 1993]). Points a–c can be accomplished via catastrophic burial (obrution), for example from ash falls and sediment avalanches on land and from various sedimentary processes in water (Baird et al. 1986; Demko and Gastaldo 1992; Wing et al. 1993; Crowley et al. 1994; Rolfe et al. 1994; Yang and Yang 1994; Downing and Park 1998; Brett et al. 1999; Feldmann et al. 1999; Hughes and Cooper 1999; Labandeira and Smith 1999).

However, contrary to stereotypes, enclosed water bodies having acidic, hypersaline, or anoxic conditions are highly effective environments for preservation *without* unusual sediment burial events, and in fact this is a more common means of konservat-lagerstätte formation (e.g., Seilacher et al. 1985; Whittington and Conway Morris 1985; Martill 1988; Barthel et al. 1990; Briggs and Crowther 1990; Brett and Seilacher 1991; Schaal and Ziegler 1992; Bartley 1996; Bartels et al. 1998). Heat and chemical transformation of volatiles to more refractory forms (e.g., charcoal, kerogen, graphite [Butterfield 1990; Lupia 1995; Vaughan and Nichols 1995]), before or after burial, is one means to lengthen the window of opportunity for mineralization (point c above). Alternative paths are acidity, which is antimicrobial, and pickling (subaqueous dehydration via salt); both can retard decomposition sufficiently for slowly polymerizing silica to replace subcel-

lular to tissue-grade structures with high fidelity (e.g., Knoll 1985; Scott 1990). Anoxia is a highly efficient agent of konservat-lagerstätte formation in aquatic systems when it affects overlying waters (accomplishing points a–c above). This may result from elevated temperature or organic matter overload (e.g., a phytoplankton bloom), and is the cause rather than the effect of metazoan mortality (e.g., Stachowitsch 1984). Although catastrophic death is important (for point a), it turns out that *mass* death is not a prerequisite for superb preservation of multiple individuals in one sedimentary layer, as evidenced by the wide spacing of metazoan specimens within classic konservat-lagerstätten (Seilacher et al. 1985). Even in overall aerobic environments, single carcasses can be highly effective in depleting oxygen from the immediate environment within sediment or under microbial mats, creating their own locally anaerobic conditions favorable to diagenetic mineralization (e.g., Schäfer 1972; Spicer 1980; Martill 1985; Baird et al. 1986; Allison et al. 1991b).

Taphonomic Feedback.—Both soft and hard biological remains can develop positive feedback systems that significantly enhance their own chances for preservation, especially where remains are densely concentrated. For example, skeletal hardparts can alter water flow dynamics, promote trapping and binding of sediment, and increase the erosion resistance of seafloors (Kidwell and Jablonski 1983; Seilacher 1985; Behrensmeyer 1990). Also, unusually high inputs of carcasses can overwhelm the capacity of normal recycling processes (e.g., scavengers faced with a surfeit of carcasses; oxygen depletion by dead organic matter in aquatic systems). Concentrations of remains also can create favorable diagenetic conditions, "self-buffering" local porewaters to reduce overall hardpart dissolution or promoting replacement of associated remains (e.g., Kotler et al. 1992; Schubert et al. 1997). Concentrations can have negative effects as well. For example, shell-rasping grazers become a major destructive force only where dead shells reach a critical abundance in tidal channels (Cutler 1989), and drought concentrations of animals around water holes focuses death and skeletal input but also may increase physical destruction and exhumation via trampling and digging (Haynes 1985, 1988, 1991).

Preservation of Mineralized and other Refractory Tissues.—Most aspects of this subject—the supply side, pre-burial effects, rates of burial, early diagenesis, and permanent incorporation into the stratigraphic record—have received heightened attention in the last 15 years, and we refer the reader to several excellent review volumes for details (Donovan 1991; Allison and Briggs 1991a; Lyman 1994; Martin 1999; Martin et al. 1999a).

A major focus for work on the postmortem sedimentology and biology of hardparts (i.e., biostratinomy) has been paleoenvironmental analysis (taphofacies analysis, sensu Speyer and Brett 1986; Parsons and Brett 1991), and these efforts are proving to have direct as well as indirect value to paleobiology. For benthic systems, for example, analysis of styles of fossil preservation and concentration reveal bed-by-bed and facies-level differences in key ecological factors such as frequency of storm reworking and oxygen levels that would be undetectable from inorganic matrix alone (e.g., Norris 1986; Parsons et al. 1988; Meyer et al. 1989; Brett et al. 1993; Ausich and Sevastopulo 1994). Determinations of the extent to which different marine and continental environments bear distinctive "taphonomic signatures" also provide a means of recognizing exotic, out-of-habitat material in fossil assemblages (e.g., Davies et al. 1989; Miller et al. 1992), or material reworked from older deposits that are ecologically or evolutionarily irrelevant to the host deposits (e.g., Argast et al. 1987; Plummer and Kinyua 1994; Trueman 1999).

Insights from hardpart condition derived from analysis of death and fossil assemblages increasingly are complemented by actualistic experiments on rates and controls on modification. This work provides insights into recycling processes themselves, for example, the huge importance of organisms as agents of skeletal transport and modification in modern systems (described above; thus affecting the limits to back-extrapolation over geologic time) and of factors intrinsic to the hardpart producers themselves (i.e., the roles of body

construction and ecology in determining different hardpart fates in the same environment; "comparative taphonomy" of Brett and Baird 1986, with implications for differential representation of taxa sensu Johnson 1960). For example, in modern shallow marine settings, mollusk shell fragmentation commonly varies independently of water energy or bears direct evidence of being the product of predators and scavengers rather than physical environment itself (Feige and Fürsich 1991; Cadée 1994; Cate and Evans 1994; Best and Kidwell 2000a), and branching colony form among scleractinian corals significantly increases postmortem disintegration relative to massive and encrusting forms (Greenstein and Moffat 1996; Pandolfi and Greenstein 1997a; and for bryozoans see Smith and Nelson 1994).

Most actualistic studies of this type previously focused on variation among taxa within a single major group, or variation among environments for a single taxon, but now include benthic foraminifera (Martin et al. 1999b), gastropods (Walker 1989, 1995; Taylor 1994; Walker and Voight 1994), bivalves (Parsons 1989; Meldahl and Flessa 1990; Parsons and Brett 1991; Cutler and Flessa 1995; Best and Kidwell 2000b), echinoids (Greenstein 1993; Nebelsick 1995), crinoids (Meyer and Meyer 1986; Llewellyn and Messing 1993; Silva de Echols 1993; Baumiller et al. 1995), brachiopods (Daley 1993; Kowalewski 1996a), and various shell-encrusters (Bishop 1989; Walters and Wethey 1991; Lescinsky 1993, 1995; McKinney 1996). In continental settings, intensive work on rates of litter decomposition (Boulton and Boon 1991; Ferguson 1995) and on sources and signatures of macrofloral material in deltas and other organic-rich coastal environments (Gastaldo et al. 1987) provides a valuable basis for comparison with the stratigraphic record. Lab and field investigations have also targeted arthropods (e.g., Wilson 1988b; Henwood 1992a; Martinez-Delclos and Martinell 1993; Labandeira and Smith 1999; Wilf and Labandeira 1999; Smith 2000; Labandeira et al. in press), fish, birds, and other lower vertebrates (Elder and Smith 1984; Smith et al. 1988; Wilson 1988c; Oliver and Graham 1994; Blob 1997; Stewart et al. 1999; Llona et al. 1999), and mammals including humans (Frison and Todd 1986; Haynes 1988, 1991; Fiorillo 1989; Andrews 1990; Blumenschine 1991; Behrensmeyer 1993; Kerbis Peterhans et al. 1993; Sept 1994; Tappen 1994a; Haglund and Sorg 1997; Cruz-Uribe and Klein 1998; Cutler et al. 1999). Zooarcheologists and paleoanthropologists have contributed important actualistic research linking damage patterns to taphonomic processes in their efforts to distinguish human from nonhuman bone modification and assemblage formation. Over the past 15 years, zooarcheologists have made important advances in characterizing bone modification patterns for specific taphonomic agents and developing more accurate methods for analyzing skeletal-part ratios (e.g., inclusion of limb shaft fragments, which were formerly omitted from such analyses, has a significant impact on archeological inferences [Bartram and Marean 1999]). This work is featured in some major volumes (Bonnichsen and Sorg 1989; Solomon et al. 1990; Hudson 1993; Lyman 1994; Oliver et al. 1994) as well as individual field and laboratory studies of hyenas (Blumenschine 1986, 1988, 1991; Marean 1992), lions (Dominguez-Rodrigo 1999), predatory birds and small mammals (Andrews 1990; Cruz-Uribe and Klein 1998; Stewart et al. 1999), and other pre- and postdepositional processes (Lyman 1985, 1994; Noe-Nygaard 1987; Marean et al. 1991, 1992; Tappen 1994b).

Actualistic studies of hardpart modification are also determining the security—and pitfalls—of "traditional" paleontologic inferences about spatial resolution and time-averaging of skeletal assemblages (and see next section). Among the questions amenable to experimentation and measurement are, How far are fossilizable materials transported outside the original life habitat? What proportion of material is moved (how great is the dilution factor for indigenous material in the ultimate host deposit)? Over what periods can biological materials survive in various environments, how different are those periods, and to what extent can these periods of potential time-averaging be interpreted from fossil condition? These questions have generated research on possible "taphonomic clocks" of damage accrual (for individual specimens or assemblages overall), how such clocks vary among groups, and how they behave with elapsed time-since-death (do rates of deterioration decrease, increase, or remain steady for a specimen held under "constant" postmortem conditions)?

Comparisons of death assemblages with local live communities are one powerful means of assessing out-of-habitat transport and time-averaging that has been applied to many marine and continental groups (reviewed by Kidwell and Flessa 1995; and next section). Direct dating of mollusk shells in death assemblages is increasingly used to explore time-averaging and taphonomic clocks in marine systems (e.g., Powell and Davies 1990; Flessa et al. 1993; Flessa and Kowalewski 1994; Martin et al. 1996; Meldahl et al. 1997a; Kowalewski et al. 1998), and the results (1) settled disputes on scales of time-averaging (commonly thousands of years even for intertidal and shallow subtidal assemblages, and tens of thousands on the open shelf, contrary to rapid rates of individual shell destruction that can be measured experimentally); (2) established the highly probabilistic and unsteady rather than monotonic accrual of damage with elapsed time-since-death (owing to erratic exposure to taphonomic agents); and (3) established the probabilistic nature of down-core stratigraphic ordering in shell ages (linked to relative rates of sediment aggradation and physical and bioadvection). A very promising direction of new research involves comparisons of major co-occurring taxa, such as mollusks versus benthic foraminifera (Martin et al. 1996; Anderson et al. 1997) and lingulid brachiopods (Kowalewski 1996a,b), where bioclasts have disparate postmortem durabilities and thus high potential for "disharmonious" scales of time-averaging.

For the paleobiologist collecting in the field, one of the most obvious taphonomic aspects of the record is the concentration of fossils in select beds or horizons and the nonrandom quality of fossil preservation. Work on this topic continues to be primarily stratigraphic rather than actualistic, and such studies consider (1) how concentrations are distributed with respect to gradients in biological input, sediment reworking, and net sediment accumulation; (2) whether such concentrations can be utilized for basin analysis (stratigraphic applications, including error-bars in biostratigraphy [see Holland this volume]); and (3) whether diverse concentration types have implications for the qualities of paleontologic information (e.g., positive versus negative effects of hiatuses in sedimentation, likely scales of time-averaging, and selective preservation). Much of this work is framed in a sequence-stratigraphic context and encompasses a range of continental (Behrensmeyer 1987, 1988; Dodson 1987; Eberth 1990; Behrensmeyer and Hook 1992; Gastaldo et al. 1993a; Rogers 1993; Smith 1993; Badgley and Behrensmeyer 1995; Smith 1995; Wilf et al. 1998; Rogers and Kidwell 2000) and marine settings (Kidwell 1991, 1993; Doyle and Macdonald 1993; Fürsich and Oschmann 1993; Ausich and Sevastopulo 1994; Brett 1995; Rivas et al. 1997; Kondo et al. 1998; Fernández-López 2000; also various papers in Kidwell and Behrensmeyer 1993). This and other research in the stratigraphic record is yielding new evidence for particular modes of accumulation, including predation (Wilson 1988c; Fernández-Jalvo et al. 1998; Andrews 1990), trapping (Richmond and Morris 1996), fluvial reworking (Schmude and Weege 1996; Smith and Kitching 1996), and drought-related or other types of mass death (Sander 1989; Rogers 1990; Fiorillo 1991; see also Eberth et al. 1999). Actualistic studies based on core samples in modern environments (e.g., Gastaldo and Huc 1992), as well as studies that track hardpart reworking under different energy and net-sedimentation conditions, would be valuable additions to continuing stratigraphic efforts.

In terms of future directions, marine studies have continued to focus on midlatitude settings, but attention to fully tropical settings is increasing. This includes both reefs and associated pure carbonate sediments (Miller 1988; Parsons 1989; Miller et al. 1992; Dent 1995; Perry 1996, 1999; Stoner and Ray 1996; Zuschin and Hohenegger 1998) as well as siliciclastic and mixed composition seafloors (Best and Kidwell 2000a), which rival carbonate sediments volumetrically on modern tropical shelves. In addition, many field surveys now include time-lapse experimental arrays over multiple years (Walker 1988; Callender et al. 1994; Best and Kidwell 1996; Walker et al. 1998; Parsons et al. 1999; Kennish and Lutz 1999) and in situ assessment of porewater geochemistry (Goldstein et al. 1997; Walker et al. 1997; Best et al. 1999). Rather than deducing early diagenetic conditions or taphonomic consequences, these can be measured directly, and the new multiyear rate information provides more explicit links to radiocarbon-calibrated studies of skeletal deterioration.

Finally, the microscopic modification and breakdown of mineralized microstructures before and during shallow burial—that is, "weathering" and early diagenesis—still receive relatively little attention, notwithstanding their huge importance in recycling biological materials. These relatively ordinary and pervasive processes are a counterpart to the extraordinary processes that preserve soft tissues, and deserve the same highly focused level of geologic, geochemical, and geomicrobiological analysis. Although the signatures of such processes may be less obvious than other kinds of damage, and require SEM to fully identify (e.g., Cutler and Flessa 1995), skeletal materials in all environments are subject to attack from some combination of the following: physical oxidation, hydrolysis, and UV light (especially in continental settings); microboring (by algae, fungi, larvae, etc. everywhere); microbial maceration (of microstructural organic matrix, in both aerobic and anaerobic conditions); and dissolution (of mineral phase within hardparts; including back-precipitation and recrystallization of minerals, which may reset isotopic ratios [Budd and Hiatt 1993]). Limited actualistic work to date indicates strong environmental differences in rates and specific pathways, but except in cases of rapid permineralization (e.g., Downing and Park 1998), hardparts generally become less resistant to destruction during reworking and time-averaging. For example, compared with bones in dry, highly seasonal, savannah settings (Behrensmeyer 1978a; Lyman and Fox 1989), those in rainforests appear to weather more slowly but are soft and spongy from the activities of bioeroders such as fungi (Kerbis Peterhans et al. 1993; Tappen 1994b; and see Cadée 1999 for supratidal example). Degradation in temperate and arctic settings generally is slow (Noe-Nygaard 1987; Andrews and Cook 1989; Sutcliffe 1990), indicating that bones on the surface have a longer opportunity for burial in cold environments. Bone weathering stages based on actualistic studies (e.g., Behrensmeyer 1978a) have been applied to the fossil record (e.g., Potts 1986; Fiorillo 1988; Cook 1995) with some success, although distinguishing primary weathering damage from similar features (e.g., cracking) acquired after burial or during diagenesis can be problematic.

In marine settings, there is growing evidence that microbial processes are at least as important as physico-chemical ones in the "dissolution" of molluscan shell both on the seafloor and during shallow burial, preferentially attacking organic-rich microstructures and proceeding at similar rates in both anaerobic and aerobic settings (Poulicek et al. 1988; Cutler and Flessa 1990; Glover and Kidwell 1993; Clark 1999; for brachiopods see Emig 1990; Daley 1993; Daley and Boyd 1996). Body size clearly has a strong effect on the preservation of macrobenthic shells (as also among continental bone assemblages [e.g., Behrensmeyer and Dechant Boaz 1980]), and there is growing evidence that rate of shell disintegration declines over time during early diagenesis (Cummins et al. 1986b; Glover and Kidwell 1993). Thus the dynamics of "loss budgets" may be complex. For example, in an innovative and highly influential set of field experiments in Texas lagoons, Cummins et al. (1986b) documented taphonomic "half-lives" as short as 60 days for mm-scale postlarval shells, suggesting very high rates of carbonate shell recycling (and see Staff et al. 1985, 1986). However, they subsequently calculated that nearly all shell carbonate produced in those sediments must be preserved to obtain the observed shell content in the long-term record; that is, virtually all of the larger shells that constitute the bulk of the skeletal biomass produced by the live community must survive (Powell et al. 1992). Other actualistic and stratigraphic evidence shows that the molluscan fossil record is time-averaged but relatively high-fidelity (various live/dead studies and paleontologic analyses based on the $> \sim 2$-mm

size fraction; for review and synthesis see Kidwell and Bosence 1991), and the Texas study supports this by illuminating how biological information can be captured in shelly death assemblages, even under conditions that may seem unfavorable on the basis of short-term loss rates.

Spatial and Temporal Resolution

Postmortem import and export of remains to an accumulation site, and the mixing of multiple generations of organisms and/or communities during time-averaging, determine the spatial and temporal resolving power of a fossil assemblage. Along with the differential destruction of species that occurs during these processes, space- and time-averaging of organic input also affect compositional fidelity of a fossil assemblage (Fig. 2). In this paper, "fidelity" refers to how closely (faithfully, accurately, truthfully) the fossil record captures original biological information, be it spatial patterning or the presence/absence and relative abundances of species; and "resolution" refers to the acuity or sharpness of that record, i.e., the finest temporal or spatial bin into which the fossil remains can confidently be assigned.

Although much more work is required for a full taxonomic and environmental picture, a taphonomic highlight of the past 15 years has been the tremendous advance in quantifying the magnitudes and selectivities of postmortem transport and time-averaging, both in modern systems and the stratigraphic record, using a diverse array of scientific methods for different groups (Figs. 3, 4). Some key hypotheses of paleontologic reconstruction, for example order-of-magnitude estimates of time-averaging, down-core stratigraphic mixing of cohorts, and how damage accrues over elapsed time, have been tested directly via radiometric and other dating of modern death assemblages (particularly the series of papers on molluscan assemblages of the Gulf of California; citations in preceding section). Both time-averaging and its relationship to "spatial averaging" also have been explored productively via probabilistic modeling (various authors in Kidwell and Behrensmeyer 1993).

The expanding baseline of information on

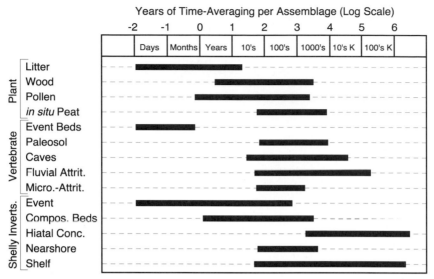

FIGURE 3. Estimated limits on time-averaging of selected types of continental plant tissues and vertebrate and marine invertebrate assemblages. The different categories (tissues versus deposits) reflect the fact that paleobotanists regard tissue type as playing the most important role in time-averaging for plant remains, while paleozoologists regard depositional environment or process as more important. Modified from Kidwell and Behrensmeyer 1993.

modern and ancient systems also is fostering conceptual models, such as the reciprocal nature between the durability of remains and their likely temporal and spatial acuity (Kowalewski 1997), and how the attributes of temporal, spatial, and compositional fidelity vary independently. For example, a mass-burial event from which mobile species and adults escaped can produce an assemblage with high time- and space-resolution but low compositional fidelity (e.g., Fig. 5), whereas if the live community is transported en masse during the fatal event (e.g., avalanching, turbidity currents), the temporal resolution and ecological fidelity of the assemblage may be high but spatial fidelity very low. Alternatively, a time-averaged assemblage in which hundreds or thousands of years of input are mixed (relatively low temporal resolution) may nonetheless contain virtually all *preservable* species that lived in the area over that time, and perhaps even in fairly accurate proportions (thus facies-level spatial resolution, and high ecological fidelity of a durable subset of the original community) (Fig. 5).

Spatial Fidelity.—Although the presence of a taxon in a fossil assemblage suggests that it occupied that site, particularly if "rooted" in life position (the highest possible spatial fidelity and resolution), biological remains can be transported out of their original life habitats, thereby becoming allochthonous and potentially problematic from the standpoint of paleocommunity reconstruction. Allochthonous or "exotic" wind-dispersed spores and pollen can account for high proportions of taxa in some samples, especially in areas with little local vegetation (e.g., middle of large lakes, offshore marine environments, and ice [Farley 1987; Calleja et al. 1993; Traverse 1994]). In contrast, animal-pollinated pollen, leaves, and other macroscopic phytodebris are relatively heavy and their records tend to have quite high spatial fidelity, although depositional context must be considered. For example, on temperate and tropical forest floors, actualistic tests indicate that litter sampled at any one point is derived largely from the surrounding 1000–3000 m² of vegetation (Burnham et al. 1992; and see Gastaldo et al. 1987; Burnham 1989, 1993, 1994; Meldahl et al. 1995; and for pollen see Jackson 1994). Such easily degraded material must be buried quickly to be preserved, but careful sampling of preserved spatial associations of taxa can capture extremely high-resolution macrofloral records

A. Continental Depositional Settings

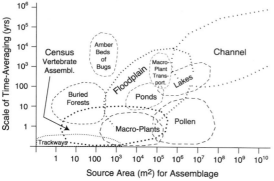

B. Benthic Marine Depositional Settings

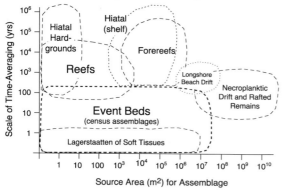

FIGURE 4. Spatial and temporal representation in fossil assemblages for different major groups of organisms, in continental and benthic marine depositional settings. A, Continental settings: dotted lines show areas of the time/space plot occupied by vertebrate remains, dashed lines plant remains; estimate for pollen excludes trees because certain morphotypes can be transported hundreds of miles by water (e.g., Farley 1987) or thousands of miles by wind (e.g., Calleja et al. 1993) prior to settling from the water or air column, respectively. B, Benthic marine settings include shelly macroinvertebrates and exclude nektonic and planktonic contributions to the fossil assemblage, because spatial resolution of these components can depend upon current drift.

do et al. 1987; Jackson 1989; Traverse 1990; Burnham 1990; Thomasson 1991; Webb 1993).

The skeletal hardparts of vertebrates and benthic invertebrates almost always are preserved out of life position, but actualistic studies indicate that out-of-life-habitat transport generally affects relatively few individuals in a given fossil assemblage (see reviews by Rich 1989; Allen et al. 1990; Behrensmeyer 1991; Kidwell and Bosence 1991; Behrensmeyer and Hook 1992; Kidwell and Flessa 1995; also specific studies by Behrensmeyer and Dechant Boaz 1980; Miller 1988; Miller et al. 1992; Nebelsick 1992; Greenstein 1993; Stoner and Ray 1996; Anderson et al. 1997; Flessa 1998; Cutler et al. 1999). Again, depositional context is crucial in determining spatial fidelity, and in providing warning flags for highly biased assemblages (Fig. 4). For example, in settings dominated by gravity-driven or surge transport of normal sediments, bioclasts may be *entirely* exotic in origin (e.g., in washover fans, tidal channels and their deltas, turbidites, base-of-slope settings). Organisms can be important transporters of biological remains, but there is great variation in the magnitude of this transport: some predators leave debris at the kill site; others concentrate it in a den or midden within the prey's life habitat (e.g., hyenas, crabs, fish, most owls), although in some cases prey remains end up outside of their life habitat (wolf dens, diving seabirds). Finally, the sprinkling onto seafloors of rocky intertidal shells rafted by seaweed (Bosence 1979), vertebrate debris from necroplanktic carcasses ("bloat and float" [Schäfer 1972]), downstream transport of bones (Behrensmeyer 1982; Dechant Boaz 1994; Aslan and Behrensmeyer 1996), and wind-transport of remains on land (Oliver and Graham 1994) can be highly effective modes of transport, but can be taphonomically subtle in terms of recognition and impact on the composition of fossil assemblage.

Natural history observations contribute to our conception of the possible, and net effects of transport on composition have been investigated by lab and field experiments (Behrensmeyer 1982; Frison and Todd 1986; Lask 1993; Prager et al. 1996; Blob 1997; and others previously cited; see also many live/dead com-

sufficient for detailed reconstructions of diversity and community interrelationships (Wing and DiMichele 1995; Gastaldo et al. 1993b, 1996, 1998; Davies-Vollum and Wing 1998). Moss polsters and small-diameter ponds collect pollen (both wind- and animal-pollinated types) from smaller areas of source vegetation than do large ponds (Jackson 1994). Fluvial channels and river deltas typically include—but usually are not dominated by—plant remains from upstream parts of their drainage basin (e.g., Scheihing and Pfefferkorn 1984; and see Collinson 1983; Gastal-

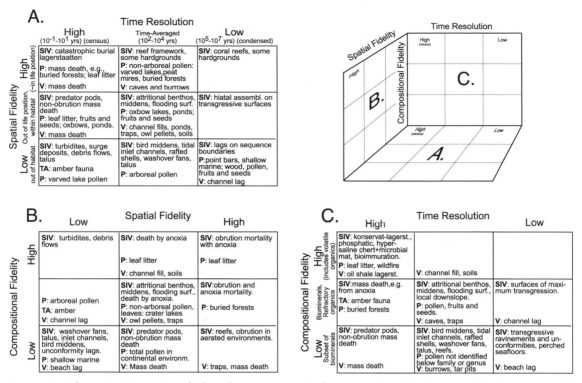

FIGURE 5. Schematic diagram in which each axis represents the summed results of preservational processes affecting time resolution, spatial fidelity, and compositional fidelity relative to the living community, showing that virtually all combinations of these aspects of record quality are possible. Note that we are restricting Compositional Fidelity to the accuracy by which the fossils represent the species present, species abundances, and population structure of the original community. In addition to the major sources of taphonomic bias represented in this diagram, some organic parts, such as pollen, wood, and bovid teeth, have inherent limitations with respect to taxonomic assignment, and these restrict the attainable level of fidelity even when preservation is excellent. SIV = shelly invertebrates, P = plants, V = vertebrates, TA = terrestrial arthropods (mostly insects).

parisons discussed in next section). It is difficult to retrodict or to reconstruct the effects of extinct organisms as agents, but overall, out-of-habitat postmortem transport does not appear to be an overwhelming taphonomic problem in ordinary depositional settings. Biological remains do not become homogenized in composition across broad environmental gradients either in modern or in ancient settings (see review by Kidwell and Flessa 1995).

Time-Averaging.—Because population turnover rates of individual taxa are less (often much less) than net rates of sediment accumulation, the biological remains of successive, noncontemporaneous populations of organisms may be admixed within a single bed, a concept first articulated by Walker and Bambach (1971). Multiple lines of evidence indicate that the *degree of time-averaging* within any

assemblage, i.e., the period of time represented by the biological components of any fossil assemblage, varies over many orders of magnitude (from virtually zero to millions of years; Fig. 3) and depends upon many factors (see papers in Kidwell and Behrensmeyer 1993; Kidwell 1998). These include (1) tissue types, (2) the habitat and specifically the frequency of burial events and exhumation events, and (3) the depth of bioturbation within the sedimentary column relative to net sediment accumulation rates (Fürsich and Aberhan 1990; Kidwell and Bosence 1991; Behrensmeyer and Hook 1992; Cutler 1993; Martin 1993; Kowalewski 1997). Robust hardparts can survive multiple reworking events, even within slowly accumulating sedimentary records (stratigraphic condensation), and can also survive exhumation and incorporation into quite different younger sedimentary deposits (stratigraphic leakage) (e.g., Cadée 1984; Henderson and Frey 1986). Highly disparate ages of co-occurring fossils generally are linked to settings of erosion (such as modern coastlines of Pleistocene or Tertiary strata [Wehmiller et al. 1995]), or prolonged low net sedimentation (e.g., modern sediment-starved continental shelves, where shells from 20,000 years ago to present co-occur in thin sedimentary veneers from most recent marine transgression [Kidwell and Bosence 1991; Flessa and Kowalewski 1994; Flessa 1998]). Such examples involve tissues that are particularly durable at death (mollusk shells, vertebrate teeth, pollen and spores) or have been made more durable by diagenesis during temporary burial (much vertebrate material associated with marine lags; steinkerns or concretions of delicate shells or refractory skeletons [Fürsich 1978; Kidwell 1991]).

Does time-averaging significantly impact our understanding of paleobiological systems? The answer to this question depends on the scale of time-averaging and the quality of information required to answer the paleobiologic question(s) at hand (Paul 1998). Time-averaged blurring of critical paleobiological events, such as the demise of the dinosaurs (Rigby et al. 1987; Argast et al. 1987), can have an obvious effect on evolutionary reconstructions. But there are many more subtle consequences of time-averaging. Catastrophic burial events such as volcanic ash falls (e.g., Voorhies 1992) can capture instantaneous samples of landscapes and organisms caught in the "wrong place at the wrong time." From a preservational perspective, the trapped invertebrates, vertebrates, and macroflora may provide highly "correct" spatial and proportional representations of the community at that spot and instant in time. However, these catastrophically trapped organisms may be admixed with (or at least deposited within very close stratigraphic proximity to) seeds, palynological components, and bones already present in the soil. Depending upon the type of soil and its maturity (Retallack 1990), these pre-event remains may represent populations that existed in the area prior to the catastrophic event, and/or populations that never lived at the locality (e.g., some wind-transported

pollen). Moreover, mass mortalities of animals rarely capture complete communities because agents of death often are taxon or age-class specific (Greenstein 1989), and organism size and mobility also are important factors (Krantz 1974). In aquatic systems, mass mortalities tend to capture unusual communities or communities in unusual states (e.g., anoxia is more common in small shallow bodies of water than in large open ones; mass death of single-species aggregations may occur following spawning events [Brett and Seilacher 1991]).

Because of the diverse ecological and taphonomic scenarios that are possible, time-averaging can have a number of effects on the species diversity and composition of fossil assemblages. However, for organisms that produce durable materials, the *usual* effect is to inflate diversity compared with what an ecologist would measure ("census") at any single moment (Fürsich and Aberhan 1990). For example, among 81 different data sets comparing live marine mollusks and dead shells from the same sediments, all contain two to ten times more dead species than species censused alive in the same habitat, even when the numbers of live *individuals* outnumber dead individuals (Kidwell in press; same phenomenon for vertebrates, see Behrensmeyer 1993). When additional live-censuses are taken and their species pooled, the known live fauna begins to more closely resemble the richness of the death assemblage. This demonstrates that, in contrast to any single, instantaneous census taken by an ecologist or captured by a catastrophic mortality, time-averaged death assemblages are fundamentally different types of samples of communities, summing biological input over longer periods (Peterson 1977; other examples in Kidwell and Bosence 1991). Other studies of marine and continental biomineralizing groups indicate that the probability of incorporation into a time-averaged death assemblage declines with tissue durability—from ~95% for shelled mollusks to ~75% for echinoids and land mammals and ~50% for marine decapods (Kidwell and Flessa 1995). These general relationships contrast with time-averaged assemblages of low-durability organisms, in which species richness may be significantly undersampled relative to the actual number of species that occur in the living community (e.g., leaf assemblages [Wing and DiMichele 1995]).

In very specific depositional settings, it is possible to use stratigraphic evidence or biological inference to constrain the absolute time over which biological remains have accumulated. The best circumstances are where there is high temporal resolution based on radiometric dates (e.g., Potts et al. 1999), or a natural cyclicity within the biotic system (e.g., seasonal deciduousness [Gastaldo et al. 1996]) or in sediment deposition (e.g., lacustrine varves [Bell et al. 1987; Wilson 1993; Wilson and Barton 1996; Briggs et al. 1998]). Variability in bone weathering stages has been used as an indicator of time-averaging (e.g., Potts 1986), and geochemical signals acquired early in diagenesis show promise as a way of calibrating relative degrees of time-averaging in vertebrate accumulations (Trueman 1999). Generally, however, paleontologists estimate the absolute and relative durations of time-averaging by a process of elimination (see papers in Kidwell and Behrensmeyer 1993; Kidwell 1998). Assemblages with a high proportion of life-positioned and/or articulated specimens, and especially those incorporating nonmineralized tissues with known rates of decay, can be categorized as snapshot-type census assemblages with minimal time-averaging (but see discussion above) (Fig. 2), whereas the opposite extreme of highly condensed or lag material may be recognized by the highly disparate diagenetic styles or biostratigraphic ages of co-occurring material and (usually) close association with a significant stratigraphic discontinuity surface (Fürsich 1978). Interpretation of material of intermediate-scale time-averaging, which accounts for the vast majority of land vertebrate and shelly invertebrate assemblages, depends on depositional context (Kidwell and Bosence 1991; Behrensmeyer and Hook 1992) and, less confidently, state of fossil preservation (see below). The less durable the material, the shorter the window for time-averaging and accrual of progressive damage. In fact, most assemblages of non-woody plant material and nonmineralized invertebrates have probably under-

gone very little time-averaging. In leaf assemblages it is even possible to infer greater time-averaging than is actually the case. For example, plant debris resting on a volcanic ash may represent canopy leaves shed as an immediate response to ash loading (Burnham and Spicer 1986), but could be mistaken for litter from a longer-term recolonization of the ash-fall deposit.

Paleobiologists have hoped to find a signature of degrees of time-averaging in the state of fossil preservation, but so far this has proved elusive. Although old shells are more consistently in poor condition than young shells (Powell and Davies 1990; Flessa et al. 1993; Meldahl et al. 1997a,b) and shell ages tend to increase down-core (Kershaw et al. 1988; Cutler and Flessa 1990; Flessa et al. 1993), neither the taphonomic grade (e.g., degree of abrasion or encrustation) nor the precise relative stratigraphic positions of skeletal remains in the sedimentary column are infallible criteria for reconstructing the ages of individual elements within molluscan assemblages. Individual shells within the same intertidal assemblage can vary in ^{14}C ages by more than 1000 years, quantifying the scale of time-averaging within a "bed," and the age range increases to ~20,000 years for assemblages from offshore subtidal areas (Flessa and Kowalewski 1994; Kowalewski et al. 1998). This is a consequence of overall robustness of molluscan shells compared with other shelly macroinvertebrates (Kidwell and Behrensmeyer 1993; Kidwell and Flessa 1995; Kowalewski 1996b) and can result in an "over-complete" record when net sedimentation rates are low—i.e., time represented by fossils is greater than that represented by sediment (Kowalewski 1996b; condensed assemblages of Fürsich 1978; Kidwell and Bosence 1991).

This contrasts with the relatively low durability of weathered bone material, which is less likely than fresh bone to survive to become fossilized in continental environments (Behrensmeyer 1978a). Most transported and/or attritional fossil bone assemblages consist of durable, unweathered elements such as teeth, jaws, and fragmentary limb parts, and the average state of fragmentation or disintegration is a poor index of the duration of surface exposure or the degree of time-averaging. However, high variability in weathering state, fragmentation, or abrasion in a single assemblage can indicate a complex taphonomic history, which should, on average, correlate with greater time-averaging. The relationship between bone damage variability and time interval of accumulation needs testing via comparative studies in both modern environments and the stratigraphic record. For example, weathering or abrasion features could be examined in concert with new chemical approaches to time-averaging in bone deposits, which suggest that variability in rare-earth elements is correlated with mixed spatial bone sources, hence greater time-averaging (Trueman and Benton 1997; Trueman 1999). The mixed preservational quality of a single type of shell also is taken as the best criterion for time-averaging within marine assemblages (Johnson 1960; Fürsich 1978; Kidwell and Aigner 1985; papers in Kidwell and Behrensmeyer 1993).

Such extrinsic and intrinsic time-averaging factors, along with analytical time-averaging (i.e., postcollection pooling of specimens from different sites or stratigraphic intervals [Behrensmeyer and Hook 1992]), reduce the resolving power of fossil assemblages for many paleobiological questions, especially those concerning species interactions, community composition, and fine-scale patterns of evolution, compared with what is possible in studies of modern biotas or fossil records dominated by census assemblages (i.e., macrofloral and nonmineralizing animals). Given the thousands of years of time-averaging that are apparently common within modern molluscan assemblages, for example, Kowalewski (1996b) has concluded that many paleobiological questions below a millennial timescale cannot be addressed (and see limits on reconstructing environmental change by Roy et al. 1996; Behrensmeyer 1982; Olszewski 1999; and see Martin et al. 1999b). Anderson et al. (1998) believe that it is possible to isolate shorter-term preservational and community trends in time, but the hardpart assemblage must be the product of episodic rather than continuous time-averaging.

Relationship between Space and Time.—Does time-averaging capture the long-term spatial variability of populations in an area? In other words, does time-averaging equal spatial-averaging? Given an environment characterized by time-averaged death assemblages, would one expect to find within a single-point sample a record of almost all the preservable taxa that ever occupied the environment (McKinney 1991). Time-space equivalence would depend on two conditions. One is that, over the period of time-averaging, the physico-chemical properties of the sample site must vary sufficiently to permit colonization by the entire range of organisms in the community. This condition will generally be met only for sites that are large relative to the size of the organism, for example hundreds of square meters for sessile invertebrates or plants. A second condition is that, following burial, all components of the spatially variable faunas or floras must have an equal chance for preservation. This condition could be met for taxa with similar kinds of organic or biomineralized remains. It is not possible at present to provide an answer to the time-space equivalency question, although the possibility is tantalizing for paleoecologists with good vertical but poor lateral exposures. Multiple spatial samples of time-equivalent fossil assemblages are needed to test this hypothesis; evaluating diversity over the sample area thus provides a way to assess potential spatial completeness. For example, Bennington and Rutherford (1999) used multiple, small samples across the exposure and then calculated cluster confidence intervals to estimate completeness.

Compositional Fidelity of Fossil Assemblages

In the last 15 years, taphonomists have applied a variety of research approaches to evaluating the compositional fidelity of fossil assemblages, i.e., the quantitative faithfulness of the record of morphs, age classes, species richness, species abundances, trophic structure, etc. to the original biological signal (e.g., Fig. 2). Research has included (1) extrapolations from laboratory and field measurements of rates of destruction of tissue types in modern systems; (2) deductive analyses of fossil assemblages, in which the preservational quality of individual specimens and sedimentary context are used to infer likely postmortem modification of taxonomic composition (informed by point 1); and (3) actualistic live/dead studies, in which the composition of a death assemblage (shells, bones, leaf litter, pollen) is compared with the local living community. These empirical approaches are complemented by probabilistic models and computer simulations aimed at testing both taphonomic and ecological (supply-side) controls on the nature of the record (Cutler and Flessa 1990; Miller and Cummins 1990, 1993; Behrensmeyer and Chapman 1993; Cutler 1993; and see Roopnarine et al. 1999 and Roopnarine 1999 for simulations of taphonomic effects on speciation patterns).

Most live/dead tests of fidelity have focused on single taxonomic groups in a limited suite of environments—as in ecological studies, there are logistical limits to the scope of an investigation. Methodological differences can make comparisons difficult across taxonomic and environmental divides (e.g., single versus multiple pooled censuses of the live community; visual survey versus sieving of upper sedimentary column for dead hardparts; methods of estimating individuals from collections of disarticulated and discarded body parts). However, we are beginning to develop a clear sense of how the construction and life habits of organisms and their postmortem environment combine to determine death assemblage fidelity for several systems (e.g., various papers in Martin et al. 1999a). There also is an increasing number of comparative taphonomic studies across taxonomic and environmental boundaries (e.g., Jackson and Whitehead 1991; Martin et al. 1996; Anderson et al. 1997). Virtually all of these live/dead studies have been concerned with the species compositions of assemblages, rather than with trophic group, age-class and morph composition, or population size (species abundance) (but see Cummins 1986a; Palmqvist 1991, 1993; Behrensmeyer 1993), and with numerical rather than biomass metrics (but see Behrensmeyer and Dechant Boaz 1980; Staff et al. 1985).

One of the clearest contrasts in fossil preservation—and thus in the fidelity of paleon-tological information—is between organisms having mineralized or highly refractory tissues and those lacking such materials ("soft-bodied taxa") (Fig. 2). Soft-bodied taxa have negligible preservation potential under ordinary environmental conditions, such as oxygenated seafloors and lake beds, and land surfaces characterized by moist and/or warm conditions (see earlier section), and the destruction of these organisms can represent a substantial loss in biological information. In marine sands and muds, for example, such taxa constitute 30–100% of species (Schopf 1978; Staff et al. 1986; Kidwell and Bosence 1991; Massé 1999; and for hardground example, see Rasmussen and Brett 1985), and in the macroflora a large proportion of the non-woody (herbaceous) species can be missing from litter samples (Scheihing 1980; Burnham 1989; Burnham et al. 1992). Thus, unless based on konservat-lagerstätten with census-level time resolution, most "reconstructions" of food webs and energy flows by paleoecologists differ fundamentally from those of living communities, and are useful only for comparison with similarly preserved (isotaphonomic) assemblages (Scott 1978; Behrensmeyer and Hook 1992) or simulations based on living communities (e.g., Behrensmeyer and Chapman 1993; Miller and Cummins 1993).

One clear pattern from existing studies is that there is tremendous variance in fidelity even among "preservable" groups, linked to the durability of their hardparts (for review see Kidwell and Flessa 1995). In the marine realm, this is a function of hardpart construction: mollusks and nonagglutinating benthic forams appear to have approximately equal durabilities and high fidelities, in terms of species representation, and are followed in decreasing order by scleractinian corals, echinoids, decapods, and agglutinating foraminifera (and for freshwater mollusks see Briggs et al. 1990; Warren 1991; Cummins 1994). There are few actualistic data for the postmortem durability of brachiopods (but see Daley 1993; Kowalewski 1996a) and bryozoans (but see Smith and Nelson 1994; Hageman et al. 2000), and no live/dead comparisons or direct age-dating for these phyla to our knowledge. Hence information for these and other groups remains largely based on inferences from the fossil record (e.g., lithology-specific diagenetic selection against small specimens of trilobites [Chatterton and Speyer 1997]).

Research on reef corals is expanding from analysis of damage styles (Scoffin 1992; Pandolfi and Greenstein 1997a) to evaluations of ecological fidelity (Greenstein and Pandolfi 1997; Pandolfi and Greenstein 1997b; Greenstein et al. 1998). Results so far are mixed: deepwater settings yield high live/dead taxonomic agreements like those for mollusks, whereas in shallow water, environmental zonation is preserved but taxonomic congruence is low, with strong underrepresentation of massive growth forms and overrepresentation of (rapidly growing) branching forms among the dead. Reef systems present quite different conditions for live/dead analysis than soft sediments: (1) dead specimens are commonly overgrown and thus more difficult to detect than live (see also this problem for reef-encrusting and -boring bivalves, where dead richness is *lower* than live richness, contrary to unlithified seafloors [Zuschin et al. 2000]); (2) resolution of corallite skeletal anatomy is essential for species-level identification of coral death assemblages, and thus systems with greater time-averaging (and thus potential for taphonomic modification) or higher proportions of fragile forms (e.g., Indo-Pacific versus Caribbean) will yield lower taxonomic fidelities; (3) among colonial organisms, the percentage of an "individual" that is alive or dead must be estimated rather than simply scored live/dead, and decisions must be made about how to count fragments on the seafloor, if at all; and (4) similarly, decisions must be made about whether dead material sieved from sedimentary pockets should be counted and how best to integrate this with live/dead data for in situ corals based on stretched-line scuba transects.

In the continental realm, fidelity among land mammals is strongly affected by body-size distributions within habitats, agent of accumulation, and climate. Natural-history observations and the few existing live/dead studies of bones on open land surfaces suggest that cool temperatures associated with high latitudes and altitudes promote longer

bone survival compared with low latitudes, and within each of these zones, dry land surfaces can be more favorable than moist ones (Behrensmeyer 1978a; Noe-Nygaard 1987; Kerbis Peterhans et al. 1993; Sept 1994; Tappen 1994a,b; Elias et al. 1995; Stewart et al. 1999). Thus, taxonomic fidelity should be greater in cool and/or dry climates where bones have higher preservation potential. For the continental macroplant record, Burnham (1989, 1993) demonstrated differences in systematic representation of taxa within modern forest-floor leaf litters and channel deposits, and also found that different depositional settings within the same regime provide dissimilar fidelities. Within the subtropical fluvial regime, for example, channel and channel margin (forebank) accumulations of leaves represent 13–47% and 38–48% of the riparian vegetation, respectively. Counterintuitively, autochthonous levee and floodplain settings adjacent to these primarily allochthonous assemblages may provide a fidelity record with as little as 29% of local vegetation represented (range from 29% to 58% depending upon sample site). Because arborescent plants along the river margin act as a barrier to the lateral movement of canopy materials, there is very little mixing among microhabitats. Different climatic regimes are characterized by different levels of fidelity, and using comparative work in subtropical, tropical, and temperate climates, Burnham developed ways to extrapolate and calculate credible values of standing taxonomic richness by applying the appropriate climatic factor (Burnham 1993; see also Gastaldo and Staub 1999).

Tests of the relative fidelity of macrofloral and pollen records underscore the limitations of any single type of paleoecological record and the benefits of a comparative approach (Gastaldo and Ferguson 1998). For example, Gastaldo et al. (1998) incorporated megafloral, carpological (fruits and seeds), palynological, and biogeochemical data to evaluate a late Oligocene abandoned fluvial channel. They demonstrated that leaf fossils recorded deciduous riparian plants; fruits and seeds not only confirmed the presence of riparian elements but increased alpha diversity nearly threefold because these body parts represented understory and herbaceous ground-cover plants that were not preserved as wood or leaves; palynological and palynofacies debris confirmed the presence of some, but not all, riparian taxa and added evidence for other local (algae) and regional components; and the biogeochemical data reflected variations in megafloral contribution to the channel. For other examples see Jackson and Whitehead 1991 and Ferguson 1995.

In addition to the need for quantitative estimates of fidelity in more environments and taxonomic groups, existing actualistic data sets could be examined for taphonomic "rules of thumb" applicable to the fossil record. For example, do fidelity levels improve as data are pooled from increasingly large geographic areas (i.e., within sample, within facies/habitat, within basin, within province) (Kidwell and Bosence 1991; Cutler 1991; Wing and DiMichele 1995; Olszewski and West 1997; Hadly 1999)? Temporally nested studies and simulations could reveal how stable death-assemblage composition is during the first few hundreds of thousands to one million years of burial (i.e., live versus dead in modern environments, uplifted Holocene strata, and/or Pleistocene fossils). The degree of fidelity might be expected to decline because of the cumulative wear and tear of diagenesis, limited outcrop areas for sampling different facies, changes in biogeographic range and community structure, and, eventually, extinction. However, data so far indicate that the agreement between live and dead floras/faunas can remain rather high over periods of a million years or more (e.g., Wolff 1975; Damuth 1982; Valentine 1989; Greenstein and Moffat 1996), so much so that it permits recognition of the uniqueness of recent environmental degradation (Greenstein et al. 1998). For example, Valentine (1989) reports that Pleistocene marine faunas in California include 77% of the living mollusk species from the Californian province, with most "live-only" species being numerically rare and restricted to deeper-water habitats that are not well represented in onshore Pleistocene outcrops. Data from other marine and continental groups could be similarly tested for sensitivity to geographic and temporal scale of analysis. All of these results underscore the potential value of death assemblages for environmental impact and other studies bearing on conservation biology (Powell et al. 1989; Davies 1993; and see review by Kidwell and Flessa 1995).

Several more fundamental difficulties still challenge our application of these insights to assessing fidelity in the older metazoan fossil record. One is the problem of evolutionary ecology: not only have the durabilities of animal hardparts changed over time (e.g., with changes in mineralogy, microstructure, body size, skeletal robustness), but organisms that interact with skeletal hardparts also have evolved. Such organisms are some of the most important agents of postmortem destruction in modern systems (e.g., shell and bone-crushing predators and scavengers, various bioeroders, sediment-irrigating bioturbators, and, of course, fungi and other microbes, whose roles and histories as biological recyclers may be important but still are difficult to assess [Robinson 1990, 1991]). How reasonable is it—and how far back into the past is it reasonable—to extrapolate present-day estimates of death-assemblage fidelity into the paleoecological past (i.e., taphonomic uniformitarianism)? The Cenozoic and Cretaceous may be within the reach of modern estimates of molluscan and scleractinian fidelity, but what do we do with older records? To move beyond arguments based on taphonomic uniformitarianism, for both Paleozoic and post-Paleozoic material, it is essential to determine whether death assemblages that appear to be highly biased in composition have distinctive damage patterns—i.e., to link taphofacies studies of damage (e.g., dead specimens of Species X are in poor condition, but those of Species Y are in good condition) to information on live/dead agreement. Greenstein (1999) has begun such work on reef corals, and this should be incorporated into other live/dead investigations.

Abundance of species in modern ecosystems is a key variable for characterizing diversity and various measures of dominance, and reconstructing such information from the fossil record is important for investigating the history of biodiversity. Taphonomic processes have the potential to alter abundances in many significant ways—for example, via differential destruction during time-averaging and because of different population turnover rates of local species—even if the import of exotic species is minor. Many live/dead investigations have generated adequate data to test agreement in rank order and relative abundance, but the numbers of studies are still too few for most groups to provide a credible basis for using abundance data in paleobiological reconstructions. For marine mollusks, meta-analysis indicates high live/dead agreements (Kidwell 1999; and for freshwater mollusks, see Briggs et al. 1990; Warren 1991; Cummins 1994). Quaternary lake deposits also provide a firm basis for assessing reliability in palynological abundance data. For the wind-pollinated part of plant communities, pollen assemblages are faithful recorders of plant relative abundances in the source area, especially when the forest is relatively homogeneous, but animal-pollinated plants are almost always grossly underrepresented because of low pollen yield per tree and because most of this large and heavy pollen falls very near the source tree (see reviews by Jackson 1994 and by Jackson and Overpeck this volume). In contrast, vertebrate paleontologists regard both relative abundances and rank-ordering of species as suspect (i.e., guilty of bias unless proven otherwise) (e.g., Badgley 1982, 1986; Barry et al. 1991; but see Behrensmeyer and Dechant Boaz 1980).

Finally, morphological fidelity of fossil populations is also a taphonomic concern, but has received relatively little work. Preserved morphologic variance might be affected in several ways, for example from time-averaging of multiple generations (broadening variance) and from differential destruction of fragile or small morphs (skewing variance or limiting recognition of true polymorphism) (Kidwell and Aigner 1985). Bell et al. (1987, 1989) provide a powerful empirical example where, because of the occurrence of both mass-mortality (census) and time-averaged assemblages in the same stratigraphic unit, the effects of time-averaging on morphological variance and character association could be evaluated entirely with fossil evidence. Capturing true pic-

tures of morphologic variability in fossil material, and distinguishing between taphonomic, sample-size, and biological controls on this variability, bears on issues of numbers of species and their stability in taxonomy. Thus, these considerations are critical to biostratigraphy, evolutionary analysis, and estimates of species richness (Hughes and Labandeira 1995). Increasing numbers of morphometric studies are based on taphonomically astute sampling, for example restricting samples to single bedding planes or horizons of constant and (hopefully) known time-resolution, and keeping close track of lithologic context to control for or assess ecopheny; but the question of bias in morphologic representation at present is still largely a qualitative assessment.

Megabiases

"Megabias" refers to bias in relatively large-scale paleobiologic patterns, such as changes in diversity and community structure over tens of millions of years, and variation in the quality of the record between mass and background extinction times or among different climate states, biogeographic provinces, and tectonic settings (Behrensmeyer and Kidwell 1985; and see treatments by Kowalewski and Flessa 1996; Martin 1999). Baseline information accumulated since then has stimulated new thinking on two reciprocal fronts: (1) broad-scale changes in climate, plate tectonics, ocean-atmosphere chemistry, and biological evolution as likely drivers of secular change in taphonomic processes (Fig. 6); and (2) the probable impact of such changes on the quality of paleontologic evidence used to reconstruct and parameterize geological and biological phenomena.

Given that Earth history can be divided into periods with different atmospheric and surface conditions, and given that the history of life also presents intervals with distinct bodyplans and life habits, it seems plausible that the geologic record would be characterized by a series of discernable "taphonomic domains" (Fig. 6). Reflecting secular changes in the nature of life and environments on Earth at a global scale, these domains would constitute the broadest-scale biases in the quality of paleobiological information. Superimposed on them would be province-scale and/or shorter-term secular and cyclical variation in taphonomic processes, and intervals of unique taphonomic conditions, e.g., those associated with regional or global mass extinctions and/or major perturbations in Earth's environmental condition. We refer to all such broad-scale taphonomic patterns and trends—affecting paleontologic analysis at provincial to global levels, over timescales of >10 m.y., or among major taxonomic groups—as megabiases.

Important taphonomic shifts may result from the evolution of organic form and behavior that makes organisms intrinsically more or less likely to fossilize, and from changes in extrinsic biotic and abiotic controls on preservation (Fig. 6). Examples of intrinsic changes include evolution in the composition and structure of mineralized skeletons, body size, mobility/life habit including burrowing behavior and pollination, deployment of life forms into new environments, and (for plants) evolution of deciduous versus perennial growth habits (for references, see discussion below). Examples of extrinsic biotic changes include the increasing depth of bioturbation through the Phanerozoic (Thayer 1983; Retallack 1990; Droser and Bottjer 1993; Buatois et al. 1998; McIlroy and Logan 1999), the evolution of more effective shell and bone crushers/ingestors (Vermeij 1977, 1987; Behrensmeyer and Hook 1992) and other biodegraders such as fungi (Robinson 1990, 1991), and the shift to detritivore and herbivore dominance on land (DiMichele and Hook 1992; Labandeira 1998) (Fig. 6). Extrinsic physical changes include fluctuations in the temperature and geochemistry of Earth's atmosphere and oceans (Berner and Canfield 1989; Maliva et al. 1989; Berner 1991; Martin 1995; Malinky and Heckel 1998; Stanley and Hardie 1998), and tectonic and climatic effects on the original extent and preservation of particular environments through geological time (e.g., the "wetlands bias" in the global plant record, variation in total sedimentary rock volume, the proportion of tropical continents, and lagerstätte-preserving lithographic limestone basins [Tardy et al. 1989; Sepkoski et al. 1991;

Gastaldo 1992; Allison and Briggs 1993a; Oost and de Boer 1994).

Although the potential impact of such large-scale processes has been recognized for some time (Efremov 1940), they remain largely unexplored aside from some aspects of the marine record. Various case studies illustrate the continuing debate over the relative importance of taphonomic versus biologic signals, and of intrinsic versus extrinsic taphonomic effects. For example, in the fossil record of unique events, such as the profound biotic changes at the Precambrian/Cambrian boundary, did multicellular life really "explode" about 550 m.y. ago, or are we simply seeing the opening of a new taphonomic window? This particular change appears to represent a linked shift in metazoan evolution and organic recycling. Given the extensive field work on late Precambrian deposits over the past two decades, it is very unlikely that shelly organisms existed in any abundance prior to the end of the Precambrian. The evolution of biomineralization indeed represents a major taphonomic event in the intrinsic preservation potential of multicellular organisms. On the other hand, debate continues over the role of taphonomic processes in the concomitant disappearance of the globally distributed soft-bodied Ediacaran fauna (Fedonkin 1994; McIlroy and Logan 1999). Researchers have hypothesized that, given the overall aerobic environments of deposition, some Ediacaran organisms had tougher body construction or ways of life that enhanced their intrinsic preservational characteristics, and that the geological disappearance is a signal of actual biological extinction (e.g., Seilacher 1984, 1994), whereas others have inferred a sharp decline at this time in the extrinsic environmental conditions that permitted the development and early cementation of microbial mats (Fedonkin 1994). Capable of preserving soft-bodied organisms as "death masks," this taphonomic mode existed in late Precambrian seas as long as effective grazers and bioturbators ("grave robbers") were absent, but disappeared when such organisms invaded this environmental zone (Gehling 1999). A comparable taphonomic mode persists in Recent hypersaline tidal flats and lagoons where most metazoans are excluded, but in normal marine environments

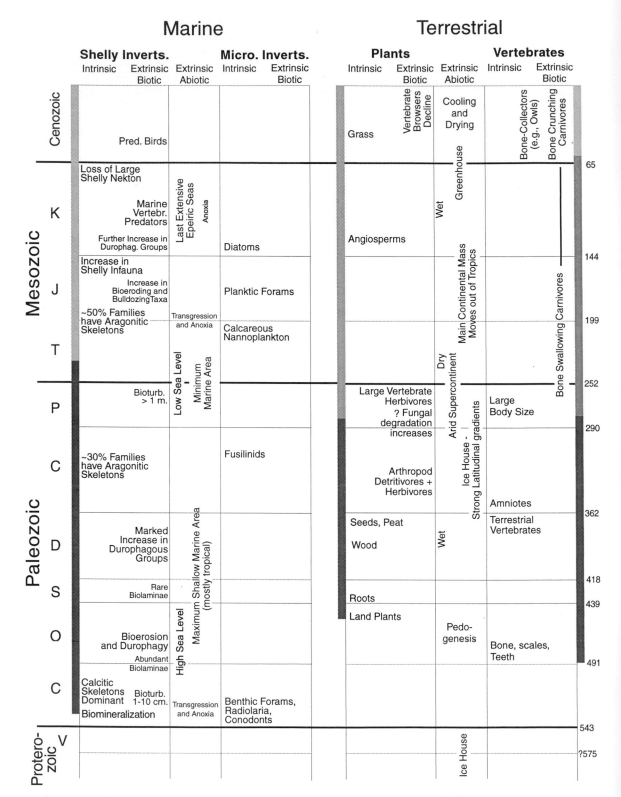

FIGURE 6.   Intrinsic and extrinsic changes with the potential for major effects on taphonomic processes and organic preservation over geologic time. This chart provides a preliminary framework for examining hypotheses concerning changing "taphonomic domains" through the fossil record for the marine and continental realms as well as possible

it became extinct as a means of preserving soft-bodied multicellular organisms at the beginning of the Cambrian (for related studies, see Sepkoski et al. 1991; Knoll et al. 1993; Knoll and Sergeev 1995; Kah and Knoll 1996).

Through the Phanerozoic, other proposed megabiases in the marine realm relate to both intrinsic and extrinsic biotic and abiotic factors. At the intrinsic end of the spectrum, styles of echinoid preservation and thus qualities of data appear to have diversified as echinoid constructional morphology diversified (Greenstein 1992); the frequency of lingulid brachiopod preservation has declined, possibly because of decreasing biomineralization during life (Kowalewski and Flessa 1996); and increases in the physical scale, taphonomic complexity and probable time-averaging of shell beds is linked to shelly macrobenthos acquiring biomechanically tougher hardparts and expanding into more energetic environments (Kidwell and Brenchley 1994, 1996; Li and Droser 1997; Simões et al. 2000b; and see Ausich 1997 for intrinsic factors in encrinites). In addition, aragonitic and calcitic biominerals may confer different degrees of resistance to predation as well as to postmortem destruction (Stanley and Hardie 1998). In contrast, primarily extrinsic factors have been invoked to explain the changing frequency of metazoan konservat-lagerstätten over the Phanerozoic (bioturbation, basin type, clay mineralogy) (Aronson 1992; Allison and Briggs 1993b; Butterfield 1995; Oost and de Boer 1994), and both intrinsic and extrinsic factors appear to have played a role in changing patterns of marine mineralization (both replacement of hardparts and early cementation of sediments) over Precambrian and Phanerozoic time (ocean saturation states, ecology and biomineralogy of target taxa, abundance of other organisms as elemental sources) (Walker and Diehl 1985; Knoll et al. 1993; Grotzinger

1994; Kah and Knoll 1996; Schubert et al. 1997).

The histories of sedimentary basins on timescales of 10^6–10^7 years could impart significant trends in the quality of fossil records, within the broader domains described above. Continental depositional systems could exhibit several distinct phases of organic preservation. For example, as a foreland basin changed from underfilled to overfilled, physical and chemical conditions should favor plant preservation in early phases (high water table, low oxidation) and vertebrate preservation in later phases (paleosols with $CaCO_3$, concentration of vertebrate remains through fluvial reworking) (e.g., as suggested in Demko et al. 1998; for analogous tectonic and sequence-stratigraphic variation in the quality of marine fossil records, see Kidwell 1991, 1993; Brett 1995). Climate change also should impose major shifts in the quality of the fossil record, with wet cool conditions favoring plant preservation, drier warmer times bone phosphate, and fluctuating CCD levels governing the preservation of deep-sea microfauna (Martin 1999; and for possible storm-bed effects, see Brandt and Elias 1989). This potentially affects the fossil record at a wide range of timescales, from regular fluctuations in preservation caused by orbital cycling, to longer-term shifts in latitudinal gradients that modify the extent of habitable and preservable biotic space, to the drift of continents across major global climate belts. Moreover, processes and circumstances favorable to preservation of one major group often are less favorable for others, resulting in potential temporal disjunctions between the marine and continental record, and the plant and land vertebrate record.

Taphonomic processes and circumstances associated with mass extinction events could constitute a recurring set of phenomena with a different set of biases relative to the intervals

---

←

links between these two realms. The points of inception of potentially important changes in intrinsic and extrinsic variables are indicated on the chart (see text for references). Shading variations on the vertical bars indicate possible taphonomic domains for shelly invertebrates based on depth of bioturbation, for plants based on the development of relatively refractory tissues and evolution within biodegraders such as fungi, and for vertebrates based on body size and changes in bone-processing capabilities of predators and scavengers. Revised dates for period boundaries from D. Erwin (personal communication 2000).

between such events. For instance, after a mass extinction event, a depleted array of consumers could lead to reduced biological recycling, allowing better preservation during the period of biotic recovery (Hunter 1994; Williams 1994; Cutler and Behrensmeyer 1996; D'Hondt et al. 1998). It is also possible that climatic or chemical stress prior to a mass extinction could affect taphonomic processes, either directly with increasing frequency of mass deaths or indirectly through the organisms, with poorer preservation indicating increased competition for scarce organic or biomineral resources. The record of mass death events also can be affected by long-term changes in taphonomic processes that control destruction and permanent burial. In many marine and continental settings where bioturbators or physical reworking can mix sediments over 10 to 100+ cm, instantaneous inputs from local mass death may have no net stratigraphic signature because the debris is rapidly homogenized with background attritional input (as in Greenstein 1989; and see Badgley 1982; Meldahl 1990; Behrensmeyer and Chapman 1993; Cutler and Behrensmeyer 1996). Through the Phanerozoic, the stratigraphic frequency of documented mass mortalities thus may be partly controlled by evolutionary changes in bioturbation depth and intensity.

Strategies for Addressing Taphonomic Biases

One of the goals of taphonomy is to establish patterns, and preferably quantitative magnitudes and causes, of bias in the fossil record. Paleontologists are developing new ways to evaluate shortcomings both in the record itself and in our sampling of it, and many researchers now approach these issues rigorously and proactively. Generally speaking, approaches to biases, both real or hypothesized, have ranged from assuming or reasoning that the record itself—or our knowledge of it—is too poor for biological analysis, to assuming or reasoning that the record can be taken at face value. These usually reflect different starting points for analysis—guilty of overwhelming bias until proven otherwise, or innocent until proven biased. Paleontologists who take this latter approach are not neces-

sarily being willfully blasé, but may reason (geologically, statistically) that taphonomic bias is relatively small and thus a second-order effect, or conclude that bias is random relative to the variable under study and thus unlikely to create artificial patterns. In other situations, the strategy is to normalize the data or adjust the metric to compensate for probable or known biases (e.g., rarefaction, comparisons of data trends with unevenness in sampling intensity, which may be a function of rock availability as well as taphonomic processes), use of taphonomic control taxa, Lazarus and other gap analyses (see discussion below; also Sepkoski and Koch 1996; Foote this volume; Holland this volume). A variant of this strategy is analytic time- and space-averaging (Behrensmeyer and Hook 1992), whereby paleontologists group (bin) data more coarsely than nature and thereby reduce noise introduced by variation at lower scales; this is axiomatic in many macroevolutionary and global analyses. Another strategy, once the quality of the record has been evaluated, is simply to sidestep biased or incomplete information by rephrasing the hypothesis or shifting the emphasis of the study to suit the quality of the data (see discussion in Paul 1992).

The construction and analysis of paleobiological data sets usually entails many taphonomic assumptions, and several points must be kept in mind in designing strategies. One is the precise meaning of the terms "bias" and "incompleteness." *Bias* by definition is neither uniform nor random (occurring unpredictably), but instead is a skewing of information in some systematic way. This is a different concern from the *incompleteness* of data, which, as used in paleontology, refers to the extent of knowledge—that is, how fully the available pool of information has been sampled and thus how stable and detailed our picture of that system is thought to be (Paul 1992, 1998). Incomplete information can provide a fair and true (unbiased) sampling of reality, and confidence intervals can be calculated for it (e.g., Sadler 1981; Strauss and Sadler 1989; Marshall 1990, 1994; McKinney 1991; see also various papers in Gilinsky and Signor 1991). Collection curves and other growth-of-information curves are longstanding examples (see Paul 1992). But if gaps in information preferentially fall, for whatever reason, within particular biota, segments of time, or environments within the scope of the analysis, then incompleteness can be transformed into the more serious problem of bias.

A second point is that bias and incompleteness can both be either *natural* or *analytical* in origin. For example, gaps can result from hiatuses and barren intervals in the stratigraphic record, but gaps also are generated analytically by coarse sampling schemes and by variation in monographic effort. Natural taphonomic bias includes the tendency for small-bodied individuals or species in a group to be underrepresented, and for transgressive records to be thin and/or faunally condensed relative to regressive deposits. Potential sources of analytic bias include relying heavily on North American and European records in the construction of *global* data sets, and interpreting at face value species richness data from samples with disparate scales of time-averaging or positions on growth-of-knowledge curves.

A third point to be clear on, whether devising a strategy or evaluating one used by others, is the reality that is being targeted—the actual history of life, the fossil record of life (a taphonomically filtered subset of information), the known fossil record (an analytic subset influenced by geopolitics), or a data set based on some analytic subset of the known fossil record (published data, unpublished data, and/or new fieldwork). Data that are complete at one scale (e.g., compendia from the published fossil record) may be neither complete nor unbiased at another more inclusive scale (e.g., the actual fossil record). Conversely, data that are incomplete and biased at a fine scale (e.g., major gaps in the local record because of facies controls on the original presence or postmortem preservation of species) might yield adequate information at a broader scale of analysis (e.g., presence of the higher taxon in the region anytime within a coarser time interval).

Finally, some metrics will be less sensitive to *incomplete* data than others—for instance, medians rather than absolute minima and maxima, ratios (e.g., predator–prey) rather than absolute numbers of taxa, rates and patterns of change rather than specific trajectories, evenness rather than total species richness (e.g., Foote this volume; Alroy et al. this volume). Factoring out *bias*, on the other hand, requires analytic dissection of the data at hand. For example, from an innocent-until-proven-guilty stance, does evolutionary rate, geographic range, or numbers of species per genus actually covary with preservation categories, such that using samples of diverse preservational quality in a single data set would yield a misleading paleobiologic interpretation (e.g., Jablonski 1988; Foote and Raup 1996)? From a guilty-until-proven-innocent stance, do observed spikes of taxonomic first and last appearances in stratigraphic sequences rise above levels expected at those horizons because of stratigraphic truncation and condensation (e.g., Holland 1995, 1996; Holland and Patzkowsky 1999), or because of collector-induced variation in sample size?

By virtue of their size and scope, data sets on broad-scale paleobiological patterns entail a particular set of taphonomic issues. For example, the marine metazoan diversity pattern (Sepkoski 1978, 1993), the continental plant record (Knoll et al. 1984; Kendricks and Crane 1997), and major branching points in vertebrate evolution (Maxwell and Benton 1990; Benton 1998) are interpreted to reflect biological history at a global, stage-sized bin level, and this may be correct at that scale of analysis, although as yet there has been no comprehensive analysis of biases. Some major potential sources of bias, such as those pertaining to variation in outcrop area and volume, the "pull of the Recent" (intensive sampling of extant fauna extends the stratigraphic ranges of poorly sampled fossils), monographic effort per geologic period, and effect of hyper-rich lagerstätten (involving taxa that occur in a single interval), have been addressed, at least coarsely (e.g., for the marine record, see Sepkoski et al. 1981). Moreover, the basic patterns (e.g., trends in numbers of families over time) have remained stable in spite of expanding knowledge of the fossil record (Sepkoski 1993; Benton 1998). Such "growth of knowledge" curves (Paul 1992, 1998) support confidence in the adequacy of information on the *known* fossil record for trends at this scale, and this type of analysis can be applied at many levels in the taxonomic, temporal, and spatial hierarchies (see examples in Donovan and Paul 1998; and for morphospace occupation, see Foote 1997).

Research is now focusing on biodiversity patterns and biases at finer geographic scales, with the effects of environmental (facies) and biogeographic variation on numbers of recorded taxa and range limits (e.g., Raymond and Metz 1995; Vrba 1995; Wing and DiMichele 1995; Alroy 1996; Behrensmeyer et al. 1997), including the effects of nonrandom patterns in facies and hiatuses through stratigraphic sequences (Holland 1995). To what extent, for example, does the acknowledged weighting of published fossil data toward North America and Europe influence the perceived global pattern? Determining whether subregions have distinctly different biodiversity patterns will indicate the extent to which uneven sampling across the globe might bias our perception to date of global trends. When diversity through time within particular geological intervals is examined closely, different patterns do emerge for different regions, and ongoing research is testing the relative roles of intrinsic biologic factors, biological response to regional environmental conditions, and taphonomic issues linked to regional environments (Miller and Foote 1996; Miller 1997; Waisfeld et al. 1999). Databases that track the environmental context and taphonomic character of paleontologic occurrences—that is, that are not done in a lithologic vacuum—are an important next step in evaluating both natural (taphonomic) and anthropogenic (analytic) sampling bias with respect to global biodiversity patterns.

In many kinds of studies, but especially in evolutionary ecology, rarefaction (Sanders 1968; Raup 1975; Miller and Foote 1996) has been widely used as a means to standardize samples (e.g., to compare species richness in samples of disparate size). To infer that differences in rarefied diversity over time or space have biological explanations, however, one must assume that the samples are isotaphonomic, i.e., that information for each bin is de-

rived from an equivalent suite of natural sampling conditions. One strategy is to limit analysis to samples from a specific habitat or suite of habitats or facies (e.g., Bambach 1977); this assures greater taphonomic equivalence, although it also limits the universality of the results. The longer the interval of time, the more important it is to sample within a single taphonomic domain: if natural taphonomic regimes have shifted through time, and thereby altered the proportional preservation of bodyplans, growth stages, or habitats (megabias), then a major assumption behind biological interpretation of rarefied data is potentially violated, notwithstanding the within-habitat design. For example, unconsolidated lower shoreface facies from the Cenozoic record, which include faunally condensed shell gravels, are likely to yield higher species richnesses for both taphonomic and analytic reasons than predominantly lithified units from the Mesozoic (Kidwell and Jablonski 1983; Kidwell and Brenchley 1996), and, owing to bioturbation, neither of these records preserves as high a proportion of discrete storm-bed concentrations as the Paleozoic (Brandt 1986; Sepkoski et al. 1991). Thus, for a variety of reasons, taphonomic biases might be expected to inflate the raw alpha diversity of benthic communities in each erathem to a greater degree than the one before it, making a biological interpretation of diversity increase ambiguous unless sampling is standardized. Cross-time analysis of comparable samples, even if piecemeal—e.g., lithified Mesozoic with lithified Cenozoic, Paleozoic storm-beds with Mesozoic storm-beds—is one obvious next step.

Another approach would be to characterize taphonomic regimes and study their effects on apparent taxonomic diversity using sampling standardization within each regime and in combined samples. At basinal scales, nonrandom geographic and temporal variations in completeness and taphonomy can in fact account for a large part of apparent faunal turnover patterns through time (Brett 1995; Holland 1995; Behrensmeyer et al. 1997), and thus scaled-up versions of similar biases might affect regional to global patterns. Benton (1998), for example, suggests that some marked shifts in Phanerozoic vertebrate diversity through time are not evidence of evolutionary events but are an artifactual "lagerstätten effect" of pooling data from scattered horizons of superb preservation with information from deposits of more ordinary fossil preservation (this has been rejected as insignificant in the periodicity of marine metazoan extinctions [Sepkoski 1990; and see Foote this volume and Alroy et al. this volume]).

At both large and small scales of investigation, establishing taphonomic equivalence (isotaphonomy) has become an important research goal. The aim is to achieve meaningful biological comparisons across space and time, and in particular to reduce dependency on modern analogues. Isotaphonomy is particularly critical for establishing the credibility of species abundance trends and other diversity measures derived from the fossil record. Criteria for defining isotaphonomic assemblages should include as many lines of evidence as possible without prohibitively limiting sample size: geological setting, general climatic regime, lateral and vertical scale of the fossiliferous unit, paleogeochemistry, body-part representation, and other indicators of taphonomic processes including time-averaging. This approach is relatively new and several strategies are possible. Benton (1998; also Briggs and Clarkson 1990) suggests comparing similar types of lagerstätten at different times as snapshots of "true" diversity, although such deposits commonly record unusual conditions of mortality and environment that limit their fidelity with respect to true or average diversity (see previous sections). Moreover, and as a general caveat, there are huge pitfalls (commonly unacknowledged) to extrapolating regional or global signals from single locales. An alternative starting point is to establish broader equivalence in depositional environments or taphofacies; hence, vertebrate faunas from channel fill versus levee versus floodplain paleosol settings might be compared through time, and all are combined for a representation of the diversity of the fluvial system as a whole. Clyde and Gingerich (1998) take such an approach in examining mammalian community response to environmental change in the late Paleocene using isotaphonomic samples from floodplain paleosol settings. Many macroplant assemblages are from specific types of wetlands environments (e.g., wet floodplain, proximal channel, abandoned channel, and channel), making them somewhat isotaphonomic by default (ditto the records of soft-bodied animals), but water chemistry, rates of aggradation, and climate can vary in ways that affect that preservation (Gastaldo 1994; Demko et al. 1998; Gastaldo and Staub 1999). "Next-generation" analysis of macroevolutionary trends requires such critical appraisal, both as a means of estimating confidence limits and as a means of testing environmental forcing factors. Although challenging, this is a logical next step toward integrating taphonomic advances into mainstream paleobiology.

An additional challenge in paleobiologic analysis, especially of broad-scale patterns, is the significance of missing taxa, and the use of "taphonomic control taxa" has been proposed as one way to determine when absences are meaningful (Bottjer and Jablonski 1988; Jablonski et al. 1997). These are biologically abundant taxa with hardparts that are comparably or less robust than those of the target taxon, and preferably relatively close taxonomically. The reasoning is that if the control taxon is present, then the target should have also been preserved if it co-occurred in that unit; moreover, the author who reports the control taxon would have been likely to report the target taxon if present. Thus, cyclostomes may serve as control taxa for tracking the environmental and evolutionary expansion of cheilostome bryozoans, isocrinids for millericrinid crinoids, and other small infaunal veneroids for tellinid bivalves. Likewise, teeth of *Hipparion* have been used as a control for the appearance datum of the similarly sized horse *Equus* in Africa (Behrensmeyer 1978b), and turtles as a control for crocodiles in an analysis of climatic effects in the Cretaceous/Tertiary extinction (Markwick 1998).

Foote et al. (1999) used similar taphonomic reasoning to challenge molecular evidence for an Early Cretaceous origin of modern Eutherian mammal orders, arguing instead that these groups did not arise much before their earliest known geologic record in the latest Cretaceous or Paleocene. The frequent lack of ancestral taxa in the fossil record often is attributed to evolution in stratigraphically underrepresented peripheral or poorly preserved habitats (e.g., for "uplands" habitats for Cretaceous angiosperms and vertebrates, see Olson 1966 and Retallack and Dilcher 1986; and for Mesozoic plant taxa now recognized in the Upper Carboniferous and Permian, see DiMichele et al. in press). However, Foote et al. (1999) take an important step beyond this by quantifying preservation rates for other Cretaceous mammals and arguing that, unless the taphonomy of the earliest eutherians differed radically from other mammal taxa in the Mesozoic (i.e., body sizes, habitats/facies, life habits) and in such a way that severely reduced the quality of their record, then the probability of missing modern eutherian orders throughout the Cretaceous is very low (barring, they note, a "Garden of Eden" in which all these orders both originated and remained in an undersampled region for tens of millions of years).

This approach touches upon a problem that needs resolution if we are to calibrate taphonomic effects on appearance events—namely, the effect of a taxon's live abundance, spatial distribution, and temporal range on its probability of preservation. Rare, localized, and geologically short-lived species, such as might initiate a major new lineage, could be particularly vulnerable to taphonomic bias (see earlier discussion of Valentine 1989), and the record of continental taxa might suffer disproportionately for both taphonomic (patchy preservation) and biologic (e.g., greater importance of endemism in origination?) reasons. For example, detailed study of Cenozoic vertebrate assemblages indicates that distinct faunal communities can exist in adjacent basins or portions of basins for long periods of time (10^6–10^7 yr) (Behrensmeyer 1978b; Bown and Beard 1990). Whether this segregation is taphonomic or ecologic in origin, it indicates how a taphonomic control approach might impart misleading results on vertebrate presence/absence at this temporal and spatial scale, and raises the issue of how such facies control might "scale up" to global patterns over tens to hundreds of millions of years and at higher taxonomic levels (DiMichele and

Aronson 1992). These concerns would apply to other taxa that lack transitional forms in the fossil record; credible evidence that a taxon is absent for evolutionary rather than taphonomic reasons requires a good understanding of both the completeness of sampling and the possible taphonomic bias in the fossil-bearing deposits (and see discussion on bias relevant to bio-events in Sepkoski and Koch 1996).

In contrast to inverse-type models, where one works back from paleontologic patterns ("what bias could generate this pattern?"), quantitative forward modeling of taphonomic processes and effects from the input perspective is in its infancy. Existing studies illustrate the great potential of this strategy for addressing biases and limits to the resolution of the fossil record. This modeling can be conducted at many scales and provides an important means of bridging the gap between actualistic data and stratigraphic patterns at the assemblage level. Behrensmeyer and Chapman (1993), for example, take this approach in using computer simulations to create artificial time-averaged vertebrate assemblages, based on a modern African assemblage and known rates of bone input. They show that hundreds to thousands of years of time-averaging is needed to capture all of the major taxa in potential fossil localities (i.e., time-averaging is good, and even necessary, for producing an accurate portrayal of species presence and rank order) (also see Miller and Cummins 1990, 1993 on marine taxa). Building upon empirical evidence that modern molluscan death assemblages are dominated by the shells of recent cohorts (Meldahl et al. 1997a), Olszewski (1999) models time-averaging and predicts the sample size necessary to ensure retrieval of specimens from each component time segment in the source assemblage, i.e., a complete sample of the entire span of time-averaging. There have also been immense strides in conceptualizing and testing the effects of incomplete preservation and stratigraphic gaps on evolutionary patterns in the broadest sense, including tempo and mode of speciation, and in using estimates of phylogeny to infer the quality of the fossil record. For recent entries to this large literature, see Carroll 1997, Roopnarine et al. 1999, and Wagner

2000a; also chapters by Alroy et al., Foote, Holland, and Wagner in this volume.

Focal Areas for the Future

Taphonomy's continuing challenge is to evaluate the prolific but problematic fossil record for systematic patterns in preservation that may constitute bias in information quality, and to develop accurate ways to measure and use such patterns in paleobiologic analysis. Over the last few decades, paleontologists have become more sanguine about the quality of paleontologic data. There are two aspects to this: (1) an appreciation that all data, paleontologic or otherwise, are incomplete, and that the critical question is whether they are adequate to address the question at hand (Paul 1998); and (2) a realization that taphonomic comparability or noncomparability of samples across time and space must be taken into consideration in deriving biological patterns from paleontological data. This means that, ideally, samples used to examine temporal and spatial trends should be from comparable depositional contexts and preservational states, even if absolute scales of time-averaging or spatial fidelity cannot be specified. Alternatively, when the point is to compare or combine biological factors such as diversity across environments, regions, and geological domains, an opposite approach is required that takes into account and compensates for clearly different qualities of data. Today this is largely done by taphonomic uniformitarianism, i.e., extrapolating modern-day rates and error estimates back in time. A challenge for next-generation research is to assess the very real limits imposed by secular changes in fossilization through the geological record.

There are particular focal areas for taphonomy that are likely to generate important contributions in the next several decades. Comparison across plants, invertebrates and vertebrates is a promising growth area for taphonomy's broader contributions to understanding geobiological processes and to developing a theoretical basis for the field. Enough now is known about each major group to suggest some common denominators, such as the effects of bioturbators and bioeroders, as well as some contrasts in approaches and problems, including the wide differences in susceptibility to time-averaging of major taxonomic groups. Is it possible to develop a general index of relative "preservation potential," including likely degrees of time-averaging, for different body-plans, life strategies, and ecological settings? Can we establish a basis for recognizing "abnormal preservation," indicating profound shifts in taphonomic regimes, such as post-extinction differences in shelly faunas? It should be possible to take assemblage-level processes and biases, and develop hypotheses about how these operate at a larger scale as a basis for defining "normal" circumstances of preservation for individual taxa and for different types of communities.

To advance these aims, and to summarize our preceding highlights of the discipline, we recommend the following key focal areas for future taphonomic research relevant to paleobiology:

1. Field and lab experiments on the budget of input and permanent burial and on the rates, agents, pathways, and conditions of *recycling* of biological materials, especially the relatively subtle geochemical and geomicrobiological aspects of "weathering" on and just below the depositional interface.

2. Quantification of *time-averaging* for a broader array of taxonomic groups and depositional settings, including the relative contributions of successive cohorts of material. In part, such work can test and amplify the hypothesis that time-averaged assemblages are dominated by the most recent cohorts, as suggested by recent empirical work on marine mollusks. Recognizing scales of time-averaging via damage levels and other tangible clues in surviving fossil material is a key aspect of this research.

3. Actualistic estimates of the *compositional fidelity* (for species richness, abundances, age groups, etc.) of assemblages for a broader array of taxonomic groups and depositional settings, including explicit attention to whether fidelity can be inferred from observed levels of damage (i.e., taphofacies information) and to how fidelity varies as a function of geographic scale of investigation and the geologic aging of an assemblage.

4. Long-term *stratigraphic trends* in the qualities (Table 1) of the fossil record, including use of the Pleistocene or Neogene as a reflection of the Recent in order to investigate taphonomic modification associated with longer periods of time/space averaging, lithification, and other aspects of the ''permanent'' fossil record.

5. A major initiative in *probabilistic and other quantitative modeling* as a means of testing existing hypotheses and formulating new hypotheses to test in the stratigraphic record and Recent systems.

After several decades of intensive research, taphonomists now visualize the fossil record of taxa with mineralized or highly refractory tissues as dominated by time-averaged assemblages, with widely spaced horizons and intervals bearing higher resolution records of taxa and paleocommunities. This contrasts with groups lacking readily preserved tissues: macroplant and soft-bodied animal records clearly are subject to much less time-averaging per assemblage than is true for pollen, marine shelly faunas, or land vertebrates, and preservation is limited to a narrower range of environmental conditions. The result is a series of geographically and temporally narrow windows of high anatomical and temporal resolution, relatively widely separated in space and time, producing a historical record with many gaps. Hence the fundamental trade-off now recognized in taphonomy: the better the preservation of individual organisms and the finer the temporal resolution of individual samples, the less likely these are to be repeated at close and regular intervals through geological time. The tendency for the taphonomically most robust groups to exhibit the greatest time-averaging (and thus spatial averaging) (reciprocal model of Kowalewski 1997) is a key corollary of this pattern. Taphonomy's agenda for the future revolves around better understanding the genesis and fidelity of these different types of records (time-averaged and time-specific), how their attributes are affected by local to global-scale tectonic, climatic, and biotic conditions, and how these taphonomic differences affect our assessment and understanding of paleobio-

logic phenomena such as evolutionary rates and diversity through time in marine versus continental organisms.

Recent and future contributions of taphonomy are relevant to an array of *paleobiologic* issues including the following (Jablonski 1999):

1. *Paleocommunity structure and composition, and how this changes through time in response to environmental perturbations, especially climate shifts.* Establishing paleocommunity structure depends heavily on studies in recent ecosystems, but neontologic and paleontologic views of communities differ in their focus on what controls species distribution and behavior (live versus dead), their selective treatment of particular taxonomic groups, and especially their degrees of temporal sampling. More exchange between neo- and paleoecologists, with an emphasis on collecting new types of field data and modeling fossil assemblages using actualistic data, could generate new insights and a stronger foundation for reconstructing paleocommunities. As a basis for this, we need more quantitative information on the potential spatial fidelity, temporal mixing, and compositional fidelity (percent living and preservable species) for a wider array of environments and taxonomic groups. By empirically linking levels of bias (inferred qualities) to damage profiles (observed states of preservation of species) and depositional modes, we also can develop criteria for isotaphonomic equivalence across space and time, both between Recent and ancient biological systems and through comparative work within the stratigraphic record.

2. *The history of biodiversity dynamics at different scales, from individual assemblages to global tallies of diversity (numbers of species, genera, families, etc.).* There is a clear need for better understanding of taphonomic effects on Phanerozoic (and Precambrian) diversity patterns. Large-scale shifts in taphonomic regime (links between organic preservation and the chemical and physical states of the earth as well as faunal/floral evolution) or between recurring taphonomic states (e.g., due to climate) as suggested in Figure 6, may be contributing confounding patterns to the diversity curves. These megabiases will not necessarily be elim-

inated by approaches such as rarefaction, and modeling what happens to diversity measures over stratigraphic shifts in taphofacies could help to clarify the effects of taphonomy on rarefaction "universes." Controlling taphonomic biases using isotaphonomic approaches may permit us to develop robust Phanerozoic diversity patterns for particular environments, and this should be complemented by greater efforts to develop nonactualistic means of estimating and compensating for bias when isotaphonomy is impossible. And the explosion of understanding of the chemical and physical aspects of fossil preservation (soft and hard parts) provides a framework for assessing the stratigraphic patterning of census assemblages (including konservat-lagerstätten) and judging what we are missing in other parts of the fossil record.

3. *Rates of evolutionary events (originations, radiations, extinctions, and rebounds after extinctions), including major periods of faunal and floral change at the end of the Permian, the K/T boundary, and the Pleistocene.* Establishing rates depends on accurate biostratigraphic records of taxonomic presence, absolute dating of these records, and comparisons that are matched for the durations over which change was measured. An important goal for taphonomy is to develop more rigorous measures of "preservation potential" for different types of organisms and provide alternative tests of biostratigraphic range limits for comparison with those based on abundance patterns and gap analysis. There should be a search for sub-Recent and Plio-Pleistocene analogues in which known extinctions or appearances are recorded in stratigraphic sequences with high temporal resolution to provide comparisons for the more distant geological record. New understanding of processes of preservation and destruction at the molecular to sequence stratigraphic scales could feed into simulations of real versus apparent records of taxonomic ranges. Likewise, increased appreciation of the limits to resolution provided by space- and time-averaging can help to provide reality checks and quantification of error-bars in correlating environmental change with major biotic events in Earth history.

4. *Correspondence of the macroevolutionary history of the biotic system to secular and cyclic geochemical and geophysical changes in Earth and its atmosphere.* Taphonomy's continuing role is to characterize sampling biases that affect macroevolutionary reconstructions, but it contributes an additional perspective on macroevolution through its focus on the recycling of organic and inorganic materials. Such processes have undoubtedly responded to and also affected environmental changes on Earth. Thus, improved information on geologic intervals and settings where physical, chemical, and biological recycling has been particularly effective or particularly ineffective in breaking down organic materials is essential. Investigation of the macrotaphonomic history of the biotic system will involve integrating different scales of evidence for plants, invertebrates, and vertebrates, and developing hypotheses about how taphonomic patterns through time relate to Earth's physical and chemical history.

Beyond the paleobiologic issues discussed in this review, taphonomy has much to contribute to the fields of ecology, biogeochemistry, sedimentary geology and stratigraphy, paleoanthropology, and conservation biology. We look forward to even greater cross-disciplinary collaborations of information and scientific methods to apply to taphonomic data and questions. Facts—the next generation of data acquisition and analysis of individual taxa and assemblages—will continue to be fundamentally important to all aspects of taphonomy in the coming decades. But there also should be much more attention to synthesis, at the local, regional, and global levels. Next-generation research can target data acquisition that feeds into the search for larger patterns and provides tests for interim hypotheses about global-scale changes in taphonomic regimes and megabiases affecting the largest-scale paleobiological interpretations of the history of life.

Acknowledgments

We thank the editors of this special issue for their encouragement and suggestions during the writing of this article, and A. I. Miller, R. R. Rogers, and D. Jablonski for helpful reviews.

Literature Cited

Allen, J. R. L., R. A. Spicer, and A. K. Behrensmeyer. 1990. Transport—hydrodynamics. Pp. 227–235 *in* Briggs and Crowther 1990.

Allison, P. A. 1986. Soft-bodied animals in the fossil record: the role of decay in fragmentation during transport. Geology 14: 979–981.

———. 1988. Konservat-Lagerstätten: cause and classification. Paleobiology 14:331–344.

Allison, P. A., and D. E. G. Briggs, eds. 1991a. Taphonomy, releasing the data locked in the fossil record. Plenum, New York.

———. 1991b. The taphonomy of soft-bodied animals. Pp. 120–140 *in* Donovan 1991.

———. 1993a. Paleolatitudinal sampling bias, Phanerozoic species diversity, and the end-Permian extinction. Geology 21: 65–68.

———. 1993b. Exceptional fossil record: distribution of soft-tissue preservation through the Phanerozoic. Geology 21:527–530.

Allison, P. A., and K. Pye. 1994. Early diagenetic mineralization and fossil preservation in modern carbonate concretions. Palaios 9:561–575.

Allison, P. A., C. R. Smith, H. Kukert, J. W. Deming, and B. Bennett. 1991. Deep-water taphonomy of vertebrate carcasses: a whale skeleton in the bathyal Santa Catalina Basin. Paleobiology 17:78–89.

Alroy, J. 1996. Constant extinction, constrained diversification, and uncoordinated stasis in North American mammals. Palaeogeography, Palaeoclimatology, Palaeoecology 127:285–311.

Alroy, J., P. L. Koch, and J. C. Zachos. 2000. Global climate change and North American mammalian evolution. *In* D. H. Erwin and S. L. Wing, eds. Deep time: *Paleobiology's* perspective. Paleobiology 26(Suppl. to No. 4):259–288.

Anderson, L. C., B. K. Sen Gupta, R. A. McBride, and M. R. Byrnes. 1997. Reduced seasonality of Holocene climate and pervasive mixing of Holocene marine section: northeastern Gulf of Mexico shelf. Geology 25:127–130.

Anderson, L. C., R. A. McBride, M. J. Taylor, and M. R. Byrnes. 1998. Late Holocene record of community replacement preserved in time-averaged molluscan assemblages, Louisiana chenier plain. Palaios 13:488–499.

Andrews, P. 1990. Owls, caves, and fossils. University of Chicago Press, Chicago.

Andrews, P., and J. Cook. 1989. Natural modifications to bones in a temperate setting. Man 20:675–691.

Archer, M., S. J. Hand, and H. Godthelp. 1991. Riversleigh: the story of animals in ancient rainforests of inland Australia. Reed Books, Balgowlah, New South Wales.

Argast, S., J. O. Farlow, R. M. Gabet, and D. L. Brinkman. 1987. Transport-induced abrasion of fossil reptilian teeth: implications for the existence of Tertiary dinosaurs in the Hell Creek Formation, Montana. Geology 15:927–930.

Aronson, R. B. 1992. Decline of the Burgess Shale fauna: ecologic or taphonomic restriction? Lethaia 25:225–229.

Aslan, A., and A. K. Behrensmeyer. 1996. Taphonomy and time resolution of bone assemblages in a contemporary fluvial system: the East Fork River, Wyoming. Palaios 11:411–421.

Ausich, W. I. 1997. Regional encrinites: a vanished lithofacies. Pp. 509–519 *in* C. E. Brett and G. C. Baird, eds. Paleontological events: stratigraphic, ecological and evolutionary implications. Columbia University Press, New York.

Ausich, W. I., and G. D. Sevastopulo. 1994. Taphonomy of Lower Carboniferous crinoids from the Hook Head Formation, Ireland. Lethaia 27:245–256.

Bada, J. L., X. S. Wang, and H. Hamilton. 1999. Preservation of key biomolecules in the fossil record: current knowledge and future challenges. Philosophical Transactions of the Royal Society of London B 354:77–87.

Badgley, C. E. 1982. How much time is represented in the present? The development of time-averaged modern assemblages as models for the fossil record. *In* B. Mamet and M. J. Copeland, eds. Proceedings of the Third North American Paleontological Convention 1:23–28.

———. 1986. Counting individuals in mammalian fossil assemblages from fluvial environments. Palaios 1:328–355.

Badgley, C. E., and A. K. Behrensmeyer. 1995. Preservational, paleoecological and evolutionary patterns in the Paleogene of Wyoming–Montana and the Neogene of Pakistan. Palaeogeography, Palaeoclimatology, Palaeoecology 115:319–340.

Baird, G. C., S. D. Sroka, C. W. Shabica, and G. J. Kuecher. 1986. Taphonomy of Middle Pennsylvanian Mazon Creek area fossil localities, northeast Illinois: significance of exceptional fossil preservation in syngenetic concretions. Palaios 1:271–285.

Bambach, R. K. 1977. Species richness in marine benthic habitats through the Phanerozoic. Paleobiology 3:152–167.

Barry, J. C., M. E. Morgan, A. J. Winkler, L. J. Flynn, E. H. Lindsay, L. L. Jacobs, and D. Pilbeam. 1991. Faunal interchange and Miocene terrestrial vertebrates of southern Asia. Paleobiology 17:231–245.

Bartels, C., D. E. G. Briggs, and G. Brassel. 1998. The fossils of the Hunsrück Slate: marine life in the Devonian. Cambridge University Press, Cambridge.

Barthel, K. W., N. H. M. Winburne, and S. Conway Morris. 1990. Solnhofen: a study in Mesozoic paleontology. Cambridge University Press, Cambridge.

Bartley, J. K. 1996. Actualistic taphonomy of cyanobacteria: implications for the Precambrian fossil record. Palaios 11:571–586.

Bartram, L. E., Jr., and C. W. Marean. 1999. Explaining the "Klasies pattern": Kua ethnoarchaeology, the Die Kelders Middle Stone Age archaeofauna, long bone fragmentation and carnivore ravaging. Journal of Archaeological Science 26:9–29.

Baumiller, T. K., G. Llewellyn, C. G. Messing, and W. E. Ausich. 1995. Taphonomy of isocrinoid stalks: influence of decay and autonomy. Palaios 10:87–95.

Behrensmeyer, A. K. 1978a. Taphonomic and ecological information from bone weathering. Paleobiology 4:150–162.

———. 1978b. Correlation in Plio-Pleistocene sequences of the northern Lake Turkana Basin: a summary of evidence and issues. Pp. 421–440 *in* W. W. Bishop, ed. Geological background to fossil man. Scottish Academic Press, Edinburgh.

———. 1982. Time resolution in fluvial vertebrate assemblages. Paleobiology 8:211–227.

———. 1987. Miocene fluvial facies and vertebrate taphonomy in northern Pakistan. *In* F. G. Ethridge, R. M. Flores, and M. D. Harvey, eds. Recent developments in fluvial sedimentology. SEPM Special Publication 39:169–176.

———. 1988. Vertebrate preservation in fluvial channels. Palaeogeography, Palaeoclimatology, Palaeoecology 63:183–199.

———. 1990. Transport/hydrodynamics of bones. Pp. 232–235 *in* Briggs and Crowther 1990.

———. 1991. Terrestrial vertebrate accumulation. Pp. 291–335 *in* Allison and Briggs 1991.

———. 1993. The bones of Amboseli: bone assemblages and ecological change in a modern African ecosystem. National Geographic Research 9:402–421.

Behrensmeyer, A. K., and R. E. Chapman. 1993. Models and simulations of time-averaging in terrestrial vertebrate accumulations. Pp. 125–149 *in* Kidwell and Behrensmeyer 1993.

Behrensmeyer, A. K., and D. E. Dechant Boaz. 1980. The Recent bones of Amboseli Park, Kenya, in relation to East African paleoecology. Pp. 72–92 *in* A. K. Behrensmeyer and A. P. Hill, eds. Fossils in the making. University of Chicago Press, Chicago.

Behrensmeyer, A. K., and R. W. Hook. 1992. Paleoenvironmental contexts and taphonomic modes. Pp. 15–136 in A. K. Behrensmeyer, J. D. Damuth, W. A. DiMichele, R. Potts, H.-D. Sues, and S. L. Wing, eds. Terrestrial ecosystems through time. University of Chicago Press, Chicago.

Behrensmeyer, A. K, and S. M. Kidwell. 1985. Taphonomy's contributions to paleobiology. Paleobiology 11:105–119.

Behrensmeyer, A. K., N. E. Todd, R. Potts, and G. E. McBrinn. 1997. Late Pliocene faunal turnover in the Turkana Basin, Kenya and Ethiopia. Science 278:1589–1594.

Bell, M. A., M. S. Sadagursky, and J. V. Baumgartner. 1987. Utility of lacustrine deposits for study of variation within fossil samples. Palaios 2:455–466.

Bell, M. A., C. E. Wells, and J. A. Marshall. 1989. Mass-mortality layers of fossil stickleback fish: catastrophic kills of polymorphic schools. Evolution 43:607–619.

Bengtson, S. 1994. Early life on Earth (Nobel Symposium No. 84). Columbia University Press, New York.

Bennington, J. B., and S. D. Rutherford. 1999. Precision and reliability in paleocommunity comparisons based on cluster-confidence intervals: how to get more statistical bang for your sampling buck. Palaios 14:506–515.

Benton, M. J. 1998. The quality of the fossil record of the vertebrates. Pp. 269–300 in S. K. Donovan and C. R. C. Paul, eds. The adequacy of the fossil record. Wiley, New York.

Berner, R. A. 1991. A model for atmospheric CO_2 over Phanerozoic time. American Journal of Science 291:339–376.

Berner, R. A., and D. E. Canfield. 1989. A new model for atmospheric oxygen over Phanerozoic time. American Journal of Science 289:333–361.

Best, M. M. R., and S. M. Kidwell. 1996. Bivalve shell taphonomy in tropical siliciclastic environments: preliminary experimental results. In J. E. Repetski, ed. Sixth North American paleontological convention, Abstracts of papers. Paleontological Society Special Publication 8:34.

———. 2000a. Bivalve taphonomy in tropical mixed siliciclastic-carbonate settings. I. Environmental variation in shell condition. Paleobiology 26:80–102.

———. 2000b. Bivalve taphonomy in tropical mixed siliciclastic-carbonate settings. II. Effect of bivalve life habits and shell types. Paleobiology 26:103–115.

Best, M. M. R., S. M. Kidwell, T. C. W. Ku, and L. M. Walter. 1999. The role of microbial iron reduction in the preservation of skeletal carbonate: bivalve taphonomy and porewater geochemistry in tropic siliciclastics vs. carbonates. Geological Society of America Abstracts with Programs 31:419–420.

Bishop, J. D. D. 1989. Colony form and the exploitation of spatial refuges by encrusting bryozoa. Biological Reviews 64:197–218.

Blob, R. W. 1997. Relative hydrodynamic dispersal potentials of soft-shelled turtle elements: implications for interpreting skeletal sorting in assemblages of non-mammalian terrestrial vertebrates. Palaios 12:151–164.

Blumenschine, R. 1986. Carcass consumption sequences and the archaeological distinction of scavenging and hunting. Journal of Human Evolution 15:639–659.

———. 1988. An experimental model of the timing of hominid and carnivore influence on archaeological bone assemblages. Journal of Archaeological Science 15:483–502.

———. 1991. Hominid carnivory and foraging strategies and the socio-economic function of early archaeological sites. Philosophical Transactions of the Royal Society of London B 334:211–221.

Bonnichsen, R., and M. H. Sorg. 1989. Bone modification. Institute for Quaternary Studies, University of Maine, Orono.

Bosence, D. W. J. 1979. Live and dead faunas from coralline algal gravels, Co. Galway. Palaeontology 22:449–478.

Bottjer, D. J., and D. Jablonski. 1988. Paleoenvironmental patterns in the evolution of post-Paleozoic benthic marine invertebrates. Palaios 3:540–560.

Boulton, A. J., and P. I. Boon. 1991. A review of methodology used to measure leaf litter decomposition in lotic environments: time to turn over an old leaf? Australian Journal of Marine and Freshwater Research 42:1–43.

Bown, T. M., and K. C. Beard. 1990. Systematic lateral variation in the distribution of fossil mammals in alluvial paleosols, lower Eocene Willwood Formation, Wyoming. Geological Society of America Special Paper 243:135–151.

Brandt, D. S. 1986. Preservation of event beds through time. Palaios 1:92–96.

Brandt, D. S., and R. J. Elias. 1989. Temporal variation in tempestite thickness may be a geologic record of atmospheric CO_2. Geology 17:951–952.

Brett, C. E. 1995. Sequence stratigraphy, biostratigraphy, and taphonomy in shallow marine environments. Palaios 10:597–616.

Brett, C. E., and G. C. Baird. 1986. Comparative taphonomy: a key to paleoenvironmental interpretation based on fossil preservation. Palaios 1:207–227.

Brett, C. E., and A. Seilacher. 1991. Fossil Lagerstätten: a taphonomic consequence of event sedimentation. Pp. 283–297 in G. Einsele, W. Ricken, and A. Seilacher, eds. Cycles and events in stratigraphy. Springer, Berlin.

Brett, C. E., A. J. Boucot, and B. Jones. 1993. Absolute depths of Silurian benthic assemblages. Lethaia 26:25–40.

Brett, C. E., T. E. Whiteley, P. A. Allison, and E. L. Yochelson. 1999. The Walcott-Rust quarry: Middle Ordovician trilobite Konservat-Lagerstätten. Journal of Paleontology 73:288–305.

Briggs, D. E. G. 1993. Fossil biomolecules. Bulletin of the Biochemical Society 15:8–12.

———. 1995. Experimental taphonomy. Palaios 10:539–550.

Briggs, D. E. G., and E. N. K. Clarkson. 1990. The late Palaeozoic radiation of malacostracan crustaceans. In P. D. Taylor and G. P. Larwood, eds. Major evolutionary radiations. Systematics Association Special Volume 42:165–186. Clarendon, Oxford.

Briggs, D. E. G., and P. R. Crowther, eds. 1990. Palaeobiology, a synthesis. Blackwell Science, Oxford.

Briggs, D. E. G., and A. J. Kear. 1994a. Decay of Branchiostoma: Implications for soft-tissue preservation in conodonts and other primitive chordates. Lethaia 26:275–287.

———. 1994b. Decay and mineralization of shrimps. Palaios 9:431–456.

Briggs, D. J., D. D. Gilbertson, and A. L. Harris. 1990. Molluscan taphonomy in a braided river environment and its implications for studies of Quaternary cold-stage river deposits. Journal of Biogeography 17:623–637.

Briggs, D. E. G., P. R. Wilby, B. P. Pérez-Moreno, J. L. Sanz, and M. Fregenal-Martínez. 1997. The mineralization of dinosaur soft tissue in the Lower Cretaceous of Las Hoyas, Spain. Journal of the Geological Society, London 154:587–588.

Briggs, D. E. G., B. A. Stankiewicz, D. Meischner, A. Bierstedt, and R. P. Evershed. 1998. Taphonomy of arthropod cuticles from Pliocene late sediments, Willershausen, Germany. Palaios 13:386–394.

Briggs, D. E. G., R. P. Evershed, and M. J. Lockheart. 2000. The molecular paleontology of continental fossils. In D. H. Erwin and S. L. Wing, eds. Deep time: Paleobiology's perspective. Paleobiology 26(Suppl. to No. 4):169–193.

Buatois, L. A., A. G. Mángano, J. F. Genise, and T. N. Taylor. 1998. The ichnologic record of the continental invertebrate invasion: evolutionary trends in environmental expansion, ecospace utilization, and behavioral complexity. Palaios 13:217–240.

Budd, D. A., and E. E. Hiatt. 1993. Mineralogical stabilization of high-magnesium calcite: geochemical evidence for intracrystal recrystallization within Holocene porcellaneous foraminifera. Journal of Sedimentary Petrology 63:261–274.

Burnham, R. J. 1989. Relationships between standing vegetation and leaf litter in a paratropical forest: implications for paleobotany. Review of Palaeobotany and Palynology 58:5–32.

———. 1990. Paleobotanical implications of drifted seeds and fruits from modern mangrove litter, Twin Cays, Belize. Palaios 5:364–370.

———. 1993. Reconstructing richness in the plant fossil record. Palaios 8:376–384.

———. 1994. Patterns in tropical leaf litter and implications for angiosperm paleobotany. Review of Palaeobotany and Palynology 81:99–113.

Burnham, R. J., and R. A. Spicer. 1986. Fossil litter preserved by volcanic activity at El Chicón, Mexico: a potentially accurate record of the pre-eruption vegetation. Palaios 1:158–161.

Burnham, R. J., S. L. Wing, and G. G. Parker. 1992. The reflection of deciduous forest communities in leaf litter: implications for autochthonous litter assemblages from the fossil record: Paleobiology 18:30–49.

Butterfield, N. J. 1990. Organic preservation of non-mineralizing organisms and the taphonomy of the Burgess Shale. Paleobiology 16:272–286.

———. 1995. Secular distribution of Burgess Shale-type preservation. Lethaia 28:1–13.

Cadée, G. C. 1984. Macrobenthos and macrobenthic remains on the Oyster Ground, North Sea. Netherlands Journal of Sea Research 18:160–178.

———. 1991. The history of taphonomy. Pp. 3–21 in Donovan 1991.

———. 1994. Eider, shelduck, and other predators, the main producers of shell fragments in the Wadden Sea—paleoecological implications. Palaeontology 37:181–202.

———. 1999. Bioerosion of shells by terrestrial gastropods. Lethaia 32:253–260.

Calleja, M., M. Rossignol-Strick, and D. Duzer. 1993. Atmospheric pollen content off West Africa. Review of Palaeobotany and Palynology 79:335–368.

Callender, W. R., E. N. Powell, and G. M. Staff. 1994. Taphonomic rates of molluscan shells placed in authochthonous assemblages on the Louisiana contental slope. Palaios 9:60–73.

Carroll, R. L. 1997. Patterns and processes of vertebrate evolution. Cambridge University Press, Cambridge.

Cate, A. S., and I. Evans. 1994. Taphonomic significance of the biomechanical fragmentation of live molluscan shell material by a bottom-feeding fish (Pogonias cromis) in Texas coastal bays. Palaios 9:254–274.

Chafetz, H., and C. Buczynski. 1992. Bacterially induced lithification of microbial mats. Palaios 7:277–293.

Chatterton, B. D. E., and S. E. Speyer. 1997. Ontogeny. Pp. 173–247 in H. B. Whittington et al. Arthropoda 1, Trilobita, revised. Part O of R. C. Moore and C. Teichert, eds. Treatise on invertebrate paleontology. Geological Society of America and University of Kansas, Boulder, Colo.

Claassen, C. 1998. Shells. Cambridge Manuals in Archaeology. Cambridge University Press, Cambridge.

Clark, G. R., II. 1999. Organic matrix taphonomy in some molluscan shell microstructures. Palaeogeography, Palaeoclimatology, Palaeoecology 149:305–312.

Clyde, W. C., and P. D. Gingerich. 1998. Mammalian community response to the latest Paleocene thermal maximum: an isotaphonomic study in the northern Bighorn Basin, Wyoming. Geology 26:1011–1014.

Collinson, M. E. 1983. Accumulations of fruits and seeds in three small sedimentary environments in southern England and their palaeoecological implications. Annals of Botany 52: 583–592.

Cook, E. 1995. Taphonomy of two non-marine Lower Cretaceous bone accumulations from southeastern England. Palaeogeography, Palaeoclimatology, Palaeoecology 116:263–270.

Crowley, S. S., D. A. Dufek, R. W. Stanton, and T. A. Ryer. 1994. The effects of volcanic ash disturbances on a peat-forming environment: environmental disruption and taphonomic consequences. Palaios 9:158–174.

Cruz-Uribe, K., and R. G. Klein. 1998. Hyrax and hare bones from modern South African eagle roosts and the detection of eagle involvement in fossil bone assemblages. Journal of Archaeological Science 25:135–147.

Cummins, H. 1994. Taphonomic processes in modern freshwater molluscan death assemblages: implications of the freshwater fossil record. Palaeogeography, Palaeoclimatology, Palaeoecology 108:55–73.

Cummins, H., E. N. Powell, R. J. Stanton, and G. M. Staff. 1986a. The size frequency distribution in palaeoecology: effects of taphonomic processes during formation of molluscan death assemblages in Texas bays. Palaeontology 29:495–518.

———. 1986b. The rate of taphonomic loss in modern benthic habitats: how much of the potentially preservable community is preserved? Palaeogeography, Palaeoclimatology, Palaeoecology 52:291–320.

Cunningham, C. R., H. R. Feldman, E. K. Franseen, R. A. Gastaldo, G. Mapes, C. G. Maples, and H.-P. Schultze. 1993. The Upper Carboniferous Hamilton fossil Lagerstätte in Kansas: a valley fill, tidally influenced deposit. Lethaia 26:225–236.

Cutler, A. H. 1989. Shells survive—loss, persistence and accumulation of hardparts in shallow marine sediments. Geological Society of America Abstracts with Programs 21:A71.

———. 1991. Nested faunas and extinction in fragmented habitats. Conservation Biology 5:496–505.

———. 1993. Mathematical models of temporal mixing in the fossil record. Pp. 169–187 in Kidwell and Behrensmeyer 1993.

Cutler, A. H., and A. K. Behrensmeyer. 1996. Models of vertebrate mass mortality events at the K/T Boundary. In G. Ryder, D. Fastovsky, and S. Gartner, eds. The Cretaceous-Tertiary Event and other catastrophes in earth history. Geological Society of America Special Paper 307:375–380.

Cutler A. H., and K. W. Flessa. 1990. Fossils out of sequence: computer simulations and strategies for dealing with stratigraphic disorder. Palaios 5:227–235.

———. 1995. Bioerosion, dissolution and precipitation as taphonomic agents at high and low latitudes. Senckenbergiana Maritima 25:115–121.

Cutler, A. H., A. K. Behrensmeyer, and R. E. Chapman. 1999. Environmental information in a recent bone assemblage: roles of taphonomic processes and ecological change. Palaeogeography, Palaeoclimatology, Palaeoecology 149:359–372.

Daley, G. M. 1993. Passive deterioration of shelly material: a study of the Recent eastern Pacific articulate brachiopod Terebratalia transversa Sowerby. Palaios 8:226–232.

Daley, R. L., and D. W. Boyd. 1996. The role of skeletal microstructure during selective silicification of brachiopods. Journal of Sedimentary Research 66:155–162.

Damuth, J. 1982. Analysis of the preservation of community structure in assemblages of fossil mammals. Paleobiology 8: 434–446.

Davies, D. J. 1993. Taphonomic analysis as a tool for long-term community baseline delineation: taphoanalysis in an environmental impact statement (EIS) for proposed human seafloor disturbances, Alabama continental shelf. Geological Society of America Abstracts with Programs 25:A459.

Davies, D. J., E. N. Powell, and R. J. Stanton. 1989. Taphonomic signature as a function of environmental process: shells and shell beds in a hurricane-influenced inlet on the Texas coast.

Palaeogeography, Palaeoclimatology, Palaeoecology 72:317–356.

Davies-Vollum, K. S., and S. L. Wing. 1998. Sedimentological, taphonomic, and climatic aspects of Eocene swamp deposits (Willwood Formation, Bighorn Basin, Wyoming). Palaios 13:28–40.

Davis, P. G., and D. E. G. Briggs. 1998. The impact of decay and disarticulation on the preservation of fossil birds. Palaios 13:3–13.

Dawson, J. W. 1882. On the results of recent explorations of erect trees containing animal remains in the coal formation of Nova Scotia. Philosophical Transactions of the Royal Society of London B 173:621–659.

Dechant Boaz, D. 1994. Taphonomy and the fluvial environment. Pp. 377–413 in R. S. Corruccini and R. L. Ciochon, eds. Integrative paths to the past: paleoanthropological advances in honor of F. Clark Howell. Prentice-Hall, Englewood Cliffs, N.J.

Demko, T. M., and R. A. Gastaldo. 1992. Paludal environments of the Lower Mary Lee coal zone, Pottsville Formation, Alabama: stacked clastic swamps and peat mires. International Journal of Coal Geology 20:23–47.

Demko, T. M., R. F. Dubiel, and J. T. Parrish. 1998. Plant taphonomy in incised valleys: implications for interpreting paleoclimate from fossil plants. Geology 26:1119–1122.

Dent, S. R. 1995. A taphofacies model of the Recent South Florida continental shelf: a new perspective for a classic, exposed carbonate environment. Ph.D. dissertation. University of Cincinnati, Cincinnati, Ohio.

D'Hondt, S., P. Donaghay, J. C. Zachos, D. Luttenberg, and M. Lindinger. 1998. Organic carbon fluxes and ecological recovery from the Cretaceous-Tertiary mass extinction. Science 282:276–279.

DiMichele, W. A., and R. B. Aronson. 1992. The Pennsylvanian-Permian vegetational transition: a terrestrial analogue to the onshore-offshore hypothesis. Evolution 46:807–824.

DiMichele, W. A., and R. W. Hook. 1992. Paleozoic terrestrial ecosystems. Pp. 205–325 in A. K. Behrensmeyer, J. D. Damuth, W. A. DiMichele, R. Potts, H.-D. Sues, and S. L. Wing, eds. Terrestrial ecosystems through time. University of Chicago Press, Chicago.

DiMichele, W. A., S. H. Mamay, D. S. Chaney, R. W. Hook, and W. J. Nelson. In press. An early Permian flora with Late Permian and Mesozoic affinities from north-central Texas. Journal of Paleontology.

Dodson, P. 1987. Microfaunal studies of dinosaur paleoecology, Judith River Formation of southern Alberta (Canada). Palaeogeography, Palaeoclimatology, Palaeoecology 10:21–74.

Dominguez-Rodrigo, M. 1999. Flesh availability and bone modifications in carcasses consumed by lions: palaeoecological relevance in hominid foraging patterns. Palaeogeography, Palaeoclimatology, Palaeoecology 149:373–388.

Donovan, S. K., ed. 1991. The processes of fossilization. Columbia University Press, New York.

Donovan, S. K., and C. R. C. Paul, eds. 1998. The adequacy of the fossil record. Wiley, New York.

Downing, K. F., and L. E. Park. 1998. Geochemistry and early diagenesis of mammal-bearing concretions from the Sucker Creek Formation (Miocene) of southeastern Oregon. Palaios 13:14–27.

Doyle, P., and D. I. M. Macdonald. 1993. Belemnite battlefields. Lethaia 26:65–80.

Droser, M. L., and D. J. Bottjer. 1993. Trends and patterns of Phanerozoic ichnofabrics. Annual Review of Earth and Planetary Sciences 21:205–225.

Duncan, I. J., D. E. G. Briggs, and M. Archer. 1998. Three-dimensionally mineralized insects and millipedes from the Tertiary of Riversleigh, Queensland, Australia. Palaeontology 41:835–851.

Eberth, D. A. 1990. Stratigraphy and sedimentology of vertebrate microfossil localities in uppermost Judith River Formation (Campanian) of Dinosaur Provincial Park, south-central Alberta, Canada. Palaeogeography, Palaeoclimatology, Palaeoecology 78:1–36.

Eberth, D., R. Rogers and T. Fiorillo, convenors. 1999. Bonebeds: genesis, analysis, and paleoecological significance (program for a symposium). Journal of Vertebrate Paleontology 19(Suppl. to No. 3):7A.

Efremov, J. A. 1940. Taphonomy: new branch of paleontology. Pan American Geologist 74:81–93.

Elder, R. L., and G. R. Smith. 1984. Fish taphonomy and paleoecology. Geobios Mémoire Spécial 8:287–291.

Elias, S. A., T. R. Van Devender, and R. De Baca. 1995. Insect fossil evidence of late glacial and Holocene environments in the Bolson de Mapimi, Chihuahuan Desert, Mexico: comparisons with the paleobotanical record. Palaios 10:454–464.

Emig, C. C. 1990. Examples of post-mortality alteration in Recent brachiopod shells and (paleo)ecological consequences. Marine Biology 104:233–238.

Evans, S., and J. A. Todd. 1997. Late Jurassic soft-bodied wood epibionts preserved by bioimmuration. Lethaia 30:185–189.

Farley, M. B. 1987. Palynomorphs from surface water of the eastern and central Caribbean Sea. Micropaleontology 33:254–262.

Fedonkin, M. A. 1994. Vendian body fossils and trace fossils. Pp. 370–388 in Bengtson 1984.

Feige, A., and F. T. Fürsich. 1991. Taphonomy of the Recent molluscs of Bahía la Choya (Gulf of California, Sonora, Mexico). Zitteliana 18:89–113.

Feldmann, R. M., T. Villamil, and E. G. Kauffman. 1999. Decapod and stomatopod crustaceans from mass mortality Lagerstätten: Turonian (Cretaceous) of Colombia. Journal of Paleontology 73:91–101.

Ferguson, D. K. 1995. Plant part processing and community reconstruction. Eclogae Geologicae Helvetiae 88:627–641.

Fernández-Jalvo, Y., C. Denys, P. Andrews, T. Williams, Y. Dauphin and L. Humphreys. 1998. Taphonomy and paleoecology of Olduvai Bed-I (Pleistocene, Tanzania). Journal of Human Evolution 34:137–172.

Fernández-López, S. 2000. Ammonite taphocycles in carbonate epicontinental platforms. Fifth international symposium on the Jurassic System (Vancouver, B.C.). GeoResearch Forum 6:293–300.

Fiorillo, A. R. 1988. Taphonomy of Hazard Homestead Quarry (Ogalalla Group), Hitchcock County, Nebraska. Contributions to Geology University of Wyoming 26:57–97.

———. 1989. An experimental study of trampling: implications for the fossil record. Pp. 61–72 in R. Bonnichsen and M. H. Sorg, eds. Bone modification. Institute for Quaternary Studies, University of Maine, Orono.

———. 1991. Taphonomy and depositional setting of Careless Creek Quarry (Judith River Formation), Wheatland County, Montana. Palaeogeography, Palaeoclimatology, Palaeoecology 81:281–311.

Flessa, K. W. 1998. Well-traveled cockles: shell transport during the Holocene transgression of the southern North Sea. Geology 26:187–190.

Flessa, K. W., and M. Kowalewski. 1994. Shell survival and time-averaging in nearshore and shelf environments: estimates from the radiocarbon literature. Lethaia 27:153–165.

Flessa, K. W., A. H. Cutler, and K. H. Meldahl. 1993. Time and taphonomy: quantitative estimates of time-averaging and stratigraphic disorder in a shallow marine habitat. Paleobiology 19:266–286.

Foote, M. 1996. On the probability of ancestors in the fossil record. Paleobiology 22:141–151.

———. 1997. Sampling, taxonomic description, and our evolving knowledge of morphological diversity. Paleobiology 23: 181–206.

———. 2000. Origination and extinction components of taxonomic diversity: general problems. In D. H. Erwin and S. L. Wing, eds. Deep time: Paleobiology's perspective. Paleobiology 26(Suppl. to No. 4):74–102.

Foote, M., and D. M. Raup. 1996. Fossil preservation and the stratigraphic ranges of taxa. Paleobiology 22:121–140.

Foote, M., J. P. Hunter, C. M. Janis, and J. J. Sepkoski Jr. 1999. Evolutionary and preservational constraints on origins of biologic groups: divergence times of Eutherian mammals. Science 283:1310–1314.

Franzen, J. L. 1985. Exceptional preservation of Eocene vertebrates in the lake deposits of Grube Messel (West Germany). Philosophical Transactions of the Royal Society of London B 311:181–186.

Frison, G. C., and L. C. Todd. 1986. The Colby Mammoth Site: taphonomy and archaeology of a Clovis kill in northern Wyoming. University of New Mexico Press, Albuquerque.

Fürsich, F. T. 1978. The influence of faunal condensation and mixing on the preservation of fossil benthic communities. Lethaia 11:243–250.

Fürsich, F. T., and M. Aberhan. 1990. Significance of time-averaging for paleocommunity analysis. Lethaia 23:143–152.

Fürsich, F. T., and W. Oschmann. 1993. Shell beds as tools in basin analysis: the Jurassic of Kachchh, western India. Journal of the Geological Society, London 150:169–185.

Gastaldo, R. A. 1988. A conspectus of phytotaphonomy. In W. A. DiMichele and S. L. Wing, eds. Methods and applications of plant paleoecology: notes for a short course. Paleontological Society Special Publication 3:14–28.

———. 1992. Taphonomic considerations for plant evolutionary investigations. The Palaeobotanist 41:211–223.

———. 1994. The genesis and sedimentation of phytoclasts with examples from coastal environments. Pp. 103–127 in A. Traverse, ed. Sedimentation of organic particles. Cambridge University Press, Cambridge.

Gastaldo, R. A., and D. K. Ferguson. 1998. Reconstructing Tertiary plant communities: introductory remarks. Review of Palaeobotany and Palynology 101:3–6.

Gastaldo, R. A., and A. Y. Huc. 1992. Sediment facies, depositional environments, and distribution of phytoclasts in the Recent Mahakam River delta, Kalimantan, Indonesia. Palaios 7:574–591.

Gastaldo, R. A., and J. R. Staub. 1999. A mechanism to explain the preservation of leaf litter lenses in coals derived from raised mires. Palaeogeography, Palaeoclimatology, Palaeoecology 149:1–14.

Gastaldo, R. A., D. P. Douglass, and S. M. McCarroll. 1987. Origin, characteristics, and provenance of plant macrodetritus in a Holocene crevasse splay, Mobile Delta, Alabama. Palaios 2:229–240.

Gastaldo, R. A., T. M. Demko, and Y. Liu. 1993a. Application of sequence and genetic stratigraphic concepts to Carboniferous coal-bearing strata: an example from the Black Warrior Basin, USA. Geologische Rundschau 82:212–226.

Gastaldo, R. A., G. P. Allen, and A. Y. Huc. 1993b. Detrital peat formation in the tropical Mahakam River delta, Kalimantan, eastern Borneo: formation, plant composition, and geochemistry. In J. C. Cobb and C. B. Cecil, eds. Modern and ancient coal-forming environments. Geological Society of America Special Paper 286:107–118.

Gastaldo, R. A., H. Walther, J. Rabold, and D. Ferguson. 1996. Criteria to distinguish parautochthonous leaves in Cenophy-

tic alluvial channel-fills. Review of Palaeobotany and Palynology 91:1–21.

Gastaldo, R. A., W. Riegel, W. Püttmann, U. H. Linnemann, and R. Zetter. 1998. A multidisciplinary approach to reconstruct the Late Oligocene vegetation in central Europe. Review of Palaeobotany and Palynology 101:71–94.

Gehling, J. G. 1999. Microbial mats in terminal Proterozoic siliciclastics: Ediacaran death masks. Palaios 14:40–57.

Gifford-Gonzalez, D. 1991. Bones are not enough: analogues, knowledge, and interpretive strategies in zooarchaeology. Journal of Anthropological Archaeology 10:215–254.

Gilinsky, N. L., and J. B. Bennington. 1994. Estimating numbers of whole individuals from collections of body parts: a taphonomic limitation of the paleontological record. Paleobiology 20:245–258.

Gilinsky, N. L., and P. W. Signor, eds. 1991. Analytical paleobiology. Short Courses in Paleontology No. 4. Paleontological Society, Knoxville, Tenn.

Glover, C. P., and S. M. Kidwell. 1993. Influence of organic matrix on the post-mortem destruction of molluscan shells. Journal of Geology 101:729–747.

Goldstein, S. T., P. Van Cappellen, A. Roychoudhury, and C. Koretsky. 1997. Preservation of salt-marsh foraminifera in experimental arrays deployed below the sediment-water interface, Sapelo Island, Georgia (USA). Geological Society of America Abstracts with Programs 30:A383.

Grayson, D. K. 1989. Bone transport, bone destruction, and reverse utility curves. Journal of Archaeological Science 16:643–652.

Greenstein, B. J. 1989. Mass mortality of the West-Indian echinoid Diadema antillarum (Echinodermata: Echinoidea): a natural experiment in taphonomy. Palaios 4:487–492.

———. 1991. An integrated study of echinoid taphonomy: predictions for the fossil record of four echinoid families. Palaios 6:519–540.

———. 1992. Taphonomic bias and the evolutionary history of the family Cidaridae (Echinodermata: Echinoidea). Paleobiology 18:50–79.

———. 1993. Is the fossil record of regular echinoids really so poor a comparison of living and subfossil assemblages? Palaios 8:587–601.

———. 1999. Taphonomy of reef-building corals II: shallow and deep reef environments of the tropical western Atlantic. Geological Society of America Abstracts with Programs 31:A420.

Greenstein, B. J., and H. A. Moffat 1996. Comparative taphonomy of modern and Pleistocene corals, San Salvador, Bahama. Palaios 11:57–63.

Greenstein, B. J., and J. M. Pandolfi. 1997. Preservation of community structure in modern reef coral life and death assemblages of the Florida Keys: implications for the Quaternary fossil record of coral reefs. Bulletin of Marine Science 61:431–452.

Greenstein, B. J., L. A. Harris, and H. Λ. Curran. 1998. Comparison of Recent coral life and death assemblages to Pleistocene reef communities: implications for rapid faunal replacement of recent reefs. Carbonates and Evaporites 13:23–31.

Grotzinger, J. P. 1994. Trends in Precambrian carbonate sediments and their implication for understanding evolution. Pp. 245–258 in Bengston 1984.

Hadly, E. A. 1999. Fidelity of terrestrial vertebrate fossils to a modern ecosystem. Palaeogeography, Palaeoclimatology, Palaeoecology 149:389–410.

Hageman, S. J., N. P. James, and Y. Bone. 2000. Cool-water carbonate production from epizoic bryozoans on ephemeral substrates. Palaios 15:33–48.

Haglund, W. D., and M. H. Sorg, eds. 1997. Forensic taphonomy, the post-mortem fate of human remains. CRC Press, New York.

Haynes, G. 1985. On watering holes, mineral licks, death, and predation. Pp. 53–71 in D. Meltzer and J. I. Mead, eds. Environments and extinctions in late glacial North America. Center for the Study of Early Man, University of Maine, Orono.

———. 1988. Mass deaths and serial predation: comparative taphonomic studies of modern large-mammal deathsites. Journal of Archaeological Science 15:219–235.

———. 1991. Mammoths, mastodonts and elephants. Cambridge University Press, Cambridge.

Henderson, S. W., and R. W. Frey. 1986. Taphonomic redistribution of mollusk shells in a tidal inlet channel, Sapelo Island, Georgia. Palaios 1:3–16.

Henwood, A. A. 1992a. Exceptional preservation of dipteran flight muscle and the taphonomy of insects in amber. Palaios 7:203–212.

———. 1992b. Soft-part preservation of beetles in Tertiary amber from the Dominican Republic. Palaeontology 35:901–912.

Holland, S. M. 1995. The stratigraphic distribution of fossils. Paleobiology 21:92–109.

———. 1996. Recognizing artifactually generated coordinated stasis: implications of numerical models and strategies for field tests. Palaeogeography, Palaeoclimatology, Palaeoecology 127:147–156.

———. 2000. The quality of the fossil record: a sequence stratigraphic perspective. In D. H. Erwin and S. L. Wing, eds. Deep time: Paleobiology's perspective. Paleobiology 26(Suppl. to No. 4):148–168.

Holland, S. M., and M. E. Patzkowsky. 1999. Models for simulating the fossil record. Geology 27:491–494.

Hudson, J. 1993. From bones to behavior. Occasional Paper No. 21. Center for Archaeological Investigations, Southern Illinois University, Carbondale.

Hughes, N. C., and D. L. Cooper. 1999. Paleobiologic and taphonomic aspects of the ''granulosa'' trilobite cluster, Kope Formation (Upper Ordovician, Cincinnati region). Journal of Paleontology 73:306–319.

Hughes, N. C., and C. C. Labandeira. 1995. The stability of species in taxonomy. Paleobiology 21:401–403.

Hunter, J. 1994. Lack of a high body count at the K-T boundary. Journal of Paleontology 68:1158.

Jablonski, D. J. 1988. Estimates of species durations. Science 240: 969.

———. 1999. The future of the fossil record. Science 284:2114–2116.

———. 2000. Micro- and macroevolution: scale and hierarchy in evolutionary biology and paleontology. In D. H. Erwin and S. L. Wing, eds. Deep time: Paleobiology's perspective. Paleobiology 26(Suppl. to No. 4):15–52.

Jablonski, D. J., S. Lidgard, and P. D. Taylor. 1997. Comparative ecology of bryozoan radiations: origin of novelties in cyclostomes and cheilostomes. Palaios 12:505–523.

Jackson, S. T. 1989. Postglacial vegetational change along an elevational gradient in the Adirondack Mountains (New York): a study of plant macrofossils. New York State Museum Bulletin 465.

———. 1994. Pollen and spores in Quaternary lake sediments as sensors of vegetation composition: theoretical models and empirical evidence. Pp. 253–286 in A. Traverse, ed. Sedimentation of organic particles. Cambridge University Press, Cambridge.

Jackson, S. T., and J. T. Overpeck. 2000. Responses of plant populations and communities to environmental changes of the late Quaternary. In D. H. Erwin and S. L. Wing, eds. Deep time: Paleobiology's perspective. Paleobiology 26(Suppl. to No. 4):194–220.

Jackson, S. T., and D. R. Whitehead. 1991. Holocene vegetation patterns in the Adirondack Mountains. Ecology 72:641–653.

Janzen, D. H. 1977. Why fruits rot, seeds mold, and meat spoils. American Naturalist 111:691–713.

Jodry, M. A., and D. J. Stanford. 1992. Stewart's Cattle Guard Site: an analysis of bison remains in a Folsom kill-butchery campsite. Pp. 101–168 in D. J. Stanford and J. S. Day, eds. Ice Age hunters of the Rockies. Denver Museum of Natural History and University Press of Colorado, Denver. 378 pp.

Johnson, R. G. 1960. Models and methods for analysis of the mode of formation of fossil assemblages. Geological Society of America Bulletin 71:105–1086.

Kah, L. C., and A. H. Knoll. 1996. Microbenthic distribution of Proterozoic tidal flats: environmental and taphonomic considerations. Geology 24:79–82.

Kendricks, P., and P. R. Crane. 1997. The origin and early diversification of land plants: a cladistic study. Smithsonian Institution Press, Washington, D.C.

Kennish, M. J., and R. A. Lutz. 1999. Calcium carbonate dissolution rates in deep-sea bivalve shells on the East pacific Rise at 21°N: results of an 8-year in-situ experiment. Palaeogeography, Palaeoclimatology, Palaeoecology 154:293–299.

Kerbis Peterhans, J. C., R. W. Wrangham, M. L. Carter, and M. D. Hauser. 1993. A contribution to tropical rain forest taphonomy: retrieval and documentation of chimpanzee remains from Kibale Forest, Uganda. Journal of Human Evolution 25: 485–514.

Kershaw, P. J., D. J. Swift, and D. C. Denoon. 1988. Evidence of recent sedimentation in the eastern Irish Sea. Marine Geology 85:1–14.

Kidston, R., and W. H. Lang. 1920. Old Red Sandstone plants showing structure, from the Rhynie Chert bed, Aberdeenshire, Part 2. Transactions of the Royal Society of Edinburgh 52: 603–627.

Kidwell, S. M. 1991. The stratigraphy of shell concentrations. Pp. 211–290 in Allison and Briggs 1991.

———. 1993. Taphonomic expressions of sedimentary hiatus: field observations on bioclastic concentrations and sequence anatomy in low, moderate and high subsidence settings. Geologische Rundschau 82:189–202.

———. 1998. Time-averaging in the marine fossil record: overview of strategies and uncertainties. Geobios 30:977–995.

———. 1999. High fidelity of species relative abundances in marine molluscan death assemblages. Geological Society of America Abstracts with Programs 31:A419.

———. In press. Ecological fidelity of molluscan death assemblages. In S. A. Woodin, J. Y. Aller, and R. C. Aller, eds. Organism-sediment interactions. Belle Baruch Institute Volume. University of South Carolina Press, Columbia.

Kidwell, S. M., and T. Aigner. 1985. Sedimentary dynamics of complex shell beds: implications for ecologic and evolutionary patterns. Pp. 382–395 in U. Bayer and A. Seilacher, eds. Sedimentary and evolutionary cycles. Springer, Berlin.

Kidwell, S. M., and T. Baumiller. 1990. Experimental disintegration of regular echinoids: roles of temperature, oxygen and decay thresholds. Paleobiology 16:247–271.

Kidwell, S. M., and A. K. Behrensmeyer, eds. 1993. Taphonomic approaches to time resolution in fossil assemblages. Short Courses in Paleontology No. 6. Paleontological Society, Knoxville, Tenn.

Kidwell, S. M., and D. W. J. Bosence. 1991. Taphonomy and time-averaging of marine shelly faunas. Pp. 115–209 in Allison and Briggs 1991.

Kidwell, S. M., and P. J. Brenchley. 1994. Patterns in bioclastic accumulation through the Phanerozoic: changes in input or in destruction? Geology 22:1139–1143.

———. 1996. Evolution of the fossil record: thickness trends in marine skeletal accumulations and their implications. Pp.

290–336 *in* D. Jablonski, D. H. Erwin, and J. H. Lipps, eds. Evolutionary paleobiology. University of Chicago Press, Chicago.

Kidwell, S. M., and K. W. Flessa. 1995. The quality of the fossil record: populations, species, and communities. Annual Review of Ecology and Systematics 26:269–299.

Kidwell, S. M., and D. Jablonski. 1983. Taphonomic feedback: ecological consequences of shell accumulation. Pp. 195–248 *in* M. J. S. Tevesz and P. L. McCall, eds. Biotic interactions in recent and fossil benthic communities. Plenum, New York.

Knoll, A. H. 1985. Exceptional preservation of photosynthetic organisms in silicified carbonates and silicified peats. Philosophical Transactions of the Royal Society of London B 311: 111–122.

Knoll, A. H., and V. N. Sergeev. 1995. Taphonomic and evolutionary changes across the Mesoproterozoic-Neoproterozoic transition. Neues Jahrbuch für Geologie und Paläontologie, Abhandlungen 195:289–302.

Knoll, A. H., K. Niklas, P. G. Gensel, and B. Tiffney. 1984. Character diversification and patterns of evolution in early vascular plants. Paleobiology 10:34–47.

Knoll, A. H., I. J. Fairchild, and K. Swett. 1993. Calcified microbes in Neoproterozoic carbonates: implications for our understanding of the Proterozoic/Cambrian transition. Palaios 8:512–525.

Kondo, Y., S. T. Abbott, A. Katamura, P. J. J. Kamp, T. R. E. Naish, T. Kamataki, and G. S. Saul. 1998. The relationship between shellbed type and sequence architecture: examples from Japan and New Zealand. Sedimentary Geology 122:109–127.

Kotler, E., R. E. Martin, and W. D. Liddell. 1992. Experimental analysis of abrasion and dissolution resistance of modern reef-dwelling Foraminifera: implications for the preservation of biogenic carbonate. Palaios 7:244–276.

Kowalewski, M. 1996a. Taphonomy of a living fossil: the lingulide brachiopod *Glottidia palmeri* Dall from Baja California, Mexico. Palaios 11:244–265.

———. 1996b. Time-averaging, overcompleteness, and the geological record. Journal of Geology 104:317–326.

———. 1997. The reciprocal taphonomic model. Lethaia 30:86–88.

Kowalewski, M., and K. W. Flessa. 1996. Improving with age: the fossil record of lingulide brachiopods and the nature of taphonomic megabiases. Geology 24:977–980.

Kowalewski, M., G. A. Goodfriend, and K. W. Flessa. 1998. High-resolution estimates of temporal mixing within shell beds: the evils and virtues of time-averaging. Paleobiology 24: 287–304.

Kranz, P. M. 1974. The anastrophic burial of bivalves and its palaeoecological significance. Journal of Geology 82:237–265.

Kristensen, E., S. I. Ahmed, and A. H. Devol. 1995. Aerobic and anaerobic decomposition of organic matter in marine sediment: which is fastest? Limnology and Oceanography 40: 1430–1437.

Labandeira, C. C. 1998. Early history of arthropod and vascular plant associations. Annual Review of Earth Planetary Sciences 28:153–193.

Labandeira, C. C., and D. M. Smith. 1999. Forging a future for fossil insects: thoughts on the First International Congress of Paleoentomology. Paleobiology 25:154–157.

Labandeira, C. C., K. R. Johnson, and P. Lang. In press. Insect herbivory across the Cretaceous/Tertiary boundary: major extinction and minimum rebound. *In* J. H. Hartman, K. R. Johnson, and D. J. Nichols, eds. The Hell Creek Formation and the Cretaceous-Tertiary boundary in the northern Great Plains—an integrated Continental record at the end of the Cretaceous. Geological Society of America Special Paper.

Lask, P. B. 1993. The hydrodynamic behavior of sclerites from the trilobite *Flexicalymene meeki*. Palaios 8:219–225.

Lescinsky, H. L. 1993. Taphonomy and paleoecology of epi-

bionts on the scallops *Chlamys hastata* (Sowerby 1843) and *Chlamys rubida* (Hinds 1845). Palaios 8:267–277.

———. 1995. The life orientation of concavo-convex brachiopods: overturning the paradigm. Paleobiology 21:520–551.

Li, X., and M. L. Droser. 1997. Nature and distribution of Cambrian shell concentrations: evidence from the Basin and Range province of western United States (California, Nevada and Utah). Palaios 12:11–1126.

Llewellyn, G., and C. G. Messing. 1993. Compositional and taphonomic variations in modern crinoid-rich sediments from the deep-water margin of a carbonate bank. Palaios 8: 554–573.

Llona, A., C. Pinto, and P. Andrews. 1999. Amphibian taphonomy and its application to the fossil record of Dolina (middle Pleistocene, Atapuerca, Spain). Palaeogeography, Palaeoclimatology, Palaeoecology 149:411–430.

Lupia, R. 1995. Paleobotanical data from fossil charcoal: an actualistic study of seed plant reproductive structures. Palaios 10:465–477.

Lyell, C., and J. W. Dawson. 1853. On the remains of a reptile (*Dendrerpeton acadianum*, Wyman and Owen), and of a land shell discovered in the interior of an erect fossil tree in the coal measures of Nova Scotia. Quarterly Journal of the Geological Society of London IX:58–63.

Lyman, R. L. 1985. Bone frequencies: differential transport, in situ destruction, and the MGUI. Journal of Archaeological Science 12:221–236.

———. 1994. Vertebrate taphonomy. Cambridge Manuals in Archaeology, Cambridge University Press, Cambridge.

Lyman, R. L., and G. L. Fox. 1989. A critical evaluation of bone weathering as an indication of bone assemblage formation. Journal of Archaeological Science 16:293–317.

Malinky, J. M., and P. H. Heckel. 1998. Paleoecology and taphonomy of faunal assemblages in gray "core" (offshore) shales in Midcontinent Pennsylvanian cyclothems. Palaios 13: 311–334.

Maliva, R. G., A. H. Knoll, and R. Siever. 1989. Secular change in chert distribution: a reflection of evolving biological participation in the silica cycle. Palaios 4:519–532.

Marean, C. W. 1991. Measuring the post-depositional destruction of bone in archaeological assemblages. Journal of Archaeological Science 18:677–694.

———. 1992. Captive hyaena bone choice and destruction, the Schlepp effect and Olduvai archaeofaunas. Journal of Archaeological Science 19:101–121.

Markwick, P. J. 1998. Crocodilian diversity in space and time: the role of climate in paleoecology and its implications for understanding K/T extinctions. Paleobiology 24:470–497.

Marshall, C. R. 1990. Confidence intervals on stratigraphic ranges. Paleobiology 16:1–10.

———. 1994. Confidence intervals on stratigraphic ranges: partial relaxation of the assumption of randomly distributed fossil horizons. Paleobiology 20:459–460.

Martill, D. M. 1985. The preservation of marine vertebrates in the Lower Oxford Clay (Jurassic) of central England. Philosophical Transactions of the Royal Society of London B 311: 155–165.

———. 1988. Preservation of fish in the Cretaceous Santana Formation of Brazil. Palaeontology 31:1–18.

———. 1990. Macromolecular resolution of fossilized muscle tissue from an elopomorph fish. Nature 346:171–172.

Martin, R. E. 1993. Time and taphonomy: actualistic evidence for time-averaging of benthic foraminiferal assemblages. Pp. 34–56 *in* Kidwell and Behrensmeyer 1993.

———. 1995. Cyclic and secular variation in microfossil biomineralization: clues to the biogeochemical evolution of Phanerozoic oceans. Global and Planetary Change 11:1–23.

———. 1999. Taphonomy, a process approach. Cambridge University Press, Cambridge.

Martin, R. E., J. F. Wehmiller, M. S. Harris, and W. D. Liddell. 1996. Comparative taphonomy of bivalves and foraminifera from Holocene tidal flat sediments, Bahía la Choya, Sonora, Mexico (Northern Gulf of California): taphonomic grades and temporal resolution. Paleobiology 22:80–90.

Martin, R. E., R. T. Patterson, S. T. Goldstein, and A. Kumar, eds. 1999a. Taphonomy as a tool in paleoenvironmental reconstruction and environmental assessment. Palaeogeography, Palaeoclimatology, Palaeoecology 149(special issue).

Martin, R. E., S. P. Hippensteel, D. Nikitina, and J. E. Pizzuto. 1999b. Artificial time-averaging and the recovery of ecological signals preserved in the subfossil record: linking the temporal scales of ecology and paleoecology. Geological Society of America Abstracts with Programs: A356.

Martinez-Delclos, X., and J. Martinell. 1993. Insect taphonomy experiments: their application to the Cretaceous outcrops of lithographic limestones from Spain. Kaupia (Darmstädter Beiträge zur Naturgeschichte) 2:133–144.

Massé, H. L. 1999. Les carbonates associés a la macrofaune des sables fins littoraux en Méditerranée nord-occidentale. Oceanologica Acta 22:413–420.

Maxwell, W. D., and M. J. Benton. 1990. Historical tests of the absolute completeness of the fossil record of tetrapods. Paleobiology 16:322–335.

McGree, H. 1984. On food and cooking: the science and lore of the kitchen. Scribner, New York.

McIlroy, D., and G. A. Logan. 1999. The impact of bioturbation on infaunal ecology and evolution during the Proterozoic-Cambrian transition. Palaios 14:58–72.

McKinney, F. K. 1996. Encrusting organisms on co-occurring disarticulated valves of two marine bivalves: comparison of living assemblages and skeletal residues. Paleobiology 222:534–567.

McKinney, M. L. 1991. Completeness of the fossil record: an overview. Pp. 66–83 in Donovan 1991.

Meldahl, K. H. 1990. Sampling, species abundance, and the stratigraphic signature of mass extinction: a test using Holocene tidal flat molluscs. Geology 18:890–893.

Meldahl, K. H., and K. W. Flessa. 1990. Taphonomic pathways and comparative biofacies and taphofacies in a Recent intertidal/shallow shelf environment. Lethaia 23:43–60.

Meldahl, K. H., D. Scott, and K. Carney. 1995. Autochthonous leaf assemblages as records of deciduous forest communities: an actualistic study. Lethaia 28:383–394.

Meldahl, K. H., K. W. Flessa, and A. H. Cutler. 1997a. Time-averaging and postmortem skeletal survival in benthic fossil assemblages: quantitative comparisons among Holocene environments. Paleobiology 23:209–229.

Meldahl, K. H., O. G. Yajimovich, C. D. Empedocles, C. S. Gustafson, M. M. Hidalgo, and T. W. Reardon. 1997b. Holocene sediments and molluscan faunas of Bahía Concepcíon: a modern analog to Neogene rift basins of the Gulf of California. Geological Society of America Special Paper 318:39–56.

Meyer, D. L., and K. B. Meyer. 1986. Biostratinomy of Recent crinoids (Echinodermata) at Lizard Island, Great Barrier Reef, Australia. Palaios 1:294–302.

Meyer, D. L., W. I. Ausich, and R. E. Terry. 1989. Comparative taphonomy of echinoderms in carbonate facies: Fort Payne Formation (Lower Mississippian) of Kentucky and Tennessee. Palaios 4:533–552.

Miller, A. I. 1988. Spatial resolution in subfossil molluscan remains: implications for paleobiological analyses. Paleobiology 14:91–103.

———. 1997. Dissecting global diversity patterns: examples from the Ordovician radiation. Annual Review of Ecology and Systematics 28:85–104.

Miller, A. I., and H. Cummins. 1990. A numerical model for the formation of fossil assemblages: estimating the amount of post-mortem transport along environmental gradients. Palaios 5:303–316.

———. 1993. Using numerical models to evaluate the consequences to time-averaging in marine fossil assemblages. Pp. 150–168 in Kidwell and Behrensmeyer 1993.

Miller, A. I., and M. Foote. 1996. Calibrating the Ordovician radiation of marine life: implications for Phanerozoic diversity trends. Paleobiology 22:304–309.

Miller, A. I., G. Llewellyn, K. M. Parsons, H. Cummins, M. R. Boardman, B. J. Greenstein, and D. K. Jacobs. 1992. Effect of Hurricane Hugo on molluscan skeletal distributions, Salt River Bay, St. Croix, U. S. Virgin Islands. Geology 20:23–26.

Mirsky, S. 1998. I shall return. Earth 7:48–53.

Morales, M., ed. 1996. The continental Jurassic. Museum of Northern Arizona Bulletin 60.

Nebelsick, J. H. 1992. Echinoid distribution by fragment identification in the northern Bay of Safaga, Red Sea, Egypt. Palaios 7:316–328.

———. 1995. Comparative taphonomy of clypeasteroids. Eclogae Geologica Helvetiae 88:685–693.

———. 1999. Taphonomy of Clypeaster fragments: preservation and taphofacies. Lethaia 32:241–252.

Noe-Nygaard, N. 1987. Taphonomy in archaeology, with special emphasis on man as a biasing factor. Journal of Danish Archaeology 6:7–62.

Norris, R. D. 1986. Taphonomic gradients in shelf fossil assemblages: Pliocene Purisima Formation, California. Palaios 1:256–270.

Oliver, J. S., and R. W. Graham. 1994. A catastrophic kill of ice-trapped coots: time-averaged versus scavenger-specific disarticulation patterns. Paleobiology 20:229–244.

Oliver, J. S., N. E. Sikes, and K. M. Stewart, eds. 1994. Early hominid behavioural ecology. Academic Press, London.

Olson, E. C. 1966. Community evolution and the origin of mammals. Ecology 47:291–302.

Olszewski, T. D. 1999. Taking advantage of time-averaging. Paleobiology 25:226–238.

Olszewski, T. D., and R. R. West. 1997. Influence of transportation and time-averaging in fossil assemblages from the Pennsylvanian of Oklahoma. Lethaia 30:315–329.

Oost, A. P., and P. L. de Boer. 1994. Tectonic and climatic setting of lithographic limestone basins. Geobios Mémoire Spécial 16.321–330.

Orr, P. J., D. E. G. Briggs, and S. L. Kearns. 1998. Cambrian Burgess Shale animals replicated in clay minerals. Science 281:1173–1175.

Palaios. 1999. Unexplored microbial worlds (theme issue). Vol. 141 Palaios. 1999. Unexplored microbial worlds (theme issue). Vol. 14, No. 1.

Palmqvist, P. 1991. Differences in the fossilization potential of bivalve and gastropod species related to their life sites and trophic resources. Lethaia 24:287–288.

———. 1993. Trophic levels and the observational completeness of the fossil record. Revista Española de Paleontología 8:33–36.

Pandolfi, J. M., and B. J. Greenstein. 1997a. Taphonomic alteration of reef coral: effects of reef environment and coral growth form. I. The Great Barrier Reef. Palaios 12:27–42.

———. 1997b. Preservation of community structure in death assemblages of deep-water Caribbean reef corals. Limnology and Oceanography 42:1505–1516.

Parsons, K. M. 1989. Taphonomy as an indicator of environment: Smuggler's Cove, St. Croix, U.S.V.I. In D. K. Hubbard, ed. Terrestrial and marine ecology of St. Croix, U.S. Virgin Islands. West Indies Laboratory Special Publication 8:135–143.

Parsons, K. M., and C. E. Brett. 1991. Taphonomic processes and biases in modern marine environments: an actualistic perspective on fossil assemblage preservation. Pp. 22–65 *in* Donovan 1991.

Parsons, K. M., C. E. Brett, and K. B. Miller. 1988. Taphonomy and depositional dynamics of Devonian shell-rich mudstones. Palaeogeography, Palaeoclimatology, Palaeoecology 63:109–140.

Parsons-Hubbard, K. M., W. R. Callender, E. N. Powell, C. E. Brett, S. E. Walker, A. L. Raymond, and G. M. Staff. 1999. Rates of burial and disturbance of experimentally-deployed molluscs: implications for preservation potential. Palaios 14:337–351.

Paul, C. R. C. 1992. How complete does the fossil record have to be? Revista Española Paleontología 7:127–133.

———. 1998. Adequacy, completeness and the fossil record. Pp. 1–22 *in* S. K. Donovan and C. R. C. Paul, eds. The adequacy of the fossil record. Wiley, New York.

Perry, C. T. 1996. The rapid response of reef sediments to changes in community composition: implications for time averaging and sediment accumulation. Journal of Sedimentary Research 66:459–467.

———. 1999. Reef framework preservation in four contrasting modern reef environments, Discovery Bay, Jamaica. Journal of Coastal Research 15:796–812.

Peterson, C. H. 1977. The paleoecological significance of undetected short-term temporal variability. Journal of Paleontology 51:976–981.

Plummer, T., and A. M. Kinyua. 1994. Provenancing of hominid and mammalian fossils from Kanjera, Kenya, using EDXRF. Journal of Archaeological Science 21:553–563.

Potts., R. 1986. Temporal span of bone accumulations at Olduvai Gorge and implications for early hominid foraging behavior. Paleobiology 12:25–31.

———. 1988. Early hominid activities at Olduvai. Aldyne de Gruyter, New York.

Potts, R., A. K. Behrensmeyer, and P. Ditchfield. 1999. Paleolandscape variation and early Pleistocene hominid activities: Members 1 and 7, Olorgesailie Formation. Journal of Human Evolution 37:747–788.

Poulicek, M., G. Goffinet, C. Jeuniaux, A. Simon, and M. F. Voss-Foucart. 1988. Early diagenesis of skeletal remains in marine sediments: a 10 years study. Actes Colloque Recherches Océanographiques en Mer Méditerranée, Université Etat, Liege 107–124.

Powell, E. N., and D. J. Davies. 1990. When is an "old" shell really old? Journal of Geology 98:823–844.

Powell, E. N., G. Staff, D. J. Davies, and W. R. Callender. 1989. Macrobenthic death assemblages in modern marine environments: formation, interpretation and application. CRC Critical Reviews in Aquatic Science 1:555–589.

Powell, E. N., R. J. Stanton Jr., A. Logan, and M. A. Craig. 1992. Preservation of Mollusca in Copano Bay, Texas. The long-term record. Palaeogeography, Palaeoclimatology, Palaeoecology 95:209–228.

Prager, E. J., J. B. Southard, and E. R. Vivoni-Gallart. 1996. Experiments on the entrainment threshold of well-sorted and poorly sorted carbonate sands. Sedimentology 43:33–40.

Pratt, B. R. 1998. Probable predation on Upper Cambrian trilobites and its relevance for the extinction of soft-bodied Burgess Shale-type animals. Lethaia 31:73–88.

Rasmussen, K. A., and C. E. Brett. 1985. Taphonomy of Holocene cryptic biotas from St. Croix, Virgin Islands: information loss and preservation biases. Geology 13:551–553.

Raup, D. M. 1975. Taxonomic diversity estimation using rarefaction. Paleobiology 1:333–342.

Raymond, A., and C. Metz. 1995. Laurussian land-plant diversity during the Silurian and Devonian: mass extinction, sampling bias, or both? Paleobiology 21:74–91.

Retallack, G. J. 1990. Soils of the past: an introduction to paleopedology. Unwin Hyman, London.

Retallack, G. J., and D. L. Dilcher. 1986. Reconstructions of selected seed ferns. Annals of the Missouri Botanical Garden 75: 1010–1057.

Rich, F. J. 1989. A review of the taphonomy of plant remains in lacustrine sediments. Review of Palaeobotany and Palynology 58:33–46.

Richmond, D. R., and T. H. Morris. 1996. The dinosaur death-trap of the Cleveland-Lloyd Dinosaur Quarry, Emery County, Utah. Pp. 533–546 *in* Morales 1996.

Rigby, J. K., Jr., K. R. Newman, J. Smit, S. Van Der Kaars, R. E. Sloan, and J. K. Rigby. 1987. Dinosaurs from the Paleocene part of the Hell Creek Formation, McCone County, Montana. Palaios 2:296–302.

Rivas, P., J. Aguirre, and J. C. Braga. 1997. *Entolium* beds: hiatal shell concentrations in starved pelagic settings (middle Liaassic, SE Spain). Eclogae Geologica Helvetiae 90:293–301.

Robinson, J. M. 1990. Lignin, land plants, and fungi. Biological evolution affecting Phanerozoic oxygen balance. Geology 15: 607–610.

———. 1991. Phanerozoic atmospheric reconstructions: a terrestrial perspective. Palaeogeography, Palaeoclimatology, Palaeoecology 97:51–62.

Rogers, R. R. 1990. Taphonomy of three dinosaur bone beds in the Upper Cretaceous Two Medicine Formation of northwestern Montana: evidence for drought-related mortality. Palaios 5:394–413.

———. 1993. Systematic patterns of time-averaging in the terrestrial vertebrate record: a Cretaceous case study. Pp. 228–249 *in* S. M. Kidwell and A. K. Behrensmeyer, eds. Taphonomic approaches to time resolution in fossil assemblages. Short Courses in Paleontology 6 (Paleontological Society).

Rogers, R. R., and S. M. Kidwell. 2000. Associations of vertebrate skeletal concentrations and discontinuity surfaces in continental and shallow marine records: a test in the Cretaceous of Montana. Journal of Geology 108:131–154.

Rolfe, W. D. I., E. N. K. Clarkson, and A. L. Panchen, eds. 1994 (1993). Volcanism and early terrestrial biotas. Transactions of the Royal Society of Edinburgh (Earth Sciences) 84.

Roopnarine, P. D. 1999. Breaking the enigma of stratophenetic series: a computational approach to the analysis of microevolutionary mode. Geological Society of America Abstracts with Programs 31(7):A42.

Roopnarine, P. D., G. Byars, and P. Fitzgerald. 1999. Anagenetic evolution, stratophenetic patterns, and random walk models. Paleobiology 25:41–57.

Roy, K., J. W. Valentine, D. Jablonski, and S. M. Kidwell. 1996. Scales of climatic variability and time averaging in Pleistocene biotas: implications for ecology and evolution. Trends in Ecology and Evolution 11:458–463.

Sadler, P. M. 1981. Sediment accumulation and the completeness of stratigraphic sections. Journal of Geology 89:569–584.

Sander, P. M. 1989. Early Permian depositional environments and pond bonebeds in central Archer County, Texas. Palaeogeography, Palaeoclimatology, Palaeoecology 69:1–21.

Sanders, H. L. 1968. Marine benthic diversity: a comparative study. American Naturalist 102:243–282.

Schaal, S., and W. Ziegler, eds. 1992. Messel: an insight into the history of life and of the Earth. Translated by M. Shaffer-Fehre. Clarendon, Oxford.

Schäfer, W. 1972. Ecology and paleoecology of marine environments. University of Chicago Press, Chicago.

Scheihing, M. 1980. Reduction of wind velocity by the forest canopy and the rarity of non-arborescent plants in the Upper Carboniferous fossil record. Augumenta Palaeobotanica 6: 133–138.

Scheihing, M. H., and H. W. Pfefferkorn. 1984. The taphonomy of land plants in the Orinoco Delta: a model for the incorporation of plant parts in clastic sediments of late Carboniferous age of Euramerica. Review of Palaeobotany and Palynology 41:205–240.

Schmude, D. E., and C. J. Weege. 1996. Stratigraphic relationship, sedimentology, and taphonomy of Meilyn, a dinosaur quarry in the basal Morrison Formation of Wyoming. Pp. 547–554 in Morales 1996.

Schopf, T. J. M. 1978. Fossilization potential of an intertidal fauna: Friday Harbor, Washington. Paleobiology 4:261–270.

Schubert, J. K., D. L. Kidder, and D. H. Erwin. 1997. Silica-replaced fossils through the Phanerozoic. Geology 25:1031–1034.

Scoffin, T. P. 1992. Taphonomy of coral reefs: a review. Coral Reefs 11:57–77.

Scott, A. C. 1990. Anatomical preservation of fossil plants. Pp. 263–266 in Briggs and Crowther 1990.

Scott, R. W. 1978. Approaches to trophic analysis of paleocommunities. Lethaia 11:1–14.

Seilacher, R. W. 1984. Late Precambrian and Early Cambrian metazoa: preservational or real extinctions? Pp. 159–168 in H. D. Holland and A. F. Trendall, eds. Patterns of change in earth evolution. Springer, Berlin.

———. 1985. The Jeram Model: event condensation in a modern intertidal environment. Pp. 336–341 in U. Bayer and A. Seilacher, eds. Sedimentary and evolutionary cycles. Springer, Berlin.

———. 1994. Early multicellular life: late Proterozoic fossils and the Cambrian extinction. Pp. 389–400 in Bengtson 1984.

Seilacher, A., W. E. Reif, and F. Westphal. 1985. Sedimentological, ecological, and temporal patterns of fossil Lagerstätten. Philosophical Transactions of the Royal Society of London B 311:5–23.

Sepkoski, J. J., Jr. 1978. A kinetic model of Phanerozoic taxonomic diversity: I. Analysis of marine orders. Paleobiology 4:223–251.

———. 1990. The taxonomic structure of periodic extinction. In V. Sharpton and P. Ward, eds. Global catastrophes in earth history. Geological Society of America Special Paper 247:33–44.

———. 1993. Ten years in the library: new data confirm paleontological patterns. Paleobiology 19:43–51.

Sepkoski, J. J., Jr., and C. F. Koch. 1996. Evaluating paleontologic data relating to bio-events. Pp. 21–34 in O. H. Walliser, ed. Global events and event stratigraphy in the Phanerozoic. Springer, Berlin.

Sepkoski, J. J., Jr., R. K. Bambach, D. M. Raup, and J. W. Valentine. 1981. Phanerozoic marine diversity and the fossil record. Nature 293:435–437.

Sepkoski, J J, Jr., R. K. Bambach, and M. L. Droser. 1991. Secular changes in Phanerozoic event bedding and the biological overprint. Pp. 298–312 in G. Einsele, W. Ricken, and A. Seilacher, eds. Cycles and events in stratigraphy. Springer, Berlin.

Sept, J. M. 1994. Bone distribution in a semi-arid riverine habitat in eastern Zaire: implications for the interpretation of faunal assemblages at early archaeological sties. Journal of Archaeological Science 21:217–235.

Silva de Echols, C. M. H. M. 1993. Diatom infestation of Recent crinoid ossicles in temperate waters, Friday harbor Laboratories, Washington: implications for biodegradation of skeletal carbonates. Palaios 8:278–288.

Simões, M. G., A. C. Marques, L. H. C. Mello, and R. P. Ghilardi. 2000a. The role of taphonomy in cladistic analysis: a case study in Permian bivalves. Revista Española de Paleontología 15:153–164.

Simões, M. G., M. Kowalewski, F. F. Torello, R. P. Ghilardi, and L. H. C. Mello. 2000b. Early onset of modern-style shell beds in the Permian sequences of the Paraná Basin: implications for the Phanerozoic trend in bioclastic accumulations. Revista Brasileira de Geociências 30:495–499.

Smith, A. M., and C. S. Nelson. 1994. Selectivity in sea-floor processes: taphonomy of bryozoans. Pp. 177–180 in P. J. Hayward, J. S. Ryland, and P. D. Taylor, eds. Biology and palaeobiology of bryozoans. Proceedings of the ninth international bryozoology conference, 1992. Olsen and Olsen, Fredensborg, Denmark.

Smith, D. M. 2000. Beetle taphonomy in a recent ephemeral lake, southeastern Arizona. Palaios 15:152–160.

Smith, G. R., R. F. Stearley, and C. E. Badgley. 1988. Taphonomic bias in fish diversity from Cenozoic floodplain environments. Palaeogeography, Paleoclimatology, Palaeoecology 63.263–273.

Smith, R. M. H. 1993. Vertebrate taphonomy of Late Permian floodplain deposits in the southwestern Karoo Basin of South Africa. Palaios 8:45–67.

———. 1995. Changing fluvial environments across the Permian-Triassic boundary in the Karoo Basin, South Africa and possible causes of tetrapod extinctions. Palaeogeography, Palaeoclimatology, Palaeoecology 117:81–104.

Smith, R. M. H., and J. Kitching. 1996. Sedimentology and vertebrate taphonomy of the Tritylodon Acme Zone: a reworked palaeosol in the Lower Jurassic Elliot Formation, Karoo Supergroup, South Africa. Pp. 531–532 in Morales 1996.

Solomon, S., I. Davidson, and D. Watson. 1990. Problem solving in taphonomy. Tempus: archaeology and material culture studies in anthropology, Vol. 2. University of Queensland, St. Lucia, Queensland, Australia.

Speyer, S. E., and C. E. Brett. 1986. Trilobite taphonomy and Middle Devonian taphofacies. Palaios 1:312–327.

———. 1991. Taphofacies controls: background and episodic processes in fossil assemblage preservation. Pp. 501–545 in Allison and Briggs 1991.

Spicer, R. A. 1980. The importance of depositional sorting to the biostratigraphy of plant megafossils. Pp. 171–183 in D. L. Dilcher and T. N. Taylor, eds. Biostratigraphy of fossil plants. Dowdon, Hutchinson, and Ross, New York.

———. 1991. Plant taphonomic processes. Pp. 71–113 in Allison and Briggs 1991.

Stachowitsch, M. 1984. Mass mortality in the Gulf of Trieste: the course of community destruction. Marine Ecology (Publicazioni della Stazione Zoologica di Napoli) 5:243–264.

Staff, G. M., E. N. Powell, R. J. Stanton Jr., and H. Cummins. 1985. Biomass: is it a useful tool in paleocommunity reconstruction? Lethaia 18:209–232.

Staff, G. M., R. J. Stanton Jr., E. N. Powell, and H. Cummins. 1986. Time averaging, taphonomy and their impact on paleocommunity reconstruction: death assemblages in Texas bays. Geological Society of America Bulletin 97:428–443.

Stankiewicz, B. A., H. N. Poinar, D. E. G. Briggs, R. P. Evershed, and G. O. Poinar Jr. 1998. Chemical preservation of plants and insects in natural resins. Proceedings of the Royal Society of London B 265:641–647.

Stanley, S. E., and L. A. Hardie. 1998. Secular oscillations in the carbonate mineralogy of reef-building and sediment-producing organism driven by tectonically forced shifts in seawater chemistry. Palaeogeography, Palaeoclimatology, Palaeoecology 144:3–19.

Stewart, K. M., L. Lebranc, D. P. Matthiesen and J. West. 1999. Microfaunal remains from a modern east African raptor roost: patterning and implications for fossil bone scatters. Paleobiology 25:483–503.

Stoner, A. W., and M. Ray. 1996. Shell remains provide clues to historical distribution and abundance patterns in a large seagrass-associated gastropod (Stombus gigas). Marine Ecology Progress Series 135:101–108.

Strauss, D., and P. M. Sadler. 1989. Classical confidence intervals and Bayesian probability estimates for ends of local taxon ranges. Mathematical Geology 21:411–427.

Sutcliffe, A. J. 1990. Rates of decay of mammalian remains in the permafrost environment of the Canadian High Arctic. Pp. 161–186 in C. R. Harrington, ed. Canada's missing dimension: science and history in the Canadian Arctic Islands, Vol. 1. Canadian Museum of Nature, Ottawa.

Tappan, M. J. 1994a. Savanna ecology and natural bone deposition: implications for early hominid site formation, hunting and scavenging. Current Anthropology 36:223–260.

———. 1994b. Bone weathering in the tropical rain forest. Journal of Archaeological Science 21:667–673.

Tardy, Y., R. N'Kounkou, and J.-L. Probst. 1989. The global water cycle and continental erosion during Phanerozoic time. American Journal of Science 289:455–483.

Taylor, P. D. 1990. Preservation of soft-bodied and other organisms by bioimmuration: a review. Palaeontology 33:1–17.

———. 1994. Evolutionary paleoecology of symbioses between bryozoans and hermit crabs. Historical Biology 9:157–205.

Thayer, C. W. 1983. Sediment-mediated biological disturbance and the evolution of marine benthos. Pp. 479–625 in M. J. S. Tevesz and P. L. McCall, eds. Biotic interactions in recent and fossil benthic communities. Plenum, New York.

Thomasson, J. R. 1991. Sediment-borne "seeds" from Sand Creek, northwestern Kansas: taphonomic significance and paleoecological and paleoenvironmental implications. Palaeogeography, Palaeoclimatology, Palaeoecology 85:213–225.

Traverse, A. 1990. Studies of pollen and spores in rivers and other bodies of water, in terms of source-vegetation and sedimentation, with special reference to Trinity River and Bay, Texas. Review of Palaeobotany Palynology 64:297–303.

———, ed. 1994. Sedimentation of organic particles. Cambridge University Press, Cambridge.

Trueman, C. N. 1999. Rare earth element geochemistry and taphonomy of terrestrial vertebrate assemblages. Palaios 14: 555–568.

Trueman, C. N., and M. J. Benton. 1997. A geochemical method to trace the taphonomic history of reworked bones in sedimentary settings. Geology 25:263–266.

Underwood, C. J., and S. M. Bottrell. 1994. Diagenetic controls on multiphase pyritization of graptolites. Geological Magazine 131:315–327.

Valentine, J. W. 1989. How good was the fossil record? Clues from the Californian Pleistocene. Paleobiology 15:83–94.

Valentine, J. W., D. Jablonski, and D. H. Erwin. 1999. Fossils, molecules, and embryos: new perspectives on the Cambrian explosion. Development 126:851–859.

Vaughan, A., and G. Nichols. 1995. Controls on the deposition of charcoal: implications for sedimentary accumulations of fusain. Journal of Sedimentary Research A 65:129–135.

Vermeij, G. J. 1977. The Mesozoic marine revolution: evidence from snails, predators and grazers. Paleobiology 2:245–258.

———. 1987. Evolution and escalation: an ecological history of life. Princeton University Press, Princeton, N.J.

Voorhies, M. R. 1992. Ashfall: life and death at a Nebraska waterhole ten million years ago. University of Nebraska State Museum, Museum Notes 81.

Vrba, E. S. 1995. On the connections between paleoclimate and evolution. Pp. 24–45 in E. S. Vrba, G. H. Denton, T. C. Partridge, and L. H. Burckle, eds. Paleoclimate and evolution, with emphasis on human origins. Yale University Press, New Haven, Conn.

Wagner, P. J. 2000a. The quality of the fossil record and the accuracy of phylogenetic inferences about sampling and diversity. Systematic Biology 49:65–86.

———. 2000b. Phylogenetic analyses and the fossil record: tests

and inferences, hypotheses and models. In D. H. Erwin and S. L. Wing, eds. Deep time: Paleobiology's perspective. Paleobiology 26(Suppl. to No. 4):341–371.

Waisfeld, B. G., T. M. Sanchez, and M. G. Carrera. 1999. Biodiversification patterns in the Early Ordovician of Argentina. Palaios 14:198–214.

Walker, K. R., and R. K. Bambach. 1971. The significance of fossil assemblages from fine-grained sediments: time-averaged communities. Geological Society of America Abstracts with Programs 3:783–784.

Walker, K. R., and W. W. Diehl. 1985. The role of marine cementation in the preservation of lower Paleozoic assemblages. Philosophical Transactions of the Royal Society of London B 311:143–153.

Walker, S. E. 1988. Taphonomic significance of hermit crabs (Anomura: Paguridea): epifaunal hermit crab—infaunal gastropod example. Palaeogeography, Palaeoclimatology, Palaeoecology 63:45–71.

———. 1989. Hermit crabs as taphonomic agents. Palaios 4:439–452.

———. 1995. Taphonomy of modern and fossil intertidal gastropod associations from Isla Santa Cruz and Isla Santa Fe, Galapagos Islands. Lethaia 28:371–382.

Walker, S. E., and J. R. Voight. 1994. Paleoecologic and taphonomic potential of deepsea gastropods. Palaios 9:48–58.

Walker, S. E., P. Van Cappellen, A. Roychoudhury, and C. Koretsky. 1997. Preservation of experimentally deployed molluscan carbonate below the sediment-water interface. Geological Society of America Abstracts with Programs 30:266.

Walker, S. E., K. Parsons-Hubbard, E. N. Powell, and C. E. Brett. 1998. Bioerosion or bioaccumulation? Shelf-slope trends for epi- and endobionts on experimentally deployed gastropod shells. Historical Biology 13:61–72.

Walter, M. R., J. J. Veevers, C. R. Calver, and K. Grey. 1995. Neoproterozoic stratigraphy of the Centralian Superbasin, Australia. Precambrian Research 73:173–195.

Walters, L. J., and D. S. Wethey. 1991. Settlement, refuges, and adult body form in colonial marine invertebrates: a field experiment. Biological Bulletin 180:112–118.

Warren, R. E. 1991. Ozarkian fresh-water mussels (Unionoidea) in the upper Eleven Point River, Missouri. American Malacological Bulletin 8:131–137.

Wayne, R. K., J. A. Leonard, and A. Cooper. 1999. Full of sound and fury: the recent history of ancient DNA. Annual Review of Ecology and Systematics 30:457–477.

Webb, T., III. 1993. Constructing the past from late-Quaternary pollen data: temporal resolution and a zoom lens space-time perspective. In Kidwell and Behrensmeyer 1993.

Wehmiller, J. F., L. L. York, and M. L. Bart. 1995. Amino acid racemization geochronology of reworked Quaternary mollusks on US Atlantic coast beaches: implications for chronostratigraphy, taphonomy, and coastal sediment transport. Marine Geology 124:303–337.

Westall, F., L. Boni, and E. Guerzoni. 1995. The experimental silicification of microorganisms. Palaeontology 38:495–528.

Whittington, H. B., and S. Conway Morris, eds. 1985. Extraordinary fossil biotas: their ecological and evolutionary significance. Philosophical Transactions of the Royal Society of London B 311:1–192.

Wilby, P. R., and D. E. G. Briggs. 1997. Taxonomic trends in the resolution of detail preserved in fossil phosphatized soft tissues. Geobios Mémoire Spécial 20:493–502.

Wilby, P. R., D. E. G. Briggs, P. Bernier, and C. Gaillard. 1996. Role of microbial mats in the fossilization of soft tissues. Geology 24:787–790.

Wilf, P., and C. Labandeira. 1999. Response of plant-insect associations to Paleocene-Eocene warming. Science 284:2153–2156.

Wilf, P., K. C. Beard, K. S. Davies-Vollum, and J. W. Norejko. 1998. Portrait of a Late Paleocene (Early Clarkforkian) terrestrial ecosystem: Big Multi Quarry and associated strata, Washakie Basin, southwestern Wyoming. Palaios 13:514–532.

Williams, M. E. 1994. Catastrophic versus noncatastrophic extinction of the dinosaurs: testing, falsifiability, and the burden of proof. Journal of Paleontology 68:183–190.

Wilson, M. V. H. 1988a. Taphonomic processes: information loss and information gain. Geoscience Canada 15:131–148.

———. 1988b. Reconstruction of ancient lake environments using both autochthonous and allochthonous fossils. Palaeogeography, Palaeoclimatology, Palaeoecology 62:609–623.

———. 1988c. Predation as a source of fish fossils in Eocene lake sediments. Palaios 2:497–504.

———. 1993. Calibration of Eocene varves at Horsefly, British Columbia, Canada, and temporal distribution of specimens of the Eocene fish *Amyzon aggregatum* Wilson. Kaupia (Darmstadter Beitrage zur Naturgeschichte) 2:27–38.

Wilson, M. V. H., and D. G. Barton. 1996. Seven centuries of taphonomic variation in Eocene freshwater fishes preserved in varves: paleoenvironments and temporal averaging. Paleobiology 22:535–542.

Wing, S. L., and W. A. DiMichele. 1995. Conflict between local and global changes in plant diversity through geologic time. Palaios 10:551–564.

Wing, S. L, L. J. Hickey, and C. C. Swisher. 1993. Implications of an exceptional fossil flora for Late Cretaceous vegetation. Nature 363:342–344.

Wolff, R. G. 1975. Sampling and sample size in ecological analyses of fossil mammals. Paleobiology 1:195–204.

Xiao, S., and A. H. Knoll. 1999. Fossil preservation in the Neoproterozoic Doushantuo phosphorite Lagerstätte, South China. Lethaia 32:219–240.

Yang, H., and S. Yang. 1994. The Shanwang fossil biota in eastern China: a Miocene Konservat-Lagerstätte in lacustrine deposits. Lethaia 27:345–354.

Zuschin, M., and J. Hohenegger. 1998. Subtropical coral-reef associated sedimentary facies characterized by molluscs (northern Bay of Safaga, Red Sea, Egypt). Facies 38:229–254.

Zuschin, M., J. Hohenegger, and F. F. Steininger. 2000. A comparison of living and dead molluscs on coral reef associated hard substrata in the northern Red Sea—implications for the fossil record. Palaeogeography, Palaeoclimatology, Palaeoecology 159:167–190.

The quality of the fossil record: a sequence stratigraphic perspective

Steven M. Holland

Abstract.—As paleobiology continues to address an ever broader array of questions, it becomes increasingly important to interpret confidently the meaning of the pattern of fossil occurrences as found in outcrop. To this end, sequence stratigraphy is an important tool for paleobiologists because it predicts the distribution of unconformities, facies changes, and changes in sedimentation rate, all factors known from numerous previous studies to affect the quality of the fossil record. Computer simulations now make it possible not only to model sequence architecture within sedimentary basins, but also to model the occurrence of fossils within those basins. These models generate predictions regarding the stratigraphic distribution of first and last occurrences, changes in species abundance, changes in species morphology, and the distribution of gaps in fossil ranges. Although confirmation of some of these predictions has been found in field studies, the extent to which these predictions describe the fossil record in general is still unknown. If the predicted patterns of fossil occurrences are found to be widespread, it will suggest that a relatively simple model of fossil occurrences in outcrops could become a new tool for solving a wide array of paleobiologic and biostratigraphic problems. With such models, paleobiologists and biostratigraphers will be able to use model data to test the accuracy of newly developed methods of analysis.

Steven M. Holland. Department of Geology, The University of Georgia, Athens, Georgia 30602-2501.
E-mail: stratum@gly.uga.edu

Accepted: 21 March 2000

Introduction

As paleontology continues to push the frontiers of understanding the patterns and processes of evolution in the fossil record, understanding the quality of the fossil record itself becomes increasingly important. The pattern of fossil occurrences in outcrops certainly preserves some temporal signal, but it is a dangerous seduction to regard those fossil occurrences as predominantly a record of the temporal sequence of events. Incompleteness of the fossil record in a variety of forms often distorts or scrambles the true record of a sequence of events.

Incompleteness of the fossil record has long been the bane of paleontology. Several decades of research have revealed numerous sources of this incompleteness (McKinney 1991; Donovan and Paul 1998). More recent advances in our understanding of how the stratigraphic record is constructed are allowing paleontologists to assemble a more holistic view of fossil preservation (Kidwell 1991a,b; Brett 1995). Computer modeling can be used to quantify many of these empirical relationships and thereby to systematically evaluate the interactions of underlying controls on the fossil record (Holland 1995b; Holland and Patzkowsky 1999). Numerical models of the fossil record have the potential to offer new interpretations of the fossil record, to suggest better sampling strategies to avoid some of the worst artifacts of the fossil record, and to develop new methods of quantifying the stratigraphic distribution of fossils.

This paper will review some of the most pervasive problems affecting the stratigraphic distribution of marine body fossils in outcrop and show how many of these patterns have been or could be quantified for computer simulations. A series of predictions made by these models will then be offered. These predictions need to be tested in field studies for a wide variety of taxonomic groups from a wide variety of geologic settings to determine the extent to which this simple model captures the fossil record as seen in outcrops. Several approaches will be offered for avoiding the worst problems imposed by these modeled sources of incompleteness. Finally, ideas for future studies and applications of the model are proposed.

     0094-8373/00/2604-0006/$1.00

Modeling the Stratigraphic Distribution of Fossils in Outcrop

Of the many controls on the occurrence of fossils in outcrop, four are singled out here as being significant in their effects and readily amenable to modeling. These include the rarity of fossils, facies control, sequence architecture, and changes in sedimentation rate. Other factors not included at this stage may be important factors to include in subsequent models; some of these are discussed below.

Rarity.—Perhaps the most immediate observation of fossil preservation in outcrops is that fossils of any given species do not occur in every layer of rock. Even when collections are made under the best conditions, in the appropriate age of rock and in the appropriate sedimentary facies, the occurrence of fossils is still erratic. Reflecting this, any species can be most simply modeled as having some probability of occurrence. This single probability subsumes all of the various factors—ecologic, biostratinomic, diagenetic, weathering, recognition, collection, identification—that affect whether a fossil is reported from any given layer. This simplest model of fossil preservation assumes that all factors governing the preservation of fossils are essentially so unpredictable or sufficiently confounding that fossil occurrence must be treated as random. Although such a model overstates our ignorance of the causes of nonpreservation and will likely be replaced with more predictive models in the future, it works quite well in many cases. For example, this simplest of models underlies most existing methods of placing confidence limits on the stratigraphic ranges of fossils (Paul 1982; Strauss and Sadler 1989; Marshall 1990).

In short, any realistic model of the fossil record must address the rarity of fossils: even under apparently optimal conditions, fossils of a given species may not be found in a given bed. This non-orderly, nonpredictable aspect of fossil preservation can be modeled by considering that each fossil species has some probability that it will be preserved, collected, and identified.

Facies Control.—It is widely recognized that most marine fossil species tend to occur more commonly in some facies than in others (Boucot 1981). Although taphonomic processes undoubtedly influence this pattern to some extent, evidence from the distribution of species in modern oceans indicates that this pattern primarily reflects the ecological preferences of organisms (Kidwell and Bosence 1991).

A range of physical, chemical, and biological factors control the geographic distribution of modern marine species. Salinity and temperature exert possibly the greatest effects on the distribution of species, but illumination, substrate consistency, water turbidity, nutrients, wave and current shear stress, dissolved oxygen concentration, water pressure, and biotic interactions can also be important (Newell 1970; Smith and Carlton 1975; Tait and Dipper 1998). Many of these factors are strongly correlated with water depth and cause organisms to be zoned with water depth. Although depth zonation is most striking in rocky intertidal areas (Smith and Carlton 1975), it has been reported across all ocean depths, including continental shelves (e.g., Pachut et al. 1995; Piepenburg and Schmid 1996; Connell and Lincoln-Smith 1999) and the deep sea (e.g., Vinogradova 1962; Grassle et al. 1975; Thompson et al. 1987; Pineda 1993). Depth zonation has been observed at all latitudes—tropical, temperate, and polar (e.g., Soto 1991; Ivany et al. 1994; Piepenburg et al. 1996). Depth zonation has been reported for a wide range of groups, including foraminifera, arthropods, molluscs, echinoderms, and vertebrates (Fig. 1) (Walton 1955; Buhl-Mortensen and Høisæter 1993; Piepenburg et al. 1997; Freeman et al. 1998; Connell and Lincoln-Smith 1999). In short, depth zonation is ubiquitous.

Because the distribution of most marine species is controlled by factors that correlate with depth (such as temperature, wave shear stress, nutrients, etc.) rather than factors that are directly caused by water depth (i.e., water pressure [see Siebenaller and Somero 1989]), few species occur in the same water depth everywhere (Murray 1991). For example, the preferred water depth of many invertebrates increases from temperate into tropical latitudes because cooler waters tend to be found at increasingly deeper depths as the tropics are ap-

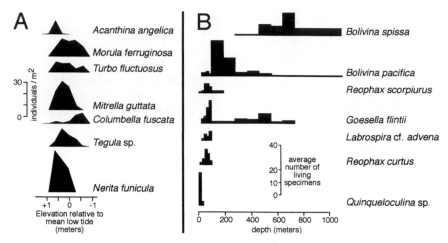

FIGURE 1.   Examples of the relationship between species abundance and water depth in modern marine organisms. A, Intertidal invertebrates from the Gulf of California (E. H. Boyer unpublished data *in* Brown 1984). B, Benthic foraminifera from Baja California (Walton 1955). Most species have a single depth range in which they occur most abundantly and their abundance decreases into greater and lesser water depths. Many of the distributions can be characterized, to a first order, by a simple symmetrical peaked distribution, such as a Gaussian distribution. Some distributions are clearly asymmetric, most commonly with a longer tail into deeper water (e.g., *Bolivina pacifica*). Rarely, the abundance distribution of a species with respect to water depth is polymodal (e.g., *Goesella flintii*) and is poorly fit by a simple peaked distribution.

proached (Gerrodette 1979). However, within limited geographic regions where the relationship of depth to important physical and chemical constraints remains relatively constant, species tend to occur within characteristic ranges of water depth (Culver 1988). Extrapolated to geologic time, if those relationships between physical and chemical constraints and water depth are relatively constant, species should be expected to show a consistent preference for sedimentary facies formed in specific bathymetric zones (e.g., Ziegler et al. 1968).

Most modern organisms show a relatively simple distribution with respect to water depth, in which each species occurs most abundantly at some water depth and becomes less abundant into shallower and deeper water (Fig. 1) (Brown 1995). This peaked relationship between abundance and depth (or probability of collection and depth) could be mathematically described in any number of ways (Fig. 2). Selecting the form of this relationship is a balance between accurately reflecting observed distributions, mathematical simplicity, and keeping the number of parameters that describe the relationship to a minimum. Because of these constraints, a Gaussian or bell-shaped relationship is most widely

used in models of the fossil record (Holland 1995b; Holland and Patzkowsky 1999; Horton et al. 1999) and in models of modern ecological distributions (Whittaker 1970; Brown 1984; Austin 1987; Minchin 1987).

With this bell-shaped distribution, the probability of collection (or abundance) of each species with respect to water depth can be described with three intuitive parameters (equations in Holland 1995b). Peak abundance (PA) is the probability of collection of a species under the best of circumstances, that is, in the preferred water depth (PD) of the species. Every species will also have some tolerance for water depths other than its preferred depth and this depth tolerance (DT) is expressed as the standard deviation of the bell curve.

Not all aspects of an organism's ecology or taphonomy are related to water depth. For example, some organisms may be adapted to either low or high siliciclastic influx, which can act independently of water depth (Bretsky 1970). Likewise, organisms may be adapted to various levels of turbidity, which might vary independently of water depth (Brett 1995). However, since the relationship of organisms is so strongly related to water depth, both in the fossil record and today, these initial models of the fossil record focus on this important

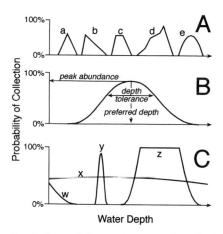

FIGURE 2. A, Several alternative ways of mathematically describing the observed distribution of modern marine species with respect to water depth, such as seen in Figure 1. Such distributions would give qualitatively the same results as the distribution actually used in these models, the bell-shaped distribution seen in B. This bell-shaped distribution allows the abundance (or probability of collection) of a species with respect to water depth to be described with three parameters: Peak abundance (PA) is the maximum probability of collection for a species, that is, the probability of collection in the water depth in which it is most likely to be collected, the preferred depth (PD). Species will also have some tolerance for other water depths, defined by the standard deviation of the curve, called depth tolerance (DT). C, This simple parameterization also allows a variety of other, less apparently bell-shaped distributions to be simulated. Species x represents an extremely eurytopic species and y represents an extremely stenotopic species. Species w has a truncated probability of collection curve because its preferred depth lies in very shallow water. Species z has a flat-topped distribution because its peak abundance is so large that the probability of collection is 100% over a range of water depths.

relationship. Future models may incorporate more complicated patterns of ecological and taphonomic relationships to facies.

Sequence Architecture.—Sequence architecture plays an important role in the stratigraphic distribution of fossils because it controls changes in facies and sedimentation rate through time and space (e.g., Kidwell 1991b; Abbott and Carter 1997; Brett 1998). Although a full explanation of the theory and application of sequence stratigraphy is not possible here, clear introductions can be found elsewhere (Van Wagoner et al. 1990; Posamentier et al. 1992; Posamentier and Allen 1993; Posamentier and James 1993; Emery and Myers 1996).

Sequence stratigraphy recognizes two types of stratigraphic cycles, the parasequence and

the depositional sequence. Parasequences are sedimentary cycles bounded at their top and base by marine flooding surfaces, across which deeper water facies abruptly overlie shallower water facies (Fig. 3A). Most commonly, parasequences are shallowing upward, but some parasequences have thin deepening-upward intervals near their base. Fully deepening-upward parasequences are rare. Facies contacts within parasequences are gradational and Waltherian, that is, each successive facies was deposited in an environment laterally adjacent to overlying and underlying facies. Although parasequences are not defined in any way by their thickness, estimated duration, or geographic extent, parasequences are most commonly one to ten meters thick and represent tens to hundreds of thousands of years (Van Wagoner et al. 1990).

Parasequences occur in bundles or parasequence sets that display net trends of shallowing, deepening, or no change in water depth (Fig. 3B). Progradational parasequence sets comprise a series of parasequences that form an overall shallowing-upward trend, with each successive parasequence composed of a shallower suite of facies than the parasequence below it. Aggradational parasequence sets are composed of a set of parasequences that shows no overall shallowing-upward trend. In this case, although each parasequence is shallowing upward, successive parasequences contain the same set of facies, such that there is no overall shallowing or deepening through the set. Retrogradational parasequence sets display an overall deepening-upward succession because, although individual parasequences are shallowing upward, each successive parasequence contains a deeper suite of facies than the one below it. Parasequence sets are controlled by the relative rates of sedimentation and accommodation (that is, the sum of the rates of eustatic sea-level rise and tectonic subsidence). Thus, under conditions of more or less constant sediment supply, progradational stacking tends to occur when rates of accommodation are relatively low, such as during eustatic stillstand, slow rise, or slow fall. Retrogradational stacking tends to develop when rates of accommodation are relatively high, such as during

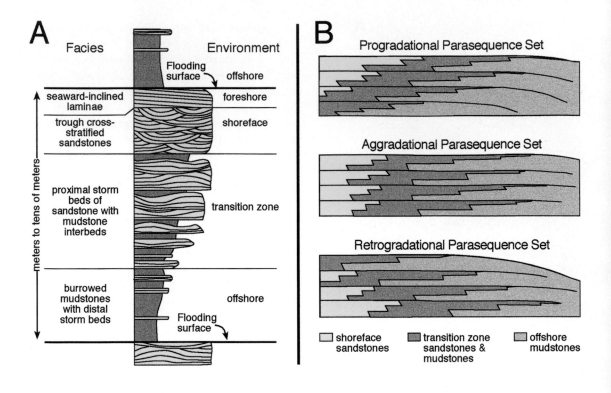

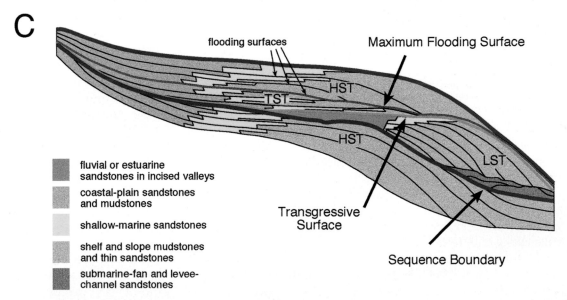

FIGURE 3. Basic elements of sequence stratigraphy, adapted from Van Wagoner et al. 1990. A, Typical parasequence of a wave-dominated sandy shoreline system. B, Parasequence stacking patterns, as expressed for facies developed in a wave-dominated sandy shoreline system. C, Type 1 depositional sequence, showing component parasequences, parasequence sets, systems tracts, and important stratigraphic surfaces. HST = highstand systems tract. TST = transgressive systems tract. LST = lowstand systems tract.

rapid rises in eustatic sea level or elevated rates of subsidence.

Depositional sequences, or sequences for short, are composed of sets of parasequences (Fig. 3C). Sequences are packages of relatively conformable strata that are bounded above and below by unconformities formed by sub-aerial exposure and erosion, their correlative submarine erosion surfaces, and their correlative conformities, which are surfaces that correlate to the age of the unconformity but record nearly continuous deposition. Internally, sequences are composed of systems tracts, which consist of sets of parasequences. The most commonly recognized type of sequence (a type 1 sequence) begins with a lowstand systems tract containing a progradational-to-aggradational set of parasequences, followed by a transgressive systems tract containing a retrogradational set of parasequences, and ending with a highstand systems tract, containing an aggradational-to-progradational set of parasequences. Any given systems tract may be present in only certain parts of a depositional basin, as the locus of deposition shifts landward and seaward during a relative cycle of sea level. As a result, the vertical succession through a sequence at any given point in the basin rarely consists of all three systems tracts.

These systems tracts are separated by important stratigraphic surfaces. The sequence boundary, which underlies the lowstand systems tract, is the unconformity as well as its correlative surfaces that separate one sequence from the next. The transgressive surface is the flooding surface (and therefore a parasequence boundary) that separates the lowstand systems tract from the overlying transgressive systems tract. The maximum flooding surface is also a flooding surface and parasequence boundary, but one that separates the transgressive systems tract from the overlying highstand systems tract. Again, because not all systems tracts occur in any given area, some of these surfaces may be merged into a single physical surface. For example, on cratons and depositionally updip portions of coastal plains, lowstand systems tracts commonly are not preserved, causing the sequence boundary and transgressive surface to be merged into one physical surface.

Sedimentation Rates.—Changes in sedimentation rate relative to the rate of bioclastic production have a pronounced effect on the stratigraphic distribution of fossils, particularly the formation of major bioclastic accumulations (Kidwell 1991a,b). Where sedimentation rates are slowed, either through starvation or sediment bypassing, shell material can accumulate. Increasing concentrations of shell material foster the growth of hard-substrate–specific shell producers (such as oysters and other epibionts), favoring the accumulation of more shell material. Furthermore, dense accumulations of shells raise their probability of preservation by buffering the pore-water system and slowing dissolution. Changes in sedimentation rate are particularly important in the expression of the fossil record in outcrop because they can profoundly distort perceptions of the tempo of evolution (MacLeod 1991).

The geometry of depositional sequences requires that sedimentation rate varies spatially and temporally; however, such variations are largely predictable from sequence stratigraphic context. Many flooding surfaces form through the temporary cessation of sediment supply, whether by trapping within coastal estuaries following a relative rise in sea level, by delta switching, or by a lag in carbonate production rates following a relative rise in sea level. As a result, most flooding surfaces display some evidence of stratigraphic condensation and commonly become sites of bioclastic accumulations (Kidwell 1991a; Brett 1995). In siliciclastic systems, such condensation is most pronounced at major flooding surfaces, such as those within the transgressive systems tract. Similarly, stratigraphic condensation also occurs within depositional sequences, most commonly near the updip terminations of strata (onlap), near the downdip terminations of strata (downlap, such as at the maximum flooding surface), and where strata thin as result of sedimentary bypass, such as in the late highstand systems tract (Kidwell 1991a). In carbonate sequences, condensation may occur in similar settings or may occur to a much lesser extent, resulting from the tendency of carbonate production to keep pace with even large rates of accommodation.

Sequence architecture and changes in sedimentation rate are easily simulated with existing sedimentary basin models. Several basin simulations are readily available, including the freeware Strata (Flemings and Grotzinger 1996), the inexpensive GEOMOD (Jervey 1988), and the commercial Sedpak (Kendall et al. 1991), as well as a variety of programs produced by individuals for their research programs (e.g., Read et al. 1986; Elrick and Read 1991; Franseen et al. 1991; Bosscher and Southam 1992; Goldhammer et al. 1993). Most stratigraphic models are based on temporal variations in accommodation and sediment flux and thereby have similar underlying principles. These models start with the controls on accommodation space, that is, eustasy and tectonic subsidence. Sediments are allowed to fill this space and it is in this aspect that models show the greatest disparity in their approaches. Some simulations use a diffusion model of sediment dispersal through the basin, others use geometric models governing the distribution of constant volumes of sand and mud, and others use local production models for carbonate sediments. Regardless of the model of sedimentation used, once sediments accumulate, compaction of underlying sediments is commonly simulated, as is isostatic subsidence as a result of changing sediment and water loads. Thus, most of these models are capable of describing changes in water depth (i.e., facies) and sedimentation rate across the basin and through time—the necessary inputs for any realistic model of fossil accumulation.

The model predictions used in this paper are based on simulations using Strata, because of its intuitive diffusion-based rules for siliciclastic sedimentation and its production-based rules for carbonate sedimentation. The model Strata is also easily modified to produce output for the model Biostrat; together, these two models are used to simulate a basin and its fossil occurrences (for details, see Holland and Patzkowsky 1999). The model Strata is first used to simulate a sedimentary basin and to generate a series of water-depth histories for multiple outcrops across the basin. Biostrat starts simulating the stratigraphic oc-

currence of fossils by building a database of species. It first uses a random-branching model to simulate the origination and extinction of species (Raup 1985), in which the probabilities of these events are held constant for the length of the model run. Biostrat keeps track of the ancestor of each species, so the model could be used in phylogenetic simulations. Whenever a new species is created, values of peak abundance, preferred depth, and depth tolerance are randomly generated for each species from within user-defined limits. Biostrat then uses the water-depth histories supplied by Strata and its own species database file to determine, at each horizon at each outcrop in the basin, the probability of collection of each species extant at that point in time. This probability of collection at each location for each species is compared with a random-number generator to determine whether the species occurs there. The net result is a file that lists every occurrence of every species at every horizon across the basin.

Predictions from Sequence Stratigraphic Models

Paleontology has focused its interest primarily on intervals of unusual change—mass extinctions, evolutionary radiations, turnover pulses between intervals of coordinated stasis, and intervals of abrupt morphological change in species. In many cases, the data for these studies are collected from one or a few relatively closely spaced (i.e., <50–100 km) outcrops, so the interpretation of the pattern of fossil occurrences in outcrops becomes paramount. It is critical to evaluate the possibility that any observed pattern is driven not by true biological changes but instead by the structure of the stratigraphic record, with its unconformities, rapid facies changes, and intervals of condensation. Numerical models can help simulate such possibilities. In these simulations, origination and extinction rates are stochastically constant, so that any abrupt changes that appear in the modeled stratigraphy must be the result of stratigraphic artifact.

These null simulations make a number of predictions about the expected appearance of the fossil record under conditions of stability in origination rates, extinction rates, and eco-

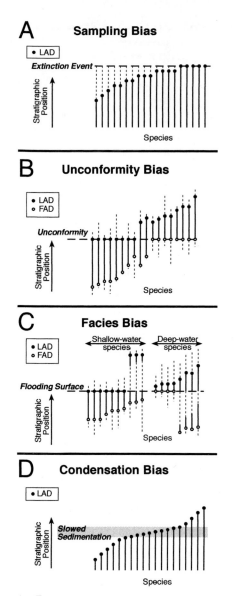

FIGURE 4. Four important biases recognized in sequence stratigraphic models of the fossil record. A, Sampling bias causes any abrupt biotic event, such as an extinction episode, to be appear gradual in a local stratigraphic column. The magnitude of sampling bias increases with the decreasing peak abundance of species and with decreased sampling intensity. The Signor-Lipps Effect is commonly equated with sampling bias but includes other types of biases as well. B, Unconformity bias causes first and last occurrences to cluster at unconformities, simply by lack of deposition and erosion during a time in which species continue to originate and go extinct elsewhere in the sedimentary basin. The magnitude of unconformity bias increases with the duration of the hiatus and the rate of faunal turnover. C, Facies bias causes first and last occurrences of species to cluster at surfaces of abrupt facies changes, such as flooding surfaces (shown here) and abrupt basinward shifts of facies. In addition, it causes gaps to appear in the ranges of other species. The magnitude of facies bias

logical structure. Some of these predictions have been made before by others without the use of numerical models. Furthermore, some of the effects of unconformities and changes in sedimentation rate are intuitive. Nonetheless, sequence stratigraphy and computer models enhance the ability to make predictions about the fossil record. Sequence stratigraphy allows changes in facies, sedimentation rates, and the distribution of unconformities in local outcrops to be predicted with greater precision than they could previously. Computer modeling allows patterns of fossil occurrence to be visualized, permits the relative effects of the underlying variables to be weighed, and lets what once was qualitative become quantitative.

Although some of these predictions have already been confirmed in the field, all need to be tested from a wider variety of taxonomic groups and geologic settings. If the model predictions hold up, then it suggests that the relatively few concepts that underlie this model sufficiently capture much of the pattern of occurrences observed in the fossil record. If so, the model may then be more confidently applied to problems in which direct observation of the fossil record might provide ambiguous answers and may be used to make more quantitative predictions about completeness of the record.

Local Range.—Under conditions of constant origination and extinction rates, clusters of first and last occurrences can arise in any stratigraphic section. Although clusters can arise purely as chance coincidence, most arise as artifacts of incompleteness imposed by facies and sequence architecture. Clusters of first or last occurrences in a local section do not necessarily indicate changes in rates of origination, migration, or extinction.

1. Concentrations of first and last occurrences are expected at major flooding surfaces

---

is increased by the magnitude of the facies change and by decreased values of depth tolerance of species. D, Condensation bias causes first or last occurrences to appear relatively clustered in zones of relatively slow sedimentation. The magnitude of condensation bias increases with the disparity between maximum and minimum sedimentation rates in the interval of interest.

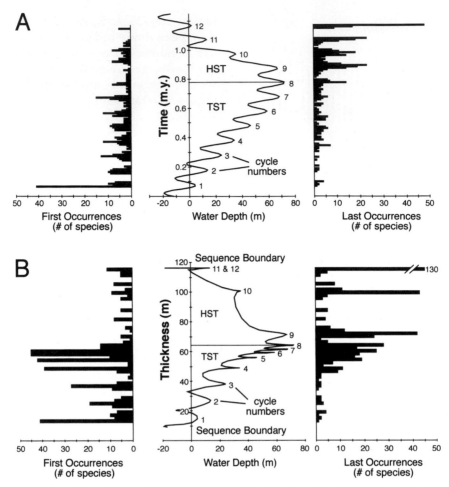

FIGURE 5. Modeled clustering of first and last occurrences of species through a single depositional sequence of 1.2 m.y. (Holland and Patzkowsky 1998). Sequence consists of a dozen numbered smaller-scale cycles of water depth. A, Section expressed as a function of time. Note bundling of first and last occurrences into spikes, with particularly large spikes of first occurrences immediately above the basal sequence boundary and last occurrences below the upper sequence boundary. B, Same section, but expressed in terms of rock thickness. Note the thinning of cycles 5–8 because of condensation near the maximum flooding surface and the thinning of cycles 11 and 12 because of sedimentary bypass in the late highstand. Such bundling dramatically increases the clustering of first and last occurrences near the maximum flooding surface, and the clustering of last occurrences near the upper sequence boundary.

(Holland 1995a,b). Facies bias (Fig. 4) causes shortening of local ranges because of the limited facies tolerance of species and the limited availability of facies in an outcrop. Where there is strong facies contrast across a flooding surface, the number of first and last occurrences of species can increase because the window of time in which is a species is not likely to be encountered increases. For example, if offshore facies occur immediately above a flooding surface, any species limited to an offshore environment will tend have its first occurrence immediately above the flooding surface, particularly if offshore facies were not developed locally for an extended period of time prior to the development of the flooding surface. Several field studies have reported such concentrations of first and last occurrences at flooding surfaces (e.g., Atrops and Ferry 1989; Armentrout 1991; Armentrout and Clement 1991).

As the degree of facies change across a flooding surface increases or the degree of facies tolerance of species decreases, the number of first and last occurrences at the flooding surface should increase (Fig. 5). In addition, the number of first occurrences should increase with the length of time since the facies above the flooding surface was last encountered. Likewise, the number of last occurrences should increase with the length of time before the facies beneath the flooding surface is encountered again. Flooding surfaces within the transgressive systems tract are commonly characterized by much greater facies change than those in the highstand systems tract. Thus, flooding surfaces within the transgressive systems tract are expected to show stronger concentrations of first and last occurrences than elsewhere within depositional sequences.

2. Concentrations of first and last occurrences are expected at sequence boundaries (Fig. 5) (Holland 1995a,b). Unconformity bias (Fig. 4) truncates the local ranges of fossils through protracted periods of nondeposition as well as erosion of previously deposited sediments. Periods of nondeposition and erosion clearly allow for more time in which background turnover can proceed and give the illusion of higher turnover rates at the unconformity (Bambach and Gilinsky 1988). Superposition of rocks of increasingly different ages allows for greater amounts of apparent faunal turnover. Many common species will have their first and last occurrences at a sequence boundary. At coarser scales of stratigraphic resolution, rarer species will also appear to have their first and last occurrences at a sequence boundary. One effect of this clustering is that zonal boundaries will commonly coincide with sequence boundaries (Mancini and Tew 1995; Ross and Ross 1995).

Although some sequence boundaries locally display almost no facies contrast, others can be characterized by significant changes in facies and should be affected by facies bias. In depositionally downdip areas, sequence boundaries can be expressed as an abrupt basinward shift in facies in which shallow-water facies are disconformably placed atop deeper-water facies (e.g., Holland 1995a). Such sequence boundaries should be characterized by the additional first occurrences of shallow-water species and last occurrences of deep-water species. In depositionally updip areas, lack of deposition during the lowstand followed by rapid onlap during the transgressive systems tract can cause deeper-water facies of the transgressive systems tract to abruptly overlie shallow-water facies of the highstand systems tract (with the two separated by the sequence boundary). In cases such as this, the sequence boundary should also contain a pulse of first occurrences of deep-water species and a pulse of last occurrences of shallow-water species. In both of these examples, facies bias increases the number of first and last occurrences over what was accumulated through unconformity bias (Fig. 4).

Clustering of first and last occurrences at sequence boundaries is commonly reported (Dockery 1986; Armentrout 1991; Gaskell 1991; Patzkowsky and Holland 1996; Harris and Sheehan 1997; Finney et al. 1999). Although some of these boundaries may represent true extinction events, the possibility that any of these surfaces simply represents a long duration of time must be weighed. This possibility might be testable in the future if high-precision radiometric dates could be obtained immediately above and below the sequence boundaries in question.

3. Concentrations of first and last occurrences are expected in intervals of stratigraphic condensation. Condensation bias (Fig. 4) alters perceptions of the relative timing of events, making events appear more closely spaced in time than they actually are (MacLeod 1991). Intervals of stratigraphic condensation are characteristically arrayed along flooding surfaces (particularly major flooding surfaces), in areas of onlap (such as the depositionally updip portions of sequence boundaries), and in areas of bypass (such as the late highstand systems tract) (Kidwell 1991b). As a result, condensation bias will tend to further increase the numbers of first and last occurrences that form at major flooding surfaces and sequence boundaries (Fig. 5).

4. Range offset, defined as the difference in age between the first appearance of a species in a local stratigraphic section and its origination within the sedimentary basin (or last appearance and extinction), is expected to vary systematically within sequences (Holland and Patzkowsky 1999). Although these simulations are in their preliminary stages, it

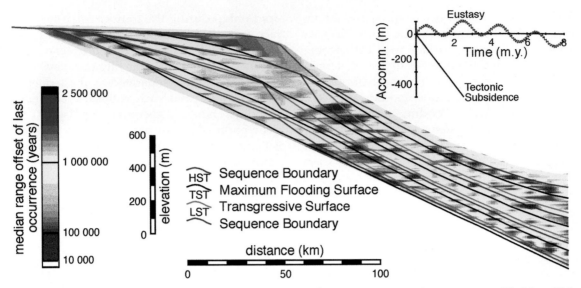

FIGURE 6.   Contours of range offset of last occurrences through several depositional sequences, modified from Holland and Patzkowsky 1999. Range offset is defined as the time between the last occurrence of a fossil in a local stratigraphic column and its time of extinction within the sedimentary basin. As such, range offset is a measure of the biostratigraphic error encountered when correlating the last occurrence of a species. Shown here is a simulation of 981 species, with the median range offset of species plotted in each of 30 stratigraphic columns across the basin. Superimposed are important sequence stratigraphic surfaces. Note that range offset reaches large values (>1 m.y.) in extreme updip areas and in the transgressive systems tract of the deep basin. Range offset also tends to increase near the shelf-slope break of the late highstand to early lowstand systems tract.

appears that range offset tends to be large within the transgressive systems tract of the deep basin, near the shelf break of the late highstand to early lowstand systems tract, and in extreme updip portions of shelves (Fig. 6). Range offset is a significant measure, particularly for biostratigraphy, because it reflects the amount of biostratigraphic error (in time) if the first or the last occurrence of a species is correlated. Zones of high range offset delineate zones in which the temporal sequence of first and last occurrences seen in outcrop should not be trusted as reflecting the true sequence of events.

Abundance.—Even under conditions of ecological stability in which species' distributions in sedimentary environments are unchanging, dramatic changes in species abundance in local outcrops can occur. Such changes can be generated easily as artifacts arising from sequence architecture and do not necessarily indicate underlying changes in the abundance or ecology of species.

1. Abrupt changes in the abundance of individual species should occur at flooding surfaces and abrupt basinward shifts of facies

(Fig. 7) (Holland 1996b). If species have a peaked distribution of abundance with respect to water depth (as in the Gaussian distribution of probability of collection versus depth), then any rapid change in water depth is likely to change the probability of occurrence of a species and therefore its abundance. The greatest changes in abundance will occur at major flooding surfaces, such as those within the transgressive systems tract, and where basinward shifts in facies are pronounced, such as at major sequence boundaries. However, if species have relatively small depth tolerances, as in relatively shallow water settings, even small flooding surfaces and minor basinward shifts in facies may suddenly change the abundance of a species. Abrupt changes in the abundance of species have been reported in field studies (e.g., Armentrout and Clement 1991; Armentrout 1996; Abbott and Carter 1997).

2. Abrupt changes in biofacies are expected at major flooding surfaces (Holland 1996b). This logically arises from the previous prediction because changes in abundance are likely to coincide for many species. Furthermore,

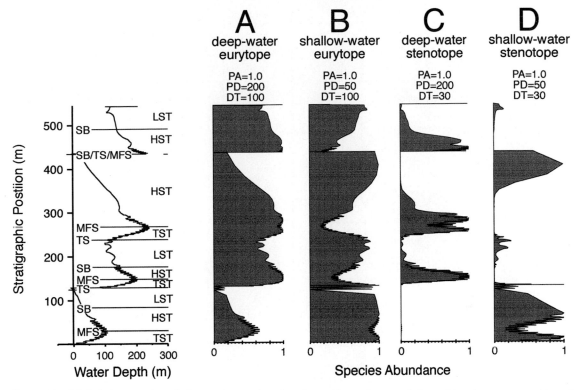

FIGURE 7. Modeled examples of change in species abundance through several depositional sequences. Abrupt changes in species abundance tend to occur at sequence boundaries and major flooding surfaces within the transgressive systems tract. Note also that changes in species abundance are synchronous near these surfaces, which will drive abrupt changes in biofacies. More gradual changes in species abundance occur within the highstand and lowstand systems tracts and should lead to more gradual biofacies replacement.

where the transgressive systems tract contains several major flooding surfaces, as it commonly does, it should be characterized by multiple repeated changes in biofacies. The highstand and lowstand systems tracts, in contrast, should be characterized by relatively gradual biofacies replacement. In the computer simulations described in this paper, species maintain their preferences for particular sedimentary environments and thereby display faunal tracking (Brett et al. 1990). The high degree of condensation commonly associated with the transgressive systems tract also promotes abrupt change in biofacies. Rapid biofacies replacement in the transgressive systems tract is therefore an expected feature of the stratigraphic record under conditions of ecological stability and does not require any ecological restructuring.

3. Abrupt changes in biofacies are expected at sequence boundaries (Holland 1996b). Where sequence boundaries are characterized by abrupt basinward shifts in facies, shallow-water biofacies should abruptly replace deep-water biofacies. Such contacts may also display evidence of marine erosion with a significant hiatus. The change in biofacies across these surfaces is simply the mirror image of that seen at flooding surfaces. In depositionally updip areas where a sequence boundary coincides with a major flooding surface, deep-water biofacies should abruptly replace shallow-water biofacies at the sequence boundary. In these depositionally updip areas, where the duration of the hiatus at the sequence boundary is likely to be long, the additional amount of time contained in the hiatus allows for prolonged background faunal turnover and accentuates the faunal change within biofacies. Again, such abrupt change in biofacies requires no change in ecological structure.

Morphological Change.—Although these simulations are in their earliest stages, it is possible to make predictions about stratigraphic

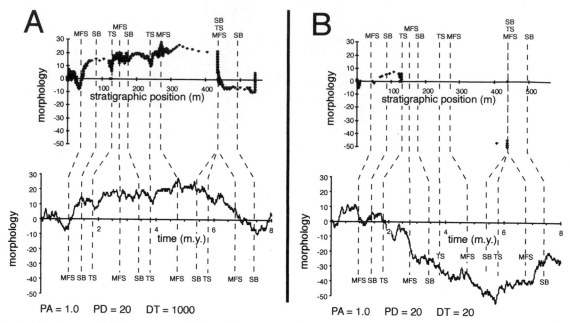

FIGURE 8.   Two modeled examples of effects of sequence architecture on observed patterns of morphologic change. Morphology in each case is simulated as a random walk of a single variable through time (lower halves of A and B). Three complete and two incomplete sequences are simulated. Facies, sequence architecture, and changes in sedimentation rate control where each species is collected within each stratigraphic section (upper halves of A and B). A, Although the random walk shows relatively gradual changes in morphology over time, the stratigraphic section preserves several abrupt changes in morphology, most strikingly shown near 450 m. Note that abrupt morphologic changes tend to occur near either a transgressive surface or a maximum flooding surface, intervals characterized by rapid facies changes and slow sedimentation. B, The random walk in morphology for this run shows a protracted trend in morphology, but because the species is relatively facies restricted (DT = 20, as opposed to DT = 1000 in A), the species is not collected over much of the section. When the species reappears in the section near 450 m, its morphology has suddenly changed, with no hint of the actual gradual morphologic transition preserved in the fossil record.

patterns of morphological changes in individual species. Here, species morphology is simulated by a single variable undergoing a random walk through geologic time. This random walk though time becomes distorted as it is filtered through the Biostrat model that incorporates rarity, facies control, sequence stratigraphic architecture, and changes in sedimentation rates. The results show examples of observed abrupt stratigraphic changes in morphology that can result when morphology simply follows a random walk.

1. Abrupt changes in morphology are expected at sequence boundaries. Not surprisingly, given the length of time contained in the hiatus at sequence boundaries, significant morphologic change can accumulate from the random walk. Longer hiatuses at sequence boundaries should be associated with greater amounts of morphologic change (Fig. 8). Such a prediction has long been recognized in the

punctuated-equilibrium-vs.-gradualism debate (Eldredge and Gould 1972). However, the cryptic expression of many sequence boundaries, particularly those that are not also a major flooding surface or characterized by an abrupt basinward shift in facies, may make the recognition of such surfaces problematic. It remains to be investigated whether reported abrupt changes in fossil morphology occur at previously undetected sequence boundaries.

2. Abrupt changes in morphology are to be expected at major flooding surfaces (Bayer and McGhee 1985; McGhee et al. 1991). In cases where the first or last occurrence of a species is associated with a flooding surface, there may be long intervals above or below the flooding surface in which the species cannot be found because suitable facies are lacking. These intervals in which the species cannot be found in a local stratigraphic section allow the

random walk in morphology to accumulate greater amounts of morphologic change (Fig. 8). The amount of morphologic change should be correlated with the amount of time between intervals in which the preferred facies of the species can be sampled.

In cases of a depth-related cline in species morphology (e.g., Cisne et al. 1982; Ludvigsen et al. 1986), flooding surfaces offer a second mechanism for achieving apparently sudden changes in species morphology. In this case, the major facies changes across flooding surfaces (and across major basinward shifts in facies as well) will cause different portions of the morphologic cline to be stratigraphically juxtaposed.

3. Rapid changes in morphology are to be expected in intervals of stratigraphic condensation (MacLeod 1991). Intervals such as onlap surfaces at the updip limit of the transgressive and highstand systems tract are commonly characterized by condensation. As these intervals directly overlie sequence boundaries, condensation here can further intensify the morphological shifts expected at sequence boundaries as a result of prolonged subaerial exposure and erosion. Similarly, stratigraphic condensation as a result of sediment bypass in the late highstand systems tract will cause the overlying sequence boundary to be preceded by an interval of condensation. In this case, such condensation will enhance the rapid morphological changes expected at the overlying sequence boundary. Sediment starvation, such as along major flooding surfaces, can also lead to stratigraphic condensation. Again, such condensation will tend to enhance the morphologic changes associated with these flooding surfaces as a result of sudden shifts in facies (Fig. 8). Finally, intervals of rapid sedimentation, such as in rapidly prograding clinoforms in the lowstand and highstand systems tracts, may cause changes in morphology that occurred quickly to appear as if they proceeded more slowly.

4. Patterns of apparent iterative evolution are expected where a given lithofacies recurs in successive sequences or parasequences and those lithofacies are separated from one another by other lithofacies. Similar species morphologies may recur in several widely spaced stratigraphic positions within a stratigraphic succession if that species is relatively facies specific and its preferred facies occurs within a small portion of several successive sequences (Fig. 9). This pattern, if taken at face value, could be interpreted as the repeated origination and extinction of a given morphology, that is, iterative evolution. This pattern will become more likely if the species is relatively rare, such that its association with a particular facies becomes less clear and the intervals between its occurrences become larger. Such interpretations of iterative morphological change have been proposed for ammonites (Bayer and McGhee 1985).

Similarly, clines in morphology can give rise to iterative or oscillating patterns in fossil morphology. Over the course of several parasequences, a given morphology would be predicted to develop repeatedly within successive parasequences and be reset to a previous morphologic state at each flooding surface. In such a case, a stable cline in morphology may have been present, but the pattern of fossil occurrences would suggest that morphology was oscillating through time. Patterns of apparently oscillating morphologies in the fossil record need to be reexamined to see if they represent the sampling of a morphological cline through several successive parasequences. Such patterns have also been recognized in ammonites (Bayer and McGhee 1985).

Gaps in Ranges.—The distribution of gap lengths between stratigraphic occurrences of fossils provides critical information on the incompleteness of the fossil record. Lazarus taxa, for example, have long been recognized as indicating incompleteness of the record, particularly in intervals of mass extinction (Jablonski 1986). The distribution of gap lengths is also important because it is a primary means for establishing confidence limits on the stratigraphic ranges of fossils (Paul 1982; Strauss and Sadler 1989; Marshall 1990). Models of the fossil record also make predictions about gap lengths.

1. When the distribution of gap lengths fails to fit the exponential distribution predicted by a truly random stratigraphic distribution of fossils, it will tend to do so because there are

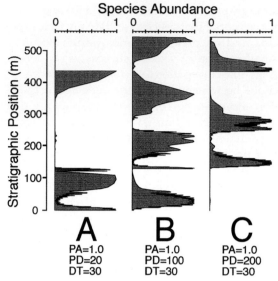

Species Abundance

A
PA=1.0
PD=20
DT=30

B
PA=1.0
PD=100
DT=30

C
PA=1.0
PD=200
DT=30

FIGURE 9. A stratigraphic mechanism for generating an artifactual pattern of iterative evolution. Shown here are three species, differing only in their values of preferred depth. Through sequence stratigraphic control on facies changes, the modeled abundance of each species increases and decreases through the stratigraphic section, such that species tend to follow one another in a characteristic pattern, with deeper-water species (C) being progressively replaced by shallower-water species (B, A). If the morphologies of these species change slightly over geologic time, a variety of taxonomic and evolutionary interpretations might arise. Each stratigraphically separated occurrence of a species might be named separately (e.g., species C might be split into three separate species). If the morphologies of each species group (A, B, C) are sufficiently distinct from one another, the repeated replacement of one species group by another could be interpreted erroneously as a pattern of iterative evolution. Some oversplitting of species, particularly in the early 1900s, may be due to this phenomenon. Water-depth history and sequence architecture are the same as that modeled in Figure 7.

too many long gaps (e.g., McKinney 1986). Long gaps are produced not only by the presence of sequence-bounding unconformities, but also by the facies specificity of taxa. Thus, highly facies-specific taxa should show a greater tendency towards nonexponential gap distributions.

2. Recurrent faunas and Lazarus taxa are expected primarily as a result of facies control. In some cases, however, the facies control is not depth related. For example, many Middle Ordovician invertebrate genera disappear from the eastern United States for almost 8 m.y., and during this long absence they are demonstrably missing from the depth-related

facies in which they previously occurred (Patzkowsky and Holland 1993, 1996, 1999). Many of these same genera reappear in the eastern United States in the Late Ordovician. Their disappearance and reappearance in this region was apparently controlled by a temporary switch to cooler, more nutrient-rich and turbid water conditions that affected nearly all bathymetrically related facies. During their absence from the eastern United States, these genera can be found in warm-water, nutrient-poor, clear-water facies in the western United States and Canada. More integrated studies of paleontologic events and lithostratigraphic indicators of physical and chemical conditions are needed to recognize the range of causes of recurrent faunas and Lazarus taxa.

Summary of Predictions.—The consistent expectation of these stratigraphic models is of a fossil record that appears to be highly episodic, even when the history of life is characterized by stable origination rates, extinction rates, and ecological structure. Because much of the fossil record comes from cratons and the depositionally updip portions of coastal plains, a relatively simple unified prediction for the fossil record can be made. In these areas, lowstand systems tracts are commonly not developed and the sequence boundary is combined with the transgressive surface. This may be overlain by a relatively condensed transgressive systems tract, which is overlain by a relatively thicker highstand systems tract lacking major flooding surfaces. Because the lowstand systems tract is missing, unconformity bias will be strong at the sequence boundary. Because the transgressive surface is merged with the sequence boundary, both facies bias and condensation bias will also be strong at this surface. Thus, the sequence boundary and transgressive systems tract interval should be characterized by numerous first and last occurrences of species, widespread changes in species abundance and biofacies, and abrupt changes in fossil morphologies across many taxa. The remainder of the sequence should be characterized by relatively few first and last occurrences, gradual changes in fossil abundance and biofacies, and slow and minor changes in fossil morphologies.

Ross and Ross (1995) have named these patterns sequence evolution and sequence extinction. What is significant is that these patterns will emerge even when origination rates, extinction rates, and ecological structure are constant and fossil morphologies follow a random walk through time. In short, the fossil record should appear episodic even if the history of life is not.

Similar patterns have been described from the fossil record. Carl Brett and Gordon Baird recognized a pattern they termed coordinated stasis, which consists of several-million-year intervals of relatively minor faunal turnover, little morphological change, and minor changes in community structure (Brett et al. 1996). Intervals of relative stasis are interrupted by brief (conservative maximum of several 100 k.y.) windows of major faunal turnover, morphological evolution, and major changes in ecological structure. Of the eleven turnover pulses recognized by Brett and Baird from the Siluro-Devonian of the Appalachian Basin, five coincide with a sequence boundary and six occur within somewhat condensed intervals of the earliest highstand systems tract (Brett and Baird 1995). Patterns similar to these are seen in Cambrian biomeres, described as the abrupt extinction of shallow-water trilobites followed by the immigration and radiation of deep-water trilobites (Palmer 1965). The Pterocephaliid/Ptychaspid biomere boundary (Saltzman 1999) and the Marjumiid/Pterocephaliid biomere boundary (Osleger and Read 1993) are each closely associated with a sequence boundary and the overlying transgressive systems tract. It is critical for studies of coordinated stasis and biomeres to assess to what degree the patterns in first and last occurrences, species abundances, biofacies, and morphological shifts arise from facies bias, unconformity bias, and condensation bias.

Most interestingly, converse patterns have also been reported. For example, some faunas demonstrate persistence across sequence boundaries (Brett et al. 1996), a pattern predicted from modeling only in cases of short duration of nondeposition and of slow rates of extinction, origination, and migration. In cases such as these, if the duration of the sequence-bounding unconformity could be estimated, the realistic range of turnover rates consistent with a lack of clustering of first and last occurrences could be estimated statistically (Baumiller 1996). In addition, some turnover pulses between blocks of stability do not occur at sequence boundaries or major flooding surfaces (Brett et al. 1996). As mentioned earlier, these would indicate biologically real events rather than stratigraphic artifacts.

Dealing with Incompleteness

If these model predictions are found to be generally true of the fossil record as seen in outcrops, then the relatively simple assumptions of this model should account for much of the distortion of the fossil record. If this is so, it will become essential to develop strategies that can recognize the relative effects of sampling bias, facies bias, unconformity bias, and condensation bias in any given study. It will also become critical to develop methods that explicitly compensate for each of these biases. Currently, though, these artifacts can be mitigated in several ways.

Where paleontological data have already been collected, perhaps without sequence stratigraphic control, the first defense is to place confidence limits on fossil ranges (Paul 1982; Strauss and Sadler 1989; Marshall 1990). Most existing methods of confidence limits assume that gap lengths between occurrences follow an exponential distribution, one that would be expected if fossil occurrences were strictly randomly placed. Before such confidence limits can be placed, gap lengths must be tested to see if their distribution is demonstrably nonexponential (McKinney 1986; Marshall 1994, 1997). If they do follow an exponential distribution, then these simple confidence limits can be applied. The confidence limits can then be used as a check on interpretations, preventing one from overreaching the quality of the data imposed by the incompleteness of the record.

In many cases, the distributions of gap lengths may be nonexponential and more advanced methods of estimating confidence limits must be applied. Such methods usually assume an underlying knowledge of the probability of collecting a given species throughout a section, which will likely depend on understanding facies changes and sequence architecture in the section (Marshall 1994, 1997). These methods are in their early stages of development. Priority should be placed on ways of estimating the probability of collecting a fossil as a function of facies. One possibility would be to estimate the three parameters defining the distribution of each species—peak abundance, preferred depth, and depth tolerance—and to estimate water depths through an outcrop. If these variables could be estimated, it would be relatively simple to define the probability of occurrence of a species throughout a section and to have confidence limits tailored to each species (Marshall 1994).

The other current option for dealing with these biases is to collect better data, and there appear to be two main approaches for doing so. The first is to develop a time-environment approach to sampling (Holland 1997; Patzkowsky and Holland 1999). Environmental divisions should be based on field analysis of sedimentary facies defined on lithologic criteria rather than biofacies or benthic assemblages (Boucot 1981). The temporal divisions in this approach should be based on depositional sequences, such that the principal gaps in the record define the boundaries of the temporal sampling units, rather than lying within them. The goal would be to fill each cell of the time-environment matrix with multiple samples, although availability of outcrop will generally greatly limit one's ability to do so. The degree to which a continuous record can be obtained in any environment will dictate the resolution of analyses of the data. In essence, this method sacrifices temporal resolution for greater environmental coverage. In this approach, the depositional sequence becomes the limit of temporal resolution. Given that third-order (i.e., 1–10 m.y.) sequences are the most widely recognized and correlated scale of depositional sequence, this approach will be best suited for paleobiological phenomena that take place over several to tens of millions of years. This approach will not be able to resolve paleobiologic changes on timescales shorter than a depositional sequence, usually 1–10 m.y., and as such, will not be useful for

studying short-term phenomena such as speciation events.

The second method for collecting better data is to use a high-resolution approach to sampling. In this approach, correlations are made at a very high resolution, perhaps through the correlation of meter-scale parasequences or even individual event beds. These units should be traced sufficiently far depositionally updip and downdip so that at least one facies can be followed throughout the interval of interest. Paleobiologic changes can then be followed in that facies. Because this method is so time intensive, it necessarily limits environmental coverage and the span of geologic time to be studied, although it is essentially the only approach to documenting paleobiologic changes at the outcrop scale for short spans of time (<1 m.y.). Because outcrop belts frequently only expose a small portion of the depositional shelf or ramp, such that outcrops even several kilometers apart expose essentially the same facies succession, the ability to track many separate facies across a given boundary will be greatly hindered. Consequently, this approach is ideal for resolving the dynamics of brief paleobiological events within single sedimentary environments, such as the turnover pulses separating postulated intervals of coordinated stasis (Brett et al. 1996), or the dynamics during and following brief and catastrophic mass extinctions.

Prospects for the Future

Model Improvements.—Recent advances in desktop computing power make it increasingly possible to quantify patterns in fossil occurrences and to formulate hypotheses regarding null expectations about fossil occurrences where no unusual biological phenomena are occurring (mass extinctions, radiations, punctuated morphological change, etc.). Three priorities for the future are to develop alternative models, to test the predictions of these models, and to develop strategies for dealing explicitly with the incompleteness that models suggest underlies much of our data.

Although a model linking the rarity of fossils, depth-dependent facies control, sequence-stratigraphic architecture, and changes in sedimentation rate integrates many of the factors known to affect fossil distributions, it clearly does not include all factors that affect the completeness of the fossil record. Once these effects are clearly understood, future models need to be developed that explore the effects of additional variables. It will be critical at each stage in this process to determine whether the additional predictive ability of the model justifies the added complexity and the consequent increase in the number of parameters to be estimated. The facies control aspect of the model is clearly one area that could be made significantly more complicated, although it is not known yet whether such complications are necessary or justified. For example, facies control that is not depth related could be included, such as hard-substrate versus soft-substrate assemblages that occur in different systems tracts (Fürsich et al. 1991). The facies characteristics of each species could also change over time, either independently of other species or in concert with them. The effect of correlations among the three preservation parameters (PA, PD, DT) should also be explored, because there is some evidence from modern organisms that preferred depth and depth tolerance are positively correlated (Pineda and Caswell 1998). The effect of correlations between these variables and origination and extinction rates should also be examined because abundant species may have broader environmental tolerances and lower extinction rates (Brown 1995). Changes in extinction and origination rates through time should be explored to determine our ability to recognize such changes when they do occur and to distinguish artifactually produced changes in turnover (e.g., Holland 1996a).

As discussed previously, all of the predictions described here need to be tested in the field. These tests need to include a wide range of taxa and a wide range of geologic settings to determine the extent to which the model successfully predicts the fossil record in outcrops. Through these tests, it will be possible to determine if the model needs to be more complex to achieve satisfactory predictions. It is likely that complications will be needed for some purposes but not others, just as simple methods of estimating confidence limits are sufficient in some cases but not others (Marshall and Ward 1996).

Model Calibration and Application.—If the model accurately describes the fossil record as seen in the outcrop, then methods must be developed for dealing with sampling bias, facies bias, unconformity bias, and condensation bias. It may be possible to develop countermeasures for each of these biases and to use them as needed in particular situations. To do this, it will also be necessary to develop methods for measuring the relative effects of these four biases in any given field situation. At some point, this will likely require methods of quantifying the three preservation parameters (PA, PD, DT) for each species. For example, transfer functions have been used to quantify for foraminifera and diatoms parameters equivalent to preferred depth and depth tolerance (Horton et al. 1999; Zong and Horton 1999).

The ultimate goal of these models is for them to be useful for understanding both the fossil record and the methods used to analyze the fossil record, particularly in cases where it may not be possible or practical to know the true sequence of events. For example, the model could be used to test the relative accuracy of biostratigraphic methods. Knowing the true relative accuracy of the methods currently requires some external method of correlation known to be true, whereas in a model, true time correlations are immediately accessible. Similarly, these models could be used to test methods of estimating confidence limits on fossil ranges and to evaluate more advanced methods of analyzing the fossil record, such as FreqRat, which simultaneously estimates extinction rate and preservation probability from the fossil record (Foote and Raup 1996).

Finally, these models might allow paleobiologists to realize answers to long-standing questions, such as the role of environmental and sea-level change on the ecology and evolution of organisms. With an honest appraisal of the limitations of the stratigraphic record, artifactual patterns in the fossil record could be weeded out from patterns that are more informative about ecological and evolutionary processes, much as Jablonski (1980) argued

two decades ago. A first step in this is to recognize paleontological patterns that cannot be
explained by sequence stratigraphic or other
biases. For example, faunal stasis across major
sequence boundaries (Brett et al. 1996) is not
predicted by these models and suggests low-
er-than-normal turnover rates. Similarly,
some changes in faunal abundance are inex-
plicable in terms of sequence stratigraphic architecture (Gräfe 1999). In this way, it will become possible, for example, to recognize the
effects of changing oceanographic conditions
(Patzkowsky and Holland 1993) or the opening of oceanographic barriers (Fernández-López and Meléndez 1996) and distinguish them
from the often more striking effects of local facies changes.

Acknowledgments

I acknowledge the shared insights of the
many that have discussed these topics with me;
in particular, I thank M. Patzkowsky, A. Miller,
S. Kidwell, M. Foote, and C. Marshall. In addition, C. Brett and J. Grotzinger improved the
manuscript with their perceptive insights. P.
Flemings and A. Deshpande assisted in modifying the model Strata so that its output could
be used by Biostrat, the model simulating the
occurrence of fossils. The model Strata is available as freeware from http://hydro.geosc.
psu.edu/Sed_html/strata_front_page.html.
Biostrat is available as freeware from http://
www.uga.edu/~strata; source code is available from SMH. The biostratigraphic applications of this modeling have been supported by
National Science Foundation grants EAR-
9705732 to S. M. H. and EAR-9705829 to M.
Patzkowsky.

Literature Cited

Abbott, S. T., and R. M. Carter. 1997. Macrofossil associations
from Mid-Pleistocene cyclothems, Castlecliff Section, New
Zealand: implications for sequence stratigraphy. Palaios 12:
188–210.
Armentrout, J. M. 1991. Paleontologic constraints on depositional modeling: examples of integration of biostratigraphy
and seismic stratigraphy, Pliocene-Pleistocene, Gulf of Mexico. Pp. 137–170 in P. Weimer and M. H. Link, eds. Seismic facies and sedimentary processes of submarine fans and tur-
bidite systems. Springer, New York.
———. 1996. High resolution sequence biostratigraphy: examples from the Gulf of Mexico Plio-Pleistocene. In J. A. Howell
and J. F. Aitken, eds. High resolution sequence stratigraphy:
innovations and applications. Geological Society of London
Special Publication 104:65–86.
Armentrout, J. M., and J. F. Clement. 1991. Biostratigraphic calibration of depositional cycles: a case study in High Island–
Galveston–East Breaks areas, offshore Texas. Pp. 21–51 in J.
M. Armentrout and B. F. Perkins, eds. Sequence stratigraphy
as an exploration tool: concepts and practices. Gulf Coast Section SEPM Foundation, eleventh annual research conference,
Program and abstracts.
Atrops, F., and S. Ferry. 1989. Sequence stratigraphy and changes in the ammonite fauna (Upper Jurassic, S-E France). Pp. 7–
9 in Mesozoic record of Western Tethyan margin. 2ème Con-
grés Français de Sédimentologie Lyon, Résumé.
Austin, M. P. 1987. Models for the analysis of species' response
to environmental gradients. Vegetatio 69:35–45.
Bambach, R. K., and N. L. Gilinsky. 1988. Artifacts in the apparent timing of macroevolutionary "events." Geological Society of America Abstracts with Programs 20:A104.
Baumiller, T. K. 1996. Exploring the pattern of coordinated stasis: simulations and extinction scenarios. Palaeogeography,
Palaeoclimatology, Palaeoecology 127:135–146.
Bayer, U., and G. R. McGhee. 1985. Evolution in marginal epi-
continental basins: the role of phylogenetic and ecologic fac-
tors (Ammonite replacements in the German Lower and Mid-
dle Jurassic). Pp. 164–220 in U. Bayer and A. Seilacher, eds.
Sedimentary and evolutionary cycles. Springer, New York.
Bosscher, H., and J. Southam. 1992. CARBPLAT—a computer
model to simulate the development of carbonate platforms.
Geology 20:235–238.
Boucot, A. J. 1981. Principles of benthic marine paleoecology.
Academic Press, New York.
Bretsky, P. W. 1970. Late Ordovician ecology of the central Ap-
palachians. Peabody Museum of Natural History Bulletin 34:
1–150.
Brett, C. E. 1995. Sequence stratigraphy, biostratigraphy, and taphonomy in shallow marine environments. Palaios 10:597–
616.
———. 1998. Sequence stratigraphy, paleoecology, and evolution: biotic clues and responses to sea-level fluctuations. Palaios 13:241–262.
Brett, C. E., and G. C. Baird. 1995. Coordinated stasis and evolutionary ecology of Silurian to Middle Devonian faunas in
the Appalachian Basin. Pp. 285–315 in D. H. Erwin and R. L.
Anstey, eds. New approaches to speciation in the fossil record. Columbia University Press, New York.
Brett, C. E., K. B. Miller, and G. C. Baird. 1990. A temporal hierarchy of paleoecologic processes within a middle Devonian
epeiric sea. In W. Miller III, ed. Paleocommunity temporal dynamics: the long term development of multispecies assem-
blies. Paleontological Society Special Publication 5:178–209.
Brett, C. E., L. C. Ivany, and K. M. Schopf. 1996. Coordinated
stasis: an overview. Palaeogeography, Palaeoclimatology, Palaeoecology 127:1–20.
Brown, J. H. 1984. On the relationship between abundance and
distribution of species. American Naturalist 124:255–279.
———. 1995. Macroecology. University of Chicago Press, Chicago.
Buhl-Mortensen, L., and T. Høisæter. 1993. Mollusc fauna along
an offshore-fjord gradient. Marine Ecology Progress Series
97:209–224.
Cisne, J. L., G. O. Chandlee, B. D. Rabe, and J. A. Cohen. 1982.
Clinal variation, episodic evolution, and possible parapatric
speciation: the trilobite Flexicalymene senaria along an Ordovician depth gradient. Lethaia 15:325–341.
Connell, S. D., and M. P. Lincoln-Smith. 1999. Depth and the
structure of assemblages of demersal fish: experimental
trawling along a temperate coast. Estuarine, Coastal and Shelf
Science 48:483–495.
Culver, S. J. 1988. New foraminiferal depth zonation of the
northwestern Gulf of Mexico. Palaios 3:69–85.

Dockery, D. T., III. 1986. Punctuated succession of Paleogene mollusks in the northern Gulf Coastal Plain. Palaios 1:582–589.

Donovan, S. K., and C. R. C. Paul. 1998. The adequacy of the fossil record. Wiley, Chichester, England.

Eldredge, N., and S. J. Gould. 1972. Punctuated equilibria: an alternative to phyletic gradualism. Pp. 82–115 in T. J. M. Schopf, ed. Models in paleobiology. Freeman, Cooper, San Francisco.

Elrick, M., and J. F. Read. 1991. Cyclic ramp-to-basin carbonate deposits, Lower Mississippian, Wyoming and Montana: a combined field and computer modeling study. Journal of Sedimentary Petrology 61:1194–1224.

Emery, D., and K. J. Myers. 1996. Sequence stratigraphy. Blackwell Scientific, Oxford.

Fernández-López, S., and G. Meléndez. 1996. Phylloceratina ammonoids in the Iberian Basin during the Middle Jurassic: a model of biogeographical and taphonomic dispersal related to relative sea-level changes. Palaeogeography, Palaeoclimatology, Palaeoecology 120:291–302.

Finney, S. C., W. B. N. Berry, J. D. Cooper, R. L. Ripperdan, W. C. Sweet, S. R. Jacobson, A. Soufiane, A. Achab, and P. J. Noble. 1999. Late Ordovician mass extinction: a new perspective from stratigraphic sections in central Nevada. Geology 27:215–218.

Flemings, P., and J. P. Grotzinger. 1996. STRATA: freeware for analyzing classic stratigraphic problems. GSA Today 6:1–7.

Foote, M., and D. M. Raup. 1996. Fossil preservation and the stratigraphic ranges of taxa. Paleobiology 22:121–140.

Franseen, E. K., W. L. Watney, C. G. S. C. Kendall, and W. Ross, eds. 1991. Sedimentary modeling: computer simulations and methods for improved parameter definition. Kansas Geological Survey Bulletin 233.

Freeman, S. M., C. A. Richardson, and R. Seed. 1998. The distribution and occurrence of Acholoë squamosa (Polychaeta: Polynoidae), a commensal with the burrowing starfish Astropecten irregularis (Echinodermata: Asteroidea). Estuarine, Coastal and Shelf Science 47:107–118.

Fürsich, F. T., W. Oschmann, A. K. Jaitly, and I. B. Singh. 1991. Faunal response to transgressive-regressive cycles: example from the Jurassic of western India. Palaeogeography, Palaeoclimatology, Palaeoecology 85:149–159.

Gaskell, B. A. 1991. Extinction patterns in Paleogene benthic foraminiferal faunas: relationship to climate and sea level. Palaios 6:2–16.

Gerrodette, T. 1979. Equatorial submergence in a solitary coral Balanophyllia elegans, and the critical life stage excluding the species from shallow water in the south. Marine Ecology Progress Series 1:227–235.

Goldhammer, R. K., P. J. Lehmann, and P. A. Dunn. 1993. The origin of high-frequency platform carbonate cycles and third-order sequences (Lower Ordovician El Paso Group, west Texas): constraints from outcrop data and stratigraphic modeling. Journal of Sedimentary Petrology 63:318–359.

Gräfe, K.-U. 1999. Foraminiferal evidence for Cenomanian sequence stratigraphy and palaeoceanography of the Boulonnais (Paris basin, northern France). Palaeogeography, Palaeoclimatology, Palaeoecology 153:41–70.

Grassle, J. F., H. L. Sanders, R. R. Hessler, G. T. Rowe, and T. McLellan. 1975. Pattern and zonation: a study of the bathyal megafauna using the research submersible Alvin. Deep-Sea Research 22:457–481.

Harris, M. T., and P. M. Sheehan. 1997. Carbonate sequences and fossil communities from the Upper Ordovician–Lower Silurian of the Eastern Great Basin. Brigham Young University Geology Studies 42:105–128.

Holland, S. M. 1995a. Sequence stratigraphy, facies control, and their effects on the stratigraphic distribution of fossils. Pp. 1–

23 in B. U. Haq, ed. Sequence stratigraphy and depositional response to eustatic, tectonic and climatic forcing. Kluwer Academic, Dordrecht, Netherlands.

———. 1995b. The stratigraphic distribution of fossils. Paleobiology 21:92–109.

———. 1996a. Guidelines for interpreting the stratigraphic record of extinctions: distinguishing pattern from artifact. Pp. 174 in J. E. Repetski, ed. Sixth North American paleontological convention, Abstracts of papers. Paleontological Society Special Publication 8:174.

———. 1996b. Recognizing artifactually generated coordinated stasis: implications of numerical models and strategies for field tests. Palaeogeography, Palaeoclimatology, Palaeoecology 127:147–156.

———. 1997. Using time/environment analysis to recognize faunal events in the Upper Ordovician of the Cincinnati Arch. Pp. 309–334 in C. E. Brett and G. C. Baird, eds. Paleontological events: stratigraphic, ecological and evolutionary implications. Columbia University Press, New York.

Holland, S. M., and M. E. Patzkowsky. 1998. Depositional sequences and the stratigraphic distribution of fossils: isolating the effects of condensation. Pp. A307(1–4) in 1998 AAPG Annual Convention Extended Abstracts, Vol. 1. American Association of Petroleum Geologists, Salt Lake City.

———. 1999. Models for simulating the fossil record. Geology 27:491–494.

Horton, B. P., R. J. Edwards, and J. M. Lloyd. 1999. A foraminiferal-based transfer function: implications for sea-level studies. Journal of Foraminiferal Research 29:117–129.

Ivany, L. C., C. R. Newton, and H. T. Mullins. 1994. Benthic invertebrates of a modern carbonate ramp: a preliminary survey. Journal of Paleontology 68:417–433.

Jablonski, D. 1980. Apparent versus real biotic effects of transgressions and regressions. Paleobiology 6:397–407.

———. 1986. Background and mass extinctions: the alternation of macroevolutionary regimes. Science 231:129–133.

Jervey, M. T. 1988. Quantitative geological modelling of siliciclastic rock sequences and their seismic expression. Pp. 47–69 in C. K. Wilgus, B. S. Hastings, C. G. S. C. Kendall, H. W. Posamentier, C. A. Ross, and J. C. Van Wagoner, eds. Sea-level changes: an integrated approach. Society of Economic Paleontologists and Mineralogists, Tulsa, Okla.

Kendall, C. G. S. C., J. Strobel, R. Cannon, J. Bezdak, and G. Biswas. 1991. The simulation of the sedimentary fill of basins. Journal of Geophysical Research 96:6911–6929.

Kidwell, S. M. 1991a. Condensed deposits in siliciclastic sequences: expected and observed features. Pp. 682–695 in G. Einsele, W. Ricken and A. Seilacher, eds. Cycles and events in stratigraphy. Springer, Berlin.

———. 1991b. The stratigraphy of shell concentrations. Pp. 211–290 in P. A. Allison and D. E. G. Briggs, eds. Taphonomy: releasing the data locked in the fossil record. Plenum, New York.

Kidwell, S. M., and D. W. J. Bosence. 1991. Taphonomy and time-averaging of marine shelly faunas. Pp. 115–209 in P. A. Allison and D. E. G. Briggs, eds. Taphonomy: releasing the data locked in the fossil record. Plenum, New York.

Ludvigsen, R., S. R. Westrop, B. R. Pratt, P. A. Tuffnell, and G. A. Young. 1986. Dual biostratigraphy: zones and biofacies. Geoscience Canada 13:139–154.

MacLeod, N. 1991. Punctuated anagenesis and the importance of stratigraphy to paleobiology. Paleobiology 17:167–188.

Mancini, E. A., and B. H. Tew. 1995. Geochronology, biostratigraphy and sequence stratigraphy of a marginal marine to marine shelf stratigraphic succession: Upper Paleocene and Lower Eocene, Wilcox Group, eastern Gulf Coastal Plain, U.S.A. In W. A. Berggren, D. V. Kent, M.-P. Aubry and J. Hardenbol, eds. Geochronology, Time Scales and Global Stratigraphic Correlation. SEPM Special Publication 54:281–293. Tulsa, Okla.

Marshall, C. R. 1990. Confidence intervals on stratigraphic ranges. Paleobiology 16:1–10.

———. 1994. Confidence intervals on stratigraphic ranges: partial relaxation of the assumption of randomly distributed fossil horizons. Paleobiology 20:459–469.

———. 1997. Confidence intervals on stratigraphic ranges with nonrandom distributions of fossil horizons. Paleobiology 23:165–173.

Marshall, C. R., and P. D. Ward. 1996. Sudden and gradual molluscan extinctions in the latest Cretaceous of Western-European Tethys. Science 274:1360–1363.

McGhee, G. R., Jr., U. Bayer, and A. Seilacher. 1991. Biological and evolutionary responses to transgressive-regressive cycles. Pp. 696–708 in W. Ricken and A. Seilacher, eds. Cycles and events in stratigraphy. Springer, Berlin.

McKinney, M. L. 1986. Biostratigraphic gap analysis. Geology 14:36–38.

———. 1991. Completeness of the fossil record: an overview. Pp. 66–83 in S. K. Donovan, ed. The processes of fossilization. Columbia University Press, New York.

Minchin, P. R. 1987. Simulation of multidimensional community patterns: towards a comprehensive model. Vegetatio 71:145–156.

Murray, J. W. 1991. Ecology and palaeoecology of benthic foraminifera. Wiley, New York.

Newell, R. C. 1970. Biology of intertidal animals. Elsevier, New York.

Osleger, D., and J. F. Read. 1993. Comparative analysis of methods used to define eustatic variations in outcrop: late Cambrian interbasinal sequence development. American Journal of Science 293:157–216.

Pachut, J. F., R. J. Cuffey, and D. R. Kobluk. 1995. Depth-related associations of cryptic-habitat bryozoans from the leeward fringing reef of Bonaire, Netherlands Antilles. Palaios 10:254–267.

Palmer, A. R. 1965. Biomere—a new kind of biostratigraphic unit. Journal of Paleontology 39:149–153.

Patzkowsky, M. E., and S. M. Holland. 1993. Biotic response to a Middle Ordovician paleoceanographic event in eastern North America. Geology 21:619–622.

———. 1996. Extinction, invasion, and sequence stratigraphy: patterns of faunal change in the Middle and Upper Ordovician of the eastern United States. In B. J. Witzke, G. A. Ludvigsen and J. E. Day, eds. Paleozoic sequence stratigraphy: views from the North American craton. Geological Society of America Special Paper 306:131–142.

———. 1999. Biofacies replacement in a sequence stratigraphic framework: Middle and Upper Ordovician of the Nashville Dome, Tennessee, USA. Palaios 14:301–323.

Paul, C. R. C. 1982. The adequacy of the fossil record. Pp. 75–117 in K. A. Joysey and A. E. Friday, eds. Problems of phylogenetic reconstruction. Academic Press, New York.

Piepenburg, D., and M. K. Schmid. 1996. Distribution, abundance, biomass and mineralization potential of the epibenthic megafauna of the Northeast Greenland Shelf. Marine Biology 125:321–332.

Piepenburg, D., N. V. Chernova, C. F. von Dorrien, J. Gutt, A. V. Neyelov, E. Rachor, L. Saldanha, and M. K. Schmid. 1996. Megabenthic communities in the waters around Svalbard. Polar Biology 16:431–446.

Piepenburg, D., J. Voß, and J. Gutt. 1997. Assemblages of sea stars (Echinodermata: Asteroidea) and brittle stars (Echinodermata: Ophiuroidea) in the Weddell Sea (Antarctica) and off Northeast Greenland (Arctic): a comparison of diversity and abundance. Polar Biology 17:305–322.

Pineda, J. 1993. Boundary effects on the vertical ranges of deep-sea benthic species. Deep-Sea Research I 70:2179–2192.

Pineda, J., and H. Caswell. 1998. Bathymetric species-diversity patterns and boundary constraints on vertical range distributions. Deep-Sea Research II 45:83–101.

Posamentier, H. W., and G. P. Allen. 1993. Variability of the sequence stratigraphic model: effects of local basin factors. Sedimentary Geology 86:91–109.

Posamentier, H. W., and D. P. James. 1993. An overview of sequence-stratigraphic concepts: uses and abuses. Pp. 3–18 in H. W. Posamentier, C. P. Summerhayes, B. U. Haq, and G. P. Allen, eds. Sequence stratigraphy and facies associations. Blackwell Scientific, Oxford.

Posamentier, H. W., G. P. Allen, D. P. James, and M. Tesson. 1992. Forced regressions in a sequence stratigraphic framework: concepts, examples, and exploration significance. American Association of Petroleum Geologists Bulletin 76:1687–1709.

Raup, D. M. 1985. Mathematical models of cladogenesis. Paleobiology 11:42–52.

Read, J. F., J. P. Grotzinger, J. A. Bova, and W. F. Koerschner. 1986. Models for generation of carbonate cycles. Geology 14:107–110.

Ross, C. A., and J. R. P. Ross. 1995. Foraminiferal zonation of late Paleozoic depositional sequences. Marine Micropaleontology 26:469–478.

Saltzman, M. R. 1999. Upper Cambrian carbonate platform evolution, Elvinia and Taenicephalus Zones (Pterocephaliid-Ptychaspid biomere boundary), northwestern Wyoming. Journal of Sedimentary Research 69:926–938.

Siebenaller, J. F., and G. N. Somero. 1989. Biochemical adaptation to the deep-sea. Reviews in Aquatic Sciences 1:1–25.

Smith, R. I., and J. T. Carlton. 1975. Light's manual: intertidal invertebrates of the central California coast. University of California Press, Berkeley.

Soto, L. A. 1991. Faunal zonation of the deep-water brachyuran crabs in the straits of Florida. Bulletin of Marine Science 49:623–637.

Strauss, D., and P. M. Sadler. 1989. Classical confidence intervals and the Bayesian probability estimates for the ends of local taxon ranges. Mathematical Geology 21:411–427.

Tait, R. V., and F. A. Dipper. 1998. Elements of marine ecology. Butterworth-Heinemann, Oxford.

Thompson, B. E., G. F. Jones, J. D. Laughlin, and D. T. Tsukada. 1987. Distribution, abundance, and size composition of echinoids from basin slopes off southern California. Bulletin of Southern California Academy of Sciences 86:113–125.

Van Wagoner, J. C., R. M. Mitchum, K. M. Campion, and V. D. Rahmanian. 1990. Siliciclastic sequence stratigraphy in well logs, cores, and outcrops. American Association of Petroleum Geologists Methods in Exploration Series, No. 7. Tulsa, Okla.

Vinogradova, N. G. 1962. Vertical zonation in the distribution of deep-sea benthic fauna in the ocean. Deep-Sea Research 8:245–250.

Walton, W. R. 1955. Ecology of living benthonic foraminifera, Todos Santos Bay, Baja California. Journal of Paleontology 29:952–1018.

Whittaker, R. H. 1970. Communities and ecosystems. Macmillan, New York.

Ziegler, A. M., R. M. Cocks, and R. K. Bambach. 1968. The composition and structure of Lower Silurian marine communities. Lethaia 1:1–27.

Zong, Y., and B. P. Horton. 1999. Diatom-based tidal-level transfer functions as an aid in reconstructing Quaternary history of sea-level movements in the UK. Journal of Quaternary Science 14:153–167.

The biomolecular paleontology of continental fossils

Derek E. G. Briggs, Richard P. Evershed, and Matthew J. Lockheart

Abstract.—The preservation of compounds of biological origin (nucleic acids, proteins, carbohydrates, lipids, and resistant biopolymers) in terrigenous fossils and the chemical and structural changes that they undergo during fossilization are discussed over three critical stratigraphic levels or "time slices." The youngest of these is the archeological record (e.g., <10 k.y. B.P.), when organic matter from living organisms undergoes the preliminary stages of fossilization (certain classes of biomolecule are selectively preserved while others undergo rapid degradation). The second time slice is the Tertiary. Well-preserved fossils of this age retain diagenetically modified biomarkers and biopolymers for which a product–precursor relationship with the original biological materials can still be identified. The final time slice is the Carboniferous. Organic material of this age has generally undergone such extensive diagenetic degradation that only the most resistant biopolymers remain and these have undergone substantial modification. Trends through time in the taphonomy and utility of ancient biomolecules in terrigenous fossils affect their potential for studies that involve chemosystematic and environmental data.

Derek E. G. Briggs. Department of Earth Sciences, University of Bristol, Wills Memorial Building, Queen's Road, Bristol BS8 1RJ, United Kingdom. E-mail: D.E.G.Briggs@bristol.ac.uk
Richard P. Evershed and Matthew J. Lockheart. School of Chemistry, University of Bristol, BS8 1TS, United Kingdom. E-mail: R.P.Evershed@bristol.ac.uk and E-mail: Matthew.Lockheart@bristol.ac.uk

Accepted: 26 June 2000

Introduction

Just as the preservation potential of body fossils depends on the resistance to decay and degradation of their constituent parts, so too does the preservation potential of their chemical constituents. A range of molecules of biological origin is preserved in the sedimentary record: nucleic acids (DNA, RNA), proteins, polysaccharides, and lipids, as well as biomacromolecules that make up the structural tissues in plants, e.g., algaenan, lignin, sporopollenin, and occasionally cellulose and cutin (Briggs and Eglinton 1994). Tegelaar et al. (1989) considered the preservation potential of biomacromolecules as ranging from those that are subject to extensive degradation "under any depositional conditions" (e.g., nucleic acids) to those that suffer "no degradation under any depositional conditions" (e.g., lignin, algaenan, suberan). Identifying the impact of chemical degradation during fossilization relies on structural analyses using a range of chemical and instrumental techniques. In some cases it is possible to speculate which chemical structures will be preserved; e.g., where only spores and pollen are preserved, the decay-resistant sporopollenin is likely to

survive. Biomolecules may be listed in rank order of decay resistance (e.g., lipid > carbohydrate > protein > nucleic acid) and, by implication, preservation potential, but this is an oversimplification because the preservation potential of a molecule may be increased if it becomes incorporated into structural tissues. Thus the jaws of polychaetes, which are composed largely of collagen strengthened by cross-linking, are considerably more decay resistant than the cuticle, which is similar in composition but lacks the cross-linking (Briggs and Kear 1993). The walls of fruits and seeds are composed of cellulose and lignin, a mixture significantly more resistant to decay than cellulose in isolation (van Bergen et al. 1994b–d, 1995). The preservation of biomolecules, e.g., collagen and noncollagenous proteins, may also be enhanced by their co-occurrence with biominerals, e.g., in bones, shells, and teeth.

Biomarkers

Organic compounds tend to be transformed diagenetically (from biomolecule to geomolecule) to more stable products with a higher preservation potential, which are known as biomarkers. The biomarker approach, when

0094-8373/00/2604-0007/$1.00

applied to the study of sedimentary environments, focuses on the identification of molecular fossils and determination of their precursor biomolecules. Aspects of the source, preservation, diagenesis, and effects of pre- and postdepositional environmental conditions on biomarkers have been extensively reviewed (e.g., Eglinton and Logan 1991; de Leeuw and Largeau 1993; Collinson et al. 1994; van Bergen et al. 1995; de Leeuw et al. 1995). The traditional organic geochemical approach involves solvent-extraction of biomarkers from ancient sedimentary rocks and profiling them using a combination of chromatographic separation and identification by mass spectrometric techniques. There are many applications of the geochemical data generated: (1) determination of the source of kerogen (at the simplest level, carbon chain lengths can indicate algal or bacterial versus higher-plant sources; specific biomarkers have been identified as diagnostic of certain source organisms; most hopanoids, for example, are derived from bacteria); (2) assessment of depositional setting, which is usually reflected in biomarkers derived from specific types of organism (e.g., marine versus fresh-water); (3) proxies for ancient environmental conditions (e.g., ocean temperature, oxygenation level); and (4) interpretation of maturation history. These applications have been extensively reviewed elsewhere (Mackenzie et al. 1982; Engel and Macko 1993; Killops and Killops 1993; Eganhouse 1997).

An approach to biomarker research largely neglected by organic geochemists has been the investigation of appropriate intact macrofossils, as opposed to relatively homogeneous sedimentary matter, as a potential means of establishing the origin and fate of organic matter in sediments. In this case the biomarkers identified can be assigned more confidently to a *known* source organism regardless of the extent of modifications by decay and diagenesis. Furthermore, the potential exists to perform analyses on the macrofossil by alternative or nonchemical methods to obtain more information about pre- and postdepositional conditions (e.g., stomatal counting on leaf surfaces).

The organism approach is proving a fruitful avenue of collaboration between organic geochemists and paleontologists, as it allows a wider range of questions to be addressed. The agenda includes issues such as (1) the relationship between chemical and macrofossil preservation—how chemical composition determines which tissues, and indeed which taxa, are preserved; (2) the relationship between the quality of morphological preservation and the chemistry of the organism—fidelity of external morphology does not necessarily imply that the chemistry is intact; (3) the variables that determine the quality of chemical preservation (e.g., depositional environment, maturation history, time); (4) the extent to which the chemistry of a fossil is diagnostic of that organism (the chemosystematic potential of different biomarkers). These are considerations of fundamental importance to paleontology, given that a significant proportion of the fossil record (of terrestrial biotas, in particular) is dominated by organisms that have no biomineralized tissues. The list includes the diversity of plants, insects, and many other arthropods, as well as graptolites (Briggs et al. 1995). Although morphological details may, in certain circumstances, be preserved as a result of mineral replication (or charcoalification in the case of plants), in the majority of cases the conservation of micro- and macroscopic structural features results from the preservation of macromolecular material.

DNA, Proteins, Carbohydrates

A great deal of interest was generated in the 1980s and 1990s by the possibility that DNA and proteins preserved in fossils might preserve sequence information that could be applied in phylogenetic analysis. Sequencing of fragments of DNA was made possible by application of the polymerase chain reaction (PCR). Unfortunately, however, both DNA and protein are highly susceptible to hydrolysis. Proteins normally hydrolyze within 100,000 to 1 million years (Bada et al. 1999), and DNA degrades even more rapidly (Poinar 1998). Thus, survival of these molecules tends to occur only where special conditions prevail; for example, remnants of plant or animal nucleic acids may be preserved in desiccated remains at cool or dry archeological sites or frozen in glaciers

(Eglinton and Logan 1991). Investigations of total nucleic acid extracts of desiccated archeological (up to 1400 years old) seeds from the site of Qasr Ibrim, Egypt, by means of GC/MS (see Analytical Methods, below) using selected ion monitoring (SIM) and MS-MS, provided the first direct mass spectrometric evidence for the presence of nucleic acid bases in ancient materials (O'Donoghue et al. 1994). Further liquid chromatography/MS-SIM analyses confirmed the presence of the corresponding deoxynucleosides (O'Donoghue et al. 1996a). Detailed studies by means of hybridization, PCR, and sequence analyses confirmed that the DNA was endogenous (O'Donoghue et al. 1996b).

DNA sequences have proved useful, to a degree, in investigations of archeological remains (Krings et al. 1997; papers in Jones et al. 1999), but reports of DNA in older fossils have proved rare and controversial (see Briggs 1999). Even in fossil specimens encased in amber, where hydrolysis and oxidation might have been prevented by the exclusion of air, the lack of evidence for the preservation of other, more decay-resistant biomacromolecules is inconsistent with the survival of DNA (Stankiewicz et al. 1998c; but see Bada et al. 1999). Structural proteins (e.g., collagen strengthened by cross-linking) are more decay resistant, as are intracrystalline proteins (in bones and shells), which are afforded additional protection from degradation. Amino acid sequence information, however, appears to be rarely preserved, and only immunological techniques have allowed simple questions of relationship to be resolved, and then only in recent "fossils" (e.g., radioimmunoassay of skin of the quagga, an extinct relative of the zebra hunted to extinction in Africa in the late nineteenth century [Lowenstein and Scheuenstuhl 1991]). Complex carbohydrates (e.g., structural polysaccharides) such as cellulose are occasionally preserved in fossils of geological age; lower molecular weight carbohydrates such as oligo-, di- and monosaccharides are more susceptible to hydrolysis and are very rarely preserved.

Lipids

It appears that the preservation of DNA, proteins, and polysaccharides in fossils and sediments more than one million years old is very limited. Unfortunately their much vaunted potential in phylogenetic studies has proved largely unfounded. However, lipids have a much higher preservation potential. These molecules are commonly hydrocarbons, which confers upon them hydrophobic properties and low chemical reactivities. As such they are likely to be preserved in close association with the fossil remains because they do not leach readily into the enclosing matrix via dissolution in water. Hence lipids are recovered from fossil remains or archeological bones, over extended timescales, often in a relatively unaltered state. However, there is still likely to be a degree of mixing with compounds derived from the surrounding sediments.

The major classes of lipids that perpetuate in fossils are n-alkyl lipids and polyisoprenoids. The former include straight-chain (aliphatic) compounds such as n-alkanes, n-alkanols, diols, ketones, hydroxy acids, diacids, and fatty acids. These may derive from the surface waxes that protect an organism from external damage, e.g., the epicuticular waxes of leaves (Eglinton et al. 1962) or the waxes that cover the cuticle of some insects, such as termites (Brown et al. 1996). These aliphatic compounds are usually found as homologous series (e.g., n-alkanes from plants have chain lengths ranging between 23 and 35 carbon atoms). Polyisoprenoidal compounds including triterpenoids, steroids, and hopanoids may also be retained in fossils either in their original form, or more often with diagenetic modification. Triacylglycerides and phospholipids are two of the major constituents of animal tissues. However, these have a low preservation potential owing to their susceptibility to hydrolysis, free-radical-induced polymerization, and β-oxidation, and as a consequence they rarely, if ever, survive into the geological record.

Historically, lipids have usually been investigated as chemical fossils in amorphous sedimentary organic matter. Recent analyses of lipids in identifiable fossils and archeological remains, however, have provided a wealth of important data (Huang et al. 1995, 1996; Logan et al. 1995; Evershed et al. 1999). The lipid profile of a macrofossil represents a complex mixture of compounds, including some diagnostic of the species of fossil analyzed and others contributed by associated exogenous organisms (e.g., bacteria, fungi). These components may have undergone pre- and post-depositional modification as a result of decay processes and/or diagenesis.

Animal remains usually undergo rapid degradation by saprotrophic bacteria, which use endogenous triacylglycerols and phospholipids as energy sources; a significant proportion of the remnant lipids may derive from these degraders. Characteristic lipids contributed by bacteria are likely to include hopanoids (ubiquitous components of sedimentary organic matter [Ourisson et al. 1979]) and branched-chain C_{15} and C_{17} fatty acids (Cranwell et al. 1987 and references therein). Plant fossil remains may yield lipids derived from epiphytic fungi, which infest the surface of some leaves (Williams 1985). Phytotoxins, such as tannins, which are naturally present in leaves, may limit the extent of degradation by fungi or microorganisms (Spicer 1991). Rapid burial of such a fossil deposited in a lacustrine setting may ensure that these protective compounds remain in situ rather than leaching away from the organism, thus reducing the extent of lipid degradation.

Lipid biomarkers carry a diversity of evidence of ancient environments (temperature, oxygen levels) and the source of hydrocarbons (reviews in Engel and Macko 1993). In contrast to proteins and DNA, where intact sequences of amino acids or nucleotides are diagnostic, it is the carbon skeleton of the lipid molecules and the composition of their mixtures that are of concern. Thus, biomarkers have wide utility so long as sufficient quantities survive to be detectable by instrumental analytical methods.

Biopolymers

A wide range of biopolymeric molecules perform a structural role in living organisms. The size of these molecules, and the bonding within their macromolecular structure, lowers their susceptibility to microbial and diagenetic degradation relative to the more labile biological components of organisms. Hence, they

are frequently preserved in the fossil record, e.g., as cuticular remains. The major types of biopolymer are the polyphenolic lignins, polysaccharides such as cellulose and chitin, cutin and suberin polyesters, and the polymethylenic structures cutan and algaenan (de Leeuw and Largeau 1993).

Lignin forms a large proportion of the woody tissues of plants with cellulose as the second major structural component, and it occurs in the woody tissues of both gymnosperms and angiosperms (see reviews by de Leeuw and Largeau 1993; de Leeuw et al. 1995). The cuticles of plant leaves are formed from cutin and/or cutan. Roots and wound tissues in plants are formed from suberin, a material analogous to cutin. Both cutin and suberin are composed of α, ω-diacids, ω-hydroxy acids ($C_{16:0}$, $C_{18:0}$, $C_{18:1}\Delta^9$, and $C_{18:2}\Delta^{9,12}$ [Holloway 1981]) and C_{16}/C_{18} di- and trihydroxy acids (Riederer et al. 1993) whereas cutan is a nonhydrolyzable polymethylenic biopolymer. Cutin, suberin, and cellulose are more susceptible to degradation than cutan and lignin and so are rarely detected in fossils, although they are more commonly present in some younger fossil and archeological specimens under favorable conditions.

Analytical Methods

The method employed in the sampling of a fossil must eliminate "contamination" from the enclosing sediment, and other sources, as far as possible. Contamination is a particular problem where ancient DNA is concerned, owing to the sensitivity of the PCR method. Plant and animal cuticles can sometimes be removed from a sedimentary matrix using a blade or needle. Beetle cuticles and plant remains have been obtained in this way from a number of localities, by gently flaking the organic fossil residue from the enclosing sediment. Chemical methods, such as dissolving the enclosing sediment in HF or HCl, can also be used; such procedures appear to have little or no effect on the preserved organic material (Collinson et al. 1994; van Bergen et al. 1995).

The main analytical methods employed are summarized in Table 1. The first stage of most analyses of fossil organic material is to perform a lipid extraction in a high-purity solvent

TABLE 1. Analytical techniques for ancient biomolecules.

Class of biomolecule	Source within the organism	Analytical methods	Target compounds
Nucleic acids	genetic material	polymerase chain reaction (replication), base sequencing, GC/MS analysis for bases	nucleotide sequences, free bases
Proteins	enzymes, hair, and connective tissues	acid digestion of peptide chain, chromatographic separation, amino acid analysis by GC/MS and LC/MS immunoassay	individual amino acids
Carbohydrate	woody tissues, storage organs	acid hydrolysis, GC/MS	monosaccharides
Free and bound lipids	plasma membranes, protective waxes, etc.	solvent extraction, mild-base hydrolysis to release "bound" lipids analysis by GC/MS, GC/combustion/isotope ratio monitoring MS	free lipids
Resistant biopolymers	leaf cuticles (cutin/cutan), algal cell walls (algaenan), pollen and spores (sporopollenin), arthropod cuticles (chitin), woody tissues (lignin)	GC/MS analysis of chemical cleavage products or pyrolysis-GC/MS of untreated sample	pyrolysis products, e.g., *lignin monomers* (syringyl, guaiacyl, and phenols); *cutin* (ω-hydroxy acids and hydroxy α,ω-diacids); *cutan* (*n*-alkanes/enes); *chitin* (acetamido furan, pyrone, and others)

(e.g., high-performance liquid chromatography [HPLC] grade). This can be achieved either by repeated sonication and removal of supernatants or, in the case of larger specimens, by continuous extraction in a Soxhlet apparatus. A complex mixture of lipids known as the total lipid extract (TLE) is obtained after vacuum concentration of the supernatants. Subfractions can be isolated from the TLE using a chromatographic procedure involving a combination of the following: solid-phase extraction, silica column chromatography, thin-layer chromatography, molecular sieving, and urea adduction.

The components in the mixture can be quantified by capillary gas chromatography (GC), using an appropriate column and conditions, by calculating peak areas relative to an internal standard introduced at the extraction stage. Identification of the lipid components of subfractions is achieved by GC/MS: a GC is coupled to a mass spectrometer (MS) that acts as a highly specialized detector. This technique provides the opportunity for both high-resolution separation and online recording of mass spectra. It permits unambiguous structural assignments to be made, based on the characteristic mass spectral fragmentation patterns of different compounds, often without the need to resort to other techniques.

Insoluble biopolymeric materials or covalently bound lipids may remain in the residue after solvent extraction. They can be analyzed by GC or HPLC after further chemical treatment (e.g., saponification, hydrolysis, RuO_4, acid digestion; Table 1). Instrumental pyrolytic techniques may also be applied in the analysis of biopolymeric materials. Flash or sequential heating of the sample on a probe or within a quartz tube leads to the breaking of intramolecular bonds and these pyrolysis products may be analyzed by GC/MS. The compositional data obtained from the pyrolysis products can be used to reassemble the "jigsaw" in order to determine the original composition of the whole macromolecular structure. There are drawbacks, however, in that the mode of breakdown of the macromolecular structure can complicate the interpretation of the data obtained from these treatments. Pyrolysis-GC/MS (py-GC/MS) is particularly suitable for the identification of recalcitrant macromolecular material in very small samples (ca. 100 μg) and has been used to analyse many fossil and modern biological tissues (e.g., de Leeuw et al. 1991; van Bergen et al. 1995; Stankiewicz et al. 1998a). Complementary data on molecular structure can be obtained from solid state ^{13}C nuclear magnetic resonance (NMR) and Fourier-transform infrared (FTIR) spectroscopy, but sample sizes up to two orders of magnitude larger are required. Morphological changes in the cuticle can be monitored using scanning and transmission electron microscopy (SEM and TEM).

Further insight can be gained from the stable isotope compositions in individual compounds. The GC-combustion-isotope ratio mass spectrometry technique (GC/C/IRMS) was developed by Matthews and Hayes (1978) and has been widely applied in the field of organic geochemistry. This technique can be used to establish the source of various biomarkers and to infer other paleoenvironmental information (applications of this technique are discussed in more detail in later sections).

Time Slices

This review is focused on the organic chemistry of fossils, rather than on biomarkers isolated from sediments. The majority of organic fossils (e.g., plants) occur in continental sediments. Although the fossilization of macromolecules is not time dependent per se, many of the factors that control preservation are, and the nature of the chemical fossil record can therefore be illustrated by considering continental deposits at three stratigraphic levels: (1) the archeological record, which exemplifies the transition from living organism to fossil, (2) the Tertiary, which represents the limit beyond which most of the original chemical composition is lost, and (3) the Carboniferous, which yields a diverse Paleozoic terrestrial biota in which labile molecules are completely degraded with only certain modified biopolymers preserved. Some of the terrestrial biomarkers considered here also occur in marine sediments. Biomarkers in the marine realm have been reviewed by de Leeuw et al. (1995).

1. The Archeological Record

Biomolecules are preserved in a number of situations at archeological sites: in plant and animal remains, as well as in association with anthropological artifacts and sedimentary deposits. The primary focus of studies of biomolecules from such sites has been the inference of past human behavior based, for example, on residues on pottery (Evershed et al. 1997b). However, organisms preserved in archeological sites provide an opportunity to gain insights into the diagenetic and taphonomic processes that occur in the early stages of fossilization. These early diagenetic processes are fundamental to the longer-term survival (or otherwise) of fossil organisms over geological timescales. Archeological materials bridge the diagenetic time gap between laboratory and field experiments conducted in real time and observations made on fossil remains many millions of years old. Additionally, archeological materials include organisms with close extant relatives, allowing the diagenetic fate of biomolecules to be traced. This provides a basis for deducing the processes that are likely to affect extinct organisms for which biochemistries are less well understood.

The biomolecules that have received most attention in the field of archeology are DNA, proteins, lignin/cellulose, and lipids. The fundamental biomarker principle underlies the retrieval of archeological information from these biomolecules. Thus biomarkers in sediments retain enough of their basic biochemical skeleton to allow them to be linked unambiguously with biological precursor compounds and, hence, their source organism. In the case of identifiable organic remains, however, the source is clearly known. The preservation potential of different biomolecules is approximately inversely proportional to their information content.

DNA

The recovery of ancient DNA raises numerous possibilities for studying the evolutionary and functional relationships between organisms. However, the technique has not lived up to its early promise. It is now generally accepted that the majority of reports of DNA from fossils dating to millions of years are probably erroneous; DNA probably survives in an PCR-amplifiable form for at most a few tens of thousands of years (Jones et al. 1999). The major degradative reactions that act to fragment the DNA include hydrolysis of the deoxyribose-adenine or -guanine bond, followed by rapid chain breakage. The Maillard reaction may also play a part in limiting the amount of amplifiable DNA, although at least some of this "bound" DNA may be released by appropriate chemical treatments (Poinar et al. 1998). Several recent papers have emphasized the value of ancient DNA in resolving important archeological questions relating to the study of human populations (Hagelberg et al. 1999; Merriwether 1999; Sykes 1999; Stone and Stoneking 1999), domesticated animals (MacHugh et al. 1999), and plants (Brown 1999).

Proteins

Proteins in archeological materials have been studied for many years. While structurally recognizable proteins have been detected using immunological techniques (Tuross and Stathoplos 1993), protein sequencing has proved to have limited potential (Lowenstein and Scheuenstuhl 1991). Claims of the survival of intact endogenous proteins (and DNA) on the surfaces of prehistoric stone tools, for example, are now regarded with skepticism (e.g., Loy 1993). Proteins may survive with just partial degradation in plant material preserved in a desiccating environment, e.g., within a 1400-yr-old radish from Qasr Ibrim, Egypt (Bland et al. 1998). The greatest contribution of proteins to archeology has been in the field of paleodietary reconstruction, using stable isotope techniques (which first emerged in the 1970s) on bones of ancient human populations (van der Merwe and Vogel 1977). It is not the molecular structure of the protein per se that is important in this approach, but the overall isotopic ($\delta^{13}C$ and $\delta^{15}N$) signal preserved in skeletal collagen (and other proteins) (Pate 1994 and references therein). The underlying principle is that differences in the composition of foodstuffs in different ecosystems are reflected in the tissues of the consumer animals. This approach has led to a number of important findings concerning the subsistence patterns of ancient peoples (Pate 1994; Pollard 1998). For example, stable carbon isotope analysis of collagen from human bones traces the introduction of maize agri-culture into temperate North America, because an isotopically distinct C_4 cultigen was introduced into an otherwise C_3 biome (van der Merwe and Vogel 1977). Paleodiet can also be determined by the isotopic composition of human hair (Macko et al. 1999).

The collagen of archeological bone can be used not only to reconstruct paleodiet, but also to determine the age of the bone (ca. <50 Ka) through radiocarbon dating. The reliability of the isotopic information obtained depends on the quality of the molecular preservation of collagen, which is just as important for dating as for paleodietary reconstruction. The indigeneity of bone collagen is usually assessed on the basis of (1) C/N ratios, which should be in the range 2.9 to 3.6 for modern collagen (Ambrose 1990); (2) amino acid profiling by means of HPLC (Katzenberg et al. 1995); and (3) a new approach involving the enantiomeric ratios of selected amino acids (Bada et al. 1999 and references therein). Proteins appear to be somewhat more resistant to degradation than DNA. Even under favorable conditions (cool or dry deposits), however, little intact collagen remains in bones after only 10–30 k.y. The extent of amino acid racemization, and the relative amino acid composition, can be used as an index of biomolecular preservation, and therefore as an indication of the potential for retrieving genetic information from bones (Poinar et al. 1996; but see Collins et al. 1999). The preservation of the skin of Iron Age bog bodies in certain peat environments (Stankiewicz et al. 1997d) also depends on the survival of proteins.

Lipids

Lipids are widely preserved in archeological bones, plant materials, and resins, as well as within the clay fabric of potsherds (Evershed 1993). The molecular separations and high sensitivities routinely achieved by GC and GC/MS allow results to be obtained even where the amounts of sample are limited, e.g., in the case of steroidal components from fragmentary remains of plants or bones. Compound-specific isotope measurements on lipids of archeological age using GC/C/IRMS provide an additional parameter that may indicate compound origins and paleodietary information. The structural diversity of lipids is fundamental to their usefulness in archeology because it allows lipid structures and distributions to be related systematically to specific source organisms.

In Plants and Animals.—Lipids occur in skeletal and other human remains, e.g., bog bodies (Evershed and Connolly 1988, 1994), mummies (Gülaçar et al. 1990; Buckley et al. 1999), and bones and teeth (Evershed et al. 1995). Cholesterol preserved in bones and teeth provides $\delta^{13}C$ information of use in paleodietary studies (Stott and Evershed 1996; Stott et al. 1999). The indigeneity of cholesterol is less of a concern than that of collagen on account of its negligible abundance in the burial environment and the fact that structural confirmation can be routinely achieved by GC/MS. Although carbonized plant remains do not yield significant lipid, desiccated propagules recovered from the exceptionally arid site of Qasr Ibrim have yielded abundant well-preserved lipids (O'Donoghue et al. 1996b; van Bergen et al. 1997a). The distributions of sterol components in ancient propagule specimens of a range of species are virtually unaltered compared with their modern counterparts (van Bergen et al. 1997a). Although triacylglycerols are hydrolyzed to their component fatty acids, preservation of polyunsaturated components within these seeds, e.g., $C_{18:2}$ and $C_{18:3}$ fatty acids, is exceptional. Clearly, these labile substances have been protected by entrapment within the macromolecular network of structural and storage macromolecules in relatively intact subfossil seeds.

In Resinous and Bituminous Materials.—The source of resins from archeological sites can be determined by biomarker analyses. Samples of amorphous material recovered from the cellar of a house at Qasr Ibrim, Egypt, dating to around 400–500 A.D., were characterized by GC/MS (Evershed et al. 1997a; van Bergen et al. 1997b), which showed the presence of triterpenoid components present in modern frankincense resin. Comparison of the composition of samples of ancient and modern reference frankincenses by mass chromatography revealed virtually identical "fingerprint" distributions. The data confirmed that the material was frankincense by demonstrating the presence of the characteristic constituents of the fresh aromatic gum resin from modern *Boswellia* trees. The lack of thermal degradation products indicated that there had been little or no processing of the resin following its collection. The frankincense was recovered in association with pieces of diterpenoid pine resin; they may have been used together in incense-burning ceremonies.

Biomarker analyses have been applied extensively to the analysis of tars and pitches (Mills and White 1994; Pollard and Heron 1996). Such materials arise from the human need to exploit natural materials to provide glues, sealants, and coating materials (Evershed et al. 1985; Robinson et al. 1987; Beck et al. 1989, 1998; Beck and Borromeo 1990; Charters et al. 1993; Aveling and Heron 1998; Regert et al. 1998) and more precious medicinal and ritual/funerary substances (van Bergen et al. 1997b; Evershed et al. 1997a). The use of biomarker techniques, such as GC and GC/MS, offers the only reliable means of identifying the origins of such materials. Chemically altered plant resins are often present in tars and pitches that are the products of the heating of resins and wood. These constituents indicate not only the botanical sources of such materials but also the methods used in their production and processing. For example, GC/MS analyses of ancient tars recovered from a number of sites throughout Europe have shown the presence of abundant triterpenoids related to compounds produced in the bark of *Betula* sp. The detection of betulin and lupeol, together with various pyrolysis products, provides strong evidence for the use of birch bark in the manufacture of these tars (Aveling and Heron 1998; Regert et al. 1998). Examination of similar samples recovered from Romano-British excavations revealed a complex mixture of high and low molecular weight components including fatty acids characteristic of animal fats, together with the triterpenoids lupeol, lupenone, betulin, betulone, and lupa-2,20(29)-dien-28-ol (Dudd and Evershed 1999). This result indicates that animal fat was deliberately mixed with the tar to provide a material with more desirable properties. Sterane and triterpane biomarker distributions derived by GC/MS have also confirmed the use of petroleum bitumens in the prehistoric Near

East (Connan et al. 1992; Boëda et al. 1996; see also Connan 1999).

In Other Archeological Contexts.—Lipid analyses are increasingly applied in other spheres of archeology, but as biomarkers in sediments and artifacts rather than in association with fossil remains. Recent analyses have suggested, for example, that the excreted sterol and bile-acid products of mammalian gut flora provide biomarkers for manure and thus are indicators of ancient animal husbandry (Evershed and Bethell 1996; Bull et al. 1999). Chemical analysis of both absorbed and surface residues on potsherds, one of the most common classes of archeological artifact, can provide information concerning the uses of ancient pottery vessels. Fats, oils and waxes, which are ubiquitous components of plants and animals, are commonly preserved in potsherds (Evershed et al. 1997b). Biomarker analyses are more diagnostic when combined with compound-specific stable isotope measurements (δ^{13}C values) by GC/C/IRMS. Milk and adipose fats are distinguished by different δ^{13}C values of their $C_{16:0}$ and $C_{18:0}$ fatty acids, providing, for example, the first reliable method for identifying the practice of dairying in prehistoric communities (Dudd and Evershed 1998).

The same properties of lipids govern their survival in archeological contexts as in substantially older fossil materials. For example, it is the hydrophobic nature of lipids that limits their loss to pore water from the immediate vicinity of archeological remains; the same principle governs their survival in association with fossils. However, all available lipids can be degraded by saprotrophic organisms and are consumed at least until the local environment becomes nutrient limited or hostile to biological activity. The protection afforded by entrapment in the protective environment of the matrix of archeological seeds or bones attests to the importance of this phenomenon for long-term survival during burial (Evershed et al. 1995; van Bergen et al. 1997a). Analogies can be drawn between this latter mechanism and the preservation of biomarkers associated with fossils in consolidated sediments.

The development of substantial populations of microorganisms during decay may result in the introduction of a microbial lipid signature superimposed on those endogenous to the fossil. Microbes are the major degraders of lipids in fossil animal remains. Marker compounds such as hopanoids are indicative of the activity of saprotrophic bacteria, which may be of exogenous or endogenous origin. For example, bones from archeological sites have been shown to contain diploptene [hop-22(29)-ene], which derives from the invasion of bone marrow by aerobic bacteria (Evershed et al. 1995). Studies of numerous specimens of animal bone in a similarly good preservational state rarely, if ever, show acylglycerol lipids (e.g., triacylglycerols and phospholipids) in appreciable quantities even though these are the major lipid components of the bones of extant mammals. This lack of acylglycerol lipids is further evidence of the activities of degradative organisms such as bacteria. Such acyl lipids yield fatty acids which are readily consumed by microorganisms, presumably largely via β-oxidation. Hydrolyzed lipids are also seen in desiccated archeological seeds, where free fatty acids persist in high concentrations. The hydrolysis in this situation is presumed to be autolytic, while the survival of fatty acids is due the apparent lack of activity of microorganisms in such environments. Most vertebrates and invertebrates also contain populations of symbiotic microorganisms. For example, the significant quantities of coprostanol in the tissues of a bog body must have been produced from the tissue cholesterol by the activities of endogenous enterohepatic bacteria invading the tissues postmortem (Evershed and Connolly 1994).

Chemical degradation becomes more significant once microbial activity has ceased. The presence of 7-oxocholesterol in archeological and subfossil animal bones provides clear evidence of chemical degradation reactions mediated by free oxygen, probably via its singlet state (Evershed et al. 1995). Free-radical oxidation of unsaturated lipids also occurs in desiccated human mummies, as indicated by the presence of complex mixtures of mono- and dihydroxy fatty acids and dicarboxylic acids (Buckley et al. 1999). Oxidation of the unsaturated compounds is also evidenced by a reduction in the concentration of polyunsaturated compounds in desiccated seeds, i.e., of the $C_{18:2}$ and $C_{18:3}$ components, although the survival of these compounds at all attests to the exceptional preservation environment of these archeological remains (O'Donoghue et al. 1996b; van Bergen et al. 1997a). The survival of these compounds may be facilitated by the presence of natural antioxidants in the seeds.

Characterization of the volatile components released upon maceration of ancient seeds has provided the first molecular evidence for the Maillard (or non-enzymatic browning) reaction occurring in buried plant matter (Evershed et al. 1997c). The presence of low molecular weight compounds, such as pyrazines, pyridines, and alkyl polysulphides, allowed operation of the reaction to be demonstrated unambiguously for the first time since it was initially suggested as a decay process in the early 1900s. The diagnostic volatile compounds would normally diffuse away from their initial site of formation during decay once the structural integrity of the plant tissue had been lost. They are, however, retained in the archeological seeds from Qasr Ibrim because the burial conditions have slowed their disaggregation, thereby ensuring preservation by encapsulation within internal networks of structural and storage macromolecules. These findings fit with the "neogenesis" model (Tissot and Welte 1984), which proposes that sedimentary organic matter can be produced by the random recombination of small labile molecules (such as simple sugars and amino acids), released from organic remains to produce refractory geopolymers.

Lignin/Cellulose

The lignin/cellulose complex is one of the most important biopolymer combinations as it constitutes the bulk of vascular plant tissues. Lignin has a heterogeneous polymeric structure composed of hydroxy propyl benzene units (C_9) linked together in three-dimensional arrangements by various oxygen linkages and carbon–carbon bonds. There are three major types of lignin: (1) in gymnosperms (particularly conifers), where it is composed solely of 2-methoxyphenol (guaiacyl) units;

(2) in dicotyledonous angiosperms, where it also contains 2,6-dimethoxyphenols (syringyl) units; and (3) in legumes and monocotyledons, where it also contains p-hydroxyphenyl units. Chemical variation among monocotyledon fruits and seeds indicates that there is potential for chemosystematic investigations of lignin in ancient material (van Bergen et al. 1994a, 1995). However, even fresh lignin is notoriously difficult to study structurally due to its heteropolymeric nature. The polysaccharide cellulose, on the other hand, is a massive homogeneous polymer of acetal linked D-glucose units.

The principal approaches to the study of modern and degraded woods include py-MS or py-GC/MS, solid state ^{13}C NMR (Hedges et al. 1985), and chemolytic methods based on oxidation of the polyphenolic lignin structure (Logan and Thomas 1987; Opsahl and Benner 1995) and on hydrolysis to degrade cellulose. Degradation of lignin involves mainly oxidation at double bonds and demethylation or demethoxylation reactions (important degradative pathways may remain obscure in view of the difficulty of studying lignin). While the cellulose component is readily lost from buried wood, the lignin frequently survives (van Bergen et al. 2000). As with proteins, the isotopic composition is the most useful type of information preserved within lignin/cellulose. Radiocarbon analyses of subfossil wood and charcoal are commonly used to date archeological sites, while stable isotope analyses of $\delta^{13}C$, $\delta^{18}O$, and D/H, have the potential to yield environmental information, e.g., past climate recorded in the cellulose of annual tree rings (Leavitt and Long 1984; Pollard 1998; Switsur and Waterhouse 1998). Van de Water et al. (1994) examined a large sample set of Quaternary pine needles from pack rat middens spanning the past 30,000 years. They found a significant correlation between $\delta^{13}C$ of cellulose and stomatal density of the needles during the Holocene (12 Ka to 0). Furthermore, they observed a 17% decrease in stomatal density as CO_2 levels rose by 30% (according to ice core records) during the period of deglaciation from 15 Ka to 12 Ka. Clearly a combination of biogeochemical and morphological data provides a more robust environmental indicator than either in isolation.

Lignin is also an important component of the decay-resistant walls of plant propagules (van Bergen et al. 1995). An exception is the sclerotic fruit wall of the water lily *Nelumbo*, which consists of a tannin-polysaccharide complex. Tannins are rarely found in the fossil record, although they have been reported from fossil bark in brown coals (Wilson and Hatcher 1988) The lack of lignin-cellulose is interpreted as the reason that the propagule of *Nelumbo* is not normally preserved in the fossil record, in contrast to other genera whose propagules have lignin-cellulose walls (van Bergen et al. 1996). Although some of the occurrences of biopolymers described above are not associated with archeological sites, they are included here because of their age and because the processes and preservation described are applicable to archeological samples.

2. The Tertiary

Exceptional preservation occurs extensively in Tertiary lacustrine settings (see Allison and Briggs 1991). Lake deposits are more prevalent in the Tertiary, mainly because they are less likely to have been subjected to deformation and erosion than those in more ancient sequences. Deep lakes become stratified, and the colder anoxic bottom conditions inhibit decay (Behrensmeyer and Hook 1992). The effects of diagenesis on Tertiary organic material are less pronounced than in older deposits. Nonetheless, because of the low preservation potential of DNA, proteins and smaller carbohydrates (e.g., oligosaccharides), the focus here is on lipids and structural biopolymers. The Eocene Green River Shales and the Messel Oil Shales have attracted considerable attention from organic geochemists, but investigations of individual fossils from Tertiary deposits have been carried out primarily on the Miocene Clarkia deposit of northwestern Idaho and the Oligocene deposit at Enspel, Germany.

The Clarkia deposit has long been renowned for its assemblage of fossil plants (Smiley 1985; Smiley and Rember 1985b). Investigations of its detailed biogeochemistry were prompted not least by reports in the early 1990s of DNA preservation in the leaves (a discovery that is now widely discounted [see Briggs 1999]). The lacustrine sequence at Enspel was targeted in the search for the earliest traces of chitin (Stankiewicz et al. 1997b).

The deposit at Clarkia comprises the sediments of a late Miocene (17–20 Ma) lake that was formed as a result of damming of a river by basalt lava flows. The sequence consists of 7.6 m of finely laminated layers of clay and silt with occasional thick layers of volcanic ash. It is thought to have accumulated over a relatively short period of perhaps 1000 years as a result of runoff from the land during violent rainstorms. Plant leaves, organs, and fruits, as well as insects, fish, and mollusks, are preserved as compression fossils. The flora and fauna at the deposit have been well documented (Smiley 1985 and references therein).

Lipids

The degree to which biomolecules within a macrofossil are altered reflects the maturity of a deposit and the thermal history of the sediments. This can be assessed through examination of the stereochemical configuration of the hydrogens at the 17 and 21 positions in the carbon skeleton of bacterial hopanes associated with the fossils (Mackenzie et al. 1980). Specifically, the ratio of the biological configuration $17\beta(H)$, $21\beta(H)$ compounds will dominate over the diagenctically altered $17\alpha(H)$, $21\beta(H)$ and $17\beta(H)$, $21\alpha(H)$ components in immature sediments. The dominance of $17\beta(H)$, $21\beta(H)$ hopanoids in Clarkia fossils is consistent with geological evidence that the Clarkia sediments are immature (Smiley and Rember 1985a; Huang et al. 1996). Gas chromatographic analyses of lipid fractions from the Clarkia fossil leaves reveal that phytosterols derived from the plants have been completely degraded (with the exception of small quantities of 24-ethyl-5α-cholestenes). Most triterpenols also appear to have been degraded or transformed; only β-amyrin is preserved intact (Huang et al. 1995; Logan et al. 1995). The steroidal A-ring has been cleaved in some cases, e.g., lupane has been converted to des-A-lupane (Logan and Eglinton 1994; Huang et al. 1995), and aromatization is also apparent (Logan 1992).

Some Clarkia leaves retain coloration, which may be due to the presence of degradation products of photosynthetic pigments such as chlorins. Niklas and Giannasi (1985) reported phaeophorbide, deoxophylloerythroaetioporphyrin (DPEP) and aetioporphyrin in acetone/methanol extracts of fossil *Betula*, *Platanus*, and *Hydrangea* by mass spectrometry (although spectra were not presented). Logan (1992) analyzed sediments from the Clarkia site and detected intact chlorins, pyrophaeophytins and pyrophaeophorbides, but no porphyrins; these compounds were perhaps derived from phytoplankton within the lake. Fossil insects have not been extensively analyzed for pigments. The iridescent coloration occasionally observed in fossil insects, e.g., from Messel, results from the refraction of light at the surface of the cuticle rather than chemical pigmentation.

Biopolymeric Materials

Microscopy of leaves from Clarkia reveals a remarkable degree of structural integrity (Niklas et al. 1985). The chemistry of structural biopolymeric material associated with the leaf remains has been analyzed using pyrolysis-MS (Logan et al. 1993). Evidence for the preservation of the highly resistant biopolymer lignin in the Clarkia fossils was provided by the detection of guaiacyl and syringyl lignin monomers using Py-GC/MS. Although the acetal-linked glucose units of cellulose are more susceptible to degradation, pyrolysis products of cellulose were detected in the side vein of a *Castanea* leaf from the Clarkia site (Logan 1992). These compounds have been observed in the pyrolysis products of Miocene seed coats of *Stratiotes* (van Bergen et al. 1995). They also occur in seeds and cone scales of *Pinus* and *Sequoia* specimens from Poland, although the pyrograms were dominated by lignin-derived guaiacyl analogues (Stankiewicz et al. 1997a). Cellulose has also been obtained from Miocene wood (Lücke et al. 1999). Evidence for cellulose as part of a lignin-cellulose biopolymer has been reported from Oligocene fruits (Boon et al. 1989) and Eocene woods (Spiker and Hatcher 1987). In contrast, cellulose is no longer present in Eocene seed coats,

and lignin is largely transformed to polyphenol macromolecules (van Bergen et al. 1994c; but see van Bergen et al. 1994a: Fig. 3b), although the gross morphology of the seed coat remains intact.

Cutin is almost always lost or altered during diagenesis. Two of the putative hydrolytic products of cutin, α,ω-diacids, and ω-hydroxy acids, have been detected in sediments at Clarkia, although at higher concentrations than in the plant fossils, perhaps suggesting a source other than leaf-derived cutin (Huang et al. 1996). Cutan is more commonly found in fossilized cuticle remains and has been identified in fossil gymnosperms and angiosperms from a number of sites ranging in age from Miocene to Permian (Tegelaar et al. 1991). Cutan-type materials have been shown to exist in some extant plants. However, it has been speculated that the fossil "cutan" might form as a result of polymerization reactions between aliphatic lipids, e.g., epicuticular waxes, during sedimentary diagenesis (Collinson et al. 1998; Stankiewicz et al. 1998d).

Arthropod cuticles frequently occur in the fossil record. They consist of a resistant macromolecular structure of 2-acetamido-2-deoxy-D-glucopyranose chains cross-linked via N-acetylglucosamine bonds to protein chains (Stankiewicz et al. 1996). Tertiary insects may yield traces of this original chitin-protein complex; the oldest known to do so are from the Oligocene lake deposits at Enspel in Germany (Stankiewicz et al. 1997b, 1998b). Although proteins are normally degraded within one million years (Bada et al. 1999), protection from degradation is afforded when they are constituents of the chitin-protein complex. Analysis of the products of chemical hydrolysis by GC/MS has shown that D-glucosamine, the monomeric unit of chitin, is released from arthropod cuticles of Oligocene age and younger (Flannery et al. in press), thereby providing evidence for the preservation of the chitin polysaccharide-protein complex and supporting the findings of py-GC/MS of the same fossils. Pyrolysis of older fossil arthropod cuticles indicates that the chitin-protein complex is altered during diagenesis and suggests that the alkyl-based wax components polymerize into an aliphatic biopolymer (Stankiewicz et al. 1997c). Thermal maturation experiments that simulate this transformation have been performed in the laboratory (Stankiewicz et al. 2000).

Chemotaxonomy

The molecular information preserved in fossil leaves may be taxonomically diagnostic provided that the biomarkers are not significantly modified diagenetically or contaminated from the surrounding sediment or other organisms (e.g., epiphytic fungi or bacteria). The potential exists to compare the lipid profiles of different fossils in order to establish taxonomic relationships. Furthermore, the lipid compositions of Tertiary or younger specimens may be compared with extant relatives.

Fossil leaves have greater potential for the preservation of lipids than fossil invertebrates. The reasons for this are twofold: (1) lipids with a reasonable preservation potential are present in abundance in epicuticular leaf waxes, e.g., n-alkyl lipids, and (2) the major constituents of animals, phospho- and glycerol lipids, are very susceptible to decay. Two approaches have been used for chemosystematic comparisons. The most specific approach is to identify biomarker compounds that are confined to a restricted group of plants or animals. Gymnosperms (conifers), for example, produce a wide range of di- and sesquiterpenoid compounds as resins and components of their needles. Otto and coworkers (1997) examined an Oligocene oxbow lake deposit in the Weisselster Basin, Germany, where the dominant macrofossil remains are conifer needles of the genus *Taxodium*. Diterpenoids with pimarane skeletons, phyllocladane isomers, and norabietane were detected in sediments from this deposit (although lipid analyses of individual fossils from this site have not been reported). Compounds with abietane-type skeletons also occur in extant *Taxodium distichum* (Kupchan et al. 1969). A similar association of diterpenoid markers and increased incidence of coniferous remains has been noted at the Clarkia P-33 site, in particular in the upper sediments (Lockheart 1997). However, analyses of individual fossils of *Taxodium*, *Metasequoia*, and *Sequoia* from these

Clarkia sediments did not reveal significantly high concentrations of the same diterpenoids.

Biomarker compositions were surveyed in a total of 89 leaf specimens (including the gymnosperms mentioned above) from the fossil flora from site P-33 at Clarkia (Lockheart et al. in press). A number of compounds of higher-plant origin were present in diagenetically altered form. These included triterpenoids with an open or lacking A-ring (Logan 1992; Huang et al. 1995, 1996). However, there was no apparent relationship between the terpenoid and steroid distributions and the genus under examination (Lockheart 1997).

Simple aliphatic compounds with a range of chain lengths are found in the epicuticular waxes of extant leaves, and these can display characteristic distributions for different genera/species. The potential for using the distributions of homologues of n-alkanes and other related n-alkyl lipids was first explored in modern leaves of the family Crassulaceae, subfamily Sempervivoideae (Eglinton et al. 1962). Subsequent studies demonstrated that n-alkyl lipid distributions allow some, but not all, genera of extant plants to be distinguished on the basis of their epicuticular wax composition (e.g., Dyson and Herbin 1968; Herbin and Robins 1969; Osborne et al. 1989; Maffei 1994; Mimura et al. 1998; Skorupa et al. 1998). Lipid profiles of fossil leaves, however, might be expected to be affected by degradation and/or contributions from other source organisms and the enclosing sediments.

At Clarkia it was found that many of the 89 fossil specimens examined showed distinct n-alkyl lipid signatures. These distribution patterns of n-alkanes and n-alkanols were often consistent among specimens of the same genus but distinctive for many of the nine different genera examined (Lockheart et al. in press). Application of principal component analysis to the data set served to further emphasize the chemosystematic utility of these lipid classes in fossils. Remarkably, specimens of oak and an extinct relative of beech exhibited preserved lipid distributions similar in appearance to the leaf waxes of their morphologically related modern counterparts (Fig. 1) (Lockheart 1997). These modern relatives are indigenous to areas with a mesic climate sim-

Quercus robur
(English common oak)

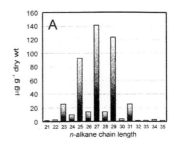

Quercus rubra
(American red oak)

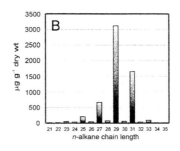

Quercus nigra
(Black oak)

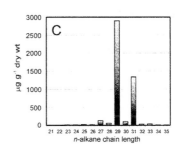

Quercus species
Miocene deposit Clarkia, Idaho

Fossil

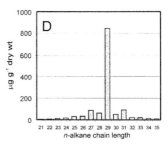

Sediment

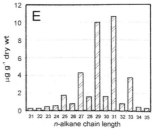

ilar to that which likely prevailed during the Miocene at Clarkia. The Clarkia oaks morphologically resemble certain species of red oaks, which are indigenous to the Appalachian Mountain region of the southeastern United States, and share with the fossils a lipid distribution with a high concentration of C_{29} n-alkane. This distribution contrasts with the broader distribution and lower concentrations of n-alkane homologues detected in waxes of white oaks, e.g., the English common oak of northern Europe, which are native to colder climates (Fig. 1).

While lipids demonstrate considerable promise for chemotaxonomic comparisons between Tertiary fossils, the chemosystematic utility of preserved biopolymeric material appears to be restricted to determining the extent of preservation of a fossil and differentiating between different organs of a plant (Stankiewicz et al. 1997a). Analysis of modern and fossil (Miocene, 6 Ma) conifers (*Pinus* and *Sequoia* scales and seeds) by py-GC/MS showed the presence of cellulose markers in modern specimens to be a key feature that distinguished them from fossil samples where cellulose had been completely degraded. Guaiacyl lignin markers were dominant in the pyrograms of the fossils, and phenols and benzenediols were also evident, probably derived from biodegraded lignin. To date, only widely separated taxa have shown appreciable differences in the pyrolysates produced by py-GC/MS, e.g., water lily seed coats (van Bergen et al. 1996) and conifer seeds and cone scales (Stankiewicz et al. 1997a). The potential of biopolymeric material to distinguish between members of the same phylogenetic family remains to be fully explored (van Bergen et al. 1995).

Environmental Indicators

Local changes in environment may be reflected in the chemistry of macrofossil remains. The seral change of vegetation over the period of deposition at a site could potentially be traced by analyses of biomarkers in fossils and sediments, in conjunction with statistical analysis of macrofossil and pollen abundance. Biomarkers for a fossil taxon may remain in the sediments where macrofossil remains have been lost through degradation. At Clarkia, lipid contributions from two dominant plant communities bordering the lake were apparent: (1) diterpenoid markers of coniferous origin increased in the upper sediments (the infilling stages of the lake) associated with a higher incidence of needles and branchlets of Taxodiaceae, and (2) the n-alkane profile of the sediments was similar to that of the most numerically dominant macrofossil, *Pseudofagus idahoensis,* an extinct relative of the beech family.

The stable carbon isotope composition of fossil lipids and bulk remains provides an additional tool with which to assess the paleoenvironmental conditions under which a plant was growing (see Arens et al. 2000). Numerous studies have demonstrated that $\delta^{13}C$ values of lipids and tissues from modern leaves are sensitive to changes in growth conditions. Fractionation against ^{13}C occurs during assimilation of CO_2 and other biosynthetic processes. The extent of fractionation depends on photosynthetic rates and on water use and availability (Farquhar et al. 1989; Meinzer et al. 1992). Concentrations of CO_2 and isotope composition of the paleoatmosphere also influence the $\delta^{13}C$ of the biosynthesized materials (see Arens et al. 2000).

Increased photosynthetic rates in leaves ex-

←

FIGURE 1. Chemosystematic classification of fossil leaves based on lipid distributions. n-Alkanes with carbon chain lengths between 21 and 35 carbons derived from the original epicuticular leaf waxes are preserved in fossil remains at Clarkia, Idaho and in some cases characteristic patterns were observed between genera (Lockheart 1997). n-Alkane distributions of *Quercus* fossil leaves were very distinctive and usually dominated by the nC_{29} alkane, which was present in high concentration (D). Compare with sediment (E). In comparisons with extant species, the fossils had a lipid distribution similar to that of several modern "red" oak species, e.g., *Q. rubra* (B) and *Q. nigra* (C). These species were similar in appearance to the fossils; i.e., they had deep-toothed margins with aristae at the end of secondary veins. However, modern "white" oaks whose morphology substantially differed from the fossils, e.g., *Q. robur* (A), showed a broader distribution with lower concentration.

posed to the sun lead to a reduced discrimination against the ^{13}C isotope relative to shaded leaves (Pearcy and Pfitsch 1991; Waring and Silvester 1994; Lockheart et al. 1997). Fortunately, stomatal counting allows sun leaves to be differentiated from shade leaves, even in fossils, by virtue of their higher stomatal indices (Poole et al. 1996). Kürschner (1996) observed a preservational bias in favor of sun leaves (90%) in Miocene and Pliocene fossil leaf assemblages from the Lower Rhine Embayment (owing to their preferential windborne transportation to favorable depositional sites). Hence, it should be possible to measure the δ^{13}C values of lipids or bulk remains on fossils that are directly comparable, e.g., sun leaves of the same species. There are, as yet, no reports of δ^{13}C values and stomatal counts in Tertiary fossils.

3. The Carboniferous

The initial controls on the preservation of macromolecules are environmental, i.e., the context in which the organism is buried. Rates of decay vary between molecules, and are influenced by external factors; selective preservation determines which molecules survive in sediment. In Tertiary rocks the effects of temperature and pressure are of limited significance. However, even the slowest chemical reactions may proceed on a longer timescale under apparently stable "environmental" conditions. On this basis it would appear that organic macrofossils of Carboniferous age should not retain their original composition. Nonetheless, it has been argued that certain structural biopolymers are sufficiently degradation resistant to survive without alteration for hundreds of millions of years. This is the basis for the model of kerogen formation by selective preservation (Nip et al. 1986a,b; Tegelaar et al. 1989; Largeau et al. 1990; Derenne et al. 1991; de Leeuw et al. 1991). The major categories of biopolymer that have been reported from Carboniferous rocks (mainly on the basis of pyrolysis) are resins, sporopollenins, algaenans, cutans, and suberans (those considered by Tegelaar et al. [1989] to have the highest preservation potential); evidence for lignins and tannins remains equivocal. In addition, organically preserved arthropod cuti-

cles are commonly associated with plants in Carboniferous deposits (Bartram et al. 1987). Their decay resistance is a product of the cross-linking of the chitin-protein complex of which they are composed.

Resins and Tannins

Resins and tannins are plant secretions. Resins have been analyzed from individual petioles of *Myeloxylon*, a medullosan pteridosperm from Upper Carboniferous coal balls of Indiana. They yielded a pyrolysate of homologous series of *n*-alk-1-enes and *n*-alkanes, as well as alkylphenols, similar to that of sub-bituminous coalified wood (Collinson et al. 1994). Resin rods have also been isolated from Upper Carboniferous Shales from Swillington, England. On pyrolysis they yielded *n*-alkanes and minor quantities of *n*-alkenes. The interpretation of both pyrolysates is problematic (van Bergen et al. 1997c), and the composition of the original resin and its diagenetic history are unknown.

Algaenans

Algaenans occur in algal cell walls. A range of examples has been documented in living algae (review in de Leeuw and Largeau 1993), and these highly aliphatic substances are assumed to be selectively preserved. The presence of microscopic algal remains in Type I kerogens (e.g., Messel Oil shale) confirms an algal source, and recognizable cell wall material (assignable to various taxa) in rocks as old as Paleozoic yields algaenan-like biopolymers. These include *Gleocapsomorpha prisca*, which occurs in marine kerogens from the Ordovician of Estonia, although the distribution pattern of alkenes and alkanes differs from that in extant algaenans (see Collinson et al. 1994). The related nonmarine *Botryococcus* and *Botryococcus*-like fossils also range from the Paleozoic to the present day; they occur in torbanites and boghead coals, where they can be identified microscopically and produce the biomarker botryococcane. Algaenans from *Botryococcus* are an important contributor to Type I kerogens because of their degradation resistance (de Leeuw and Largeau 1993; Collinson et al. 1994).

Sporopollenins

Sporopollenin is the macromolecular constituent that accounts for the decay-resistant nature of the walls of spores and pollen. There appear to be two different types of sporopollenin, one dominated by oxygenated aromatics, the other by aliphatics (see van Bergen et al. 1995 for review). There is little difference in the composition of sporopollenin in the spores of different Tertiary water ferns—no chemosystematic information survives (van Bergen et al. 1993). In Carboniferous material the chemical composition is diagenetically altered, particularly where subjected to temperature or pressure (Hemsley et al. 1996). The relative amounts of unsaturated carbon species are lower in Carboniferous gymnosperms than in lycopods, however, and as this also applies to living examples, it may indicate the retention of a chemosystematically diagnostic signature (Hemsley et al. 1995). The composition of sporopollenins, and the degree to which they are transformed chemically through time, remains to be determined in detail. A combination of microscopic, spectroscopic, chemolytic, and pyrolytic techniques needs to be applied to a range of appropriate material (van Bergen et al. 1995).

Lignins

Lignin-containing plant remains are thought to provide a major constituent of coals (Hatcher et al. 1994) as a precursor for vitrinite. Chemical evidence for the preservation of lignin in the Carboniferous is equivocal (see van Bergen et al. 1997c for a critical review). An early investigation of the likely survival of lignin in Carboniferous lycophytes, using cupric oxide oxidation, revealed significant lignin derivatives in *Sigillaria* compression fossils (Logan and Thomas 1987). This may reflect a thick-walled inner layer of periderm, or the derivatives may have been a product of the chemolytical technique used (P. F. van Bergen personal communication 2000). Other lycophyte periderms have yielded alkylbenzene and naphthalene derivatives in addition to aliphatic components (Collinson et al. 1994). Carboniferous lignin-based tissues have yet to be investigated by pyrolysis, but they are likely to have been highly altered by diagenesis (van Bergen et al. 1994c).

Suberans

Suberans are aliphatic biomacromolecules that occur in the periderm (i.e., bark) of higher plants such as *Betula* (see de Leeuw and Largeau 1993; Tegelaar et al. 1995). A similar compound, which yields a homologous series of *n*-alk-1-enes and *n*-alkanes upon pyrolysis, has been discovered in the lycophyte stem *Diaphoradendron* and root *Stigmaria* from the Upper Carboniferous of the United Kingdom (Collinson et al. 1994). Suberan is thought to be an important component of bark that occurs as a coal maceral and may, like lignin, contribute to the structural strength of certain plant tissues (see van Bergen et al. 1995).

Cutan

Cutan was first reported in the cuticles of leaves and stems of higher plants by Nip et al. (1986a,b). Tegelaar et al. (1991) analyzed the leaves of a range of fossil and modern plants and demonstrated not only that the fossil leaves they analyzed all contained cutan, but also that modern leaves lacking cutan are rarely preserved in the fossil record. This provided important evidence to support the model of kerogen formation by selective preservation (Tegelaar et al. 1989). More recent analyses, however, have cast doubt on the very existence of cutan in living species, suggesting that it may form in fossils as a product of diagenetic alteration (Mösle et al. 1997; Collinson et al. 1998; Stankiewicz et al. 1998).

Pyrolysis of cuticles of Carboniferous cordaites and pteridosperms yielded a highly aliphatic signature, similar to that normally attributed to cutan. The pyrolysates were dominated by homologous series of C_6 to C_{30} *n*-alk-1-enes and *n*-alkanes (C_{10} to C_{14} homologues the most abundant [Stankiewicz et al. 1998d]). However, Recent relatives (*Araucaria, Ginkgo*) of these Carboniferous plants yielded no cutan, but fatty acids (pyrolysis products diagnostic of cutin), together with minor products derived from polysaccharide and lignin (Mösle et al. 1997; Stankiewicz et al. 1998d). Thus, assuming a similar original composition, the so called "cutan" in the fossils cannot be a

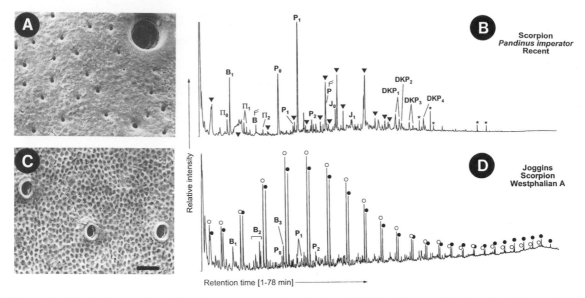

FIGURE 2.   Morphology and total ion chromatograms (pyrolysis at 610°C for 10 seconds) of cuticle of Recent Emperor scorpion (internal surface) (A, B) and scorpion from Upper Carboniferous, Joggins, Nova Scotia, Canada (C, D). Scale bar 50 μm. Carbon number for alkene/alkane pairs ranges from 6 to 30. ○ = n-alk-1-enes, ● = n-alkanes, ▼ = pyrolysis products directly related to chitin polymer; DKP = 2,5-diketopiperazines, pyrolysis products indicative of protein moieties (where DKP_1 = pro-ala, DKP_2 = pro-gly, DKP_3 = pro-val, DKP_4 = pro-pro); $\Pi(n)$ = alkylpyrroles; $J(n)$ = alkylindoles; $B(n)$ = alkylbenzenes; $P(n)$ = alkylphenols, where (n) indicates the extent of alkyl substitution (i.e., 1 = methyl, 2 = dimethyl, or ethyl, etc.); * = contaminants. (After Stankiewicz et al. 1998d.)

product of selective preservation but must be the result of diagenetic modification.

Arthropod Cuticles

Arthropod cuticles (from eurypterids and scorpions) were analyzed from the same Carboniferous localities as those from which the cordaites and pteridosperms were obtained. They, too, yielded an aliphatic signature, but with a lower proportion of C_{20} to C_{30} hydrocarbons. The cuticle of modern scorpion, on the other hand, is composed of a chitin-protein complex (Stankiewicz et al. 1998d) and lacks an aliphatic biopolymer (Fig. 2). While selective preservation of more decay-resistant compounds may have been critical to the initial survival of these Carboniferous cuticles, traces of chitin have not been detected in arthropod cuticles older than Tertiary (Stankiewicz et al. 1997c, 1998b; Briggs et al. 1988a). Thus their final composition must be a product of diagenetic modification. The differences that remain in the pyrolysates of different Carboniferous arthropods and plants suggest that there was no mixing or external contamination: the process involved was in situ po-

lymerization. This is borne out by similar transformations of insect and leaf cuticles encapsulated in amber (Stankiewicz et al. 1998b).

Molecular Diagenesis and Morphological Preservation

The analysis of associated plant and animal cuticles from the Carboniferous of North America (Stankiewicz et al. 1998d) demonstrated that selective preservation alone is not an adequate explanation of the composition of the fossils. Clearly, diagenesis has played an important role, and primarily through the transformation of components of the fossil itself, in situ polymerization. While this process usually destroys the internal structure of the cuticles of both plants and animals (Stankiewicz et al. 1998d; but see van Bergen et al. 1995: Fig. 3d), the external morphology remains largely intact. Thus, the quality of preservation of surface detail in Paleozoic macrofossils (as revealed by the scanning electron microscope) is not paralleled by the retention of the original cuticle chemistry. Evidence for the role of diagenetic transformation in the gen-

TABLE 2. Distribution of biomolecules in organic fossils through time.

Biomolecule	Source organism	In archaeological remains	In Tertiary fossils	In Paleozoic fossils
Nucleic acids				
DNA/RNA	all organisms: encapsulation (e.g., in bone) may enhance preservation	up to 10^5 yr	none confirmed, but see Bada et al. 1999	none
Proteins	animals (shell better than bone); hydrolyze to amino acids that have some applications to dating and paleotemperature determination	up to 10^5–10^6 yr in shell and bone; apparently survives longer in arthropod cuticles (but detection by Py-GC MS)	present in insect cuticles as far back as Oligocene (but sequences not analyzed	none
Carbohydrates				
Cellulose	vascular plants, some fungi	present	altered, even where it occurs as lignin-cellulose in seed coats	none
Chitin	arthropods, fungi, algae	present in variable amounts, depending on setting, and in Quaternary beetles	traces present in insect cuticles as far back as Oligocene	none
Lipids				
Glycolipids	plants, algae, eubacteria, animals	present	wax and cuticle sources, bacterial overprint	?
Lipopolysaccharides	gram-negative eubacteria	present	present?	?
Resistant biopolymers				
Algaenans	algae	present	present	diagenetically modified
Resins	vascular plants	present	present	diagenetically modified
Lignins	vascular plants	present	present, but greater alteration in the early Tertiary	traces, diagenetically modified
Sporopollenin	vascular plants	present	present, but in an altered state	diagenetically modified
Cutan	vascular plants	occurs in modern *Agave*, but existence in other taxa doubtful (Collinson et al. 1998)		
Suberan	vascular plants	present	?	diagenetically modified

eration of aliphatic biopolymers is presently confined to cutan (Collinson et al. 1998) and to the chitin-protein complex of arthropod cuticles (Briggs et al. 1998a; Briggs 1999). The details of the polymerization process have yet to be unraveled. However, the evidence that aliphatic macromolecules can be generated in this way opens the possibility that a similar process was involved in the preservation of other biopolymeric materials (although its impact may have been less significant where a substantial aliphatic component occurs in the original molecule). Thus, polymerization of decay-resistant material may be a critical factor in explaining the diversity of organically preserved fossils preserved in the pre-Tertiary continental fossil record.

Conclusions

Table 2 highlights the pattern of preservation of biomolecules in the fossil record. While preservation is initially determined by resistance to biological decay, physico-chemical processes become increasingly important in more ancient rocks. Examples of the full suite of fossil biomolecules are present in appropriate archeological materials up to 10^5 years old, preserved in favorable conditions, although nucleic acids, proteins, and lower molecular weight polysaccharides have undergone varying degrees of degradation. In the Tertiary there are no reliable records of DNA preservation, and proteins and lower molecular weight polysaccharides normally survive only within biominerals or where they are protected by cross-linking. Lipids and biopolymers are widespread. Only the most degradation-resistant macromolecules survive in Carboniferous and older rocks; the preservation of lipids has not been demonstrated in fossils of this age.

Degradation and diagenesis reduce the utility of molecular fossils in successively older contexts. Sequence information, from DNA or proteins, for example, and detailed paleodietary information, from stable isotopic investigations of proteins or biomarker approaches using lipids, have been obtained only from materials of Pleistocene or younger age. In Tertiary fossils the degree and nature of molecular degradation can be used to determine climate and environment of deposition, and subsequent thermal history. Lipids may still retain significant chemotaxonomic information, because the chemistry of fossils usually shows a clear relationship to that of the once-living organism. The composition of Paleozoic fossils is much less diagnostic of taxonomic position. However, even diagenetically altered signatures may retain chemosystematic information at a high taxonomic level. The isotopic compositions of preserved compounds are also likely to be retained, although this has yet to be explored in detail. The chemistry of fossils provides important clues to likely sources and diagenetic history of sedimentary organic matter (kerogen).

The factors controlling the transformation of organic macromolecules in fossils are incompletely known. Time plays a role; the speed of alteration depends on reaction rates. The susceptibility of nucleic acids, proteins, and carbohydrates to hydrolysis and oxida-tion explains why they are normally eliminated within thousands of years. Biopolymeric materials may persist for millions of years, but even they are affected by diagenesis. Traces of chitin and proteins are evident in arthropod cuticles from a small number of Tertiary biotas (Stankiewicz et al. 1997c), whereas others, even from the same sequence of beds, display a predominantly aliphatic signature. Lignin may survive relatively unaltered in seed coats from the early Tertiary (van Bergen et al. 1995), whereas contemporaneous samples may yield quite different chemical signatures. The chitin-protein complex of arthropod cuticles is completely altered in pre-Tertiary fossils, and other biopolymeric materials are likely to be similarly affected. Thus the early Tertiary appears to represent some kind of threshold beyond which alteration is ubiquitous. Factors controlling the rate of diagenesis, however, are poorly understood. The nature of the organism, the depositional environment (Briggs et al. 1998b), the enclosing lithology, and metamorphism clearly play important roles, but their relative importance is unknown. It has been reported, for example, that ligno-cellulose is better preserved in coarse-grained rather than fine-grained sediments (van Bergen et al. 1994a), a counterintuitive result given the higher porosity of sands. Organic-rich deposits such as peats and coals might be expected to promote the preservation of macromolecules owing to high productivity and low levels of microbial activity within the peat environment.

Much research remains to be done on the chemistry of organic macrofossils. Further decay experiments on modern organisms are needed to address the relative preservation potential of different chemical constituents under different environmental conditions. This can also be documented through the archeological record. The first unambiguous evidence for the Maillard reaction in the decay of the protein and carbohydrate components of buried plant matter has only recently been reported through studies of archeological seeds (Evershed et al. 1997c). Maturation experiments can help to constrain likely pathways to the diagenetic products found in fossil material. The influence of lithology on the preservation of biomolecules in plant fossils has yet to be investigated in detail. Assemblages need to be documented in terms of the nature of plant and animal remains (morphology and ultrastructure, macromolecular composition) and associated variables such as sedimentary setting, organic productivity and content, and composition of enclosing medium. Further research of this kind will allow the processes involved in the preservation and transformation of the macromolecular constituents of animals and plants to be modeled, thus improving our understanding of the taphonomy of organic fossils. It will also identify the characteristics of sediments that favor the preservation of different types of organic macrofossils and their intrinsic chemical data, allowing us to access the chemosystematic data in fossil organic matter. This area of investigation also has wider implications in advancing the identification of sources of organic matter in kerogen.

Acknowledgments

Our research has been supported extensively in the form of grants, studentships and mass spectrometry facilities by the Natural Environment Research Council. P. F. van Bergen provided very helpful critical comments on an earlier version of the paper.

Literature Cited

Allison, P. A., and D. E. G. Briggs. 1991. The taphonomy of soft-bodied animals. Pp. 120–140 *in* S. K. Donovan, ed. The processes of fossilization. Belhaven, London.

Ambrose, S. H. 1990. Preparation and characterization of bone and tooth collagen for isotopic analysis. Journal of Archeological Science 17:431–451.

Arens, N. C., A. H. Jahren, and R. Amundsen. 2000. Can C3 plants faithfully record the carbon isotopic composition of atmospheric carbon dioxide? Paleobiology 26:137–155.

Aveling, E. M., and C. Heron. 1998. Identification of birch bark tar at the Mesolithic site of Star Carr. Ancient Biomolecules 2: 69–80.

Bada, J. L., X. S. Wang, and H. Hamilton. 1999. Preservation of key biomolecules in the fossil record: current knowledge and future challenges. Philosophical Transactions of the Royal Society of London B 354:77–87.

Bartram, K. M., A. J. Jeram, and P. A. Selden. 1987. Arthropod cuticles in coal. Journal of the Geological Society, London 144: 513–517.

Beck, C. W., and C. Borromeo. 1990. Ancient pine pitch: technological perspectives from a Hellenistic shipwreck. Pp. 51–58 *in* A. R. Biers and P. E. McGovern, eds. Organic contents of ancient vessels: materials analysis and archeological investigation. MASCA Research Papers in Science and Archaeolo-

gy, Vol. 7. University Museum of Archaeology and Anthropology, University of Pennsylvania, Philadelphia.

Beck, C. W., C. J. Smart, and D. J. Ossen. 1989. Residues and linings in ancient Mediterranean transport amphoras. ACS Symposium Series 220:369–380.

Beck, C. W., E. C. Stout, and P. A. Janne. 1998. The pyrotechnology of pine tar and pitch inferred from quantitative analyses by gas chromatography/mass spectrometry and carbon-13 nuclear magnetic resonance spectroscopy. Pp. 181–190 *in* W. Brzezi'nski and W. Peotrowski, eds. Proceedings of the first international symposium on wood tar and pitch, pp. 181–190. Domu Wydawniczym Pawla Dabrowskiego, Warsaw.

Behrensmeyer, A. K., and R. W. Hook. 1992. Paleoenvironmental contexts and taphonomic modes. Pp. 15–136 *in* A. K. Behrensmeyer, J. D. Camuth, W. A. DiMichele, R. Potts, H.-D. Sues, and S. L. Wing, eds. Terrestrial ecosystems through time. University of Chicago Press, Chicago.

Bland, H. A., P. F. van Bergen, J. F. Carter, and R. P. Evershed. 1998. Early diagenetic transformations of proteins and polysaccharides in archaeological plant remains. Pp. 113–131 *in* Stankiewicz and van Bergen 1998.

Boëda, E., J. Connan, D. Dessort, S. Muhesen, N. Mercier, H. Valadas, and N. Tisnerat. 1996. Bitumen as a hafting material on Middle Palaeolithic artefacts. Nature 380:336–338.

Boon, J. J., S. A. Stout, W. Genuit, and W. Spackman. 1989. Molecular paleobotany of *Nyssa* endocarps. Acta Botanica Neerlandica 38:391–404.

Briggs, D. E. G. 1999. Molecular taphonomy of animal and plant cuticles: selective preservation and diagenesis. Philosophical Transactions of the Royal Society of London B 354:7–17.

Briggs, D. E. G., and G. Eglinton. 1994. Chemical traces of ancient life. Chemistry in Britain 31:907–912.

Briggs, D. E. G., and A. J. Kear. 1993. Decay and preservation of polychaetes: taphonomic thresholds in soft-bodied organisms. Paleobiology 19:107–135.

Briggs, D. E. G., A. J. Kear, M. Baas, J. W. de Leeuw, and S. Rigby. 1995. Decay and composition of hemichordate *Rhabdopleura*: implications for the taphonomy of graptolites. Lethaia 28:15–23.

Briggs, D. E. G., R. P. Evershed, and B. A. Stankiewicz. 1998a. The molecular preservation of fossil arthropod cuticles. Ancient Biomolecules 2:135–146.

Briggs, D. E. G., B. A. Stankiewicz, D. Meischner, A. Bierstedt, and R. P. Evershed. 1998b. Taphonomy of arthropod cuticles from Pliocene lake sediments, Willershausen, Germany. Palaios 13:386–394.

Brown, T. A. 1999. How ancient DNA may help in understanding the origin and spread of agriculture. Philosophical Transactions of the Royal Society of London B 354:89–98.

Brown, W. V., J. A. L. Watson, and M. J. Lacey. 1996. A chemotaxonomic survey using cuticular hydrocarbons of some species of the Australian harvester termite genus Drepanotermes (Isoptera: Termitidae). Sociobiology 27:199–221.

Buckley, S. A., A. W. Stott, and R. P. Evershed. 1999. Studies of organic residues from ancient Egyptian mummies using high temperature-gas chromatography-mass spectrometry and sequential thermal desorption-gas chromatography-mass spectrometry and pyrolysis-gas chromatography-mass spectrometry. Analyst 124:443–452.

Bull, I. D., I. A. Simpson, P. F. van Bergen, and R. P. Evershed. 1999. Muck 'n' molecules: organic geochemical methods for detecting ancient manuring. Antiquity 73:86–87.

Charters, S., R. P. Evershed, L. J. Goad, C. Heron, and P. W. Blinkhorn. 1993. Identification of an adhesive used to repair a Roman jar. Archaeometry 35:91–101.

Collins, M. J., E. R. Waite, and A. C. T. van Duin. 1999. Predicting protein decomposition: the case of aspartic-acid racemization kinetics. Philosophical Transactions of the Royal Society of London B 354:51–64.

Collinson, M. E., P. M. van Bergen, A. C. Scott, and J. W. de Leeuw. 1994. The oil-generating potential of plants from coal and coal-bearing strata through time: a review with new evidence from Carboniferous plants. *In* A. C. Scott and A. J. Fleet, eds. Coal and coal-bearing strata as oil-prone source rocks. Geological Society of London Special Publication 77: 31–70.

Collinson, M. E., B. Mösle, P. Finch, A. C. Scott, and R. Wilson. 1998. The preservation of plant cuticle in the fossil record: a chemical and microscopical investigation. Ancient Biomolecules 2:251–265.

Connan, J. 1999. Use and trade of bitumen in antiquity and prehistory: molecular archaeology reveals secrets of past civilizations. Philosophical Transactions of the Royal Society of London B 354:33–50.

Connan, J., A. Nissenbaum, and D. Dessort 1992. Molecular archaeology: export of Dead Sea asphalts to Canaan and Egypt in the Chalcolithic-early bronze age (4th–3rd millennium BC). Geochimica Cosmochimica Acta 56:2743–2759.

Cranwell, P. A., G. Eglinton, and N. Robinson. 1987. Lipids of aquatic organisms as potential contributors to lacustrine sediments. II. Organic Geochemistry 11:513–527.

de Leeuw, J. W., and C. Largeau. 1993. A review of macromolecular organic compounds that comprise living organisms and their role in kerogen, coal and petroleum formation. Pp. 23–72 *in* Engel and Macko 1993.

de Leeuw, J. W., P. F. van Bergen, B. G. K. van Aarssen, J.-P. L. A. Gatellier, J. S. Sinninghe Damsté, and M. E. Collinson. 1991. Resistant biomacromolecules as major contributors to kerogen. Philosophical Transactions of the Royal Society of London B 333:329–337.

de Leeuw, J. W., N. L. Frewin, P. F. van Bergen, J. S. Sinninghe Damsté, and M. E. Collinson. 1995. Organic carbon as a palaeoenvironmental indicator in the marine realm. *In* D. W. J. Bosence and P. A. Allison, eds. Marine palaeoenvironmental analysis from fossils. Geological Society of London Special Publication 83:43–71.

Derenne, S., C. Largeau, E. Casadevall, J. F. Raynaud, C. Berkaloff, and B. Rousseau. 1991. Chemical evidence of kerogen formation in source rocks and oil shales via selective preservation of thin resistant outer walls of microalgae: origin of ultralaminae. Geochimica Cosmochimica Acta 55:1041–1050.

Dudd, S. N., and R. P. Evershed. 1998. Direct demonstration of milk as an element of archaeological economies. Science 282: 1478–1481.

———. 1999. Unusual triterpenoid fatty acyl ester components of archaeological birch bark tars. Tetrahedron Letters 40:359–362.

Dyson, W. G., and G. A. Herbin. 1968. Studies on plant cuticular waxes. IV. Leaf wax alkanes as a taxonomic discriminant for cypresses grown in Kenya. Phytochemistry 7:1339–1344.

Eganhouse, R. P. 1997. Molecular markers in environmental geochemistry. ACS Symposium Series No. 671. American Chemical Society, Washington D.C.

Eglinton, G., and G. A. Logan. 1991. Molecular preservation. Philosophical Transactions of the Royal Society of London B 333:315–328.

Eglinton, G., A. G. Gonzalez, R. J. Hamilton, and R. A. Raphael. 1962. Hydrocarbon constituents of the wax coatings of plant leaves: a taxonomic survey. Phytochemistry 1:89–102.

Engel, M. H., and S. A. Macko. 1993. Organic geochemistry. Plenum, New York.

Evershed, R. P. 1993. Biomolecular archaeology and lipids. World Archaeology 25:74–93.

Evershed, R. P., and P. H. Bethell. 1996. Application of multimolecular biomarker techniques to the identification of faecal material in archaeological soils and sediments. ACS Symposium Series 625:157–172.

Evershed, R. P., and R. C. Connolly. 1988. Lipid preservation in Lindow Man. Naturwissenschaften 75:143–145.

———. 1994. Post-mortem transformations of sterols in bog body tissues. Journal of Archaeological Science 21:577–583.

Evershed, R. P., K. Jerman, and G. Eglinton. 1985. Pine wood origin for pitch from the Mary-Rose. Nature 314:528–530.

Evershed, R. P., G. Turner-Walker, R. E. M. Hedges, N. Tuross, and A. Leyden. 1995. Preliminary results for the analysis of lipids in ancient bone. Journal of Archaeological Science 22: 277–290.

Evershed, R. P., P. F. van Bergen, T. M. Peakman, E. C. Leigh-Firbank, M. C. Horton, D. Edwards, M. Biddle, B. Kjolbye-Biddle, and P. A. Rowley-Conwy. 1997a. Archaeological frankincense. Nature 390:667–668.

Evershed, R. P., H. R. Mottram, S. N. Dudd, S. Charters, A. W. Stott, G. J. Lawrence, A. M. Gibson, A. Conner, P. W. Blinkhorn, and V. Reeves. 1997b. New criteria for the identification of animal fats preserved in archaeological pottery. Naturwissenschaften 84:402–406.

Evershed R. P., H. A. Bland, P. F. van Bergen, J. F. Carter, M. C. Horton, and P. A. Rowley-Conwy. 1997c. Volatile compounds in archaeological plant remains and the Maillard reaction during decay of organic matter. Science 278:432–433.

Evershed, R. P., S. N. Dudd, S. Charters, H. Mottram, A. W. Stott, A. Raven, P. F. van Bergen, and H. A. Bland. 1999. Lipids as carriers of anthropogenic signals from prehistory. Philosophical Transactions of the Royal Society of London B 354:19–31.

Farquhar, G. D., J. R. Ehleringer, and K. T. Hubick. 1989. Carbon isotope discrimination and photosynthesis. Annual Review of Plant Physiology and Molecular Biology 40:503–537.

Flannery, M. F., A. W. Stott, D. E. G. Briggs, and R. P. Evershed. In press. Chitin in the fossil record: identification and quantification of D-glucosamine. Organic Geochemistry.

Gülaçar, F. O., A. Susini, and M. Koln. 1990. Preservation and post-mortem transformations of lipids in samples from a 4000-year-old Nubian mummy. Journal of Archaeological Science 17:691–695.

Hagelberg, E., M. Kayser, M. Nagy, L. Roewer, H. Zimdahl, M. Krawczak, P. Lió, and W. Schiefenhövel. 1999. Molecular genetic evidence for the human settlement of the Pacific: analysis of mitochondrial DNA, Y chromosome and HLA markers. Philosophical Transactions of the Royal Society of London B 354:141–152.

Hatcher, P. G., K. A. Wenzel, and G. D. Cody. 1994. Coalification reactions of vitrinite derived from coalified wood: transformations to rank of bituminous coal. *In* P. K. Mukhopadhyay and W. G. Dow, eds. Reevaluation of vitrinite reflectance as a maturity indicator. ACS Symposium Series 570:112–135. American Chemical Society, Washington, D.C.

Hedges J. I., G. L. Cowie, J. R. Ertel, R. J. Barbour, and P. G. Hatcher. 1985. Degradation of carbohydrates and lignins in buried woods. Geochimica. Cosmochimica Acta 49:701–711.

Hemsley, A. R., P. J. Barrie, and A. C. Scott. 1995. ^{13}C solid state NMR spectroscopy of fossil sporopollenins: variation in composition independent of diagenesis. Fuel 74:1009–1012.

Hemsley, A. R., A. C. Scott, P. J. Barrie, and W. G. Chaloner. 1996. Studies of fossil and modern spore wall biomacromolecules using ^{13}C solid state NMR. Annals of Botany 78:83–94.

Herbin, G. A., and P. A. Robins. 1969. Patterns of variation and development in leaf wax alkanes. Phytochemistry 8:1985–1998.

Holloway, P. J. 1981. The chemical constitution of plant cutins. Pp. 1–32 *in* D. F. Cutler, K. L. Alvin, and C. E. Price, eds. The plant cuticle. Academic Press, London.

Huang, Y., M. J. Lockheart, J. W. Collister, and G. Eglinton. 1995. Molecular and isotopic biogeochemistry of the Miocene

Clarkia Formation: hydrocarbons and alcohols. Organic Geochemistry 23:785–801.

Huang Y., M. J. Lockheart, G. A. Logan, and G. Eglinton. 1996. Isotope and molecular evidence for the diverse origins of carboxylic acids in leaf fossils and sediments from the Miocene Lake Clarkia deposit, Idaho, U.S.A. Organic Geochemistry 24: 289–299.

Jones, M. K., D. E. G. Briggs, G. Eglinton, and R. Hagelberg, eds. 1999. Molecular information and prehistory. Philosophical Transactions of the Royal Society of London B 354:1–159.

Katzenberg, M. A., H. P. Schwarcz, M. Knyf, and F. J. Melbye. 1995. Stable isotope evidence for maize horticulture and palaeodiet in southern Ontario, Canada. American Antiquity 60: 335–350.

Killops, S. D., and V. J. Killops. 1993. An introduction to organic geochemistry. Longman Group, Harlow, England.

Krings, M., A. Stone, R. W. Schmitz, H. Krainitzki, M. Stoneking, and S. Pääbo. 1997. Neandertal DNA sequences and the origin of modern humans. Cell 90:19–30.

Kupchan, S. M., A. Karim, and C. Marcks. 1969. Tumour inhibitors. XL VIII. Taxodione and taxodone, two novel diterpenoid quinone methide tumour inhibitors from *Taxodium distichum*. Journal of Organic Chemistry 34:3912–3918.

Kürschner, W. M. 1996. Leaf stomata as biosensors of palaeoatmospheric CO_2 levels. Ph.D. dissertation. LPP Foundation, Utrecht.

Largeau, C., S. Derenne, E. Casadevall, C. Berkaloff, M. Corolleur, B. Lugardon, J. F. Raynaud, and J. Connan. 1990. Occurrence and origin of "ultralaminar" structures in "amorphous" kerogens of various source rocks and oil shales. Organic Geochemistry 16:889–895.

Leavitt S. W., and A. Long. 1984. Sampling strategy for stable carbon isotope analysis of tree rings in pine. Nature 311:145–147.

Lockheart, M. J. 1997. Isotope compositions and distributions of individual compounds as indicators for environmental conditions: comparisons between contemporary and Clarkia fossil leaves. Ph.D. thesis. University of Bristol, Bristol, U.K.

Lockheart, M. J., P. F. van Bergen, and R. P. Evershed. 1997. Variations in the stable carbon isotope compositions of individual lipids from the leaves of modern angiosperms: implications for the study of higher land plant derived sedimentary organic matter. Organic Geochemistry 26:137–153.

Lockheart, M. J., P. F. van Bergen, and R. P. Evershed. In press. Chemotaxonomic classification of fossil leaves from the Miocene Clarkia lake deposit, Idaho, USA, based on n-alkyl lipid distributions and principal component analyses. Organic Geochemistry.

Logan, G. A. 1992. Biogeochemistry of the Miocene lacustrine deposit, Clarkia, northern Idaho, U.S.A. Ph.D. thesis, University of Bristol, Bristol, England.

Logan, G. A., and G. Eglinton. 1994. Biogeochemistry of the Miocene lacustrine deposit at Clarkia, northern Idaho, U.S.A. Organic Geochemistry 21:857–870.

Logan, G. A., J. J. Boon, and G. Eglinton. 1993. Structural biopolymer preservation in Miocene leaf fossils from the Clarkia site, northern Idaho. Proceedings of the National Academy of Sciences USA 90:2246–2250.

Logan, G. A., C. J. Smiley, and G. Eglinton. 1995. Preservation of fossil leaf waxes in association with their source tissues, Clarkia, northern Idaho, USA. Geochimica Cosmochimica Acta 59:751–763.

Logan, K. J., and B. A. Thomas. 1987. The distribution of lignin derivatives in fossil plants. New Phytologist 105:157–173.

Lowenstein, J. M., and G. Scheuenstuhl. 1991. Immunological methods in molecular palaeontology. Philosophical Transactions of the Royal Society of London B 333:375–380.

Loy, T. H. 1993. The artifact as site: an example of the biomolecular analysis of organic residues on prehistoric tools. World Archaeology 25:44–63.

Lücke, A., G. Helle, G. H. Schleser, I. Figueiral, V. Mosbrugger, T. P. Jones, and N. P. Rowe. 1999. Environmental history of the German Lower Rhine Embayment during the Middle Miocene as reflected by carbon isotopes in brown coal. Palaeogeography, Palaeoclimatology, Palaeoecology 154:339–352.

MacHugh, D. E., C. S. Troy, F. McCormick, I. Olsaker, E. Eythórsdóttir, and D. G. Bradley. 1999. Early medieval cattle remains from a Scandinavian settlement in Dublin: genetic analysis and comparison with extant breeds. Philosophical Transactions of the Royal Society of London B 354:99–109.

Mackenzie A. S., R. L. Patience, J. R. Maxwell, M. Vandenbroucke, and B. Durand. 1980. Molecular parameters of maturation in the Toarcian shales, Paris Basin, France. I. Changes in the configuration of the acyclic isoprenoid alkanes, steranes and triterpanes. Geochimica Cosmochimica Acta 45:1345–1355.

Mackenzie A. S., S. C. Brassell, G. Eglinton, and J. R. Maxwell. 1982. Chemical fossils: the geological fate of steroids. Science 217:491–504.

Macko, S. A., M. H. Engel, V. Andrusevich, G. Lubec, T. C. O'Connell, and R. E. M. Hedges. 1999. Documenting the diet in ancient human populations through stable isotope analysis of hair. Philosophical Transactions of the Royal Society of London B 354:65–76.

Maffei, M. 1994. Discriminant analysis of leaf wax alkanes in the Lamiaceae and four other plant families. Biochemical Systematics and Ecology 22:711–728.

Matthews, D. E., and J. M. Hayes. 1978. Isotope-ratio monitoring gas chromatography mass spectrometry. Analytical Chemistry 50:1465–1473.

Meinzer, F. C., P. W. Rundel, G. Goldstein, and M. R. Sharifi. 1992. Carbon isotope composition in relation to leaf gas exchange and environmental conditions in Hawaiian *Metrosideros polymorpha* populations. Oecologia 91:305–311.

Merriwether, D. A. 1999. Freezer anthropology: new uses for old blood. Philosophical Transactions of the Royal Society of London B 354:121–130.

Mills, J. S., and R. White. 1994. The organic chemistry of museum objects, 2d ed. Butterworths, London.

Mimura, M. R. M., M. L. F. Salatino, A. Salatino, and J. F. A. Baumgratz. 1998. Alkanes from foliar epicuticular waxes of *Huberia* species: taxonomic implications. Biochemical Systematics and Ecology 26:581–588.

Mosle, B., P. F. Finch, M. E. Collinson, and A. C. Scott. 1997. Comparison of modern and fossil plant cuticles by selective chemical extraction monitored by flash pyrolysis-gas chromatography-mass spectrometry and electron microscopy. Journal of Analytical and Applied Pyrolysis 40–41:585–597.

Niklas K. J., and D. E. Giannasi. 1985. The paleobiochemistry of fossil angiosperm floras. Part II: Diagenesis of organic compounds with particular reference to steroids. Pp. 175–183 in Smiley 1985.

Niklas K. J., J. R. Brown, and R. Santos. 1985. Ultrastructural states of preservation in Clarkia angiosperm leaf tissues: implications on modes of fossilization. Pp. 143–159 in Smiley 1985.

Nip, M., E. Tegelaar, J. W. de Leeuw, P. A. Schenck, and P. J. Holloway. 1986a. Analysis of modern and fossil plant cuticles by Curie-point pyrolysis-gas chromatography and Curie-point pyrolysis-gas chromatography-mass spectrometry. Recognition of a new, highly aliphatic and resistant biopolymer. Organic Geochemistry 10:769–778.

———. 1986b. A new, non-saponifiable highly aliphatic and resistant biopolymer in plant cuticles. Naturwissenschaften 73: 579–585.

O'Donoghue, K., T. A. Brown, J. Carter and R. P. Evershed. 1994.

Detection of nucleotides bases in ancient seeds using gas chromatography/mass spectrometry and gas chromatography/mass spectrometry/mass spectrometry. Rapid Communications in Mass Spectrometry 8:503–508.

O'Donoghue, K., T. A. Brown, J. F. Carter, and R. P. Evershed. 1996a. Application of high performance liquid chromatography/mass spectrometry with electrospray ionisation to the detection of DNA nucleosides in ancient seeds. Rapid Communications in Mass Spectrometry 10:495–500.

O'Donoghue, K., A. Clapham, R. P. Evershed, and T. Brown. 1996b. Remarkable preservation of biomolecules in ancient radish seeds. Proceedings of the Royal Society of London B 263:541–547.

Opsahl, S., and R. Benner. 1995. Early diagenesis of vascular plant tissues: lignin and cutin decomposition and biogeochemical implications. Geochimica Cosmochimica Acta 59: 4889–4904.

Osborne, R., M. L. F. Salatino, and A. Salatino. 1989. Alkanes of foliar epicuticular waxes of the genus Encephalartos. Phytochemistry 28:3027–3030.

Otto, A., H. Walther, and W. Puttmann. 1997. Sesqui- and diterpenoid biomarkers preserved in Taxodium-rich Oligocene oxbow lake clays, Weisselster basin, Germany. Organic Geochemistry 26:105–115.

Ourisson G., P. Albrecht, and M. Rohmer. 1979. The hopanoids, palaeochemistry and biochemistry of a group of natural products. Pure and Applied Chemistry 51:709–729.

Pate, F. D. 1994. Bone chemistry and palaeodiet. Journal of Archaeological Method and Theory 1:161–209.

Pearcy, R. W., and W. A. Pfitsch. 1991. Influence of sunflecks on the $\delta^{13}C$ of Adenocaulon bicolor plants occurring in contrasting forest understory microsites. Oecologia 86:457–462.

Poinar, H. N. 1998. Preservation of DNA in the fossil record. Pp. 132–146 in Stankiewicz and van Bergen 1998.

Poinar, H. N., M. Höss, S. X. Wang, J. L. Bada, and S. Pääbo. 1996. Amino acid racemization and the preservation of ancient DNA. Science 272:864–866.

Poinar, H. N., M. Hofreiter, W. G. Spaulding, P. S. Martin, B. A. Stankiewicz, H. Bland, R. P. Evershed, G. Possnert, and S. Pääbo. 1998. Molecular coproscopy: dung and diet of the extinct ground sloth Nothrotheriops shastensis. Science 281:402–406.

Pollard, A. M. 1998. Archaeological reconstruction using stable isotopes. Pp. 285–298 in H. Griffiths, ed. Stable isotopes: integration of biological, ecological and geochemical processes. Environmental Plant Biology Series. Bios Scientific, Oxford.

Pollard, A. M., and C. Heron. 1996. Archaeological chemistry. Royal Society of Chemistry, Cambridge, England.

Poole, I., J. D. B. Weyers, T. Lawson, and J. A. Raven. 1996. Variations in stomatal density and index: implications for palaeoclimatic reconstructions. Plant, Cell and Environment 19: 705–712.

Regert, M., J.-M. Delacotte, M. Menu, P. Pétrequin, and C. Rolando. 1998. Identification of Neolithic hafting adhesives from two lake dwellings at Chalain (Jura, France). Ancient Biomolecules 2:81–96.

Riederer, M., K. Matzke, F. Ziegler, and I. Kögel-Knabner. 1993. Occurrence, distribution and fate of lipid plant biopolymers cutin and suberin in temperate forest soils. Organic Geochemistry 20:1063–1076.

Robinson, N., R. P. Evershed, W. H. Higgs, K. Jerman, and G. Eglinton. 1987. Proof of a pine wood origin for pitch from Tudor (Mary Rose) and Etruscan shipwrecks: application of analytical organic chemistry in archaeology. Analyst 112:637–644.

Skorupa, L. A., M. L. F. Salatino, and A. Salatino. 1998. Hydrocarbons of leaf epicuticular waxes of Pilocarpus (Rutaceae): taxonomic meaning. Biochemical Systematics and Ecology 26: 655–662.

Smiley, C. J., ed. 1985. Late Cenozoic history of the Pacific Northwest. American Association for the Advancement of Science, San Francisco.

Smiley, C. J., and W. C. Rember. 1985a. Physical setting of the Miocene Clarkia fossil beds, northern Idaho. Pp. 11–31 in Smiley 1985.

———. 1985b. Composition of the Miocene Clarkia flora. Pp. 95–112 in Smiley 1985.

Spicer, R. A. 1991. Plant taphonomic processes. Pp. 71–113 in P. A. Allison and D. E. G. Briggs, eds. Taphonomy: releasing the data locked in the fossil record. Plenum, New York.

Spiker, E. C., and P. G. Hatcher. 1987. The effects of early diagenesis on the chemical and stable carbon isotopic composition of wood. Geochimica et Cosmochimica Acta 51:1385–1391.

Stankiewicz, B. A., and P. F. van Bergen, eds. 1998. Nitrogen-containing molecules in the bio- and geosphere. ACS Symposium Series No. 707. American Chemical Society, Washington, D.C.

Stankiewicz, B. A., P. F. van Bergen, I. J. Duncan, J. F. Carter, D. E. G. Briggs, and R. P. Evershed. 1996. Recognition of chitin and proteins in invertebrate cuticles using analytical pyrolysis-gas chromatography and pyrolysis-gas chromatography/mass spectrometry. Rapid Communications in Mass Spectrometry 10:1747–1757.

Stankiewicz B. A., M. Mastalerz, M. A. Kruge, P. F. van Bergen, and A. Sadowska. 1997a. A comparative study of modern and fossil cone scales and seeds of conifers: a geochemical approach. New Phytologist 135:375–393.

Stankiewicz, B. A., D. E. G. Briggs, R. P. Evershed, M. B. Flannery, and M. Wuttke. 1997b. Preservation of chitin in 25-million-year-old fossils. Science 276:1541–1543.

Stankiewicz, B. A., D. E. G. Briggs, and R. P. Evershed. 1997c. Chemical composition of Paleozoic and Mesozoic fossil invertebrate cuticles as revealed by pyrolysis-gas chromatography/mass spectrometry. Energy Fuels 11:515–521.

Stankiewicz, B. A., J. C. Hutchins, R. Thomson, D. E. G. Briggs, and R. P. Evershed. 1997d. Assessment of bog-body tissue preservation by pyrolysis-gas chromatography/mass spectrometry. Rapid Communications in Mass Spectrometry 11: 1884–1890.

Stankiewicz, B. A., P. F. van Bergen, M. B. Smith, J. F. Carter, D. E. G. Briggs, and R. P. Evershed. 1998a. Comparison of the analytical performance of filament and Curie-point pyrolysis devices. Journal of Analytical and Applied Pyrolysis 45:133–151.

Stankiewicz, B. A., D. E. G. Briggs, R. P. Evershed, R. F. Miller, and A. Bierstedt 1998b. The fate of chitin in Quaternary and Tertiary strata. Pp. 211–225 in Stankiewicz and van Bergen 1998.

Stankiewicz, B. A., H. N. Poinar, D. E. G. Briggs, R. P. Evershed, and G. O. Poinar Jr. 1998c. Chemical preservation of plants and insects in natural resins. Proceedings of the Royal Society of London B 265:641–647.

Stankiewicz, B. A., A. C. Scott, M. E. Collinson, P. Finch, B. Mösle, D. E. G. Briggs, and R. P. Evershed. 1998d. The molecular taphonomy of arthropod and plant cuticles from the Carboniferous of North America. Journal of the Geological Society, London 155:453–462.

Stankiewicz, B. A., D. E. G. Briggs, R. Michels, M. E. Collinson, M. B. Flannery, and R. P. Evershed. 2000. Alternative origin of aliphatic polymer in kerogen. Geology 28:559–562.

Stone, A. C., and M. Stoneking. 1999. Analysis of ancient DNA from a prehistoric Amerindian cemetery. Philosophical Transactions of the Royal Society of London B 354:153–159.

Stott, A. W., and R. P. Evershed. 1996. ^{13}C analysis of cholesterol preserved in archaeological bones and teeth. Analytical Chemistry 24:4402–4408.

Stott, A. W., R. P. Evershed, S. Jim, V. Jones, J. M. Rogers, N. Tuross, and S. Ambrose. 1999. Cholesterol as a new source of palaeodietary information: experimental approaches and archaeological applications. Journal of Archaeological Science 26:705–716.

Switsur, R., and J. Waterhouse. 1998. Stable isotopes and tree ring cellulose. Pp. 303–316 *in* H. Griffiths, ed. Stable isotopes: integration of biological, ecological and geochemical processes. Environmental Plant Biology Series. Bios Scientific, Oxford.

Sykes, B. 1999. The molecular genetics of European ancestry. Philosophical Transactions of the Royal Society of London B 354:131–139.

Tegelaar, E. W., J. W. de Leeuw, S. Derenne, and C. Largeau. 1989. A reappraisal of kerogen formation. Geochimica Cosmochimica Acta 53:3103–3106.

Tegelaar, E. W., H. Kerp, H. Visscher, P. A. Schenk, and J. W. de Leeuw. 1991. Bias of the paleobotanical record as a consequence of variations in the chemical composition of higher vascular plant cuticles. Paleobiology 17:133–144.

Tegelaar, E. W., G. G. Hollman, P. van de Vegt, J. W. de Leeuw, and P. J. Holloway. 1995. Chemical characterization of the periderm tissue of some angiosperm species: recognition of an insoluble, non-hydrolyzable, aliphatic biomacromolecule (suberan). Organic Geochemistry 23:239–250.

Tissot, B., and D. H. Welte, eds. 1984. Petroleum formation and occurrence. Springer, Berlin.

Tuross, N., and L. Stathoplos. 1993. Ancient proteins in fossil bones. Methods in Enzymology 224:121–129.

van Bergen, P. F., M. E. Collinson, and J. W. de Leeuw. 1993. Chemical composition and ultrastructure of fossil and extant salvinialean microspore massulae and megaspores. Grana (Suppl.) 1:18–30.

van Bergen, P. F., M. E. Collinson, P. G. Hatcher, and J. W. de Leeuw. 1994a. Lithological control on the state of preservation of fossil seed coats of water plants. *In* N. Telnaes, B. van Graas, and K. Oygard, eds. Advances in organic geochemistry 1993. Organic Geochemistry 22:683–702.

van Bergen, P. F., M. E. Collinson, J. S. Sinninghe Damsté, and J. W. de Leeuw. 1994b. Chemical and microscopical characterization of inner seed coats of fossil water plants. Geochimica Cosmochimica Acta 58:231–240.

van Bergen, P. F., M. Goni, M. E. Collinson, P. J. Barrie, J. S. Sinninghe Damsté, and J. W. de Leeuw. 1994c. Chemical and microscopical characterization of outer seed coats of fossil and extant water plants. Geochimica Cosmochimica Acta 58:3823–3844.

van Bergen, P. F., A. C. Scott, P. J. Barrie, J. W. de Leeuw, and M. E. Collinson. 1994d. The chemical composition of upper Carboniferous pteridosperm cuticles. Organic Geochemistry 21:107–112.

van Bergen, P. F., M. E. Collinson, D. E. G. Briggs, J. W. de Leeuw, A. C. Scott, R. P. Evershed, and P. Finch. 1995. Resistant biomacromolecules in the fossil record. Acta Botanica Neerlandica 44:319–342.

van Bergen, P. F., M. E. Collinson, and J. W. de Leeuw. 1996. Characterization of the insoluble constituents of propagule walls of fossil and extant water lilies: implications for the fossil record. Ancient Biomolecules 1:55–81.

van Bergen P. F., H. A. Bland, M. C. Horton, and R. P. Evershed. 1997a. Chemical and morphological changes in ancient seeds and fruits during preservation by desiccation. Geochimica Cosmochimica Acta 61:1919–1930.

van Bergen P. F., T. M. Peakman, E. C. Leigh-Firbank, and R. P. Evershed. 1997b. Chemical evidence for archaeological frank incense: boswellic acids and their derivatives in solvent soluble and insoluble fractions of resin-like materials. Tetrahedron Letters 38:8409–8412.

van Bergen, P. F., M. E. Collinson, A. C. Scott, and J. W. de Leeuw. 1997c. Unusual resin chemistry from Upper Carboniferous pteridosperm resin rodlets. *In* K. B. Anderson and J. C. Crelling, eds. Amber, resinite and fossil resins. ACS Symposium Series 617:150–169. American Chemical Society, Washington, DC.

van Bergen P. F, I. Poole, T. M. A. Ogilvie, C. Caple, and R. P. Evershed. 2000. Evidence for the demethylation of syringyl moieties in archaeological wood using pyrolysis/gas chromatography/mass spectrometry. Rapid Communications in Mass Spectrometry (in press).

van der Merwe, N. J., and J. C. Vogel. 1977. [13]C content of human collagen as a measure of prehistoric diet in woodland North America. Nature 276:815–816.

van de Water, P. K., S. W. Leavitt, and J. L. Betancourt. 1994. Trends in stomatal density and the ^{13}C/^{12}C ratios of *Pinus flexilis* during last glacial-interglacial cycle. Science 264:239–243.

Waring, R. H., and W. B. Silvester. 1994. Variation in foliar $\delta^{13}C$ values within the crowns of *Pinus radiata* trees. Tree Physiology 14:1203–1213.

Williams, J. L. 1985. Miocene epiphyllous fungi from northern Idaho. Pp. 139–142 *in* Smiley 1985.

Wilson, M. A., and P. G. Hatcher. 1988. Detection of tannins in modern and fossil barks and in plant residues by high-resolution solid state ^{13}C nuclear magnetic resonance. Organic Geochemistry 12:539–546.

Responses of plant populations and communities to environmental changes of the late Quaternary

Stephen T. Jackson and Jonathan T. Overpeck

Abstract. — The environmental and biotic history of the late Quaternary represents a critical junction between ecology, global change studies, and pre-Quaternary paleobiology. Late Quaternary records indicate the modes and mechanisms of environmental variation and biotic responses at time-scales of 10^1–10^4 years. Climatic changes of the late Quaternary have occurred continuously across a wide range of temporal scales, with the magnitude of change generally increasing with time span. Responses of terrestrial plant populations have ranged from tolerance in situ to moderate shifts in habitat to migration and/or extinction, depending on magnitudes and rates of environmental change. Species assemblages have been disaggregated and recombined, forming a changing array of vegetation patterns on the landscape. These patterns of change are characteristic of terrestrial plants and animals but may not be representative of all other life-forms or habitats. Complexity of response, particularly extent of species recombination, depends in part on the nature of the underlying environmental gradients and how they change through time. Environmental gradients in certain habitats may change in relatively simple fashion, allowing long-term persistence of species associations and spatial patterns. Consideration of late Quaternary climatic changes indicates that both the rate and magnitude of climatic changes anticipated for the coming century are unprecedented, presenting unique challenges to the biota of the planet.

Stephen T. Jackson. Department of Botany, Aven Nelson Building, University of Wyoming, Laramie, Wyoming 82071. E-mail: jackson@uwyo.edu
Jonathan T. Overpeck. Institute for the Study of Planet Earth and Department of Geosciences, University of Arizona, Tucson, Arizona 85721. E-mail: jto@u.arizona.edu

Accepted: 15 June 2000

Introduction

The environmental history of the late Quaternary represents a critical bridge between ecology and pre-Quaternary paleobiology. Sites with records of past biota and climate spanning the last 25,000 years are distributed across the globe, and those records can be dated with high precision (10^0–10^3 years) using ^{14}C, U/Th, dendrochronology, varve chronology, and other methods. Thus, local to global patterns can be observed at timescales of 10^1–10^3 years. No other period in geological history permits observation of contemporaneous events and spatial patterns of temporal change at such timescales across such a wide range of spatial scales. Late Quaternary pollen and macrofossil assemblages have been studied from thousands of sites worldwide, and ecological knowledge of the species represented in the fossil record and taphonomic knowledge of relationships between modern and fossil assemblages underpin their interpretation. Records of past biota can be dated independently and compared with independent records of environmental change from ice cores, marine sediments, tree rings, and many other sources.

Environmental changes of the past 25,000 years are superimposed upon longer-term and higher-magnitude changes during the entire Quaternary (Bartlein 1997; Bradley 1999), which are in turn embedded in trends of even higher magnitude spanning the Cenozoic (McDowell et al. 1990; Crowley and North 1991; Parrish 1998). The late Quaternary record does not encompass all of the dynamics of the earth's climatic or biotic systems. Nonetheless, the spatial and temporal detail of the record, together with the ability to connect past phenomena with ongoing, observable environmental processes and biological patterns, renders the late Quaternary a rich source of models, hypotheses, and exemplars for pre-Quaternary paleobiologists (Valentine and Jablonski 1993; Jablonski and Sepkoski 1996; Roy et al. 1996).

Modern ecological patterns and processes are superimposed on late Quaternary dynam-

 0094-8373/00/2604-0008/$1.00

ics. Ecologists and resource managers are recognizing that environmental variability at timescales of 10^1–10^4 years plays a major role in governing spatial patterns and temporal dynamics of populations, communities, and ecosystems (Swetnam and Betancourt 1998; Parsons et al. 1999). Furthermore, the magnitude of climatic change during the past 25,000 years matches the magnitude of predicted climate change over the next few centuries (Overpeck et al. 1991; Wright et al. 1993; Houghton et al. 1996), so late Quaternary records provide contexts for assessing biotic response to global change and for developing ecological management strategies in the face of environmental change, whether natural or anthropogenic (Hunter et al. 1988; Davis 1990; Huntley 1990d, 1999; Overpeck et al. 1991; Webb 1992; Overpeck 1993; Huntley et al. 1997).

In this paper, we discuss responses of terrestrial plant populations and communities to environmental changes of the late Quaternary. First, we present a conceptual scheme showing expected responses of plant species and communities to environmental changes. Our conceptual model, based on ecological niche theory, builds on some ideas presented by Good (1931) and Webb (1987, 1988), and parallels recent conceptual schemes of Huntley (1996, 1999). Second, we discuss the multivariate nature of climatic change, using insights from climate modeling and paleoclimate studies. Third, we review the broad spectrum of population responses and the narrow spectrum of community responses to environmental changes of the late Quaternary. In that review, we emphasize examples from North America and Europe, which are the best-studied parts of the globe. We then discuss whether the patterns observed in terrestrial plant populations and communities can be generalized to other organisms and habitats, speculate on evolutionary responses of niches, and briefly note implications for ongoing and future climate changes.

A Conceptual Model of Biotic Response to Environmental Variation

Ecological Niche Theory and the Nature of Environmental Space

Ecological niche theory is central to understanding how environmental change affects species abundance patterns. However, modern niche theory is inadequate for understanding the responses of plant populations and communities to environmental changes characteristic of the late Quaternary. Hutchinson (1958, 1978) conceived of the niche as a multidimensional conceptual space whose n dimensions are defined by the environmental factors[1] that influence fitness of individuals of a species (Appendix). Most discussions of niche theory implicitly assume that sites exist in the natural world that correspond to all possible combinations of niche variables (but see Austin 1990, 1992). Hutchinson (1958) noted possible exceptions but did not explore their implications. As Griesemer (1992) has observed, Hutchinson's conception of the niche is static and overlooks temporal changes in environment and population response. Environmental change, however, involves emergence and disappearance of combinations of niche variables.

The *fundamental niche* of a species comprises a subset of the *environmental space* defined by the n dimensions, consisting of the suite of combinations of variables that permit survival and reproduction of individuals (Hutchinson 1978). Maguire (1973) noted that the fundamental niche could be envisioned as a fitness response-surface, with an outer boundary delineating the absolute limit of population viability, and inner contours representing increasing fitness. In classical niche theory, the actual niche space occupied by a species comprises a subset of the fundamental niche. This subset, the *realized niche*, is constrained by biotic factors, which may prevent individuals of a species from occupying part of its fundamental niche. The realized niche can be portrayed as a response-surface of population density or biomass (Bartlein et al. 1986; Austin et al. 1990, 1997).

Large portions of environmental space are

---

[1] These factors include resources consumed by individuals and nonresource factors that affect individuals. We focus on nonresource factors, defining the niche similar to the *requirement niche* of Leibold (1996), and roughly equivalent to the *habitat hyperspace* (Whittaker et al. 1973), the *scenopoetic niche* (Hutchinson 1978), the *Grinnellian niche* (James et al. 1984), and the *environmental niche* (Austin et al. 1990, 1997).

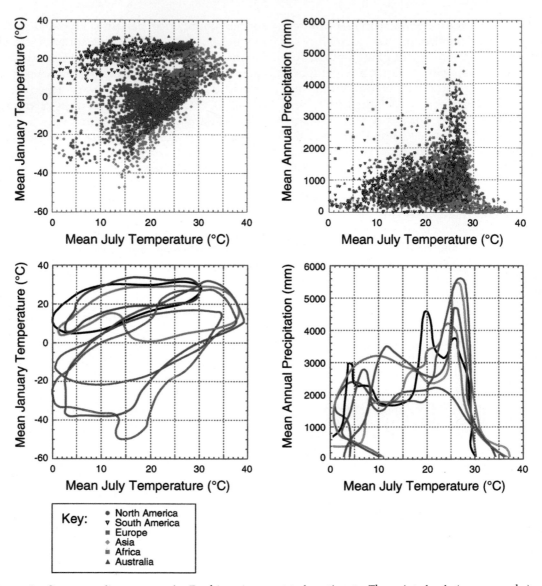

FIGURE 1.   Summary climate space for Earth's major vegetated continents. The point clouds (upper graphs) represent the particular observed combinations of mean July temperature, mean January temperature, and mean annual precipitation that were realized at 8516 climate stations (Vose et al. 1992) on six continents in the late twentieth century (averaged for the period 1951 to 1980). The station counts for each continent are: North America (3364), South America (776), Europe (864), Asia (1010), Africa (1665), and Australia (837). Lower graphs show generalized outlines of the climate space for each continent as defined by the point clouds.

empty (i.e., unrepresented in the realized world) at any given time, and temporal environmental change inevitably leads to portions of environmental space alternating between empty and full. Empty environmental space arises for two reasons. First, many niche variables covary. For example, summer and winter temperatures are positively correlated for each continent (Fig. 1 top left). Summer temperature and annual precipitation do not oc-

cur in all possible combinations (Fig. 1 top right). Second, spatial heterogeneities imposed by geology, topography, and other factors dictate that not all combinations of variables can occur. For example, serpentine outcrops do not occur in midcontinental North America, and therefore much of the environmental space defined by three variables—soil [Ni], summer temperature, and annual precipitation—does not exist in North America.

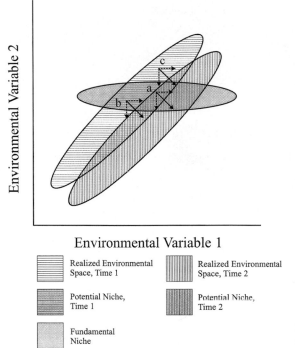

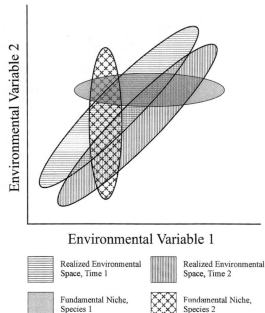

FIGURE 4.   How environmental change can affect a species population at a site. Vectors represent the direction and magnitude of change at individual sites within the environmental space. In all cases, Environmental Variable 1 increases while Environmental Variable 2 decreases; solid vector shows net change in both variables at the site. If the environment changes, but a particular site remains within the potential niche space of a species, the population can remain in place (a). However, if a site moves out of the potential niche of a species, the population at that site will undergo extinction (b). New sites may move into the potential niche space of the species (c), creating opportunities for colonization.

FIGURE 5.   How environmental change can affect species associations. At Time 1, the potential niches of species 1 and 2 overlap, and hence the species can potentially coexist at sites within that intersection. At Time 2, the potential niches of the two species do not overlap, and hence the species will not coexist in the realized world.

Nomadic Niches and Contingent Communities

Species differ in their fundamental niches, and different species respond to different environmental variables. Consequently, particular species combinations may appear and disappear as the environment changes (Fig. 5). A particular ecological community can exist only in that portion of environmental space where fundamental niches of all its constituent species overlap. In general, a particular species association will occupy a smaller portion of realized environmental space than any of its constituent species. Put another way, the amplitude of environmental change that will allow survival of a particular species will tend to be much greater than the amplitude of var-

iation that will allow continued realization of a particular collection of species (Fig. 5).

This model predicts not only that environmental change will alter composition of ecological communities, but also that it can alter the sequence of species arrayed along spatial gradients (Fig. 6). Different species combinations may arise along a spatial gradient, and species may even reverse positions along the gradient as the environment changes (Fig. 6).

The Nature of Environmental Change

Climate Change: A Redundant Expression

Recent advances in paleoclimatology have demonstrated that climate changes continually on all timescales (Overpeck 1995; Parrish 1998; Bradley 1999), and hence the term "climate change" is redundant—climate is always changing. High-resolution studies indicate significant shifts in atmospheric circulation patterns at annual to decadal scales during the last two centuries (Martinson et al. 1995; Diaz and Markgraf 2000), and changes in climate are recognized at progressively expanding

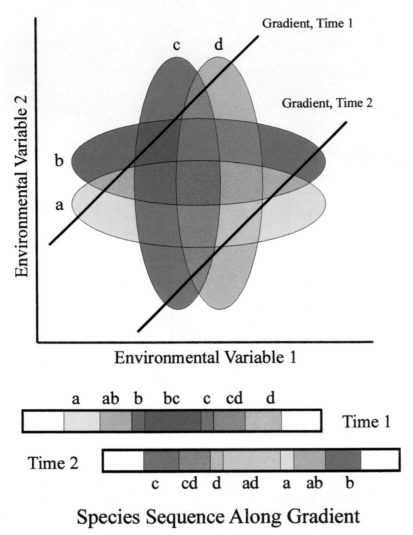

FIGURE 6. How environmental change can influence patterns of species and communities along environmental gradients. Ovals represent fundamental niches of four species (*a–d*). The diagonal lines represent a spatial gradient at two different times. For example, the gradient might be an elevational or latitudinal transect, with Environmental Variables 1 and 2 respectively representing summer and winter temperatures. Lower part of figure shows species sequence projected along the gradients. Vegetational transects along the spatial gradient would yield a different array of species associations (e.g., *b* and *c* co-occur at Time 1 but not at Time 2, whereas *a* and *d* co-occur only at Time 2). Furthermore, the sequence of species encountered along the gradient might shift. For example, in a latitudinal gradient, species *a* would be north of species *d* at Time 1, while species *d* would be north of species *a* at Time 2.

timescales, from centuries (Bradley and Jones 1992; Woodhouse and Overpeck 1998) to millennia (Wright et al. 1993) and beyond (Crowley and North 1991; Parrish 1998). In general, the magnitude of change increases with the timescale considered (e.g., Bradley 1999: Fig. 2.14). The regional climates represented by instrumental observations are just a snapshot of an ever-changing continuum, and also just a subset of the possible range of future climate change. The shape of Earth's realized environmental space is continually changing.

Climate Forcing and the Ever-Changing Shape of the Realized Climate Hyperspace

Paleoclimatology was revolutionized by the demonstration that climate variability at glacial-interglacial scales (10^4–10^5 years) is paced

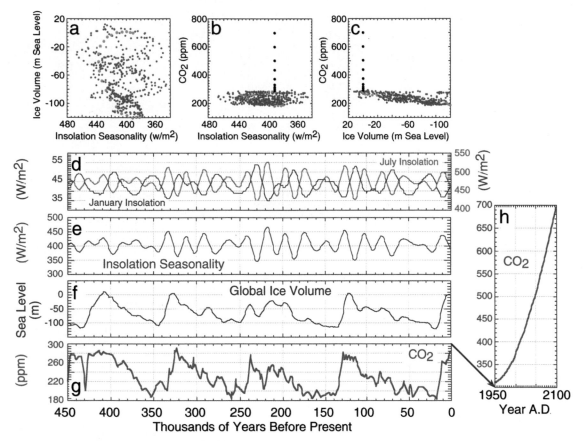

FIGURE 7.   Major climate forcing for the last 450,000 years (observed at 1000-year intervals to 1950), and for the period 1850 to 2100 (observed and projected at 25-year intervals). Scatter plots (A–C) and time series (D–H) of glacial ice volume (in meters of sea-level equivalent, see below), insolation at the top of the atmosphere (January, July, and July minus January ["seasonality"], all calculated for 60°N [Berger and Loutre 1991]), and atmospheric CO_2 concentration (Etheridge et al. 1996; Houghton et al. 1996; Petit et al. 1999; Robertson et al. 2000). Note that CO_2 levels projected for the next century (red line in H, and black dots in A–C) (Houghton et al. 1996) are plotted versus years A.D. rather than B.P. Glacial ice volumes were obtained using the global-average deep-water ("benthic") $\delta^{18}O$ (Imbrie et al. 1992) scaled to a 20,000 years B.P. (glacial) to present (interglacial) sea-level amplitude equal to the observed value of 120 m (Fairbanks 1989). The trajectory (sensu Bartlein 1997) of climate forcing over the last 21,000 years is displayed as green dots (A–C). Whereas atmospheric trace gas (e.g., CO_2) levels are expected to increase dramatically to unprecedented levels in the next century (A–C, H), both insolation and global sea level are not likely to change much relative to recent geologic variations. An interactive 3-D view of climate forcing over the period 450,000 years ago to A.D. 2100 can be viewed on the worldwide web at: *http://www.ngdc.noaa.gov/paleo/class/javatest/solidtest/SolidTest.html*

by variations in Earth's orbital geometry (Hays et al. 1976; Imbrie et al. 1992, 1993). Gravitational influences of other planets in the solar system cause long-term variations in eccentricity of Earth's orbit, precession of the equinoxes, and tilt of Earth's rotational axis. These Milankovitch variations have been calculated for the past several million years, and can be quantified in terms of insolation at the top of the atmosphere as a function of year, season, and latitude (Fig. 7) (Berger and Lou-

tre 1991). Their net effect is to alter the seasonal and latitudinal distribution of insolation, which in turn causes changes in atmospheric circulation patterns, including changes in seasonal and geographic patterns of temperature, precipitation, and other climate variables (Fig. 8) (Kutzbach and Webb 1991, 1993; Kutzbach et al. 1993, 1998; Bartlein et al. 1998).

The COHMAP (Cooperative Holocene Mapping Project) Group provided another

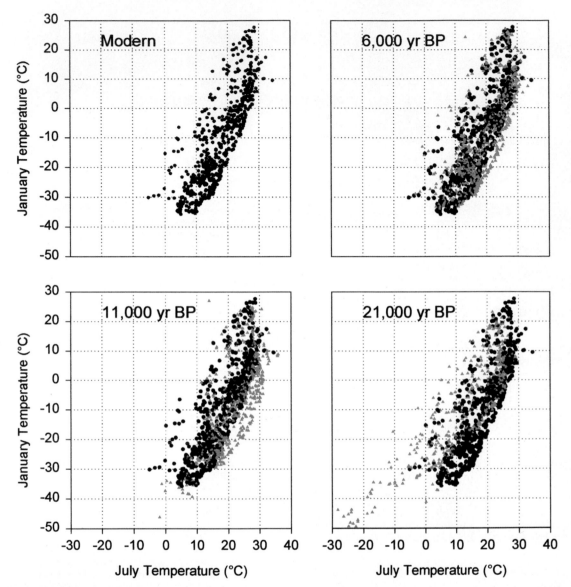

FIGURE 8.   Simulated changes in the realized environmental space of North America (Canada, United States, Mexico) since the Last Glacial Maximum. Modern January and July temperatures (black circles) are from a 25-km equal-area grid based on >8000 modern climate stations (Thompson et al. 1999). Simulated January and July temperatures for 6000, 11,000, and 21,000 years B.P. (green triangles) are from NCAR CCM1 simulations (Kutzbach et al. 1998). We used the data of Bartlein et al. (1998), in which anomalies (paleoclimate simulation minus modern control simulation) were applied to the 25-km grid of modern climate data. Because of the size of the data sets (22,690 to 23,839 points), we plotted 2% of the values for each time period, selected at random, to give a representation of the scatter patterns. The contrasting scatter patterns indicate the extent of simulated change in the shape, position, and orientation of the realized environmental space as defined by these two variables under different forcings of the last 21,000 years. The time periods differ in orbital forcing, ice-sheet extent, and atmospheric CO_2 concentration (Kutzbach et al. 1998).

key advance, detailing how the earth's climate system (broadly defined to include the atmosphere, oceans, biosphere, and cryosphere) evolved through the last glacial cycle in response to Milankovitch forcing and changes in continental ice sheets (COHMAP 1988; Wright et al. 1993). Key lessons of the COHMAP effort and modern climate theory are that climate is multivariate and that biologically relevant climate variables may change

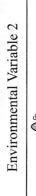

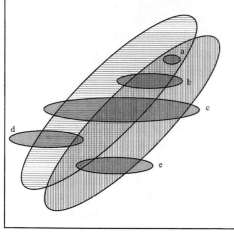

Environmental Variable 1

Realized Environmental Space, Time 1

Realized Environmental Space, Time 2

Fundamental Niche

FIGURE 9. Modes of population response to environmental change. Species a: Mode 1 (populations stay in place). Species b: Modes 1 and 2 (populations shift along local habitat gradients). Species c: Modes 1, 2, and 3 (populations migrate to distant, newly suitable sites and disappear from some former sites). Species d: Modes 1 and 4 (widespread extirpation without colonization of new territory. In this case a formerly widespread species becomes restricted in geographic range and habitat. Species e: Inverse of species d: A formerly rare species colonizes extensive new territory.

independently. For instance, summer and winter insolation vary independently in response to orbital forcing (Fig. 7) (Berger and Loutre 1991). From 14,000 to 6000 years ago[2], high-latitude Northern Hemisphere summer insolation was up to 9% higher than today, while winter insolation was up to 15% lower (Fig. 7) (see also COHMAP 1988: Fig. 2). Thus, summer and winter temperatures, although positively correlated along geographic gradients, changed in opposite directions in many

---

[2] Late Quaternary timescales are complicated by long-term variations in atmospheric ^{14}C concentrations; ^{14}C-based age estimates do not correspond precisely to calendar years. Calibrations between ^{14}C chronologies and calendar-year chronologies have been developed based on dendrochronology, annually laminated sediments, and U/Th dating (Hughen et al. 1998; Bradley 1999). For example, the Last Glacial Maximum, dated at 18,000 ^{14}C years B.P., is closer to 21,500 calendar years B.P. We use calendar-year ages except where noted.

regions during the Holocene (Fig. 8; cf. Fig. 3D). The mean values, variability, and seasonal patterns of the many biologically influential climatic variables have changed in complex ways, leading to continual changes in the realized environmental space.

In summary, climate is a dynamic system, the global climate system is always on its way elsewhere, and the particular realization of the climate hyperspace at any given time rarely repeats any previous realization (Fig. 7). For example, climates from one glacial or interglacial period do not represent analogues for previous or succeeding glacials and interglacials (Watts 1988; Imbrie et al. 1992, 1993; Webb et al 1993). The major forcing elements were different in each case (Fig. 7), and hence each interglacial, glacial, stadial, and interstadial period has differed in important respects from others. Each point in time represents a unique set of climate gradients and combinations of gradients. Atmospheric CO_2 concentration, which not only influences climate but is an environmental variable directly affecting plant ecophysiology, has also varied substantially at these timescales (Fig. 7), as have other climatically influential variables (other trace gases, stratospheric [volcanic] aerosols, tropospheric aerosols [e.g., dust]).

Case Studies from the Late Quaternary

Modes of Biotic Response to Environmental Change

Environmental change exerts its influence directly on individual organisms at the sites where they live. As noted earlier, the environment of a site may remain within the fundamental niche of a species, so individuals can continue to live and reproduce there. The local environment may pass out of the fundamental niche, leading to extirpation via mortality or lack of recruitment. And other sites may pass into the fundamental niche, so individuals can colonize if propagules reach the sites. Depending on the nature, magnitude, and rate of the environmental changes, these site-specific processes are manifested at coarser spatial scales as a spectrum of patterns. We have, somewhat arbitrarily, classified this spectrum into four distinct response modes (Fig. 9).

Mode 1: Populations continue to occupy the same sites as before the change. *Mode 2*: Populations shift locally along habitat gradients (e.g., elevation, soil texture). *Mode 3*: Populations undergo migration, colonizing suitable territory 10^1–10^3 km from the origin, often disappearing from formerly occupied territory. *Mode 4*: Populations undergo local extinction while failing to colonize new territory, resulting in range contraction.

Each of these response modes has ecological, geographic, and evolutionary consequences for the species as a whole. The particular geographic range occupied by a species is a realization of local colonization and extinction processes and biotic interactions in the context of the spatial distribution of the species' potential niche. As a species migrates in response to spatial shifts in its potential niche, it may encounter new species (competitors, facilitators, consumers) and lose contact with others as it colonizes new and abandons old territory. Such properties as the total area occupied by a species, the total population size of the species, spatial structure (e.g., continuity versus patchiness), and genetic structure (e.g., diversity, gene flow, clinal variation, hybridization) will be governed or constrained by the outcome of the individual population responses. If population extinctions are unaccompanied by migrations, the species may undergo genetic bottlenecking or extinction. Evolutionary changes resulting from selection and genetic drift will accompany all of these response modes, although magnitude and rate will vary widely.

Each of these response modes can be applied to species associations as well. Conceivably, associations can remain intact in situ, they can shift position along local habitat gradients, they can migrate, or they can simply vanish entirely. We describe each of these responses in turn, noting examples (or lack thereof) from the Quaternary record for each.

Mode 1: Remaining in Place

Species Populations.—Individual plants, particularly long-lived perennials, often tolerate environmental variation at timescales of 10^0–10^3 years. Net photosynthesis, resource allocation, seed production, growth form, and organ morphology may change, but the plants remain literally rooted in place unless the environmental change exceeds their capacity to adapt physiologically and morphologically. Such persistence of individuals is well known in the case of long-lived woody plants, but genets of some herbaceous and graminoid species may persist for hundreds to thousands of years (e.g., Steinger et al. 1996). Similarly, populations of a species may persist at a site for many generations. Populations may change in density, age structure, size structure, and growth form as the environment changes, and they may undergo recruitment or mortality pulses in response to decade- and century-scale environmental variation (Swetnam and Betancourt 1998; Swetnam et al. 1999). They may also undergo evolutionary changes resulting from selection and from immigration of genotypes better suited to the local environment (Rehfeldt et al. 1999).

Some populations have occupied single sites over several millennia. *Choisya dumosa*, *Juniperus osteosperma*, and other species have grown at some sites in the southwestern United States since the Last Glacial Maximum (Thompson 1988; Nowak et al. 1994a; Betancourt et al. 2000). Several temperate tree species grew in the lower Mississippi Valley during the Last Glacial Maximum (Delcourt et al. 1980; Givens and Givens 1987; Jackson and Givens 1994) and still grow there today. However, these populations have experienced different climatic regimes and species associations and have likely changed in density, age structure, and genetic structure.

Even though these and other populations have persisted at some sites through entire glacial/interglacial cycles, geographic ranges have changed. For example, *Juniperus osteosperma* populations in western Nevada during the Last Glacial Maximum were at the northern limit of the species range (Nowak et al. 1994b). Populations at the site are now at the southwestern end of the range, which extends >1000 km to the northeast. Most plant species well documented in the fossil record have undergone geographic range shifts during the past 25,000 years.

Species Associations.—Modes of community response to decade- and century-scale environmental change are poorly known. Ecological studies are based on permanent-plot monitoring, repeat photography, age-structure studies, and repeated floristic surveys, and interpretation is confounded by impacts of human activity and secondary succession. Studies in relatively undisturbed areas often show community stasis in a coarse sense (e.g., no species invasion or extirpation). However, many studies show changes in relative abundance of species, and wholesale conversion of vegetation physiognomy (e.g., grassland to woodland) has been documented. Several of these cases are attributable at least in part to climatic change (e.g., Weaver and Albertson 1936, 1944; Weaver 1954; Hastings and Turner 1965; Archer 1989).

Paleoecological studies of sites with relatively high spatial and/or temporal resolution provide records of vegetation dynamics at decade to century timescales before extensive human disturbance (Bernabo 1981; Campbell and McAndrews 1993; Foster and Zebryk 1993; Davis et al. 1994; Björkman and Bradshaw 1996; Clark et al. 1996; Laird et al. 1996). Interpretation can be hampered by coarse taxonomic resolution, spatial smoothing, and absence of independent records of climatic change at similar timescales. However, in cases where independent paleoclimate data are available, changes in vegetation correspond well to climatic changes. A singular exception is the mid-Holocene *Tsuga* decline in eastern North America, which is attributable to a pathogen outbreak (Davis 1981a; Allison et al. 1986; Bhiry and Filion 1996).

Persistence of particular species assemblages at individual sites for more than a few thousand years is rare. Pollen sequences typically show turnover in assemblage composition over periods greater than 2000–5000 years (Jacobson and Grimm 1986; Overpeck 1987; Overpeck et al. 1991, 1992; Foster and Zebryk 1993). Macrofossil data show that this turnover consists of not only changes in the relative abundance of plant taxa at the site but also invasion and extirpation of species (Thompson 1988, 1990; Nowak et al. 1994a; Baker et al. 1996; Jackson et al. 1997; Weng and Jackson 1999), even at sites where some species have persisted for longer periods.

The best examples of apparent community stasis are from montane forests dominated by a single species. For example, Ponderosa pine (*Pinus ponderosa*) forests have persisted for 10,000 years on the Kaibab Plateau of northern Arizona (Weng and Jackson 1999). Lodgepole pine (*P. contorta*) forests have occupied the Yellowstone Plateau of northwestern Wyoming for a similar period (Whitlock 1993). However, Holocene climate changes have forced changes in fire frequency and intensity within the lodgepole pine forests (Millspaugh et al. 2000), undoubtedly causing changes in emergent attributes such as stand density, standing crop, and understory composition. Subalpine forests in New York and New Hampshire have been dominated by balsam fir (*Abies balsamea*) throughout the Holocene, but disturbance regime and stand structure appear to have varied (Jackson 1989; Spear et al. 1994). In all these cases the forests were established in the early Holocene, replacing late-glacial tundra or spruce woodland.

Mode 2: Shifts along Local Habitat Gradients

Species Populations.—Species populations can respond to environmental changes by fine-scale adjustments in their site or habitat occupation within a region. Such shifts are documented at decade to century scales for woody species based on tree-ring demographic studies. Tree invasions of open areas are documented from demography of living populations (e.g., Archer 1989; Graumlich 1994; Hessl and Baker 1997; Mast et al. 1998), and tree-ring studies of dead wood in open areas reveal extirpation patterns (Szeicz and MacDonald 1995; Lloyd and Graumlich 1997). Comparison of aerial or ground photographs taken at different times document twentieth-century dynamics of tree lines (Mast et al. 1997) and other ecotones (Allen and Breshears 1998). Spatial shifts in individual species distributions can be very rapid (<5 yr) when climate episodes lead directly to adult mortality (Allen and Breshears 1998).

Holocene shifts of populations along local habitat gradients have been documented by paleoecological studies in many regions. Examples of shifts along elevational gradients are legion, and come from radiocarbon-dating of dead wood above tree line (Luckman et al. 1993; Kullman 1995) and from elevational transects of lake cores (Gaudreau et al. 1989; Jackson 1989; Spear et al. 1994; Anderson 1996; Weng and Jackson 1999) and packrat-midden series (Betancourt 1990; Cole 1990; Thompson 1990; Nowak et al. 1994b). In general, the magnitude of elevational displacement increases with the time span considered. Studies along other environmental gradients (soil texture, soil chemistry, bedrock type) are fewer. In a now-classic study, Brubaker (1975) showed that mesic tree species colonized coarse-textured outwash soils during moist periods but were restricted to fine-textured soils during dry periods.

Species Associations.—Shifts of species associations along habitat gradients at decade to century timescales are not as well known as population shifts, especially for nonwoody species. Movement of woody species is individualistic, but invasion or retreat of woody plants (e.g., upper or lower tree line) may induce secondary environmental changes (e.g., ground-level insolation, soil temperature and moisture, soil chemistry) important to other plants. For example, twentieth-century expansion of juniper woodlands from thin, rocky soils onto deep-soiled slopes in Wyoming is accompanied by extirpation of many steppe/grassland species and invasion of forb and graminoid species better suited to the microenvironment created by the junipers (Wight and Fisser 1968; S. T. Jackson et al. unpublished data). In a sense, the juniper woodland association is moving in toto along the edaphic gradient, but the movement is driven by invasion of the dominant species.

Holocene shifts along habitat gradients have been generally nonzonal: the shifts do not involve displacement of vegetation zones but instead lead to recombined associations and new zonation patterns. For example, the current elevational zonation of montane forests in New York and New England runs from subalpine *Abies* forest downward through *Picea/Abies* forest into mixed conifer/hardwood forests (*Acer, Betula, Fagus, Tsuga, Pinus*). In the warmer mid-Holocene (8000–4000 years B.P.), several low-elevation species (*Betula alleghe-niensis, Pinus strobus, Tsuga canadensis*) grew as

much as 300 m higher than today (Jackson 1989; Spear et al. 1994). However, they graded upward directly into subalpine *Abies* forest; no intervening *Picea*-dominated forest existed before 3000 years B.P. (Jackson and Whitehead 1991; Spear et al. 1994). Elevational gradients in western North America show similar zonal reorganization (Thompson 1988; Betancourt 1990; Van Devender 1990).

Mode 3: Migration

Species Populations.—Late Quaternary shifts in geographic distributions of species have been discussed since the recognition of a "Glacial Epoch" and accompanying climatic changes (Lyell 1832; Forbes 1846; Gray 1858; Darwin 1859), and postglacial migration patterns and pathways were proposed long ago from phytogeographic patterns (e.g., Peattie 1922; Gleason 1923; Clements 1934). Pollen and macrofossil records now allow mapping of migration patterns at scales ranging from regional (Davis and Jacobson 1985; Gaudreau 1988; Woods and Davis 1989) to continental (Davis 1976; Huntley and Birks 1983; Webb 1988; Jackson et al. 1997).

Shifts in species distributions can involve two processes, colonization of previously unoccupied territory and extirpation in formerly occupied territory. Most discussions of migration have focused on patterns and dynamics at the advancing end (Gleason 1917, 1923; Davis 1976; Huntley and Webb 1989; Clark et al. 1998).

Colonization is a slow process, at least by human timescales. It proceeds by ecological processes (dispersal, establishment, population growth, reproduction) that can be studied directly, but the aggregate consequences of these processes for migration may take decades to centuries or more to be manifested. Mapping of pollen and macrofossil data indicates that spread of many species has been far more rapid than can be accounted for by mean seed-dispersal distances, so infrequent long-distance dispersal and colonization events must play critical roles (Clark 1998; Clark et al. 1998). Such events are rarely observed directly. Although some recent "alien" plant invasions are well documented, (e.g., Mack 1981, 1986), most of our knowledge of

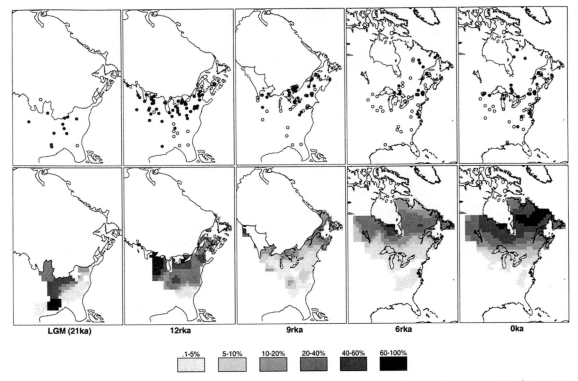

| .1-5% | 5-10% | 10-20% | 20-40% | 40-60% | 60-100% |

FIGURE 10.   Changes in the geographic distribution and abundance of spruce (*Picea* spp.) in eastern North America since the Last Glacial Maximum (21,500 calendar years B.P., approximately 18,000 ¹⁴C years B.P.), as shown by maps of macrofossil presence/absence (upper) and isopolls (lower). Closed circles in macrofossil maps denote sites where *Picea* macrofossils occurred at the specified time interval; open circles represent absence of *Picea* macrofossils. "*rka*" denotes 1000 ¹⁴C years B.P. (6 rka ≅ 6 Ka; 9rka ≅ 11 Ka; 12 rka ≅ 14 Ka). Modified from Jackson et al. 1997, 2000.

natural plant invasions comes from the paleo-ecological record.

Unfortunately, paleoecological data are in-effective at recording colonizing events and populations (Bennett 1985; Davis et al. 1991). Pollen data provide a distance-weighted in-tegration of populations within 10–50 km of a basin and are better at showing approach and expansion of population masses than estab-lishment and growth of individual popula-tions. Macrofossil studies indicate that scat-tered populations of tree species can go un-recorded by pollen data for thousands of years (Kullman 1996, 1998a,b,c). Macrofossils con-stitute a finer-scale sensor but are spatially bi-ased (toward lake margins for lake sediments and rock outcrops for packrat middens). Mac-rofossil data networks are usually not dense enough to record colonization patterns, with a few exceptions (Gear and Huntley 1991; Ly-ford et al. 2000).

Despite these limitations, the paleoecologi-cal record provides some general lessons. Mi-grations spanning hundreds to thousands of kilometers have occurred throughout the last 25,000 years. Rare, long-distance dispersal events have played a key role (Betancourt et al. 1991; Clark et al. 1998). Dispersal across geo-graphic barriers of ca. 100 km has occurred frequently (e.g., S. L. Webb 1987; Betancourt et al. 1991; Kullman 1996, 1998a,b,c). Migration rates vary in space and time (King and Her-strom 1997) and in some cases have been lim-ited more by rate of climatic change than by biotic factors such as dispersal and establish-ment (Johnson and Webb 1989). Postglacial migration of *Picea* and other taxa onto degla-ciated terrain in North America was limited only by the rate of ice retreat (Webb 1988; Jack-son et al. 1997) (Fig. 10). The migrations not only consisted of range extensions but also in-volved shifts in population mass and density, and probably changes in genetic structure as well. Rapid migrations may have been accompanied at least temporarily by selection for more effective seed dispersal (Cwynar and MacDonald 1987), decrease in genetic diversity owing to multiple founder effects (Premoli et al. 1994), and escape from pathogens and herbivores.

Population declines or extirpations at the retreating end of a species migration, although little studied, are undoubtedly diverse in pattern, rate, and mechanism. Unfavorable environmental changes may be manifested in adult mortality, which can be gradual or episodic, or in recruitment decline via lowered seed production, germination, or survival. Climatic effects may be amplified by biotic factors (e.g., competition from indigenous or newly arrived species, increased susceptibility to pathogens or herbivores). Disturbances (fire, windthrow) can induce mortality and provide recruitment opportunities for other species (Clark et al. 1996; Weng and Jackson 1999). Population extirpation is likely to be patchy. Population susceptibility will track landscape-scale heterogeneity in environment, disturbance, and population size. Relict populations may persist in locally suitable habitats.

Paleoecological studies of population retreat face many of the same challenges of scale, lag, and resolution as for colonization. Pollen sequences record regional population declines effectively, but population fragmentation and disappearance are harder to document owing to spatial imprecision of the record. Scattered remnant populations, like scattered colonizing populations, may go undetected. Nevertheless, opportunities exist for "model-system" studies in regions where suitably dense site networks with sufficient temporal resolution can be developed. Such studies, in tandem with ecological and demographic studies of extant populations and documentation of disturbance events, could greatly advance understanding of population retreat.

Plant Associations.—Early discussions of postglacial migrations (Adams 1902; Transeau 1903; Clements 1904; Gleason 1923) conceived of them as zonal, consisting of latitudinal, longitudinal, and elevational shifts in the same species associations that can be recognized today. Clements formalized this in his conception of the *clisere,* which he defined as a temporal series of associations corresponding to those found along a spatial transect (Clements 1916: Fig. 37).

Paleoecologists have long argued, on the basis of both past pollen assemblages that have no modern counterparts and independent rates and routes of migration for different species, that the clisere concept is untenable (West 1964; Davis 1976, 1981b). Jacobson et al. (1987) showed how pollen assemblages formed new and unique combinations in eastern North America during the last deglaciation, and Webb (1988) noted that most of the major vegetation associations of the region cannot be traced before the early Holocene. Overpeck et al. (1992) demonstrated this formally via application of distance metrics between modern and fossil pollen assemblages (Fig. 11). The European record shows similar patterns (Huntley 1990a,b,c). At various times, spatially extensive plant associations unlike any existing today have been established. A particularly striking example is the late-glacial and early Holocene mixed forest of *Picea* and hardwoods (*Ulmus, Ostrya/Carpinus, Fraxinus, Quercus*) in midcontinental North America (Jacobson et al. 1987; Baker et al. 1996). These floristically unique assemblages represent extensive, dense populations of several taxa (*Ulmus, Ostrya, Fraxinus nigra*) now restricted to small and scattered populations in disturbed or wet habitats.

The inference that species migrate independently does not imply that correlated migration patterns among associated species do not occur. The fundamental niches of many coexisting species are probably similar on some environmental dimensions, and hence an environmental change may induce migration by two or more species in the same general direction. Furthermore, some niche variables may be correlated with vegetation physiognomy. For example, many forest understory species require cool, moist microclimates with low insolation. Thus it is unsurprising to observe, for instance, replacement of forest by prairie accompanied by extirpation of forest-floor herbs and establishment of prairie herbs, shrubs, and graminoids (e.g., Baker et al. 1996).

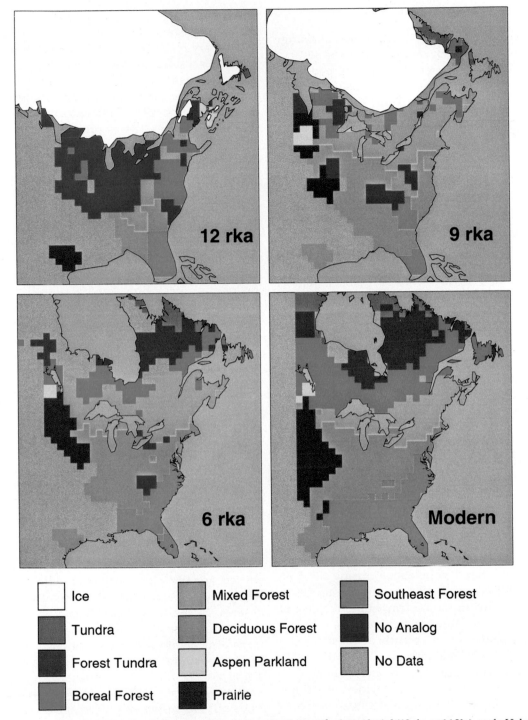

FIGURE 11. Paleovegetation maps of eastern North America during the late-glacial (12 rka ≅ 14 Ka), early Holocene (9 rka ≅ 11 Ka), mid-Holocene (6 rka ≅ 6 Ka), and modern times, based on application of multivariate distance metrics to modern and fossil pollen data. Modified from Overpeck et al. 1992.

Mode 4: Extinction/Contraction

Species Populations.—Environmental change that reduces the overall size of a species' potential niche may result in widespread extirpation of populations unaccompanied by colonization of new sites. Thus, the geographic range of the species may shrink substantially, and the species may become relegated to scattered populations in isolated habitats. Depending on the nature of environmental change relative to size and shape of the fundamental niche, species many alternate between being widespread and abundant at certain times and being scattered or isolated at others.

Numerous examples are documented in the Quaternary fossil record. Papershell pinyon (*Pinus remota*) is currently restricted to a few scattered populations in mountain ranges along the Texas/Chihuahua border. During the Last Glacial Maximum, it formed extensive woodlands on valley floors and extended 300 km to the north (Lanner and Van Devender 1998). During the glacial/interglacial transition, elm (*Ulmus*), hophornbeam (*Ostrya*), and black ash (*Fraxinus nigra*) were extensive and abundant in midcontinental North America, but they have since contracted their ranges and habitat distributions (Jacobson et al. 1987; Webb 1988).

Some species display an inverse pattern, going from rarity to abundance. Ponderosa pine (*P. ponderosa*), now widespread and abundant in the western United States, occurred as small, scattered populations during the Last Glacial Maximum (Betancourt et al. 1990). Many of the important temperate tree taxa of eastern North America occurred as small, isolated populations as recently as the early Holocene (Webb 1988; Jackson et al. 1997, 2000). Interglacial populations of many temperate species in northern Europe evidently did not migrate south during transition to glacial conditions, but underwent extinction. Residual populations in southern Europe survived the glacial periods and provided seed sources for recolonization of northern Europe during subsequent deglaciations (Bennett et al. 1991; Willis 1994; Bennett 1997).

Species undergoing widespread range contraction during periods of environmental change face some risk of extinction. One late Quaternary plant extinction has been reported to date (Jackson and Weng 1999). *Picea critchfieldii* had a widespread range (>240,000 km^2) and was a dominant component of forests in the Lower Mississippi Valley during the Last Glacial Maximum. It was probably extinct by 10,000 years B.P. (Jackson and Weng 1999). Like many vertebrates in North America, it disappeared during the last deglaciation (Martin and Klein 1984; Stuart 1991; MacPhee 1999). Unlike the vertebrates, however, its demise cannot be attributed to direct human activities. The last deglaciation was characterized by rapid and often abrupt climatic changes of high magnitude (Overpeck et al. 1991; Taylor et al. 1993; Broecker 1997), so *P. critchfieldii* may have encountered a brief period of "bad luck" in which its potential niche contracted or vanished.

Differentiation of *P. critchfieldii* from extant *Picea* species, which was required to recognize the extinction, was made possible only by detailed anatomical and morphological studies of ovulate cones and needles. The demise of *P. critchfieldii* may have been a unique event. However, the last deglaciation may have been accompanied by other species extinctions masked by the coarse taxonomic resolution of pollen data and the paucity of detailed studies of pre-Holocene floras in North America.

Species Associations.—Earlier we presented evidence that late Quaternary plant associations have not persisted in situ, nor have they shifted zonally along habitat gradients, nor have they migrated across the landscape as cliseres. In the face of late Quaternary environmental changes, plant associations were transformed into other associations different in composition, structure, and dynamics. These associations can (and usually do) share species with those existing at other time periods but are sufficiently different to warrant recognition as different entities. Webb (1988: p. 406) suggested that "plant assemblages are to the biosphere what clouds, fronts, and storms are to the atmosphere. . . .They are features that come and go."

Although plant associations are not conserved through time, an association could conceivably disappear and reappear later if environmental conditions conducive to it were repeated. For example, associations might be similar between different interglacial periods or glacial maxima. This is difficult to evaluate in view of the paucity of pre-Wisconsinan pollen and macrofossil assemblages. However, periodic reappearance of associations over glacial/interglacial timescales is probably not routine. Pollen studies spanning the past 100,000–250,000 years show continual transformation of community composition, with no repeating patterns (e.g., Whitlock and Bartlein 1997; Allen et al. 1999). Comparative floristic studies of successive interglacial periods show that no two interglacial floras are alike (West 1980), although Watts (1988) suggests some climatic and vegetational parallels between portions of successive interglacials. The global climate system follows a complex, multidimensional trajectory, never returning precisely to a preexisting state (Fig. 7), and so plant associations are unlikely to be duplicated precisely through time.

Is All the World Gleasonian[3]?

The biotic response patterns we described in the previous section can be explained as direct consequences of multivariate environmental changes. If fundamental niches vary among species in shape, orientation, and position within environmental space, then temporal changes in realized environmental space should result in predictably individualistic behavior, including disaggregation of species associations and emergence of new ones. As the magnitude of environmental change increases,

population responses should go from tolerance to habitat shift to migration and extinction. Larger environmental changes should lead to larger shifts in realized environmental space and to greater reorganization of species assemblages.

Our discussion has focused on terrestrial plant populations and communities at mid- to high latitudes and underscores the conclusion of many other authors that terrestrial plant species have behaved individualistically (e.g., Davis 1976; Webb 1987, 1988; Huntley 1988, 1991, 1996; Betancourt et al. 1990). Vegetation history of low-latitude regions has been less intensively studied, but data emerging from the Tropics shows substantial late Quaternary change in vegetation composition and species associations (e.g., Bush et al. 1990; Livingstone 1993; Colinvaux et al. 1996). Late Quaternary records of terrestrial fauna show a similar array of population responses and recombinations of species assemblages (Coope 1995; FAUNMAP 1996; Ashworth 1997; Preece 1997), which can also be explained by changes in realized environmental space. Figure 6, for example, provides an explanatory model for mammalian range reversals along latitudinal gradients during the late Quaternary (FAUNMAP 1994, 1996).

To what extent are the individualistic patterns of change observed in terrestrial plants, vertebrates, coleopterans, and gastropods representative of other environments and biotic groups? The answer for any specific group or environment will depend on the nature of the primary environmental factors that influence individuals and populations and, in particular, on how those factors and gradients change through time. Establishment, growth, and reproduction of terrestrial plants are strongly influenced by climate, a dynamic, multivariate entity. The high magnitude and complex nature of late Quaternary climate changes account for the dramatic changes in distribution and abundance of individual plant species, as well as the consequent reorganizations of terrestrial plant communities. Terrestrial animals responded not only to these climate changes but also to consequent changes in vegetation composition and structure.

In contrast, coral reef communities appear

---

[3] We define "Gleasonian" as individualistic behavior of species responding primarily to environmental variables, with stochastic factors of colonization and establishment playing restricted roles in space and time. Gleasonian communities are often confounded with random assemblages of species, evidently stemming from Gleason's (1926) hypothetical examples of identical environments reaching alternative stable communities owing to stochastic differences in colonization. Our definition here is closer to Gleason's own environmentally deterministic view at broader spatial and temporal scales, as indicated by his emphasis on physiological tolerances as constraints to community composition in the 1926 paper, and by his explicit statements in a companion paper (Gleason 1927) that has been largely overlooked.

to have maintained their overall structure through glacial/interglacial cycles, at least in terms of the dominant coral species (Jackson 1992; Pandolfi 1996). Modern coral species array themselves mainly along gradients of water depth and disturbance, and changes at millennial scales have consisted of local spatial shifts in response to sea-level changes, with water-depth zonation patterns conserved (Jackson 1992). The nature of environmental change in these habitats may be similar to that depicted in Figure 2A. The gradient in water depth (x-axis) has remained constant, while other variables (water temperature, salinity) may have increased or decreased. Thus, the realized environmental space has changed relatively little, and the essential community patterns have remained the same.

The coral reef pattern may have terrestrial counterparts. Arrays of species along certain soil-nutrient or hydrological gradients are likely to show similar patterns along those gradients under contrasting climatic regimes. Throughout the late Quaternary, lake-margin gradients in cool-temperate eastern North America have consisted of deep-water submersed aquatics (*Potamogeton, Najas*), shallow-water emergents (*Scirpus, Carex*), and lakeshore trees (*Picea*), whether they were in mixed pine/spruce/oak forest in northern Georgia 21,000 years ago (Watts 1970), in spruce/elm/ash/hornbeam forest in south-central Minnesota 11,000 years ago (Watts and Winter 1966), or in hemlock/hardwood forest in northern Wisconsin today (Curtis 1959). The realized environmental space for variables controlling upland vegetation along latitudinal and elevational gradients has behaved like that depicted in Figures 2C and 2D, while changes in the variables controlling vegetation along lakeshore gradients have been more like Figure 2A.

Some communities that are structured strongly by biotic interactions (e.g., host-specific herbivores, parasites, parasitoids, and mutualists) may be conserved in the face of environmental change. Host-specific organisms are probably under strong selection to develop environmental tolerances similar to those of their host, and in many cases the host's tissues will provide buffering from the climatic environment (except during interhost dispersal). In such cases, the associations may be highly stable through long periods of time, although examples from the Quaternary fossil record are scarce. Also, the short generation times of many host-specific consumers and mutualists may permit rapid evolutionary responses to environmental changes.

The Fundamental Niche in Time

Phylogeographic studies are revealing the impact of Quaternary history on genetic structure of species (Hewitt 1999; Sinclair et al. 1999; Mitton et al. 2000; Terry et al. 2000). We focus on the specific question of evolutionary change in the fundamental niche. To what extent do the shape, size, and position of fundamental niches change at timescales of 10^1–10^5 years?

If a species is to avoid extinction, either it must maintain a fundamental niche sufficiently broad that the fundamental niche and realized environmental space always overlap, or it must be capable of evolutionary changes sufficiently large and rapid to allow the fundamental niche to track the realized environmental space. The latter strategy will work if the evolutionary response time is short compared with the rate of environmental change. Species with relatively long evolutionary response times must track environmental change through spatial displacement rather than evolution (Good 1931; Webb 1987, 1997; Huntley and Webb 1989). Populations of such species will be subject to different selection regimes as the realized environmental space changes, but this selection is likely to be interrupted or redirected by subsequent environmental changes. However, genetic variations lost to drift or to strong selection under an environmental regime will not necessarily reappear even if favorable under a new regime.

At timescales of 10^3–10^4 years, the population mass of a species might be envisioned as sloshing from one portion of the fundamental niche to another as populations track the realized environmental space, with the fundamental niche boundaries remaining stable overall. T. Webb (1987) and Bennett (1990, 1997) have argued that millennial-scale climatic changes obliterate adaptation to local environments at shorter timescales.

Assessment of long-term evolutionary change in niches is difficult. A comparative study of *Fagus* in Europe and North America indicates little divergence in climatic tolerances since the Tertiary (Huntley et al. 1989), suggesting that the fundamental niche for *Fagus* along several climatic axes is constrained phylogenetically (by physiological and developmental pathways, leaf architecture, wood anatomy, etc.). Similar comparative studies are needed for other taxa.

Species that track the environment by spatial displacement must have fundamental niches broad enough to ensure that sites always exist in the realized world where they can establish and maintain populations. Niche breadth can be governed by phenotypic plasticity within genotypes and/or genetic variation among genotypes. A species with a narrow fundamental niche relative to the magnitude of environmental change it experiences is at high risk of eventual extinction. Many species have restricted realized niches (Austin et al. 1990; Thompson et al. 1999), which may result from currently restricted potential niches (masking broader fundamental niches) but may also represent narrow fundamental niches. Narrow fundamental niches may represent low genetic diversity resulting from genetic drift (e.g., Waters and Schaal 1991).

Broad niches may be maintained in some cases because suites of traits adaptive in one environmental setting may be useful in others. For example, many conifers have morphological and physiological traits that impart high carbon-use efficiency, which is advantageous where growing seasons are short and cool (Smith and Brewer 1994; Smith et al. 1997; Eckstein et al. 1999). These same traits may have been advantageous in the low-CO_2 environment of the Last Glacial Maximum (Jackson et al. 2000), which is nowhere replicated on Earth today.

Environmental variability across a range of timescales may also help maintain broad fundamental niches. Genetic diversity within a population can be maintained by environmental heterogeneity, and temporal variation can

have similar effects (Gillespie 1991; Mitton 1997).

Environmental Changes of the Future in the Context of the Past

History is better suited to providing cautionary tales rather than specific images of future climate and vegetation change. Past climate change does not provide strict analogues for future conditions (Webb and Wigley 1985; Crowley 1990; Webb et al. 1993). For example, the total amount, seasonal pattern, and latitudinal distribution of solar energy reaching Earth are expected to remain constant for the foreseeable future. In contrast, concentrations of atmospheric greenhouse gases are likely to rise within the next century beyond any documented in recent Earth history (Fig. 7) (Petit et al. 1999). Thus, the planet is entering a domain of the climate-forcing space that it has never experienced before.

Recent efforts at simulating observed late Quaternary climate changes indicate that, although many aspects of regional climate change (e.g., moisture-related variables) are difficult to simulate accurately, the current generation of predictive climate models are effective in estimating the mean global and hemispheric sensitivity to altered climate forcing (Webb 1998; Joussaume et al. 1999). The paleoclimate data/model comparisons lend confidence to model-based estimates of future sensitivity to altered trace-gas (e.g., CO_2) forcing, which indicate that future global warming will be significant (i.e., 1.5 to 4.5°C by the end of the twenty-first century) (Houghton et al. 1996). Regional climatic shifts could match or exceed the magnitude of those associated with the past 21,000 years (Fig. 12) (Wright et al. 1993; Cuffey and Clow 1997). In many regions, the largest changes occurred over hundreds to thousands of years (e.g., Overpeck et al. 1991), in sharp contrast to the speed at which regional climates could change in the next 100 years.

Future shifts in the realized environmental space are likely to be unprecedented both in terms of climate forcing (i.e., atmospheric greenhouse-gas levels) and in terms of the possible rates and magnitudes of climate change. A comparison of potential future tem-

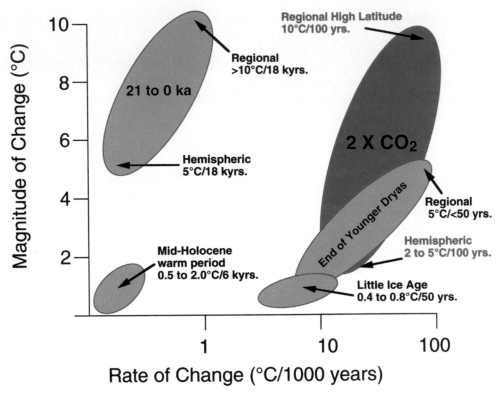

FIGURE 12.   Summary comparison of the rates and magnitudes of possible future climate change (estimated in terms of mean annual temperature) with those associated with several well-known periods of past change in regions that were vegetated. Rates of future regional temperature change could far exceed any widespread change in the late Quaternary. See text for sources, and note that the estimated warming associated with the end of the "Little Ice Age" is currently estimated to be about half of the warming observed over the last 150 years (Overpeck et al. 1997; Mann et al. 1999, 2000).

perature change with changes during well-known periods in the past (Fig. 12) indicates that biota of many regions will be faced with high-magnitude environmental changes more rapid than any experienced in a pre-industrial natural world. Although some past changes (e.g., the warming associated with the end of the Younger Dryas ca. 11,500 years ago) were both rapid and large, they were restricted to small portions of the earth (Rind et al. 1998; Hughen et al. 1996; Fawcett et al. 1997). The time-transgressive mid-Holocene warm period (Wright et al. 1993; Kerwin et al. 1999) was characterized by slow rates of change, and the late Holocene "Little Ice Age" (Overpeck et al. 1997; Mann et al. 1999, 2000) was of relatively small magnitude. Moreover, once biologically important factors other than temperature are considered (e.g., drought variability), it becomes increasingly likely that many of the earth's species will encounter changes sub-stantially different from any they have faced in the course of their evolution. Disruption and fragmentation of the natural landscape by human activities pose additional challenges. Will the mechanisms by which plants and animals have responded to environmental changes of the past be sufficient to ensure their survival in the coming centuries?

Acknowledgments

The niche-based conceptual scheme in this paper was originally presented by S. T. Jackson at the 1990 Ecological Society of America meeting, and subsequently in several seminars and workshops. He thanks the many colleagues who pestered him over the years to publish it. Discussions with T. Webb III played an important role in development of these ideas, and we appreciate his thoughtful and thorough critique of the paper. We also thank S. L. Wing, J. S. Clark, W. A. DiMichele, S. T.

Gray, and J. Norris for comments, and P. J. Bartlein and J. W. Williams for providing climate and paleoclimate data. This research was supported by the National Science Foundation (Ecology and Climate Dynamics Programs) and the National Oceanic and Atmospheric Administration (Paleoclimate Program). We are indebted to S. Stewart, K. H. Anderson, and S. T. Gray for creation of many of the illustrations, J. Keltner and J. W. Williams for help in climate data manipulation, and W. Gross for development of the interactive, three-dimensional web version of Figure 7.

Literature Cited

Adams, C. C. 1902. Postglacial origin and migrations of the life of the northeastern United States. Journal of Geography 1: 303–310, 355–357.

Allen, C. D., and D. D. Breshears. 1998. Drought-induced shift of a forest-woodland ecotone: rapid landscape response to climate variation. Proceedings of the National Academy of Sciences USA 95:14839–14842.

Allen, J. R. M., U. Brandte, A. Brauer, H.-W. Hubberten, B. Huntley, J. Keller, M. Kraml, A. Mackensen, J. Mingram, J. F. W. Negendank, N. R. Nowaczyk, H. Oberhänsli, W. A. Watts, S. Wulf, and B. Zolitschka. 1999. Rapid environmental changes in southern Europe during the last glacial period. Nature 400: 740–743.

Allison, T. D., R. K. Moeller, and M. B. Davis. 1986. Pollen in laminated sediments provides evidence for a mid-Holocene forest pathogen outbreak. Ecology 67:1101–1105.

Anderson, R. S. 1996. Postglacial biogeography of Sierra lodgepole pine (Pinus contorta var. murrayana) in California. Écoscience 3:343–351.

Archer, S. 1989. Have southern Texas savannas been converted to woodlands in recent history? American Naturalist 134:545–561.

Ashworth, A. C. 1997. The response of beetles to Quaternary climate changes. Pp. 119–127 in Huntley et al. 1997.

Austin, M. P. 1990. Community theory and competition in vegetation. Pp. 215–238 in J. B. Grace and D. Tilman, eds. Perspectives on plant competition. Academic Press, San Diego.

———. 1992. Modelling the environmental niche of plants: implications for plant community response to elevated CO_2 levels. Australian Journal of Botany 40:615–630.

Austin, M. P., A. O. Nicholls, and C. R. Margules. 1990. Measurement of the realized quantitative niche: environmental niches of five Eucalyptus species. Ecological Monographs 60: 161–177.

Austin, M. P., J. G. Pausas, and I. R. Noble. 1997. Modelling environmental and temporal niches of eucalypts. Pp. 129–150 in J. E. Williams and J. C. Z. Woinarski, eds. Eucalypt ecology: individuals to ecosystems. Cambridge University Press, Cambridge.

Baker, R. G., E. A. Bettis III, D. P. Schwert, D. G. Horton, C. A. Chumbley, L. A. Gonzalez, and M. K. Reagan. 1996. Holocene paleoenvironments of northeast Iowa. Ecological Monographs 66:203–234.

Bartlein, P. J. 1997. Past environmental changes: characteristic features of Quaternary climate variations. Pp. 11–29 in Huntley et al. 1997.

Bartlein, P. J., I. C. Prentice, and T. Webb III. 1986. Climatic response surfaces from pollen data for some eastern North American taxa. Journal of Biogeography 13:35–57.

Bartlein, P. J., K. H. Anderson, P. M. Anderson, M. E. Edwards, C. J. Mock, R. S. Thompson, R. S. Webb, T. Webb III, and C. Whitlock. 1998. Paleoclimate simulations for North America over the past 21,000 years: features of the simulated climate and comparisons with paleoenvironmental data. Quaternary Science Reviews 17:549–585.

Bennett, K. D. 1985. The spread of Fagus grandifolia across eastern North America during the last 18 000 years. Journal of Biogeography 12:147–164.

———. 1990. Milankovitch cycles and their effects on species in ecological and evolutionary time. Paleobiology 16:11–21.

———. 1997. Evolution and ecology: the pace of life. Cambridge University Press, Cambridge.

Bennett, K. D., P. C. Tzedakis, and K. J. Willis. 1991. Quaternary refugia of north European trees. Journal of Biogeography 18: 103–115.

Berger, A., and M. F. Loutre. 1991. Insolation values for the climate of the last 10 million years. Quaternary Science Reviews 10:297–317.

Bernabo, J. C. 1981. Quantitative estimates of temperature changes over the last 2700 years in Michigan based on pollen data. Quaternary Research 15:143–159.

Betancourt, J. L. 1990. Late Quaternary biogeography of the Colorado Plateau. Pp. 259–292 in Betancourt et al. 1990.

Betancourt, J. L., T. R. Van Devender, and P. S. Martin. 1990a. Synthesis and prospectus. Pp. 435–447 in Betancourt et al. 1990b.

Betancourt, J. L., T. R. Van Devender, and P. S. Martin, eds. 1990b. Packrat middens: the last 40,000 years of biotic change. University of Arizona Press, Tucson.

Betancourt, J. L., W. S. Schuster, J. B. Mitton, and R. S. Anderson. 1991. Fossil and genetic history of a pinyon pine (Pinus edulis) isolate. Ecology 72:1685–1697.

Betancourt, J. L., K. A. Rylander, C. Peñalba, and J. L. McVickar. 2000. Late Quaternary vegetation history of Rough Canyon, south-central New Mexico, USA. Palaeogeography, Palaeoclimatology, Palaeoecology (in press).

Bhiry, N., and L. Filion. 1996. Mid-Holocene hemlock decline in eastern North America linked with phytophagous insect activity. Quaternary Research 45:312–320.

Björkman, L., and R. Bradshaw. 1996. The immigration of Fagus sylvatica L. and Picea abies (L.) Karst. into a natural forest stand in southern Sweden during the last 2000 years. Journal of Biogeography 23:235–244.

Bradley, R. S. 1999. Paleoclimatology: reconstructing climates of the Quaternary, 2d ed. Academic Press, San Diego.

Bradley, R. S., and P. D. Jones. 1992. Climate since A.D. 1500. Routledge, London.

Broecker, W. S. 1997. Thermohaline circulation, the Achilles heel of our climate system: will man-made CO_2 upset the current balance? Science 278:1582–1588.

Brubaker, L. B. 1975. Postglacial forest patterns associated with till and outwash in north-central upper Michigan. Quaternary Research 5:499–527.

Bush, M. B., P. A. Colinvaux, M. C. Wiemann, D. R. Piperno, and K.-B. Liu. 1990. Pleistocene temperature depression and vegetation change in Ecuadorian Amazonia. Quaternary Research 34:330–345.

Campbell, I. D., and J. H. McAndrews. 1993. Forest disequilibrium caused by rapid Little Ice Age cooling. Nature 366:336–338.

Clark, J. S. 1998. Why trees migrate so fast: confronting theory with dispersal biology and the paleorecord. American Naturalist 152:204–224.

Clark, J. S., P. D. Royall, and C. Chumbley. 1996. The role of fire during climate change in an eastern deciduous forest at Devil's Bathtub, New York. Ecology 77:2148–2166.

Clark, J. S., C. L. Fastie, G. Hurtt, S. T. Jackson, W. C. Johnson, G. A. King, M. Lewis, J. Lynch, S. Pacala, I. C. Prentice, G. Schupp, T. Webb III, and P. Wyckoff. 1998. Dispersal theory offers solutions to Reid's paradox of rapid plant migration. BioScience 48:13–24.

Clements, F. E. 1904. The development and structure of vegetation. Contributions from the Botanical Survey of Nebraska, No. 7. University of Nebraska, Lincoln.

———. 1916. Plant succession: an analysis of the development of vegetation. Carnegie Institution of Washington Publication No. 242.

———. 1934. The relict method in dynamic ecology. Journal of Ecology 22:39–68.

COHMAP. 1988. Climatic changes of the last 18,000 years: observations and model simulations. Science 241:1043–1052.

Cole, K. L. 1990. Late Quaternary vegetation gradients through the Grand Canyon. Pp. 240–258 in Betancourt et al. 1990.

Colinvaux, P. A., P. E. De Oliveira, J. E. Moreno, M. C. Miller, and M. B. Bush. 1996. A long pollen record from lowland Amazonia: forest and cooling in glacial times. Science 274:85–88.

Coope, G. R. 1995. Insect faunas in ice age environments: why so little extinction? Pp. 55–74 in J. H. Lawton and R. M. May, eds. Extinction rates. Oxford University Press, Oxford.

Crowley, T. J. 1990. Are there any satisfactory geologic analogs for a future greenhouse warming? Journal of Climate 3:1282–1292.

Crowley, T. J., and G. R. North. 1991. Paleoclimatology. Oxford University Press, New York.

Cuffey, K. M., and G. D. Clow. 1997. Temperature, accumulation, and ice sheet elevation in central Greenland through the last deglacial cycle. Journal of Geophysical Research 102:26, 383–26, 396.

Curtis, J. T. 1959. The vegetation of Wisconsin: an ordination of plant communities. University of Wisconsin Press, Madison.

Cwynar, L. C., and G. M. MacDonald. 1987. Geographical variation in lodgepole pine in relation to population history. American Naturalist 134:668–673.

Darwin, C. 1859. On the origin of species by means of natural selection: or the preservation of favoured races in the struggle for life. John Murray, London.

Davis, M. B. 1976. Pleistocene biogeography of temperate deciduous forests. Geoscience and Man 13:13–26.

———. 1981a. Outbreaks of forest pathogens in Quaternary history. Fourth international conference on palynology, Lucknow (1976–1977) 3:216–228.

———. 1981b. Quaternary history and the stability of forest communities. Pp. 132–154 in D. C. West, H. H. Shugart, and D. B. Botkin, eds. Forest succession: concepts and applications. Springer, New York.

———. 1990. Climatic change and the survival of forest species. Pp. 99–110 in G. M. Woodwell, ed. The earth in transition: patterns and processes of biotic impoverishment. Cambridge University Press, Cambridge.

Davis, M. B., M. W. Schwartz, and K. Woods. 1991. Detecting a species limit from pollen in sediments. Journal of Biogeography 18:653–668.

Davis, M. B., S. Sugita, R. R. Calcote, J. B. Ferrari, and L. E. Frelich. 1994. Historical development of alternate communities in a hemlock-hardwood forest in northern Michigan, USA. Pp. 19–39 in P. J. Edwards, R. May, and N. R. Webb, eds. Large-scale ecology and conservation biology. Blackwell Scientific, Oxford.

Davis, R. B., and G. L. Jacobson Jr. 1985. Late glacial and early Holocene landscapes in northern New England and adjacent areas of Canada. Quaternary Research 23:341–368.

Delcourt, P. A., H. R. Delcourt, R. C. Brister, and L. E. Lackey. 1980. Quaternary vegetation history of the Mississippi Embayment. Quaternary Research 13:111–132.

Diaz, H. F., and V. Markgraf, eds. 2000. El Niño and the Southern Oscillation: multiscale variability and global and regional impacts. Cambridge University Press, Cambridge.

Eckstein, R. L., P. S. Karlsson, and M. Weih. 1999. Leaf life span and nutrient resorption as determinants of plant nutrient conservation in temperate-arctic regions. New Phytologist 143:177–189.

Etheridge, D. M., L. P. Steele, R. L. Langenfelds, R. J. Francey, J.-M. Barnola, and V. I. Morgan. 1996. Natural and anthropogenic changes in atmospheric CO_2 over the last 1000 years from air in Antarctic ice and firn. Journal of Geophysical Research 101(D2):4115–4128.

Fairbanks, R. G. 1989. A 17,000-year glacio-eustatic sea level record: influence of glacial melting rates on the Younger-Dryas event and deep-ocean circulation. Nature 342:637–617.

FAUNMAP Working Group. 1994. FAUNMAP: a database documenting Late Quaternary distributions of mammal species in the United States. Illinois State Museum Scientific Papers 25.

———. 1996. Spatial responses of mammals to Late Quaternary environmental fluctuations. Science 272:1601–1606.

Fawcett, P. J., A. M. Agustsdottir, R. B. Alley, and C. A. Shuman. 1997. The Younger Dryas termination and North Atlantic deep water formation: insights from climate model simulations and Greenland ice cores. Paleoceanography 12:23–28.

Forbes, E. 1846. On the connexion between the distribution of the existing fauna and flora of the British Isles, and the geological changes which have affected their area, especially during the epoch of the northern drift. Memoirs of the Geological Survey of Great Britain 1:336–432.

Foster, D. R., and T. M. Zebryk. 1993. Long-term vegetation dynamics and disturbance history of a Tsuga-dominated forest in New England. Ecology 74:982–998.

Gaudreau, D. C. 1988. Paleoecological interpretation of geographic patterns in pollen data: spruce and birch in northeastern North America. Bulletin of the Buffalo Society of Natural Sciences 33:15–29.

Gaudreau, D. C., S. T. Jackson, and T. Webb III. 1989. Spatial scale and sampling strategy in paleoecological studies of vegetation patterns in mountainous terrain. Acta Botanica Neerlandica 38:369–390.

Gear, A. J., and B. Huntley. 1991. Rapid changes in the range limits of Scots pine 4000 years ago. Science 251:544–547.

Gillespie, J. H. 1991. The causes of molecular evolution. Oxford University Press, Oxford.

Givens, C. R., and F. M. Givens. 1987. Age and significance of fossil white spruce (Picea glauca), Tunica Hills, Louisiana–Mississippi. Quaternary Research 27:283–296.

Gleason, H. A. 1917. The structure and development of the plant association. Bulletin of the Torrey Botanical Club 43:463–481.

———. 1923. The vegetational history of the Middle West. Annals of the Association of American Geographers 12:39–85.

———. 1926. The individualistic concept of the plant association. Bulletin of the Torrey Botanical Club 53:7–26.

———. 1927. Further views on the succession-concept. Ecology 8:299–326.

Good, R. O' D. 1931. A theory of plant geography. New Phytologist 30:149–171.

Graumlich, L. J. 1994. Long-term vegetation change in mountain environments: paleoecological insights into modern vegetation dynamics. Pp. 167–179 in M. Beniston, ed. Mountain environments in changing climates. Routledge, London.

Gray, A. 1858. Diagnostic characters of new species of phenogamous plants, collected in Japan by Charles Wright, Botanist of the U.S. North Pacific Exploring Expedition (Published by request of Captain John Rodgers, Commander of the Expedition). With observations upon the relations of the Japanese flora to that of North America, and of other parts of the Northern Temperate Zone. Memoirs of the American Academy of Arts and Sciences 6:377–452.

Griesemer, J. R. 1992. Niche: historical perspectives. Pp. 231–240 in E. Fox-Keller and E. A. Lloyd, eds. Keywords in evolutionary biology. Harvard University Press, Cambridge.

Hastings, J. R., and R. M. Turner. 1965. The changing mile: an ecological study of vegetation change with time in the lower mile of an arid and semiarid region. University of Arizona Press, Tucson.

Hays, J. D., J. Imbrie, and N. J. Shackleton. 1976. Variations in the Earth's orbit: pacemaker of the ice ages. Science 194:1121–1132.

Hessl, A. E., and W. L. Baker. 1997. Spruce and fir regeneration and climate in the forest-tundra ecotone of Rocky Mountain National Park, Colorado, U.S.A. Arctic and Alpine Research 29:173–183.

Hewitt, G. M. 1999. Post-glacial re-colonization of European biota. Biological Journal of the Linnean Society 68:87–112.

Houghton, J. T., L. G. Meira-Filho, B. A. Callander, N. Harris, A. Kattenberg, and K. Maskell, eds. 1996. Climate change 1995: the science of climate change. Cambridge University Press, Cambridge.

Hughen, K. A., J. T. Overpeck, L. C. Peterson, and S. Trumbore. 1996. Abrupt deglacial climatic change in the tropical Atlantic. Nature 380:51–54.

Hughen, K. A., J. T. Overpeck, S. J. Lehman, M. Kashgarian, L. C. Peterson, and R. Alley. 1998. Deglacial ^{14}C calibration, activity and climate from a marine varve record. Nature 391:65–68.

Hunter, M. L., Jr., G. L. Jacobson Jr., and T. Webb III. 1988. Paleoecology and the coarse-filter approach to maintaining biological diversity. Conservation Biology 2:375–385.

Huntley, B. 1988. Glacial and Holocene vegetation history: Europe. Pp. 341–383 in Huntley and Webb 1988.

———. 1990a. Dissimilarity mapping between fossil and contemporary pollen spectra in Europe for the past 13,000 years. Quaternary Research 33:360–376.

———. 1990b. European post-glacial forests: compositional changes in response to climatic change. Journal of Vegetation Science 1:507–518.

———. 1990c. European vegetation history: palaeovegetation maps from pollen data—13 000 yr BP to present. Journal of Quaternary Science 5:103–122.

———. 1990d. Lessons from climates of the past. Pp. 133–148 in J. Leggett, ed. Global warming: the Greenpeace report. Oxford University Press, Oxford.

———. 1991. How plants respond to climate change: migration rates, individualism and the consequences for plant communities. Annals of Botany 67(Suppl. 1):15–22.

———. 1996. Quaternary palaeoecology and ecology. Quaternary Science Reviews 15:591–606.

———. 1999. The dynamic response of plants to environmental change and the resulting risks of extinction. Pp. 69–85 in G. M. Mace, A. Balmford, and J. R. Ginsberg, eds. Conservation in a changing world. Cambridge University Press, Cambridge.

Huntley, B., and H. J. B. Birks. 1983. An atlas of past and present pollen maps for Europe: 0–13000 years ago. Cambridge University Press, Cambridge.

Huntley, B., and T. Webb III, eds. 1988. Vegetation history. Kluwer Academic, Dordrecht.

———. 1989. Migration: species' response to climatic variations caused by changes in the earth's orbit. Journal of Biogeography 16:5–19.

Huntley, B., P. J. Bartlein, and I. C. Prentice. 1989. Climatic control of the distribution and abundance of beech (Fagus L.) in Europe and North America. Journal of Biogeography 16:551–560.

Huntley, B, W. Cramer, A. V. Morgan, H. C. Prentice, and J. R. M. Allen, eds. 1997. Past and future rapid environmental changes: the spatial and evolutionary responses of terrestrial biota. Springer, New York.

Hutchinson, G. E. 1958. Concluding remarks. Cold Spring Harbor Symposia on Quantitative Biology 22:425–427.

———. 1978. An introduction to population ecology. Yale University Press, New Haven, Conn.

Imbrie, J., E. A. Boyle, S. C. Clemens, A. Duffy, W. R. Howard, G. Kukla, J. Kutzbach, D. G. Martinson, A. McIntyre, A. C. Mix, B. Molfino, J. J. Morley, L. C. Peterson, N. G. Pisias, W. L. Prell, M. E. Raymo, N. J. Shackleton, and J. R. Toggweiler. 1992. On the structure and origin of major glaciation cycles. 1. Linear responses to Milankovitch forcing. Paleoceanography 7:701–738.

Imbrie, J., A. Berger, E. A. Boyle, S. C. Clemens, A. Duffy, W. R. Howard, G. Kukla, J. Kutzbach, D. G. Martinson, A. McIntyre, A. C. Mix, B. Molfino, J. J. Morley, L. C. Peterson, N. G. Pisias, W. L. Prell, M. E. Raymo, N. J. Shackleton, and J. R. Toggweiler. 1993. On the structure and origin of major glaciation cycles. 2. The 100,000-year cycle. Paleoceanography 8:699–735.

Jablonski, D., and J. J. Sepkoski Jr. 1996. Paleobiology, community ecology, and scales of ecological pattern. Ecology 77:1367–1378.

Jackson, J. C. B. 1992. Pleistocene perspectives on coral reef community structure. American Zoologist 32:719–731.

Jackson, S. T. 1989. Postglacial vegetational changes along an elevational gradient in the Adirondack Mountains (New York): a study of plant macrofossils. New York State Museum Bulletin 465.

Jackson, S. T., and C. R. Givens. 1994. Late Wisconsinan vegetation and environment of the Tunica Hills region, Louisiana/Mississippi. Quaternary Research 41:316–325.

Jackson, S. T., and C. Weng. 1999. Late Quaternary extinction of a tree species in eastern North America. Proceedings of the National Academy of Sciences USA 96:13847–13852.

Jackson, S. T., and D. R. Whitehead. 1991. Holocene vegetation patterns in the Adirondack Mountains. Ecology 72:641–653.

Jackson, S. T., J. T. Overpeck, T. Webb III, S. E. Keattch, and K. H. Anderson. 1997. Mapped plant macrofossil and pollen records of Late Quaternary vegetation change in eastern North America. Quaternary Science Reviews 16:1–70.

Jackson, S. T., R. S. Webb, K. H. Anderson, J. T. Overpeck, T. Webb III, J. W. Williams, and B. C. S. Hansen. 2000. Vegetation and environment in eastern North America during the Last Glacial Maximum. Quaternary Science Reviews 19:489–508.

Jacobson, G. L., Jr., and E. C. Grimm. 1986. A numerical analysis of Holocene forest and prairie vegetation in central Minnesota. Ecology 76:958–966.

Jacobson, G. L., Jr., T. Webb III, and E. C. Grimm. 1987. Patterns and rates of vegetation change during the deglaciation of eastern North America. Pp. 277–288 in W. F. Ruddiman and H. E. Wright Jr., eds. North America and adjacent oceans during the last deglaciation. Geology of North America, Vol. K-3. Geological Society of America, Boulder, Colo.

James, F. C., R. F. Johnston, N. O. Wamer, G. J. Niemi, and W. J. Boecklen. 1984. The Grinnellian niche of the wood thrush. American Naturalist 124:17–47.

Johnson, W. C., and T. Webb III. 1989. The role of blue jays (Cyanocitta cristata L.) in the postglacial dispersal of fagaceous trees in eastern North America. Journal of Biogeography 16:561–571.

Joussaume, S., K. E. Taylor, P. Braconnot, J. F. B. Mitchell, J. E. Kutzbach, S. P. Harrison, I. C. Prentice, A. J. Broccoli, A. Abe-Ouchi, P. J. Bartlein, C. Bonfils, B. Dong, J. Guiot, K. Herterich,

C. D. Hewitt, D. Jolly, J. W. Kim, A. Kislov, A. Kitoh, M. F. Loutre, V. Masson, B. McAvaney, N. McFarlane, N. de Noblet, W. R. Peltier, J. Y. Peterschmitt, D. Pollard, D. Rind, J. F. Royer, M. E. Schlesinger, J. Syktus, S. Thompson, P. Valdes, G. Vettoretti, R. S. Webb, and U. Wypputta. 1999. Monsoon changes for 6000 years ago: results of 18 simulations from the Paleoclimate Modeling Intercomparison Project (PMIP). Geophysical Research Letters 26:859–862.

Kerwin, M., J. T. Overpeck, R. S. Webb, A. DeVernal, D. H. Rind, and R. J. Healy. 1999. The role of oceanic forcing in mid-Holocene Northern Hemisphere climatic change. Paleoceanography 14:200–210.

King, G. A., and A. A. Herstrom. 1997. Holocene tree migration rates objectively determined from fossil pollen data. Pp. 91–101 in Huntley et al. 1997.

Kullman, L. 1995. Holocene tree-limit and climate history from the Scandes Mountains, Sweden. Ecology 76:2490–2502.

———. 1996. Norway spruce present in the Scandes Mountains, Sweden at 8000 BP: new light on Holocene tree spread. Global Ecology and Biogeography Letters 5:94–101.

———. 1998a. Non-analogous tree flora in the Scandes Mountains, Sweden, during the early Holocene—macrofossil evidence of rapid geographic spread and response to paleoclimate. Boreas 27:153–161.

———. 1998b. Palaeoecological, biogeographical and palaeoclimatological implications of early Holocene immigration of Larix sibirica Ledeb. into the Scandes Mountains, Sweden. Global Ecology and Biogeography Letters 7:181–188.

———. 1998c. The occurrence of thermophilous trees in the Scandes Mountains during the early Holocene: evidence for a diverse tree flora from macroscopic remains. Journal of Ecology 86:421–428.

Kutzbach, J. E., and T. Webb III. 1991. Late Quaternary climatic and vegetational change in eastern North America: concepts, models, and data. Pp. 175–217 in L. C. K. Shane and E. J. Cushing, eds. Quaternary landscapes. University of Minnesota Press, Minneapolis.

———. 1993. Conceptual basis for understanding Late-Quaternary climates. Pp. 5–11 in H. E. Wright Jr., J. E. Kutzbach, T. Webb III, W. F. Ruddiman, F. A. Street-Perrott, and P. J. Bartlein, eds. 1993. Global climates since the last glacial maximum. University of Minnesota Press, Minneapolis.

Kutzbach, J. E., P. J. Guetter, P. J. Behling, and R. Selin. 1993. Simulated climatic changes: results of the COHMAP climate-model experiments. Pp. 24–93 in H. E. Wright Jr., J. E. Kutzbach, T. Webb III, W. F. Ruddiman, F. A. Street-Perrott, and P. J. Bartlein, eds. 1993. Global climates since the last glacial maximum. University of Minnesota Press, Minneapolis.

Kutzbach, J. E., R. Gallimore, S. Harrison, P. Behling, R. Selin, and F. Laarif. 1998. Climate and biome simulations for the past 21,000 years. Quaternary Science Reviews 17:473–506.

Laird, K. R., S. C. Fritz, E. C. Grimm, and P. G. Mueller. 1996. Century-scale paleoclimatic reconstruction from Moon Lake, a closed-basin lake in the northern Great Plains. Limnology and Oceanography 41:890–902.

Lanner, R. M., and T. R. Van Devender. 1998. The recent history of pinyon pines in the American Southwest. Pp. 171–182 in D. M. Richardson, ed. Ecology and biogeography of Pinus. Cambridge University Press, Cambridge.

Leibold, M. A. 1996. The niche concept revisited: mechanistic models and community context. Ecology 76:1371–1382.

Livingstone, D. A. 1993. Evolution of African climate. Pp. 455–472 in P. Goldblatt, ed. Biological relationships between Africa and South America. Yale University Press, New Haven, Conn.

Lloyd, A. H., and L. J. Graumlich. 1997. Holocene dynamics of treeline forests in the Sierra Nevada. Ecology 78:1199–1210.

Luckman, B. H., G. Holdsworth, and G. D. Osborn. 1993. Neo-

glacial glacier fluctuations in the Canadian Rockies. Quaternary Research 39:144–153.

Lyell, C. 1832. Principles of geology, Vol. 2. John Murray, London.

Lyford, M. E., J. L. Betancourt, and S. T. Jackson. 2000. Holocene vegetation and climate history of the northern Big Horn Basin, Montana. Quaternary Research, in review.

Mack, R. N. 1981. Invasion of Bromus tectorum L. into western North America: an ecological chronicle. Agro-Ecosystems 7:145–165.

———. 1986. Alien plant invasion into the Intermountain West: a case history. Pp. 191–213 in H. A. Mooney and J. A. Drake, eds. Ecology of biological invasions of North America and Hawaii. Springer, New York.

MacPhee, R. D. E., ed. 1999. Extinctions in near time: causes, contexts, and consequences. Plenum, New York.

Maguire, B. 1973. Niche response structure and the analytical potentials of its relationship to the habitat. American Naturalist 107:213–246.

Mann, M. E., R. S. Bradley, and M. K. Hughes. 1999. Northern Hemisphere temperatures during the past millennium: inferences, uncertainties, and limitations. Geophysical Research Letters 26:759–762.

Mann, M. E., E. Gille, R. S. Bradley, M. K. Hughes, J. Overpeck, F. T. Keimig, and W. Gross. 2000. Global temperature patterns in past centuries: an interactive presentation. Earth Interactions (in press).

Martin, P. S., and R. G. Klein, eds. 1984. Quaternary extinctions: a prehistoric revolution. University of Arizona Press, Tucson.

Martinson, D. G., K. Bryan, M. Ghil, M. M. Hall, T. R. Karl, E. S. Sarachik, S. Sorooshian, and L. D. Talley, eds. 1995. Natural climate variability on decade-to-century time scales. National Academy Press, Washington, D.C.

Mast, J. N., T. T. Veblen, and M. E. Hodgson. 1997. Tree invasion within a pine/grassland ecotone: an approach with historic aerial photography and GIS modeling. Forest Ecology and Management 93:181–194.

Mast, J. N., T. T. Veblen, and Y. B. Linhart. 1998. Disturbance and climatic influences on age structure of ponderosa pine at the pine/grassland ecotone, Colorado Front Range. Journal of Biogeography 25:743–755.

McDowell, P. F., T. Webb III, and P. J. Bartlein. 1990. Long-term environmental change. Pp. 143–162 in B. L. Turner II, W. C. Clark, R. W. Kates, J. F. Richards, J. T. Mathews, and W. B. Meyer, eds. The earth as transformed by human action. Cambridge University Press, Cambridge.

Millspaugh, S. H., C. Whitlock, and P. J. Bartlein. 2000. Variations in fire frequency and climate over the past 17 000 yr in central Yellowstone National Park. Geology 28:211–214.

Mitton, J. B. 1997. Selection in natural populations. Oxford University Press, Oxford.

Mitton, J. B., B. R. Kreiser, and R. G. Latta. 2000. Glacial refugia of limber pine (Pinus flexilis James) inferred from the population structure of mitochondrial DNA. Molecular Ecology (in press).

Nowak, C. L., R. S. Nowak, R. J. Tausch, and P. E. Wigand. 1994a. A 30 000 year record of vegetation dynamics at a semi-arid locale in the Great Basin. Journal of Vegetation Science 5:579–590.

———. 1994b. Tree and shrub dynamics in northwestern Great Basin woodland and shrub steppe during the Late-Pleistocene and Holocene. American Journal of Botany 81:265–277.

Overpeck, J. T. 1987. Pollen time series and Holocene climate variability of the midwest United States. Pp. 137–143 in W. H. Berger and L. D. Labeyrie, eds. Abrupt climatic change. D. Reidel, Dordrecht.

———. 1993. The role and response of continental vegetation in the global climate system. Pp. 221–238 in J. A. Eddy and H.

Oeschger, eds. Global changes in the perspective of the past. Wiley, New York.

———. 1995. Paleoclimatology and climate system dynamics. Reviews of Geophysics 33:863–871.

Overpeck, J. T., P. J. Bartlein, and T. Webb III. 1991. Potential magnitude of future vegetation change in eastern North America: comparisons with the past. Science 254:692–695.

Overpeck, J. T., R. S. Webb, and T. Webb III. 1992. Mapping eastern North American vegetation changes of the past 18 ka: no-analogs and the future. Geology 20:1071–1074.

Overpeck, J., K. Hughen, D. Hardy, R. Bradley, R. Case, M. Douglas, B. Finney, K. Gajewski, G. Jacoby, A. Jennings, S. Lamoureux, A. Lasca, G. MacDonald, J. Moore, M. Retelle, S. Smith, A. Wolfe, and G. Zielinski. 1997. Arctic environmental change of the last four centuries. Science 278:1251–1256.

Pandolfi, J. M. 1996. Limited membership in Pleistocene reef coral assemblages from the Huon Peninsula, Papua New Guinea: constancy during global change. Paleobiology 22: 152–176.

Parrish, J. T. 1998. Interpreting pre-Quaternary climate from the geologic record. Columbia University Press, New York.

Parsons, D. J., T. W. Swetnam, and N. L. Christensen, eds. 1999. Historical variability concepts in ecosystem management. Ecological Applications 94:1177–1277.

Peattie, D. C. 1922. The Atlantic Coastal Plain element in the flora of the Great Lakes. Rhodora 24:57–88.

Petit J. R., J.-M. Barnola, I. Basile, M. Bender, J. Chappellaz, M. Davis, G. Delaygue, M. Delmotte, V. M. Kotlyakov, M. Legrand, V. Y. Lipenkov, C. Lorius, L. Pepin, C. Ritz, E. Saltzman, M. Stievenard, J. Jouzel D. Raynaud, and N. I. Barkov. 1999. Climate and atmospheric history of the past 420,000 years from the Vostok ice core, Antarctica. Nature 399:429–436.

Preece, R. C. 1997. The spatial response of non-marine Mollusca to past climate changes. Pp. 163–177 in Huntley et al. 1997.

Premoli, A., S. Chischilly, and J. B. Mitton. 1994. Genetic variation and the establishment of new populations of pinyon pine. Biodiversity and Conservation 3:331–340.

Rehfeldt, G. E., C. C. Ying, D. L. Spittlehouse, and D. A. Hamilton Jr. 1999. Genetic responses to climate for Pinus contorta in British Columbia: niche breadth, climate change, and reforestation. Ecological Monographs 69:375–407.

Rind, D., D. Petcet, W. S. Broecker, A. McIntyre, and W. F. Ruddiman. 1986. Impact of cold North Atlantic sea surface temperatures on climate: implications for the Younger Dryas cooling (11–10k). Climate Dynamics 1:1–33.

Robertson, A. D., J. T. Overpeck, D. Rind, E. Mosley-Thompson, G. A. Zielinski, J. L. Lean, D. Koch, J. E. Penner, I. Tegen, and R. Healy. 2000. Hypothesized climate forcing time series for the last 500 years. Journal of Geophysical Research (in press).

Roy, K., J. W. Valentine, D. Jablonski, and S. M. Kidwell. 1996. Scales of climatic variability and time averaging in Pleistocene biotas: implications for ecology and evolution. Trends in Ecology and Evolution 11:458–463.

Sinclair, W. T., J. D. Morman, and R. A. Ennos. 1999. The postglacial history of Scots pine (Pinus sylvestris) in western Europe: evidence from mitochondrial DNA variation. Molecular Ecology 8:83–88.

Smith, W. K., and C. A. Brewer. 1994. The adaptive importance of shoot and crown architecture in conifer trees. American Naturalist 143:165–169.

Smith, W. K., T. C. Vogelmann, E. H. DeLucia, D. T. Bell, and K. A. Shepherd. 1997. Leaf form and photosynthesis. BioScience 47:785–793.

Spear, R. W., M. B. Davis, and L. C. K. Shane. 1994. Late Quaternary history of low- and mid-elevation vegetation in the White Mountains of New Hampshire. Ecological Monographs 64:85–109.

Steinger, T., C. Körner, and B. Schmid. 1996. Long-term persistence in a changing climate: DNA analysis suggests very old ages of clones of alpine Carex curvula. Oecologia 105:94–99.

Stuart, A. J. 1991. Mammalian extinctions in the Late Pleistocene of Northern Eurasia and North America. Biological Reviews 66:453–562.

Swetnam, T. W., and J. L. Betancourt. 1998. Mesoscale disturbance and ecological response to decadal climatic variability in the American Southwest. Journal of Climate 11:3128–3147.

Swetnam, T. W., C. D. Allen, and J. L. Betancourt. 1999. Applied historical ecology: using the past to manage for the future. Ecological Applications 94:1189–1206.

Szeicz, J. M., and G. M. MacDonald. 1995. Recent white spruce dynamics at the subarctic alpine treeline of north-western Canada. Journal of Ecology 83:873–885.

Taylor, K. C., G. W. Lamorey, G. A. Doyle, R. B. Alley, P. M. Grootes, P. A. Mayewski, J. W. C. White, and L. K. Barlow. 1993. The "flickering switch" of late Pleistocene climate change. Nature 361:432–436.

Terry, R. G., R. S. Nowak, and R. J. Tausch. 2000. Variation in chloroplast and nuclear ribosomal DNA in Utah juniper (Juniperus osteosperma, Cupressaceae): Evidence for interspecific gene flow. American Journal of Botany 87:250–258.

Thompson, R. S. 1988. Western North America. Vegetation dynamics in the western United States: modes of response to climatic fluctuations. Pp. 415–458 in Huntley and Webb 1988.

———. 1990. Late Quaternary vegetation and climate in the Great Basin. Pp. 200–239 in Betancourt et al. 1990.

Thompson, R. S., K. H. Anderson, and P. J. Bartlein. 1999. Atlas of relations between climatic parameters and distributions of important trees and shrubs in North America. United States Geological Survey Professional Paper 1650.

Transeau, E. N. 1903. On the geographic distribution and ecological relations of the bog plant societies of North America. Botanical Gazette 36:401–420.

Valentine, J. W., and D. Jablonski. 1993. Fossil communities: compositional variation at many time scales. Pp. 341–349 in R. E. Ricklefs and D. Schluter, eds. Species diversity in ecological communities: historical and geographical perspectives. University of Chicago Press, Chicago.

Van Devender, T. R. 1990. Late Quaternary vegetation and climate of the Sonoran Desert, United States and Mexico. Pp. 134–163 in Betancourt et al. 1990.

Vose, R. S., R. L. Schmoyer, P. M. Steurer, T. C. Peterson, R. Heim, T. R. Karl, and J. K. Eischeid. 1992. The global historical climatology network: long-term monthly temperature, precipitation, sea level pressure, and station pressure data. Data Set NDP-041, Carbon Dioxide Information Analysis Center, Oak Ridge National Laboratory, Oak Ridge, Tenn.

Waters, E. R., and B. A. Schaal. 1991. No variation is detected in the chloroplast genome of Pinus torreyana. Canadian Journal of Forest Research 21:1832–1835.

Watts, W. A. 1970. The full-glacial vegetation of northwestern Georgia. Ecology 51:17–33.

———. 1988. Europe. Pp. 155–192 in Huntley and Webb 1988.

Watts, W. A., and T. C. Winter. 1966. Plant macrofossils from Kirchner Marsh, Minnesota—a paleoecological study. Geological Society of America Bulletin 77:1339–1360.

Weaver, J. E. 1954. North American prairie. Johnsen, Lincoln, Nebr.

Weaver, J. E., and F. W. Albertson. 1936. Effects of the great drought on the prairies of Iowa, Nebraska, and Kansas. Ecology 17:567–639.

———. 1944. Nature and degree of recovery of grassland from the great drought of 1933 to 1940. Ecological Monographs 14: 393–479.

Webb, S. L. 1987. Beech range extension and vegetation history:

pollen stratigraphy of two Wisconsin lakes. Ecology 68:1993–2005.

Webb, T., III. 1987. The appearance and disappearance of major vegetational assemblages: long-term vegetational dynamics in eastern North America. Vegetatio 69:177–187.

———. 1988. Eastern North America. Pp. 385–414 in Huntley and Webb 1988.

———. 1992. Past changes in vegetation and climate: lessons for the future. Pp. 59–75 in R. L. Peters and T. E. Lovejoy, eds. Global warming and biological diversity. Yale University Press, New Haven, Conn.

———. 1997. Spatial response of plant taxa to climate change: a palaeoecological perspective. Pp. 55–71 in Huntley et al. 1997.

———, ed. 1998. Late Quaternary climates: data synthesis and model experiments. Quaternary Science Reviews 17:463–688.

Webb, T., III, and T. M. L. Wigley. 1985. What past climate climates can indicate about a warmer world. Pp. 237–258 in M. C. MacCracken and F. M. Luther, eds. Projecting the climatic effects of increasing carbon dioxide. U.S. Department of Energy Report DOE/ER-0237. Washington, D.C.

Webb, T., III, T. J. Crowley, B. Frenzel, A.-K. Gliemeroth, J. Jouzel, L. Labeyrie, I. C. Prentice, D. Rind, W. F. Ruddiman, M. Sarntheim, and A. Zwick. 1993. Group report: use of paleoclimatic data as analogs for understanding future global changes. Pp. 51–71 in J. A. Eddy and H. Oeschger, eds. Global changes in the perspective of the past. Wiley, Chichester, England.

Weng, C., and S. T. Jackson. 1999. Late Glacial and Holocene vegetation history and paleoclimate of the Kaibab Plateau, Arizona. Palaeogeography, Palaeoclimatology, Palaeoecology 153:179–201.

West, R. G. 1964. Inter-relations of ecology and Quaternary paleobotany. Journal of Ecology 52(Suppl.):47–57.

———. 1980. Pleistocene forest history in East Anglia. New Phytologist 85:571–622.

Whitlock, C. 1993. Postglacial vegetation and climate of Grand Teton and Yellowstone National Parks. Ecological Monographs 63:173–198.

Whitlock, C., and P. J. Bartlein. 1997. Vegetation and climate change in northwest America during the past 125 kyr. Nature 388:57–61.

Whittaker, R. H., S. A. Levin, and R. B. Root. 1973. Niche, habitat, and ecotope. American Naturalist 107:321–338.

Wight, J. R., and H. J. Fisser. 1968. Juniperus osteosperma in northwestern Wyoming: their distribution and ecology. Science Monograph 6:1–31. Agricultural Experiment Station, University of Wyoming, Laramie.

Willis, K. J. 1994. The vegetational history of the Balkans. Quaternary Science Reviews 13:769–788.

Woodhouse, C. A., and J. T. Overpeck. 1998. 2000 years of drought variability in the central United States. Bulletin of the American Meteorological Society 79:2693–2714.

Woods, K. D., and M. B. Davis. 1989. Paleoecology of range limits: beech in the Upper Peninsula of Michigan. Ecology 70:681–696.

Wright, H. E., Jr., J. E. Kutzbach, T. Webb III, W. F. Ruddiman, F. A. Street-Perrott, and P. J. Bartlein, eds. 1993. Global climates since the last glacial maximum. University of Minnesota Press, Minneapolis.

Glossary of Terms Used in Conceptual Model of Biotic Response to Environmental Change

Environmental Space: An n-dimensional hyperspace delimited by the n environmental variables that are relevant to establishment, growth, survival, and reproduction of individuals of a species. Two-dimensional views of this space are shown in Figures 1 and 2.

Realized Environmental Space: The portion of the total n-dimensional environmental space that is actually represented on Earth (or within a specified region, such as eastern North America) at a given time t (Figs. 1, 2). The realized environmental space can be envisioned as a cloud of points within the boundaries of the environmental space, with each point representing the joint values of the n variables at some point in geographic space at time t. In a two-dimensional environmental space (e.g., Fig. 2), the cloud may appear as an ellipse, circle, blob, or some other planar shape. The point cloud may change shape and position through time as the environment changes at each point in geographic space (Figs. 3, 8).

Empty Environmental Space: The portion of the total environmental space that is not represented by any points on Earth (or within a specified region) at time t. Some portions of environmental space are permanently empty, representing combinations of variables that are physically impossible at Earth's surface (Fig. 1). Other parts of environmental space will alternate between realized and empty environmental space (Figs. 3, 8).

Fundamental Niche: The portion of n-dimensional environmental space that is capable of sustaining populations of a species. This can be visualized as a uniform solid within the n-dimensional environmental space, representing all possible combinations of the n environmental variables that permit establishment, survival, and reproduction of individuals of the species (Fig. 2 provides a two-dimensional representation). Alternatively, it can be perceived as a variable-density solid, with density corresponding to fitness or potential population density. The shape, orientation, and position of the solid are determined by the species' environmental tolerances and requirements, as governed by the genetic attributes of the species (diversity, structure, phenotypic plasticity, etc.), and hence is subject to evolutionary change.

Potential Niche: The portion of environmental space that is capable of supporting populations of a species at time t, defined as the intersection of the fundamental niche for the species with the realized environmental space for time t (Fig. 2). The potential niche will change shape, size, and position within the environmental space as the realized environmental space changes through time (Fig. 4), and as the fundamental niche changes through evolution.

Realized Niche: The portion of n-dimensional environmental space that is actually occupied by populations of a species. The realized niche can be modeled empirically given suitable information on geographic occurrences of the species and associated environmental factors. The realized niche is a subset of the potential niche and may not fill the entire potential niche space owing to dispersal limitations and biotic interactions (Fig. 2).

Life in the last few million years

Jeremy B. C. Jackson and Kenneth G. Johnson

Abstract.—The excellent fossil record of the past few million years, combined with the overwhelming similarity of the biota to extant species, provides an outstanding opportunity for understanding paleoecological and macroevolutionary patterns and processes within a rigorous biological framework. Unfortunately, this potential has not been fully exploited because of lack of well-sampled time series and adequate statistical analysis. Nevertheless, four basic patterns appear to be of general significance. First, a major pulse of extinction occurred 1–2 m.y. ago in many ocean basins, more or less coincident with the intensification of glaciation in the Northern Hemisphere. Rates of origination also increased greatly but were more variable in magnitude and timing. The fine-scale correlation of these evolutionary events with changes in climate is poorly understood. Similar events probably occurred on land but have not been tested adequately. Second, rates of origination and extinction in the oceans waned after the pulse of extinction, especially during the past 1 m.y. Thus, most marine species originated long before the Pleistocene under very different environmental circumstances, suggesting that they are "exapted" rather than adapted to their present ecological circumstances. The same may be true for many terrestrial groups, but not for the mammals or fresh-water fishes that have continued to undergo speciation throughout the Pleistocene. Third, community membership of late Pleistocene coral reef communities was more stable than expected by chance. These are the only paleoecological data adequate to test hypotheses of community stability, so that we do not know whether community structure involving other taxa or environments typically reflects more than the collective behavior of individual species distributions. Regardless, the strong evidence for nearly universal exaptation of ecological characteristics argues strongly against ideas of coevolution of species in communities. Finally, ecological communities were profoundly altered by human activities long before modern ecological studies began. Holocene paleontological, archeological, and historical data constitute the only ecological baseline for "pristine" ecological communities before significant human disturbance. Holocene records should be much more extensively used as a baseline for Recent ecological studies and for conservation and management.

Jeremy B. C. Jackson. Geosciences and Marine Biology Research Divisions, Scripps Institution of Oceanography, University of California, San Diego, La Jolla, California 92093-0244 and Smithsonian Tropical Research Institute, Apartado 2072, Balboa, Republic of Panama. E-mail: jbcj@ucsd.edu
Kenneth G. Johnson. Geosciences Research Division, Scripps Institution of Oceanography, University of California, San Diego, La Jolla, California 92093-0244. E-mail: kenjohnson@ucsd.edu

Accepted: 24 April 2000

Introduction

Understanding paleobiological events of the past few million years should be simple in comparison to earlier times. Biases of preservation are minimal and most species are still alive today or closely related to extant species so that life habits and genetic lineages can be assessed using conventional biological methods (Valentine 1961, 1989; Jackson et al. 1996a; Jackson and Cheetham 1999). Moreover, the global environmental record has been reconstructed in unprecedented detail with excellent stratigraphic control both on land and in the oceans (Shackleton et al. 1984; Webb and Bartlein 1992; Berger and Jansen 1994; Raymo 1994; Kennett 1995; de Menocal 1995). Rapid climatic changes have been accompanied by striking ecological shifts, extinctions, and origination of new species (Stanley 1986a; Crowley and North 1988; Webb and Bartlein 1992; Valentine and Jablonski 1993; Vrba et al. 1995; Jackson et al. 1996a). Thus we have all the components of what should be an ideal scenario for understanding the coupling of environmental and biological changes past, present, and future.

But a closer look reveals large gaps in understanding and widespread confusion of the documentation of patterns with the inference of processes supposedly responsible. Environments and biotas have changed so fast that it is a major task just to document what happened, and this actually has been accomplished in only a few cases (Prentice et al.

   0094-8373/00/2604-0009/$1.00

1991; Overpeck et al. 1992; Budd and Johnson 1999; Cheetham et al. 1999; Jackson et al. 1999). At the same time, inadequate data have not dampened speculation about causes and effects of ecological and evolutionary change, to the point that it is often unclear what testable hypotheses are being proposed. This is particularly true of controversies surrounding hypotheses of turnover pulse (Vrba 1995), co-ordinated stasis (Brett and Baird 1995), and the Gleasonian view of entirely individualistic responses of species (Valentine and Jablonski 1993; Jablonski and Sepkoski 1996). Moreover, all three of these concepts mean increasingly different things to different people (Ivany and Schopf 1996), so there is no common basis for communication.

Here we review some of what we believe we have learned about the biological turmoil of the past few million years. We ask four questions: (1) What are the patterns of taxonomic diversity, origination, and extinction, and are they constant, random, or correlated to produce pulses of evolutionary turnover of taxa? (2) What were the evolutionary effects of the rapidly increasing magnitude and variance of environmental changes of the past 1–2 m.y.? Did rates of speciation or extinction increase, decrease or remain unchanged? (3) Throughout all of the changes observed, is there ecological community unity? That is, do communities exist as more than just the sum of the dynamics of the individual species present? (4) What were the ecological effects of emerging *Homo sapiens*? In each case, we begin with a discussion of marine biota of tropical America that we know best. We then briefly compare these patterns with data for marine and terrestrial biotas elsewhere. We make no pretence of exhaustive coverage but hope to provide an overview of where we stand and the challenges ahead.

Patterns of Diversity, Origination, and Extinction

Caribbean Mollusks and Reef Corals.—Most paleoecological and macroevolutionary studies are based on taxonomic lists for entire faunas and range data for occurrences through space and time, rather than on actual occurrences. The fundamental problem with this

approach is lack of statistical confidence for describing patterns of distribution and abundance without replicate sampling data (Koch 1987; Koch and Morgan 1988; Strauss and Sadler 1989; Marshall 1990; Sepkoski and Koch 1995; Budd et al. 1996; Jackson et al. 1999; Hayek and Bura in press). In fact, use of occurrence data should be straightforward in most cases, because museum labels and collection numbers published in the primary literature provide unbiased criteria for assigning taxa to samples (Bambach 1977; Koch and Sohl 1983; Koch 1995, 1996); but this approach has not been widely used.

We have attempted to address all these issues of quality and consistency of data in our studies of Neogene Caribbean mollusks and reef corals based on collections of the Panama Paleontology Project and earlier work (Saunders et al. 1986; Jackson et al. 1996a; Collins and Coates 1999). Data for fossil mollusks come from 441 collections from the Limón Basin in Costa Rica and from the Bocas del Toro, Canal, and Chuqunaque Basins in Panama. Great effort was made to obtain collections from approximately similar environments (depths), with standardized taxonomy and revised age dating (Collins and Coates 1999; Jackson et al. 1999). Data are for genera and subgenera because most species are undescribed. Names and images of bivalve taxa are available on the Neogene Marine Biota of Tropical America (NMITA) database (http:// nmita.geology.uiowa.edu). Collections of fossils range in age from about 15 to 1 Ma, with the great majority from 11.6 to 1.4 Ma. Recent data are from 30 dredge collections from around Cayos Cochinos in the Gulf of Honduras, Puerto Cabezas in Nicaragua and the western San Blas Archipelago of Panama.

Ranges of molluscan genera and subgenera are shown in order of both stratigraphic appearance and extinction (Fig. 1). Originations appear to occur at a roughly constant rate whereas there is an apparent burst of extinction between 2 and 1 Ma. The same data are broken down into 1-m.y. intervals for numbers of taxa, samples, and first and last occurrences of taxa in Figure 2. Total diversity appears to increase until 3–4 Ma, then levels off until declining during the last 1 m.y. (Fig. 2A).

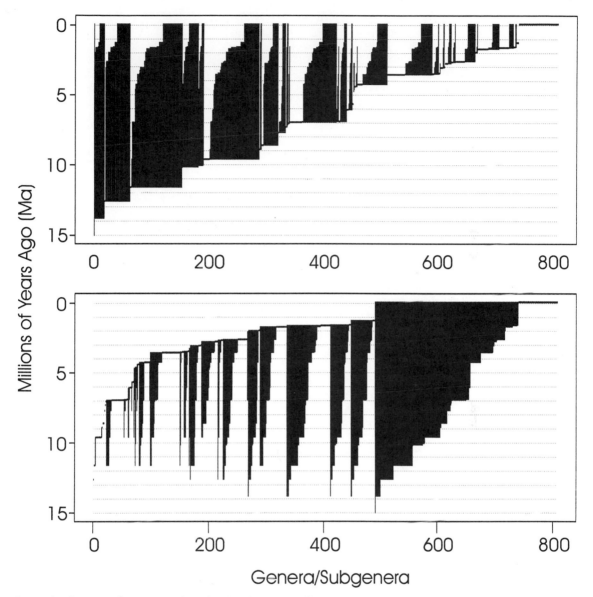

FIGURE 1. Stratigraphic ranges of 808 fossil and Recent molluscan genera and subgenera from 441 collections from the southwestern Caribbean. Ranges were estimated as extending from the midpoint of the estimated age of the sample in which the taxon first occurs to the midpoint of the age of the sample in which the taxon last occurs. In the upper diagram, the taxa are sorted by their ages of origination, and in the lower figure they are sorted by their ages of extinction.

High rates of origination occur sporadically throughout the entire interval sampled, but extinction is concentrated between 4 and 1 Ma (Fig. 2B). However, all these patterns are strongly biased by variations in numbers of samples among intervals and by problems of estimating diversity for intervals at the boundaries of available data (Signor and Lipps 1982).

We therefore evaluated effects of uneven sampling by a permutation procedure that tests the null hypothesis of no significant temporal pattern in origination or extinction (Johnson and McCormick 1999; Budd and Johnson 1999). The procedure is based on the assumption that all taxa potentially range through the entire interval sampled, as if all originations and extinctions were due to in-

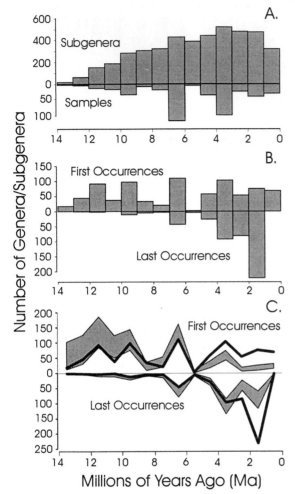

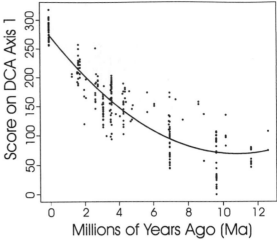

FIGURE 3. Plot of scores from detrended correspondence analysis (DCA) and median geologic ages of 278 molluscan assemblages. Assemblages comprising fewer than 20 taxa were excluded from the analysis. The line shows a second-order polynomial fit to the data ($r^2 = 0.805$, $F = 5689$, $p < 0.0001$, df = 2,275).

FIGURE 2. Patterns of taxonomic turnover of Caribbean mollusks during the past 14 million years. A, Number of genera and subgenera estimated using the range-through method plotted against numbers of samples within 1-m.y. time intervals. B, Numbers of first and last occurrences of genera and subgenera during 1-m.y. intervals. C, Results of a stratigraphic permutation test of the null hypothesis that first and last occurrences of all taxa are due to incomplete sampling. The dark lines indicate the observed numbers of first and last occurrences and the shaded region lies between the tenth and ninetieth percentiles of a test distribution constructed by repeated permutation of sample age assignments. See text for further explanation of this procedure.

complete sampling. The test distribution is generated by repeated estimates of origination and extinction based upon randomized shuffling of the ages of the assemblages while holding constant their taxonomic composition. Rates of origination and extinction are estimated for each of 1000 such permutations to build the test distribution. We accept the null

hypothesis if rates of origination or extinction are not significantly different from random expectation at some high value. However, results for mollusks were highly significant. Numbers of first occurrences exceed null expectation during the last 4 m.y., and there is a very strong pulse of extinction from 2 to 1 Ma (Fig. 2C).

Patterns of association of molluscan genera and subgenera were analyzed using the ordination procedure of detrended correspondence analysis (McCune and Medford 1997) for the 231 collections containing 20 or more taxa. Analyses were made for Recent and fossil collections combined, and for fossil collections only, to test for possible taphonomic differences between fossil and Recent dead shell assemblages (Kidwell and Flessa 1996). In general, collections are arrayed along a single axis (dimension 1) that is strongly correlated with age (Fig. 3). Recent collections were grouped well apart from fossil collections along dimension 1 but still lie well on the line of a second-order polynomial fit to the data. The excellent polynomial fit and considerable overlap of community composition over time suggest gradually escalating ecological reorganization of assemblages. Results were similar when Recent collections were not includ-

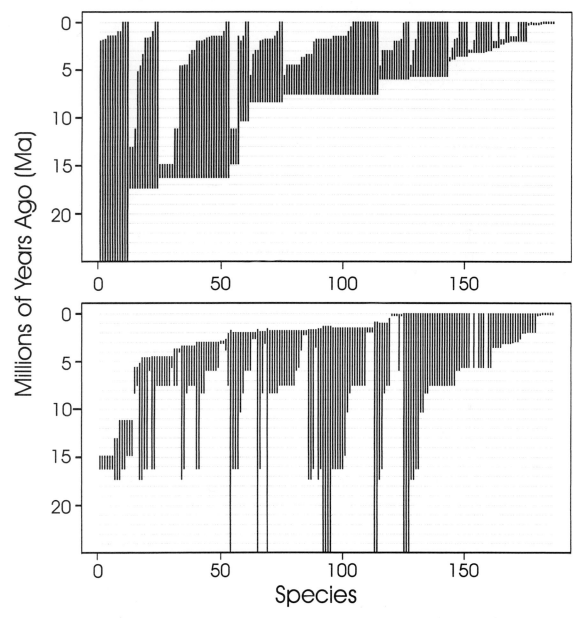

FIGURE 4. Stratigraphic ranges of 186 fossil and Recent Caribbean reef-coral species from 481 collections. Ranges were estimated as for mollusks in Figure 1.

ed, suggesting that taphonomic effects were of minor importance.

Similar data were compiled for the 186 known reef coral species from throughout the Caribbean (Budd and Johnson 1999, unpublished data). Budd and colleagues (Budd et al. 1994, unpublished data) have revised the taxonomy of all the species based on morphometrics; names, distributions, and images are available on the NMITA database. Ranges of

coral species are shown in order of stratigraphic appearance and extinction in Figure 4, and broken down into 1-m.y. intervals for numbers of species, samples, and first and last occurrences in Figure 5. Unlike in mollusks, most originations occurred between 8 and 5 Ma, whereas most extinction occurred between about 4 and 1 Ma, as for the mollusks. Most originations and extinctions occurred after 8–6 Ma, but most of the samples are also

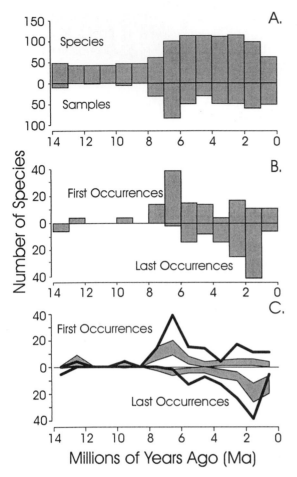

FIGURE 5. Pattern of species turnover among Caribbean reef corals. Explanation as for Figure 2.

TABLE 1. Ages of first appearances of living species of various marine invertebrates.

Taxa	Number of species	Median age (Ma)	Number of species <1 Ma	Number of unfossi- lized species
Reef corals[1]	62	4.3	7	5
Stylopoma[2]	18	6.2	4	4
Puriana[3]	11	3	2	2†
Globigerinids[4]	15	15.3*	0	0
Globorotaliids[4]	16	3.6*	1	0
Pectinids[5]	11	6	0	0

[1] Budd and Johnson 1999, unpublished data; [2] Jackson and Cheetham 1994; [3] Cronin 1987; [4] Stanley et al. 1988; [5] Smith 2000; * ages interpolated from Tables 1 and 2 and Figure 1 in Stanley et al. 1988; † except for Holocene.

Zealand that suggests high rates of origination and extinction toward the end of the Pliocene followed by much lower rates of evolution thereafter (Stanley and Campbell 1981; Stanley 1986a; Beu 1990; Allmon et al. 1993, 1996). In contrast, there was apparently very little extinction of bivalve mollusks along the coasts of California and Japan, except for scallops and shallow burrowing, non-siphonate species (Stanley 1986b; Smith and Roy 1999; Smith 2000).

Terrestrial Biotas.—Rates of origination and extinction of African antelopes, hominids, and the entire mammalian fauna of the Turkana Basin increase greatly from about 3.0 to 1.5 Ma, but so also do the intensity of sampling and numbers of collections (de Menocal 1995; Vrba 1995; Behrensmeyer et al. 1997). Moreover, all the analyses are based on first and last occurrences of species rather than on individual occurrences (i.e., samples), which are necessary to evaluate statistical significance. Thus, we do not know whether or not there was a pulse of mammalian origination or extinction.

How Old Are Species Alive Today?

Rates of speciation and extinction for most marine and terrestrial fauna except mammals decreased during the rapid, large Pleistocene changes in sea level, temperature, and other environmental conditions (Table 1) (Coope 1979, 1994; Stanley 1979; Potts 1984; Jackson 1994a). Consequently, the great majority of extant species originated before the Pleistocene under very different environmental conditions than exist today.

from this time. Nevertheless, the permutation test demonstrates that rates of origination and extinction are significantly higher than expected over this interval, with the greatest peak in origination at 7–5 Ma and of extinction at 2–1 Ma. Thus, increased rates of origination precede and are spread out over a much longer time for coral species than for molluscan genera and subgenera, regardless of similarly strong peaks in extinction between 2 and 1 Ma.

Other Marine Biotas.—Most other studies of diversity, origination, and extinction over the past few million years are based on raw counts of taxa per time interval rather than on individual collections, so that we cannot evaluate adequacy of sampling or statistical significance. Nevertheless, there is considerable evidence for mollusks from Florida and New

Caribbean Invertebrates.—The most extensive data are for reef corals revised by Budd and colleagues (Budd et al. 1994, unpublished data). The median age of first appearance for the 62 extant species in Figure 4 is 4.3 Ma. Only seven of these 62 species (11%) may have originated within the past one million years, and all but two of these are rare or lack a fossil record, so we have no good evidence for when they actually originated. A similar pattern is evident for the cheilostome bryozoan *Stylopoma,* which has undergone extensive late Neogene speciation in the tropical western Atlantic (Jackson and Cheetham 1994; Cheetham and Jackson 1995). The median age of first appearance for the 18 extant species is 6.2 Ma. Four of these 18 species have no fossil record, but the youngest species with a fossil record appeared at least 1.8 m.y. ago. The ostracod *Puriana* shows a similar pattern in tropical America (Cronin 1987). Nine of eleven living species have a fossil record extending back 3–20 m.y. and the other two species are known only from the Holocene and Recent. Combining all these data for corals, bryozoans, and ostracods, 13 of 91 (14.3%) living species may have originated within the past one million years, but only 2 of 80 (2.5%) living species with a fossil record are less than 1 m.y. old.

Other Marine Invertebrates.—Globigerinids and globorotaliids are the most abundant and diverse planktonic foraminifera in Recent seas (Stanley et al. 1988). The median age of first appearance for the 15 extant species of globigerinids is 15.3 Ma and for the 16 extant globorotaliids is 3.6 Ma. Only one globorotaliid species and no globigerinids originated less than 1.0 m.y. ago. Valentine (1989) examined the fossil record of Recent mollusks of California. There were 698 Recent species, of which 538 (77%) are known as fossils. Species unknown as fossils are mostly minute, fragile, or rare, so it is likely that very few species actually originated during the past one million years. For example, the median age of the 11 living scallop species from the same region is 6 Ma and all are known as fossils before 1 Ma (Smith 2000).

Terrestrial Biotas.—The longevity of Recent terrestrial species is much more variable. At one extreme, all of the more than 300 species of cichlids in Lake Victoria originated less than 10 k.y. ago because the lake was dry during the last glacial maximum (Johnson et al. 1996). At the opposite extreme, virtually all early Pleistocene species of beetles are alive today (Coope 1979, 1994; Elias 1994), although we found no data on the earliest fossil occurrences of entire faunas of extant beetles. Among European mammals, "at least 111 modern species appeared during the Pleistocene" (Barnosky 1997: p. 133), an unspecified large percentage of which originated less than 1 m.y. ago. Likewise, 15 of the 32 Recent African antelope species originated less than 1 m.y. ago (Vrba 1995). In contrast, the fossil records of all six extant species of *Equus* (horses and zebras) extend back well before 1 Ma (MacFadden 1992). But horses are a declining clade and antelopes are not.

Community Unity?

The central question in community paleoecology is whether associations of species are a random sample of the available pool of species capable of living in a particular environment, or a predictable subset of those species whose ecological characteristics favor their coexistence (Roughgarden 1989; Jackson 1992). In other words, do predictable associations of species persist in space and time? The short answer to this question is at least sometimes no. There are numerous, well-documented examples of assemblages of pollen, insects, and mammals without modern analogues from the end of the last glacial maximum and early Holocene (Webb 1993; Coope 1994; Elias 1994; FAUNMAP 1996; Jablonski and Sepkoski 1996; Stafford et al. 1999) and from the middle Pleistocene (Barnosky et al. 1996). These data clearly demonstrate that species do not always co-occur in assemblages of similar species composition. The same may be true for late Pleistocene to Recent assemblages of marine mollusks (Valentine and Jablonski 1993; Roy et al. 1993; Jablonski and Sepkoski 1996). However, the data for Recent species are for ranges rather than samples, so that the actual composition of local assemblages is unknown, and there are conflicting ecological interpretations (Lindberg and Lipps 1996).

Compared with climatically more stable

glacial maxima or interglacials (Webb and Bartlein 1992; Kennett 1995), when community composition also may have been less variable (Schopf and Ivany 1998), the end of the last glacial maximum was a time of extremely rapid environmental change. Moreover, the ecologically more interesting question is not whether communities vary in composition, because they all do (Bennington and Bambach 1996), but whether they vary more than expected by chance according to an appropriate null model (Connor and Simberloff 1978, 1979; Pandolfi and Jackson 1997). To our knowledge, this approach has not been attempted for any fossil communities except late Pleistocene coral reefs from New Guinea and the Caribbean (Jackson et al. 1996b; Pandolfi 1996, 1999; Pandolfi and Jackson 1997, in press).

Community Similarity through Time.—Pandolfi (1996, 1999) examined coral community composition on nine late Pleistocene reef terraces on the Huon Peninsula of Papua, New Guinea, that were formed by the interaction of local tectonic uplift and glacial sea-level fluctuations over 95,000 years. A new reef community formed after each uplift event or rise in sea level, so that each terrace records a separate ecological experiment in coral community assembly after a major disturbance. Reef crest and reef slope assemblages were sampled separately at three sites along 35 km of coast. Adequacy of sampling was tested using species effort curves (Hayek and Buzas 1997). There were no significant differences in community composition or species richness within each environment among the nine terraces, but there were persistent local differences along the coast reflecting local differences in environmental conditions.

Community membership also was more stable over the 95,000 years than expected from a random sample of the habitat-specific species pool (i.e., only species known to occur in each specific environment rather than all species from the geographic region). Restriction of the analysis to the habitat-specific species pool is essential because fewer species occur in each local habitat than in the entire faunal province. For example, only 18 species of *Acropora* occur along the Huon Peninsula to-

day, out of the approximately 125 species of *Acropora* in the western Pacific, and only 4 of these 18 are common. These same four *Acropora* species dominate Pandolfi's Pleistocene communities throughout the entire 95,000 years. Reef crest and reef slope communities were consistently dominated by fewer than 25% of the 66–92 species encountered in each environment in the Pleistocene and Recent. These dominant species were no more widely distributed than the rest. Thus, their dominance does not reflect mere differences in dispersal ability, as proposed by Hubbell (1997), and therefore requires some other ecological explanation.

Community Similarity through Space.—The Huon study was based on presence or absence data. However, similar community stability was observed using the same methods for relative abundance data from three different reef environments on the 125-thousand-year-old terrace around Curaçao (Pandolfi and Jackson 1997, in press). Community composition was significantly different among environments, but was significantly more stable than expected by chance over distances as great as 40 km within the same environments. As in New Guinea, coral communities were dominated by a small percentage of the roughly 30 species recorded from each paleocommunity, and species richness was comparable to that in the same environments today. Likewise, dominant species were no more widely distributed than rarer species in the same habitats.

Human Impacts

Mass extinction of large mammals throughout the Americas at the close of the Pleistocene about 10 Ka has been attributed to overkill by Recent human immigrants, the unique climatic events of Wisconsin deglaciation, or some combination of the two (Martin and Klein 1984; Webb and Rancy 1996). The basic problem is that increased human hunting is confounded with climatic change, and that the abrupt climatic changes at that time do not simply repeat those of earlier deglaciations (Berger 1991). However, the strikingly different histories of large mammals in the New World and the Old World are more difficult to explain by global climate change (Azzaroli 1992).

It also seems unlikely that climate change alone would affect only large mammals while other vertebrates were almost unscathed. For example, 51 of the 70 large tropical genera became extinct while smaller tropical taxa were unaffected (Webb and Rancy 1996)—this in the Tropics, where climatic changes were much less than in higher latitudes (Jackson and Budd 1996). At the same time in North America, extinction of fishes, amphibians, and reptiles ranged from 0% to 5.6%, for birds 16.0%, and for all mammals combined 36.8% (Pinsof 1996). Climate change also cannot explain the increasingly well-documented extinctions on islands that are closely correlated with the arrival of human settlers, spread out over many thousands of years so that no single climatic episode could be responsible (Diamond 1991, 1998). The problem deserves more work because the ecological consequences of the extinction were very great (Janzen and Martin 1982; Webb and Rancy 1996). If, as we believe, humans were primarily responsible, then the mass extinction of large mammals at the end of the Pleistocene is the earliest and most dramatic example of the romantic fallacy that primitive peoples lived in ecological harmony with nature.

Much less attention has been paid until very recently to the massive ecological extinction of large marine animals by overfishing and to other profound changes due to human interference that long preceded modern ecological investigations (Simenstad et al. 1978; Fisher 1987; Estes et al. 1989; Cooper and Brush 1991; Brush 1994; Hughes 1994; Dayton et al. 1995, 1998; Gallagher and Carpenter 1997; Jackson 1997; Nixon 1997; Steneck 1997). Some of these changes were due to aboriginal hunting 2–3 k.y. ago, most strikingly in the case of the colonization of the Aleutian Islands and the overfishing of sea otters and other marine mammals (Simenstad et al. 1978). However, most severe changes were due to exploitation for distant markets and to massive deforestation and runoff associated with colonization and agriculture during the past few centuries (Cooper and Brush 1991; Brush 1994; Jackson 1997; Steneck 1997).

An outstanding feature of the effects of humans in the oceans is the time lag between the initial removal of large consumers and the subsequent collapse of coral reefs, sea grasses, kelps, and other large sessile organisms that provide the living habitat structure of many benthic communities (Simenstad et al. 1978; Brush 1994; Hughes 1994; Jackson 1997). Lags of up to several centuries are geologically instantaneous but longer than the entire history of ecological science, and thus predate any opportunity for study in real time. The reasons for the delayed collapse of reefs, sea grasses, and kelps are not well understood. They presumably reflect the very long generation times of many such species and the nonlinear consequences of compounded episodes of disturbance that result in the sudden development of sometimes unrecognizably different alternative communities (Sutherland 1974; Knowlton 1992; Hughes 1994; Paine et al. 1998).

Discussion

For most of the taxa and communities reviewed here, the modern era began with the extinction of large proportions of the marine and terrestrial biota between 2 and 1 Ma. This extinction shaped the composition of living communities more than the origination of new taxa before. Since then, speciation has slackened greatly in the oceans and on the land, except for mammals; but rampant speciation and extinction have continued throughout the Pleistocene in many lakes. Community membership on Pleistocene reefs is much more predictable than expected by chance, but there are no comparable data for any other fossil communities. These paleontological data, along with archeological and historical records, typically provide the only meaningful ecological baseline to examine the effects of humans on marine ecosystems. Each of these assertions is reviewed in more detail below.

Causes of Origination and Extinction.—There is no obvious climatic mechanism for the increased latest Miocene to Pliocene origination of Caribbean mollusks and reef corals. The most likely possibility involves changes in regional current patterns associated with the rise of the Isthmus of Panama, which may have favored speciation by geographic isolation between different parts of the Caribbean as well as the eastern Pacific. In contrast, greatly increased extinction at the end of the Pliocene and beginning of the Pleistocene is correlated with accelerated cooling and changes in primary production associated with the intensification of Northern Hemisphere glaciation (Stanley 1986a; Vermeij and Petuch 1986; Allmon et al. 1993, 1996; Jackson et al. 1993). The correlation is only approximate and the timing of biotic change is not nearly as well constrained as the paleoceanography (Jackson and Budd 1996). Moreover, the strongest shifts in tropical climatic conditions occurred at about 2.5 Ma and 1.0 Ma (Berger and Jansen 1994) and do not closely coincide with major pulses of extinction between 2 and 1 Ma. There are also striking regional differences in extinction that have been attributed to variations in geographic settings that might have restricted migration in response to changes in climate (Stanley 1986a,b).

For terrestrial environments, controversy about the timing and duration of mammalian turnover in Africa and its correlation with the excellent record of African climate change seems moot pending better resolution of the faunal data. The approximately contemporaneous burst of mammalian evolution in the Americas was associated with the mingling of faunas across the newly emergent Isthmus of Panama and their subsequent rapid radiation and extinction in new habitats populated by new neighbors (Stehli and Webb 1985; Webb and Rancy 1996). Thus, any signal of turnover due to climate is confounded by biogeographic change.

Causes of Evolutionary Stability.—The taxonomy of the Caribbean corals and *Stylopoma* has been reevaluated using morphometrics and a variety of molecular methods (Jackson and Cheetham 1994; Knowlton and Budd in press). Newly discovered cryptic species are as old as all the rest, so that the great median longevity of extant species is unlikely to be an artifact of lumping cryptic species. In contrast, we know less about the correspondence between morphospecies and biological species of planktonic foraminifera, so that cryptic species may be common and species not so long

lived (Norris et al. 1996; Huber et al. 1997). Nevertheless, whatever caused increased speciation and extinction toward the end of the Pliocene appears to have filtered out the great majority of species vulnerable to the subsequent seesaw climatic changes of the Pleistocene, so that rates of both speciation and extinction declined dramatically thereafter (Valentine and Jablonski 1991; Jackson 1994a). It is also possible that rapid environmental change precluded geographic isolation for sufficiently long periods for speciation to occur (Potts 1984).

Nothing like the extremely rapid and recent speciation of cichlids in African lakes (Meyer 1993) is known for any marine taxa. Molecular data suggest that the great majority of marine fishes originated earlier than 1 Ma (e.g., Johns and Avise 1998), although fishes may indeed evolve more rapidly than invertebrates in the same environments (Stanley 1979; Knowlton in press). Extant species of beetles and some terrestrial mammals are as old as marine species, but antelopes clearly are not. We do not know why some mammalian clades evolved so rapidly during the past one million years, although behavioral as well as environmental factors were almost certainly of considerable importance (Stanley 1979, 1996; Wood and Brooks 1999).

Causes and Implications of Coral Reef Community Stability? The paleontological data demonstrate predictable patterns of reef coral community membership over large temporal and spatial scales before intense human disturbance (Jackson 1992; Pandolfi and Jackson 1997, in press). Natural disturbances, such as hurricanes and outbreaks of predators, followed by subsequent processes of recovery, alter coral community composition on a scale of decades to centuries (Connell 1978; Jackson 1991; Done 1992, 1997, 1999; Connell et al. 1997). However, growth rates of acroporid corals are so fast that, on average, competitive dominance occurs faster than the average interval between extreme disturbances (Jackson 1992; Aronson and Precht 1997; Aronson et al. 1998; Greenstein et al. 1998). In contrast, other dominant corals such as *Montastraea* and *Diploria* simply are more resistant to disturbance (Woodley et al. 1981; Hughes and Jackson

1985). Thus, the outcomes of disturbance, persistence, and recovery were sufficiently predictable over large scales of the reefscape for the development of predictable coral community composition, just as has been demonstrated for intertidal communities on rocky shores (Paine and Levin 1981).

It cannot be overemphasized that predictability of late Pleistocene reef community composition does not imply local communities tightly integrated by coevolution or strong ecological interaction (Bambach and Bennington 1996; Jablonski and Sepkoski 1996; Aronson and Precht 1997). To the contrary, extant species of *Acropora* and *Montastraea* originated during the middle to late Pliocene under very different environmental conditions from those they came to dominate in the Pleistocene (Jackson 1994b; Budd and Johnson 1999; Knowlton and Budd in press). Thus, especially if coral species originate by punctuated patterns of speciation like most other benthic marine invertebrates (Jackson and Cheetham 1999), their apparent adaptations for dominance on Pleistocene and Recent reefs are really accidents of exaptation to new physical and biological environments.

Community composition in temperate marine environments may be less stable than for coral reefs (Valentine and Jablonski 1993; Jablonski and Sepkoski 1996), if for no other reason than that species living near the equator have nowhere warmer to migrate during glacial maxima, whereas temperate species may migrate toward the Tropics when climates cool. However, without proper sampling we cannot know whether community composition is random or more stable than expected by chance.

The demonstration of predictable composition and assembly of late Pleistocene coral reef communities provides a baseline for comparison of the effects of subsequent human disturbance to coral reefs (Pandolfi and Jackson 1997, in press). Other late Pleistocene and Holocene deposits offer similar opportunities for a wide variety of marine benthic and planktonic communities. High-resolution sedimentary records (Baumgartner et al. 1992; Lange et al. 1997), uplifted or otherwise isolated deposits (Stemann and Johnson 1992; Pandolfi

1996), and archeological remains (Simenstad et al. 1978) are particularly promising because of better stratigraphic control and minimal effects of mixing up deposits from before and after human disturbance. However, with appropriate precautions, time-averaged assemblages can also produce important baseline information (Kidwell and Flessa 1996; Kowalewski et al. 1998). The combination of paleontological, archeological, and historical records is the only means of obtaining time series sufficiently long to resolve and understand natural versus anthropogenic sources of variability in marine ecosystems. Paleontologists can contribute greatly to this central problem of marine conservation and management.

Acknowledgments

Collections and stratigraphy of the Dominican Republic and Panama Paleontology Projects are the primary basis of this paper. Most of the molluscan data were compiled by J. Todd, A. Heitz, P. Jung, and M. Alvarez and standardized taxonomically by J. Todd. A. Budd provided the coral data. R. Bambach and R. Aronson reviewed the paper and provided useful comments. This research was supported by grants from the Kuglerfonds of the Naturhistorisches Museum of Basel, National Geographic Society, Scholarly Studies and Walcott Funds of the Smithsonian Institution, Schweizerischer Nationalfonds Forschung grants 21-36589.92 and 20-43229.95, National Science Foundation grants BSR90-06523, DEB-9300905, DEB-9696123, and DEB-9705289, the Naturhistorisches Museum of Basel, and the Smithsonian Tropical Research Institute.

Literature Cited

Allmon, W. D., G. Rosenberg, R. W. Portell, and K. S. Schindler. 1993. Diversity of Atlantic coastal plain mollusks since the Pliocene. Science 260.1626–1629.

———. 1996. Diversity of Pliocene-Recent mollusks in the Western Atlantic: extinction, origination, and environmental change. Pp. 271–302 in Jackson et al. 1996a.

Aronson, R. B., and W. F. Precht. 1997. Stasis, biological disturbance, and community structure of a Holocene coral reef. Paleobiology 23:326–346.

Aronson, R. B., W. F. Precht, and I. G. Macintyre. 1998. Extrinsic control of species replacement on a Holocene reef in Belize: the role of coral disease. Coral Reefs 17:223–230.

Azzaroli, A. 1992. Ascent and decline of monodactyl equids: a case for prehistoric overkill. Annals Zoologica Fennici 28:151–163.

Bambach, R. K. 1977. Species richness in marine benthic environments through the Phanerozoic. Paleobiology 3:152–167.

Bambach, R. K., and J. B. Bennington. 1996. Statistical testing for paleocommunity recurrence: are similar fossil assemblages ever the same? In L. C. Ivany and K. M. Schopf, eds. New perspectives on faunal stability in the fossil record. Palaeogeography, Palaeoclimatology, Palaeoecology 127:107–133.

Barnosky, A. D. 1997. Punctuated equilibrium and phyletic gradualism: some facts from the Quaternary mammalian record. Pp. 109–147 in H. H. Genoways, ed. Current mammalogy, Vol. 1. Plenum, New York.

Barnosky, A. D., T. I. Rouse, E. A. Hadley, D. L. Wood, F. L. Keesing, and V. A. Schmidt. 1996. Comparison of mammalian response to glacial–interglacial transitions in the middle and late Pleistocene. Pp. 16–33 in K. M. Stewart and K. L. Seymour, eds. Palaeoecology and palaeoenvironments of late Cenozoic mammals. University of Toronto Press, Toronto.

Baumgartner, T. R., A. Soutar, and V. Ferreira-Bartrina. 1992. Reconstruction of the history of northern anchovy populations over the past two millennia from sediments of the Santa Barbara Basin, California. California Cooperative Oceanic Fisheries Investigations Reports 33:24–40.

Behrensmeyer, A. K., N. E. Todd, R. Potts, and G. E. McBrinn. 1997. Late Pliocene faunal turnover in the Turkana Basin, Kenya and Ethiopia. Science 278:1589–1594.

Bennington, J. B., and R. K. Bambach. 1996. Statistical testing for paleocommunity recurrence: are fossil assemblages ever the same? Palaeogeography, Palaeoclimatology, Palaeoecology 127:83–106.

Berger, W. H. 1991. On the extinction of the mammoth: science and myth. Pp. 115–132 in D. W. Müller, J. A. McKenzie, and H. Weissert, eds. Controversies in modern geology. Academic Press, London.

Berger, W. H., and E. Jansen. 1994. Mid-Pleistocene climate shift—the Nansen connection. American Geophysical Union Monograph 84:295–311.

Beu, A. G. 1990. Molluscan generic diversity of New Zealand Neogene stages: extinction and biostratigraphic events. Palaeogeography, Palaeoclimatology, Palaeoecology 77:279–288.

Brett, C. E., and G. C. Baird. 1995. Coordinated stasis and evolutionary ecology of Silurian to Middle Devonian faunas in the Appalachian Basin. Pp. 285–315 in D. H. Erwin and R. L. Anstey, eds. New approaches to speciation in the fossil record. Columbia University Press, New York.

Brush, G. S. 1994. Case 2: the Chesapeake Bay estuarine system. Pp. 398–416 in N. Roberts, ed. The changing global environment. Blackwell, Oxford.

Budd, A. F., and K. G. Johnson. 1999. Origination preceding extinction during late Cenozoic turnover of Caribbean reefs. Paleobiology 25:188–200.

Budd, A. F., T. A. Stemann, and K. G. Johnson. 1994. Stratigraphic distributions of genera and species of Neogene to Recent Caribbean reef corals. Journal of Paleontology 68:951–977.

Budd, A. F., K. G. Johnson, and T. F. Stemann. 1996. Plio-Pleistocene turnover and extinctions in the Caribbean reef-coral fauna. Pp. 168–204 in Jackson et al. 1996a.

Cheetham, A. H., and J. B. C. Jackson. 1995. Process from pattern: tests for selection versus random change in punctuated bryozoan speciation. Pp. 184–207 in D. H. Erwin and R. L. Anstey, eds. 1995. New approaches to speciation in the fossil record. Columbia University Press, New York.

Cheetham, A. H., J. B. C. Jackson, J. Sanner, and Y. Ventocilla. 1999. Neogene and Quaternary cheilostome Bryozoa of tropical America: comparison and contrast between the Central American isthmus (Panama, Costa Rica) and the north-central Caribbean (Dominican Republic). In L. Collins and A. G. Coates, eds. The Neogene of the Isthmus of Panama: a paleobiotic survey of the Caribbean coast. Bulletins of American Paleontology 357:159–192.

Collins, L., and A. G. Coates, eds. 1999. The Neogene of the Isthmus of Panama: a paleobiotic survey of the Caribbean coast. Bulletins of American Paleontology No. 357.

Connell, J. H. 1978. Diversity in tropical rain forests and coral reefs. Science 199:1302–1310.

Connell, J. H., T. P. Hughes, and C. C. Wallace. 1997. A 30-year study of coral abundance, recruitment, and disturbance at several scales in space and time. Ecological Monographs 67:461–488.

Connor, E. F., and D. S. Simberloff. 1978. Species number and compositional similarity of the Galápagos flora and avifauna. Ecological Monographs 48:219–248.

———. 1979. The assembly of species communities: chance or competition? Ecology 60:1132–1140.

Coope, G. R. 1979. Late Cenozoic fossil Coleoptera: evolution, biogeography, and ecology. Annual Review of Ecology and Systematics 10:247–267.

———. 1994. Insect faunas in Ice Age environments: why so little extinction? Pp. 55–74 in J. H. Lawton and R. M. May, eds. Extinction rates. Oxford University Press, Oxford.

Cooper, S. R., and G. S. Brush. 1991. A 2,500-year history of anoxia and eutrophication in Chesapeake Bay. Estuaries 16:617–626.

Cronin, T. M. 1987. Evolution, biogeography, and systematics of Puriana: evolution and speciation in Ostracoda. III. Paleontological Society Memoir 21. Journal of Paleontology 61(Suppl.):1–71.

Crowley, T. J., and G. R. North. 1988. Abrupt climate change and extinction events in earth history. Science 240:996–1002.

Dayton, P. K., S. F. Thrush, M. T. Agardy, and R. J. Hofman. 1995. Viewpoint. Environmental effects of marine fishing. Aquatic Conservation: Marine and Freshwater Ecosystems 5:205–232.

Dayton, P. K., M. J. Tegner, P. B. Edwards, and K. L. Riser. 1998. Sliding baselines, ghosts, and reduced expectations in kelp forest communities. Ecological Applications 8:309–322.

de Menocal, P. B. 1995. Plio-Pleistocene African climate. Science 270:53–59.

Diamond, J. M. 1991. The rise and fall of the third chimpanzee. Vintage, London.

———. 1998. Guns, germs, and steel: the fates of human societies. Norton, New York.

Done, T. 1992. Constancy and change in some Great Barrier Reef coral communities. American Zoologist 32:655–662.

———. 1997. Decadal changes in reef-building communities: implications for reef growth and monitoring programs. Proceedings of the Eighth International Coral Reef Symposium, Panamá 1:411–416.

———. 1999. Coral community adaptability to environmental changes at the scales of regions, reefs, and reef zones. American Zoologist 39:66–79.

Elias, S. A. 1994. Quaternary insects and their environments. Smithsonian Institution Press, Washington, D.C.

Estes, J. A., D. O. Duggins, and G. B. Rathbun. 1989. The ecology of extinctions in kelp forest communities. Conservation Biology 3:252–264.

FAUNMAP Working Group. 1996. Spatial responses of mammals to late Quaternary environmental fluctuations. Science 272:1601–1606.

Fisher, D. C. 1987. Mastodont procurement by Paleoindians of the Great Lakes region: hunting or scavenging? Pp. 309–421 in M. H. Nitecki and D. V. Nitecki, eds. The evolution of human hunting. Plenum, New York.

Gallagher, R., and B. Carpenter. 1997. Human-dominated ecosystems. Science 277:485.

Greenstein, B. J., H. A. Curran, and J. M. Pandolfi. 1998. Shifting ecological baselines and the demise of *Acropora cervicornis* in the western North Atlantic and Caribbean Province: a Pleistocene perspective. Coral Reefs 17:249–261.

Hayek, L. C., and E. Bura. In press. On the ends of the taxon range problem. *In* J. B. C. Jackson, S. Lidgard, and F. K. McKinney, eds. Process from pattern in the fossil record. University of Chicago Press, Chicago.

Hayek, L. C., and M. A. Buzas. 1997. Surveying natural populations. Columbia University Press, New York.

Hubbell, S. P. 1997. Niche assembly, dispersal limitation, and the maintenance of diversity in tropical tree communities and coral reefs. Proceedings of the Eighth International Coral Reef Symposium, Panamá 1:387–396.

Huber, B. T., J. Bijma, and K. Darling. 1997. Cryptic speciation in the living planktonic foraminifer *Globigerinella siphonifera* (d'Orbigny). Paleobiology 23:33–62.

Hughes, T. P. 1994. Catastrophes, phase shifts, and large-scale degradation of a Caribbean coral reef. Science 265:1547–1551.

Hughes, T. P., and J. B. C. Jackson. 1985. Population dynamics and life histories of foliaceous corals. Ecological Monographs 55:141–166.

Ivany, L. C., and K. M. Schopf, eds. 1996. New perspectives on faunal stability in the fossil record. Palaeogeography, Palaeoclimatology, Palaeoecology 127.

Jablonski, D., and J. J. Sepkoski Jr. 1996. Paleobiology, community ecology, and scales of ecological pattern. Ecology 77: 1367–1378.

Jackson, J. B. C. 1991. Adaptation and diversity of reef corals. BioScience 41:475–482.

———. 1992. Pleistocene perspectives on coral reef community structure. American Zoologist 32:719–731.

———. 1994a. Constancy and change of life in the sea. Philosophical Transactions of the Royal Society of London B 344: 55–60.

———. 1994b. Community unity? Science 264:1412–1413.

———. 1997. Reefs since Columbus. Coral Reefs 16(Suppl.): S23–S32.

Jackson, J. B. C., and A. F. Budd. 1996. Evolution and environment: introduction and overview. Pp. 1–20 *in* Jackson et al. 1996a.

Jackson, J. B. C., and A. H. Cheetham. 1994. Phylogeny reconstruction and the tempo of speciation in cheilostome Bryozoa. Paleobiology 20:407–423.

———. 1999. Tempo and mode of speciation in the sea. Trends in Ecology and Evolution 14:72–77.

Jackson, J. B. C., P. Jung, A. G. Coates, and L. S. Collins. 1993. Diversity and extinction of tropical American mollusks and the emergence of the Isthmus of Panama. Science 260:1624–1626.

Jackson, J. B. C., A. F. Budd, and A. G. Coates, eds. 1996a. Evolution and environment in tropical America. University of Chicago Press, Chicago.

Jackson, J. B. C., A. F. Budd, and J. M. Pandolfi. 1996b. The shifting balance of natural communities? Pp. 89–122 *in* D. Jablonski, D. H. Erwin, and J. H. Lipps, eds. Evolutionary paleobiology: in honor of James W. Valentine. University of Chicago Press, Chicago.

Jackson, J. B. C., J. A. Todd, H. Fortunato, and P. Jung. 1999. Diversity and assemblages of Neogene Caribbean Mollusca of lower Central America. *In* L. S. Collins and A. G. Coates, eds. The Neogene of the Isthmus of Panama: a paleobiotic survey of the Caribbean coast. Bulletins of American Paleontology 357:193–230.

Janzen, D. H., and P. S. Martin. 1982. Neotropical anachronisms: the fruits the gomphotheres ate. Science 215:19–27.

Johns, G. C., and J. C. Avise. 1998. Tests for ancient species flocks based on molecular phylogenetic appraisals of *Sebastes* rockfishes and other marine fishes. Evolution 52:1135–1146.

Johnson, K. G., and T. McCormick. 1999. The quantitative description of faunal change using paleontological databases. Pp. 227–247 *in* D. Harper, ed. Numerical palaeobiology: computer-based modelling and analysis of fossils and their distributions. Wiley, Chichester, England.

Johnson, T. C., C. A. Scholz, M. R. Talbot, K. Kelts, R. D. Ricketts, G. Ngobi, K. Buening, I. Ssemmanda, and J. W. McGill. 1996. Late Pleistocene desiccation of Lake Victoria and rapid evolution of cichlid fishes. Science 273:1091–1093.

Kennett, J. P. 1995. A review of polar climatic evolution during the Neogene, based on the marine sediment record. Pp. 49–64 *in* E. S. Vrba, G. H. Denton, T. C. Partridge, and L. H. Burckle, eds. Paleoclimate and evolution, with emphasis on human origins. Yale University Press, New Haven, Conn.

Kidwell, S. M., and K. W. Flessa. 1996. The quality of the fossil record: populations, species, and communities. Annual Review of Earth and Planetary Science 24:433–464.

Knowlton, N. 1992. Thresholds and multiple stable states in coral reef community dynamics. American Zoologist 32:674–682.

———. In press. Molecular genetic analyses of species boundaries in the sea. Hydrobiologia.

Knowlton, N., and A. F. Budd. In press. Recognizing coral species past and present. *In* J. B. C. Jackson, S. Lidgard, and F. K. McKinney, eds. Process from pattern in the fossil record. University of Chicago Press, Chicago.

Koch, C. F. 1987. Prediction of sample size effects on the measured temporal and geographic distribution patterns of species. Paleobiology 13:100–107.

———. 1995. Sampling effects, species-sediment relationships, and observed geographic distribution: an uppermost Cretaceous bivalve example. Geobios Mémoire Spécial 18:237–241.

———. 1996. Latest Cretaceous mollusc species 'fabric' of the US Atlantic and Gulf coastal plain: a baseline for measuring biotic recovery. *In* M. B. Hart, ed. Biotic recovery from mass extinctions. Geological Society of London Special Publication 102:309–317.

Koch, C. F., and J. P. Morgan. 1988. On the expected distribution of species' ranges. Paleobiology 14:126–138.

Koch, C. F., and N. F. Sohl. 1983. Preservational effects in paleoecological studies: cretaceous mollusc examples. Paleobiology 9:26–34.

Kowalewski, M., G. A. Goodfriend, and K. W. Flessa. 1998. High-resolution estimates of temporal mixing within shell beds: the evils and virtues of time-averaging. Paleobiology 24. 287–304.

Lange, C. B., A. L. Wernheimer, F. M. H. Reid, and R. C. Thunell. 1997. Sedimentation patterns of diatoms in Santa Barbara Basin, California. California Cooperative Oceanic Fishery Investigation Reports 38:161–170.

Lindberg, D. R., and J. H. Lipps. 1996. Reading the chronicle of Quaternary temperate rocky shore faunas. Pp. 161–184 *in* D. Jablonski, D. H. Erwin, and J. H. Lipps, eds. Evolutionary paleobiology: in honor of James W. Valentine. University of Chicago Press, Chicago.

MacFadden, B. J. 1992. Fossil horses: systematics, paleobiology, and evolution of the Family Equidae. Cambridge University Press, Cambridge.

Marshall, C. R. 1990. Confidence intervals on stratigraphic ranges. Paleobiology 16:1–10.

Martin, P. S., and R. G. Klein, eds. 1984. Quaternary extinctions: a prehistoric revolution. University of Arizona Press, Tucson.

McCune, B., and M. J. Medford. 1997. PC-ORD, Multivariate analysis of ecological data, Version 3.0. MjM Software Design, Gleneden Beach, Ore.

Meyer, A. 1993. Phylogenetic relationships and evolutionary processes in East African cichlid fishes. Trends in Ecology and Evolution 8:279–284.

Nixon, S. W. 1997. Prehistoric nutrient inputs and productivity in Narragansett Bay. Estuaries 20:253–261.

Norris, R. D., R. M. Corfield, and J. Cartlidge. 1996. What is gradualism? Cryptic speciation in globorotaliid foraminifera. Paleobiology 22:386–405.

Overpeck, J. T., R. S. Webb, and T. Webb III. 1992. Mapping eastern North American vegetation change of the past 18 ka: no-analogs and the future. Geology 20:1071–1074.

Paine, R. T., and S. A. Levin. 1981. Intertidal landscapes: disturbance and the dynamics of pattern. Ecological Monographs 51:145–178.

Paine, R. T., M. J. Tegner, and E. A. Johnson. 1998. Compounded perturbations yield ecological surprises. Ecosystems 1:535–545.

Pandolfi, J. M. 1996. Limited membership in Pleistocene reef coral assemblages from the Huon Peninsula, Papua New Guinea: constancy during global change. Paleobiology 22: 152–176.

———. 1999. Response of Pleistocene coral reefs to environmental change over long temporal scales. American Zoologist 39:113–130.

Pandolfi, J. M., and J. B. C. Jackson. 1997. The maintenance of diversity on coral reefs: examples from the fossil record. Proceedings of the Eighth International Coral Reef Symposium, Panamá 1:397–404.

———. In press. Community structure of Pleistocene coral reefs of Curaçao, Netherlands Antilles. Ecology.

Pinsof, J. D. 1996. Current status of North American Sangamonian local faunas and vertebrate taxa. Pp. 156–190 in K. M. Stewart and K. L. Seymour, eds. Palaeoecology and palaeoenvironments of late Cenozoic mammals. University of Toronto Press, Toronto.

Potts, D. C. 1984. Generation times and the Quaternary evolution of reef-building corals. Paleobiology 10:48–58.

Prentice, I. C., P. J. Bartlein, and T. Webb III. 1991. Vegetation and climatic change in eastern North America since the last glacial maximum. Ecology 72:2038–2056.

Raymo, M. E. 1994. The initiation of Northern Hemisphere glaciation. Annual Review of Earth and Planetary Science 22: 353–383.

Roughgarden, J. 1989. The structure and assembly of communities. Pp. 203–226 in J. Roughgarden, R. M. May, and S. A. Levin, eds. Perspectives in ecological theory. Princeton University Press, Princeton, N.J.

Roy, K., D. Jablonski, and J. W. Valentine. 1993. Thermally anomalous assemblages revisited: patterns in the extraprovincial latitudinal range shifts of Pleistocene marine mollusks. Geology 23:1071–1074.

Saunders, J. B., P. Jung, and B. Biju-Duval. 1986. Neogene paleontology in the northern Dominican Republic. 1. Field surveys, lithology, environment, and age. Bulletins of American Paleontology 89:1–79.

Schopf, K. M., and L. C. Ivany. 1998. Scaling the ecosystem: a hierarchical view of stasis and change. Pp. 187–211 in M. L. McKinney and J. A. Drake, eds. Biodiversity dynamics: turnover of populations, taxa, and communities. Columbia University Press, New York.

Sepkoski, J. J., Jr., and C. F. Koch. 1995. Evaluating paleontological data relating to bio-events. Pp. 21–34 in O. M. Walliser, ed. Global biological events in earth history. Springer, Berlin.

Shackleton, N. J., J. Backman, H. Zimmerman, D. V. Kent, M. A. Hall, D. G. Roberts, D. Schnitker, J. G. Baldauf, A. Desprairies, R. Homrighausen, P. Huddlestun, J. P. Keene, A. J. Kaltenback, K. A. O. Krumseik, A. C. Morton, J. W. Murray, and J. Westberg-Smith. 1984. Oxygen isotope calibration of the onset of ice-rafting and history of glaciation in the North Atlantic region. Nature 307:620–623.

Signor, P. J., and J. H. Lipps. 1982. Sampling bias, gradual extinction patterns, and catastrophes in the fossil record. Geological Society of America Special Paper 190:291–296.

Simenstad, C. A., J. A. Estes, and K. W. Kenyon. 1978. Aleuts, sea otters, and alternate stable-state communities. Science 200:403–411.

Smith, J. T. 2000. The dynamics of extinction selectivity in the Late Neogene Pectinidae of California. M.S. thesis. University of California, San Diego, La Jolla.

Smith, J. T., and K. Roy. 1999. Late Neogene extinctions and modern regional species diversity: analysis using the Pectinidae of California. Geological Society of America Abstracts with Programs 31:A473.

Stafford, T. W., Jr., H. A. Semken Jr., R. W. Graham, W. F. Klippel, A. Markova, N. G. Smirnov, and J. Southon. 1999. First accelerator mass spectrometry ^{14}C dates documenting contemporaneity of nonanalog species in late Pleistocene mammal communities. Geology 27:903–906.

Stanley, S. M. 1979. Macroevolution: pattern and process. W. H. Freeman, San Francisco.

———. 1986a. Anatomy of a regional mass extinction: Plio-Pleistocene decimation of the western Atlantic bivalve fauna. Palaios 1:17–36.

———. 1986b. Population size, extinction, and speciation: the fission effect in Neogene Bivalvia. Paleobiology 12:89–110.

———. 1996. Children of the ice age. W. H. Freeman, New York.

Stanley, S. M., and L. D. Campbell. 1981. Neogene mass extinction of western Atlantic mollusks. Nature 293:457–459.

Stanley, S. M., K. L. Wetmore, and J. P. Kennett. 1988. Macroevolutionary differences between the two major clades of Neogene planktonic foraminifera. Paleobiology 14:235–249.

Stehli, F. G., and S. D. Webb. 1985. The Great American biotic interchange. Plenum, New York.

Stemann, T. A., and K. G. Johnson. 1992. Coral assemblages, biofacies, and ecologic zones in the mid-Holocene reef deposits of the Enriquillo Valley, Dominican Republic. Lethaia 25:231–241.

Steneck, R. S. 1997. Fisheries-induced biological changes to the structure and function of the Gulf of Maine ecosystem. Pp. 151–165 in G. T. Wallace and E. F. Braasch, eds. Proceedings Gulf of Maine ecosystem dynamics scientific symposium and workshop. RARGOM Report 91-1. Regional Association for Research on the Gulf of Maine, Hanover, N.H.

Strauss, D., and P. M. Sadler. 1989. Classical confidence intervals and Bayesian probability estimates for ends of taxon ranges. Mathematical Geology 21:411–427.

Sutherland, J. P. 1974. Multiple stable points in natural communities. American Naturalist 108:859–873.

Valentine, J. W. 1961. Paleoecologic molluscan geography of the California Pleistocene. University of California Publications in Geological Sciences 34:309–442.

———. 1989. How good was the fossil record? Clues from the California Pleistocene. Paleobiology 15:83–94.

Valentine, J. W., and D. Jablonski. 1991. Biotic effects of sea level change: the Pleistocene test. Journal of Geophysical Research 96:6873–6878.

———. 1993. Fossil communities: compositional variation at many time scales. Pp. 341–348 in R. E. Ricklefs and D. Schluter, eds. Species diversity in ecological communities: historical and geographical perspectives. University of Chicago Press, Chicago.

Vermeij, G. J., and E. J. Petuch. 1986. Differential extinction in tropical American mollusks: endemism, architecture, and the Panama land bridge. Malacologia 27:29–41.

Vrba, E. S. 1995. The fossil record of African antelopes. Pp. 385–424 in E. S. Vrba, G. H. Denton, T. C. Partridge, and L. H. Burckle, eds. Paleoclimate and evolution, with emphasis on human origins. Yale University Press, New Haven, Conn.

Vrba, E. S., G. H. Denton, T. C. Partridge, and L. H. Burckle, eds.

1995. Paleoclimate and evolution, with emphasis on human origins. Yale University Press, New Haven, Conn.

Webb, S. D., and A. Rancy. 1996. Late Cenozoic evolution of the Neotropical mammal fauna. Pp. 335–358 *in* Jackson et al. 1996a.

Webb, T., III. 1993. Constructing the past from late Quaternary pollen data: temporal resolution and a zoom lens space-time perspective. *In* S. M. Kidwell and A. K. Behrensmeyer, eds. Taphonomic approaches to time resolution in fossil assemblages. Short Courses in Paleontology 6:79–101. Paleontological Society, Knoxville, Tenn.

Webb, T., III, and P. J. Bartlein. 1992. Global changes during the last three million years: climatic controls and biotic responses. Annual Review of Ecology and Systematics 23:141–173.

Wood, B., and A. Brooks. 1999. We are what we ate. Nature 400: 219–220.

Woodley, J. D., E. A. Chornesky, P. A. Clifford, J. B. C. Jackson, L. S. Kaufman, N. Knowlton, J. C. Lang, M. P. Pearson, J. W. Porter, M. C. Rooney, K. W. Rylaarsdam, V. J. Tunnicliffe, C. M. Wahle, J. L. Wulff, A. S. G. Curtis, M. D. Dallmeyer, B. P. Jupp, M. A. R. Koehl, J. Neigel, and E. M. Sides. 1981. Hurricane Allen's impact on Jamaican coral reefs. Science 214:749–755.

Pelagic species diversity, biogeography, and evolution

Richard D. Norris

Abstract.—Pelagic (open-ocean) species have enormous population sizes and broad, even global, distributions. These characteristics should damp rates of speciation in allopatric and vicariant evolutionary models since dispersal should swamp diverging populations and prevent divergence. Yet the fossil record suggests that rates of evolutionary turnover in pelagic organisms are often quite rapid, comparable to rates observed in much more highly fragmented terrestrial and shallow-marine environments. Furthermore, genetic and ecological studies increasingly suggest that species diversity is considerably higher in the pelagic realm than inferred from many morphological taxonomies.

Zoogeographic evidence suggests that ranges of many pelagic groups are much more limited by their ability to maintain viable populations than by any inability to disperse past tectonic and hydrographic barriers to population exchange. Freely dispersing pelagic taxa resemble airborne spores or wind-dispersed seeds that can drift almost anywhere but complete the entire life cycle only in favorable habitats. It seems likely that vicariant and allopatric models for speciation are far less important in pelagic evolution than sympatric or parapatric speciation in which dispersal is not limiting. Nevertheless, speciation can be quite rapid and involve cladogenesis even in cases where morphological data suggest gradual species transitions. Indeed, recent paleoecological and molecular studies increasingly suggest that classic examples of "phyletic gradualism" involve multiple, cryptic speciation events.

Paleoceanographic and climatic change seem to influence rates of turnover by modifying surface water masses and environmental gradients between them to create new habitats rather than by preventing dispersal. Changes in the vertical structure and seasonality of water masses may be particularly important since these can lead to changes in the depth and timing of reproduction. Long-distance dispersal may actually promote evolution by regularly carrying variants of a species across major oceanic fronts and exposing them to very different selection pressures than occur in their home range. High dispersal in pelagic taxa also implies that extinction should be difficult to achieve except though global perturbations that prevent populations from reestablishing themselves following local extinction. High rates of extinction in some pelagic groups suggests either that global perturbations are common, or that the species are much more narrowly adapted than we would infer from current taxonomies.

Richard D. Norris. MS-23, Woods Hole Oceanographic Institution, Woods Hole, Massachusetts 02543-1541. E-mail: RNorris@whoi.edu

Accepted: 20 May 2000

Introduction

Species diversity in the marine plankton is generally acknowledged to be only a fraction of that in terrestrial environments despite the extremely large size of open-ocean habitats. For example, Briggs (1994) calculates there are some 12 million terrestrial multicellular animal and plant species (of which perhaps 10 million are insects) but only some 200,000 marine taxa. These findings are particularly surprising considering that models that analyze ecosystem size, energy flow, and stability suggest that the diversity of marine species should be substantially higher than that in terrestrial environments (Briggs 1994).

Recently, the growth of paleoecologic and molecular genetic studies of open-ocean species has begun to challenge long-standing assumptions about the amount and origins of biodiversity in the open-ocean (pelagic) environment (Norris 1999; Norris and de Vargas 2000). The challenge comes in two parts: First, the combination of molecular studies with morphological and ecological data have led to a reexamination of traditional morphospecies concepts that influence our concepts of species diversity and ecological specialization. Second, surveys of genetic variability have begun to suggest that supposed barriers to long-distance dispersal are much more permeable than once believed. In this paper, I examine molecular, biogeographic and ecological data that bear on the general problem of how pelagic biodiversity is created and maintained. I

0094-8373/00/2604-0010/$1.00

conclude that traditional species concepts based on morphologically defined taxa (morphospecies) have greatly underestimated open-ocean species richness. Models of speciation have also overestimated the role of physical and hydrographic barriers to exchange as a motor for evolution.

Reevaluation of Species Concepts in Pelagic Organisms.—A major problem in evolutionary biology is to explain the amount and structure of biodiversity in relatively homogeneous environments like the open ocean. Generally speaking, recent results based on molecular phylogenetic analyses have revealed an impressive cryptic biodiversity, which is well structured in the open ocean. Pioneer studies on population genetics of zooplankton have demonstrated that copepod and euphausid morphospecies comprise numerous sibling species that may be discriminated by a very few, subtle, morphological features (Bucklin 1986; Bucklin et al. 1996; Bucklin and Wiebe 1998). Miya and Nishida (1997) demonstrated that there is considerable genetic variability in the pelagic fish *Cyclothone alba* that suggests this morphospecies actually consists of numerous cryptic species. These genetic results for *Cyclothone* echo earlier morphological studies of mesopelagic fish in the genus *Eustomais,* which find that the eight species traditionally recognized within this taxon may actually contain over a hundred morphologically defined taxa (Gibbs 1986).

Extensive cryptic biodiversity is evident in taxa with extensive fossil records. For example, Etter et al. (1999) have shown that the variation in 16S rDNA in several "species" of deep-sea mollusks is as high as or higher than that in shallow-marine genera composed of numerous species. The deep-sea species either have considerably greater genetic variability than shallow water genera or are composed of many previously unrecognized cryptic species. Knowlton (1993) also documents many cases of cryptic or sibling species in shallow-marine invertebrates. She observes (following the prescient observations of Gibbs [1986]) that many examples of sibling species reflect inadequate study of the morphological features of living taxa and estimates that the number of shallow-marine species will increase by an order of magnitude when cryptic taxa are considered (Knowlton 2000).

Pelagic microbial diversity is probably also greatly underestimated. Cloning surveys of the planktic cyanobacteria *Prochlorococcus* and *Synechococcus*—the most important marine primary producers—reveal an unsuspected genetic microdiversity that may represent adaptive radiation of species (Fuhrman and Campbell 1998; Moore et al. 1998). Such microdiversity has also been observed for marine bacterioplankton (Giovannoni et al. 1990; Olsen 1990; DeLong et al. 1994). Surprisingly small genetic distances in the 16S rDNA ($\sim$2%) have been associated with important physiological adaptations in the case of cyanobacteria (Moore et al. 1998). Recently, Ferris and Palenik (1998) demonstrated, by using rpoC1 DNA sequences, that different planktic cyanobacterial clones from very remote localities (Pacific, Mediterranean, and Sargasso seas) are organized in homogeneous clades adapted to specific surface or deep-oceanic niches, despite considerable geographic distances between populations.

How should we interpret such results? It is hazardous to interpret the genetic results strictly in terms of speciation and taxonomic diversity in the absence of other data (Knowlton 1997; Fuhrman and Campbell 1998; Knowlton 2000). For instance, the Restriction Fragment Lengths Polymorphism procedure (RFLP) that is commonly used to survey genetic diversity does not discriminate well between closely related genotypes and can end up underestimating population variation. Errors due to misalignment of genetic sequences can produce artificial variability. Conservative parts of the genome may evolve too slowly to be useful in discriminating recently diverged taxa (de Vargas et al. in press). There is also the general problem of how to interpret relatively modest molecular differences between populations. Indeed, studies of demonstrably different pathogens have found virtual genetic identity in their rDNA (Fuhrman and Campbell 1998). Furthermore, ecologically distinctive populations of the phytoplankton *Prochlorococcus* may differ from each other in the structure of the 16S rDNA molecule by less than 0.5% (Moore et al. 1998). These results

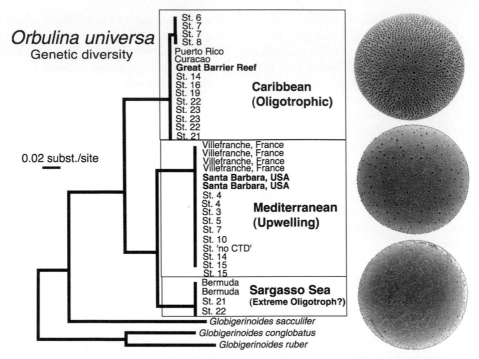

FIGURE 1. Cryptic speciation in the planktic foraminifer *Orbulina* (after de Vargas et al. 1999). Three distinct genotypes are present in the Atlantic, some of which are also present in the Mediterranean, the Caribbean, and the Pacific (records identified by station numbers along a north-south transect, from No. 1 in the north, to No. 23 in the south). In addition to the N-S transect, there were sample stations in the Mediterranean (Villefranche sur Mer), the Caribbean (Puerto Rico and Curaçao), and the Pacific (Great Barrier Reef and Santa Barbara Channel). Note the modest differences in shell porosity that appear to correlate with genotype distribution.

suggest that it is not always possible to distinguish taxa using molecular techniques, which must be combined with morphological and ecological data to make a strong case for species identification.

Among pelagic species with a fossil record, the planktic foraminifera are becoming increasingly well known from a genetic standpoint, and both biogeographic and ecological data can be brought to bear on the question of species definitions. Recent studies of planktic foraminifera have detected considerable genetic diversity that is correlated with predictable differences in ecology and subtle differences in morphology. For example, Huber et al. (1997) demonstrated that there are at least two sibling species of the planktic foraminifer *Globigerinella siphonifera*. These authors found not only that this morphospecies could be subdivided into two distinct genotypes, but that these varieties could also be distinguished by their characteristic symbionts, isotopic signature, and shell porosity. The com-

bination of molecular, ecological and morphological evidence suggests the two varieties of *G. siphonifera* are biological species. Parallel studies by Darling et al. (1999) and de Vargas et al. (1999) demonstrated cryptic diversity within the morphospecies *Orbulina universa*. Phylogenetic analysis clearly reveals the presence of three "cryptic" genotypes that each maintain genetic identity from diverse localities in the oceans (de Vargas et al. 1999) (Fig. 1). In addition, de Vargas et al. (1999) found that the three *Orbulina* genotypes have distributions correlated with total chlorophyll-a concentrations at the sea surface, suggesting that the various genotypes have become specialized on different water masses in the tropical and subtropical oceans (Fig. 2). These *Orbulina* genotypes also appear to be recognizable by their shell porosity, as is the case for the *G. siphonifera* genotypes (Fig. 1). Studies of the latitudinal cline in *Globorotalia truncatulinoides* (Lohmann and Malmgren 1983; Lohmann 1992) show that there are many genet-

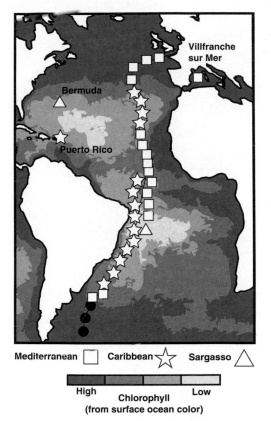

Figure 2. Geographic distribution of *Orbulina* genotypes along a north-south transect in the Atlantic compared with surface ocean color (simplified from de Vargas et al. 1999 and Norris and de Vargas 2000). The various genotypes are genuine species whose biogeographic distributions are related to nutrient concentrations. The "Caribbean" genotype (star) is found mostly in the subtropical gyres of the North and South Atlantic where chlorophyll concentrations are low, whereas the "Mediterranean" genotype (square) is present at higher chlorophyll concentrations in the equatorial Atlantic and poleward of the subtropical convergences. A third genotype, "Sargasso" (triangle), has been found only in the core of the South Atlantic subtropical gyre and at Bermuda. Black circles indicate stations where *Orbulina* is absent.

ically and morphologically distinctive taxa within this single "species" (de Vargas et al. in press). It is probable that other morphospecies can be split into several biological species based on a refined look at ecological, morphological, and genetic differences that have been traditionally regarded as ecophenotypic variation.

Clearly, a great part of the diversity has been overlooked in traditional morphological analyses of pelagic groups. In the case of planktic foraminifera, it appears that the ~50

living morphospecies (Hemleben et al. 1989) are likely to represent clusters of related taxa. For example, the diversity of distinct genotypes recognized within the polar species *Neogloboquadrina pachyderma*, *Turborotalita quinqueloba*, and *Globigerina bulloides* suggests that each taxon consists of at least three to five genetically distinctive variants that may represent discrete species (Darling et al. 2000). The deep genetic differences between some "cryptic" taxa, as in the cases of *G. siphonifera* (Huber et al. 1997) and *Orbulina universa* (de Vargas et al. 1999), suggest that some of them may represent relatively ancient splitting events (perhaps 5–15 m.y. old), belying the usual assumption that morphologically similar taxa have been produced by recent speciation (Darling et al. 1999). In other cases, it is possible that divergence is recent, within the last several hundred thousand years (de Vargas et al. in press). Moreover, speciation is evidently associated with specialization on different oceanic habitats in everything from protistans to zooplankton.

The apparent wealth of biodiversity within the planktic foraminifera, copepods, and bathypelagic fish suggests that we have to fundamentally revise our morphological species concepts for these groups. It appears that substantial genetic and ecological differences between taxa are associated with very modest morphological differences, which have been regarded as ecophenotypes in all but the most highly split traditional taxonomies (Knowlton 1997). The limited genetic studies of other pelagic groups, such as the examples from the phytoplankter *Prochlorococcus* and the fish *Cyclothone*, suggest that the underestimation of pelagic diversity may be widespread.

The Nature of Pelagic Ecosystems

Pelagic ecosystems are structured both geographically and by water depth with important consequences for evolution of open-ocean clades. There are about 18 major surface water masses in the global ocean (Emery and Meincke 1986) as well as local water masses associated with marginal seas such as the Mediterranean, Sea of Japan, and Persian Gulf among others (Fig. 3). In turn, these surface water masses can be further subdivided by

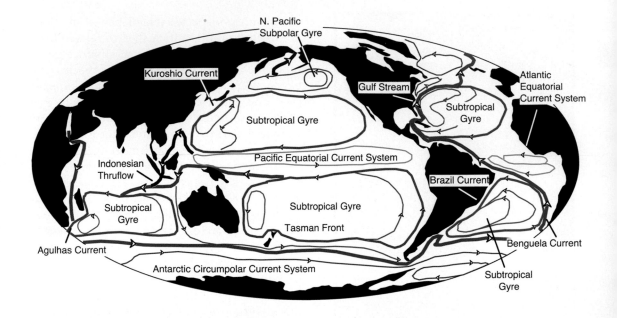

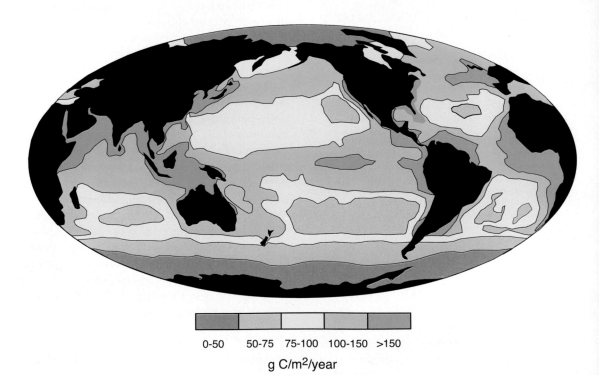

0-50   50-75   75-100   100-150   >150

g C/m²/year

FIGURE 3.   The pattern of major surface-water currents defines the boundaries of surface water masses. Heavy lines show currents that exchange water between water masses whereas thin lines illustrate circulation within current systems (after Schmitz 1996). Lower panel shows estimated yearly production of organic carbon in surface waters of the world based on satellite imagery (after Antoine et al. 1996). Note that surface production generally follows the distributions of major surface water masses.

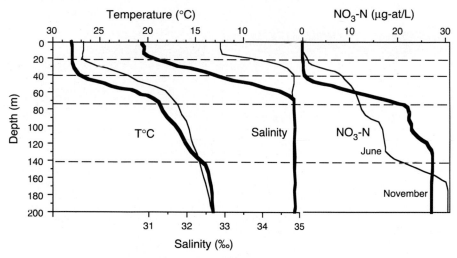

FIGURE 4. Features of upper-ocean stratification that could be used to structure populations vertically in the water column. The rapid changes in temperature (far left), salinity (center), and nutrients (right) can be sharp enough to form surfaces (shown approximately by dashed lines) against which species or their prey may congregate for growth and reproduction. Data from the Panama Basin (eastern equatorial Pacific) from Thunell et al. 1983. Thin lines are hydrographic properties in June, thick lines in November. Note the change in strength and depth of hydrographic gradients between seasons and the diversity of features that could be used to organize (or isolate) populations at different depths.

variations in seasonality, chlorophyll production rates, and degree of stratification. The water masses correspond reasonably well with the distribution of major biogeographic regions tabulated by a number of authors (e.g., Steuer 1933; McGowan 1971, 1972; McGowan and Walker 1993), suggesting that the zoogeography of many species is strongly controlled by hydrography and attendant changes in biological assemblages (e.g., Dortch and Packhard 1989). The number and character of major modern water masses is partly set by the shape of the ocean basins, the directions of surface winds, and the patterns of evaporation and precipitation over the oceans (Emery and Meincke 1986). Hence, the basic number of water masses has probably been fairly constant over the past 20–30 m.y., if not longer (McGowan 1986; McGowan and Walker 1993).

Surface water masses show considerable variation with depth in salinity, temperature, pressure, light attenuation, nutrient levels, dissolved oxygen, and biological production, all of which may play important roles in organizing species distributions and providing means for ecological and genetic subdivision of populations (Fig. 4). Vertical gradients in temperature (thermoclines) and salinity (pychnoclines) create density surfaces that can be

strong enough to accumulate sinking organic detritus and organisms (e.g., Sieburth 1986; Michaels and Silver 1988). Density surfaces are also commonly associated with maxima in chlorophyll and nitrite concentrations that reflect peaks in phytoplankton abundance and bacterial production, respectively (e.g., Fairbanks and Wiebe 1980; Fairbanks et al. 1982; Garfield et al. 1983; Sieburth 1986). Furthermore, sinking flocs of organic detritus (marine snow) can produce microenvironments hosting microbial assemblages that may be considerably more productive and species-rich than the surrounding open water (Silver and Allredge 1981; Sieburth 1986; Michaels and Silver 1988). All these structural features of the water column could serve a wide array of biological needs including concentration of nutrients, prey species, mature individuals, gametes, and free-living symbiont populations.

The vertical structure of the upper ocean also varies seasonally in virtually all open-ocean water masses and includes changes in the degree of stratification, upwelling intensity, light transmission, and biomass. For example, the subtropical gyres experience an increase in stratification in the summer owing to surface heating, followed by deep mixing and wind-driven overturning in the winter (Fig.

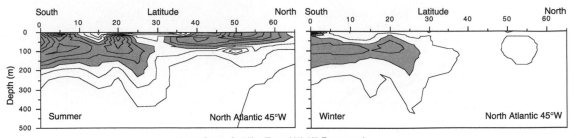

FIGURE 5.   Seasonal changes in surface stratification in the upper ocean. The surface ocean becomes well stratified by surface heating in the northern summer (left panel). Stratification breaks down in the northern subtropical gyre (between 28–40°N) and in the subpolar gyre (~40–60°N) during deep winter mixing, but remains strong in the Tropics and warm subtropics throughout the year (right panel). The variability in stratification strongly affects the supply of nutrients to the surface waters and resulting biological assemblages. Stratification given by the Brunt Väisälä frequency in cycles/second, from Bell et al. (1974).

5). Seasonal variations in surface ocean stability are evident in maps of tropical Atlantic mixed-layer depth (Hastenrath and Merle 1987), latitudinal trends in surface stratification (Lohmann 1992), and seasonal variation in chlorophyll seen in satellite imagery (Brown and Yoder 1994; Antoine et al. 1996). Seasonality is also evident in the marked variation in organic carbon fluxes, biogenic opal production, and both coccolith and foraminifer abundance seen in sediment traps from diverse localities in the oceans (e.g., Thunell et al. 1983; Thunell and Honjo 1987; Wefer 1989; Thunell and Sautter Reynolds 1992; Wefer and Fischer 1993; Honjo 1996; Honjo et al. 1999). Seasonal trends provide opportunities for speciation through changes in the timing of reproduction and the production of diverse habitats over the course of a year.

Population Structure in Plankton

Pelagic plankton have enormous population sizes and prodigious gene flow that seemingly should greatly restrict populations from becoming genetically isolated and creating new species. For zooplankton with densities of $\sim1/m^3$, the total number of individuals representing the species is in the hundreds of trillions. The total stock of euphausids—"krill"—is estimated to be a billion tons or more in the Southern Ocean (Bakun 1996), which, given an average dry weight of ~1 mg/individual (Parsons et al. 1984), suggests the genus is represented by some 10^{18} individuals in any given year. Lazarus (1983) calcu-

lated a total population of 5×10^{14} individuals for the tropical Pacific radiolarian *Amphirhopalum ypsilon*. Bucklin and Wiebe (1998) have estimated the female population size of the copepod *Nannocalanus minor*—a cool subtropical North Atlantic taxon with a geographic range about five times as large as that of *A. ypsilon*—at 2.6×10^{15} individuals. Species with global distributions within the subtropical gyres would be expected to have population sizes an order of magnitude greater than this. For example, the surface area of the world's subtropical gyres is ~192 million km^2, so a species with ~100 individuals/m^3 (the female density of *N. minor* in the North Atlantic gyre [Bucklin and Wiebe 1998]) results in a global population of $\sim1.9 \times 10^{16}$ individuals.

The large size of pelagic populations does not mean that they necessarily have high levels of genetic diversity or that they are not susceptible to population crashes. Bucklin and Wiebe (1998) show that the genetic diversity of several copepod species is significantly lower than expected from their population size. Thus, while they estimate female populations of $\sim10^{15}$ individuals for *Calanus finmarchicus*, the genetic diversity is consistent with a population size of only $\sim10^5$ individuals. The small effective population size in this species may be due to a recent bottleneck in the population created by an expansion of sea ice and colder water over most of the species' geographic range during the Pleistocene (Bucklin and Wiebe 1998). Hence, it is conceivable that even very large populations could be dramatically reduced in effective size by range contraction. A range contraction, or the isolation of a subset of a population, could set the stage for genetic divergence and speciation.

Evidence for High Dispersal

The vast oceanic ranges of most pelagic taxa imply that there are few important barriers to exchange between far-flung populations and that many oceanic species are reasonably interpreted as freely interchanging populations united by prodigious gene flow. Genetic data for pelagic cyanobacteria (Ferris and Palenik 1998), copepods (Bucklin 1986; Bucklin et al. 1996), and foraminifera (Darling et al. 1999, 2000; de Vargas et al. 1999) demonstrate that gene flow must be strong across ocean basins as well as between them. Genetic evidence suggests that Arctic and Antarctic groups of planktic foraminifera exchanged populations during the Pleistocene and may be in genetic contact today (Darling et al. 2000). Likewise, Pacific and Atlantic populations of the planktic foraminifer *Orbulina* have essentially identical sequences in the portion of the 16s rDNA fragment that was studied (de Vargas et al. 1999); this is also the case for Atlantic and Pacific *Globigerinoides siphonifera* Type II and for *Globigerinoides sacculifer* (Darling et al. 1999). Other foraminifer species, such as *Globigerinoides ruber* (white), display large differences between Caribbean and South Pacific genotypes (Darling et al. 1999), but this could reflect unrecognized species-level differences rather than within-species variation. The implication is that many foraminifer genotypes have either evolved little after they became separated into the different ocean basins or they have remained in continuous exchange. Bipolar species seem to have maintained gene flow despite the necessity to cross the seemingly inhospitable Tropics. Tropical species in the Atlantic and Pacific have also maintained genetic exchange despite formidable tectonic barriers such as the Isthmus of Panama (e.g., Norris 1999).

High dispersal coupled with large population sizes should not only connect distant populations but should reduce intraspecific genetic variability and slow speciation rates (Palumbi 1994). Paradoxically, rates of speciation can also be as high for pelagic taxa as for shallow-marine and terrestrial species (e.g., Norris 1991, 1992; Palumbi 1994; Palumbi et al. 1997) suggesting that high dispersal and large population size have not impeded species formation (Norris 1999).

Biogeographic ranges of living and fossil species provide evidence that the distributions of pelagic species are set more by environmental limits on reproduction and population maintenance than by limited dispersal (Fig. 6). In effect, many pelagic species seem to be able to easily circumvent formidable tectonic and hydrographic barriers but still fail to establish viable populations outside their main ranges. Individuals behave like wind-dispersed seeds that are able to go almost anywhere the currents take them, but able to establish themselves only where conditions are favorable. For example, the black marlin, *Makaira indica*, is present throughout the tropical and subtropical Indo-Pacific but is absent from the Atlantic save for occasional, non-spawning individuals that disperse around the Cape of Good Hope and are caught off the west coast of Africa (Mooney-Seus and Stone 1997; Norris 1999). The mesopelagic fish *Cyclothone braueri* is found throughout the Atlantic and Southern Oceans but not in the mid-latitude Indo-Pacific (Miya and Nishida 1997), whereas the tuna *Thunnus thynnus* is present in the tropical–subtropical Pacific and Atlantic but not in the Indian Ocean (Mooney-Seus and Stone 1997). Clearly the distributions of the black marlin and these other fish are controlled not by failure to disperse but by some factor that limits successful reproduction or population growth outside their core ranges. There are many examples of pelagic species ranging from fish and crustaceans to pelagic tunicates, pteropods, and planktic foraminifera that are found in only part of the oceans, despite indications that they can disperse regularly outside their home range (van der Spoel 1983; Mooney-Seus and Stone 1997; Norris 1999). These biogeographic distributions are hard to explain without a combination of high dispersal and regional limits on reproduction.

Speciation in Pelagic Environments

Our calculations of populations sizes for pelagic species and evidence of strong dispersal

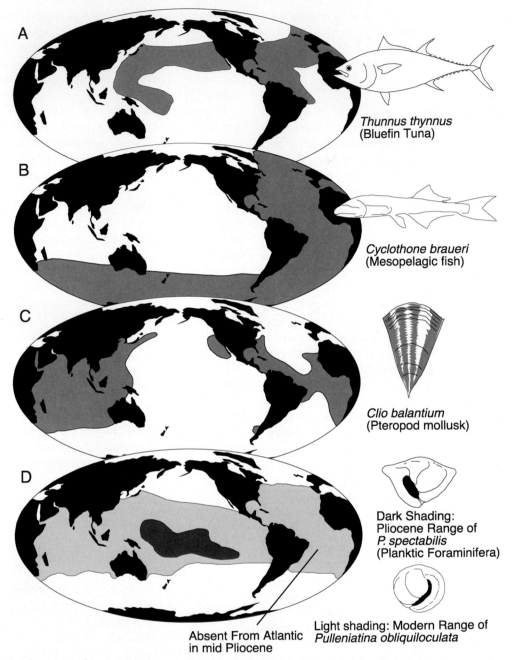

FIGURE 6. Biogeographic distributions of species with disjunct or ''endemic'' distributions. In all cases, the distributions suggest that species' ranges are controlled more by limits on a species' ability to maintain viable populations than by their ability to disperse. For example, the bluefin tuna (A) is not present in the Indian Ocean despite direct surface-water flow from the Pacific. In other cases (D), a species like *Pulleniatina spectabilis* maintains populations in the warm tropical western Pacific despite extensive surface water connections with the Indian Ocean. A close relative, *P. obliquiloculata* is episodically eliminated from the Atlantic during glacial stages, but is able to reinvade by dispersal around southern Africa (e.g., Norris 1999).

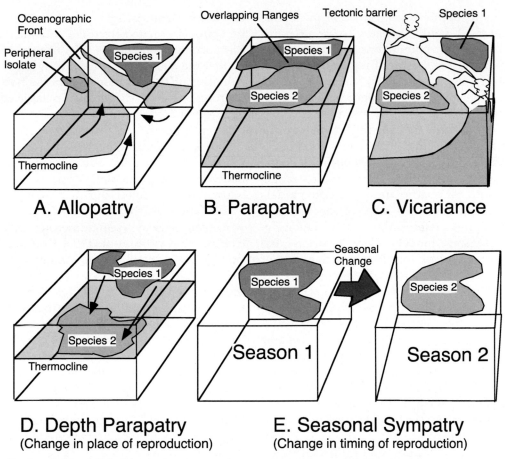

FIGURE 7. Speciation models for pelagic environments. A, Allopatry, created by divergence on either side of a pronounced hydrographic barrier. B, Parapatry, in which two species diverge along some gradual hydrographic gradient (in this case, gradually changing thermocline depth). C, Vicariance, where speciation is created by isolation of large populations on either side of an impenetrable barrier. D, Depth parapatry (or sympatry), wherein speciation occurs by a change in the place of reproduction (in this case water depth). E, Seasonal sympatry (or parapatry), in which isolation is established by change in timing of reproduction. Seasonal sympatry could be produced by changing the environmental cues used by a species to synchronize reproduction between members of a population. Evidence for high dispersal (this paper and Norris 1999) suggests that a complete shutoff in gene flow implied by models A and C is unlikely.

capabilities for many of them directly challenge models for speciation that rely on the establishment of isolation by limited dispersal in continuously floating organisms (holoplankton). Many models for speciation in terrestrial environments rely on limited dispersal capability as a means of creating genetic isolation and divergence. Such models include (1) allopatric models, in which isolation of small subsets of a population leads to divergence and speciation, and (2) vicariant models, in which oceanic populations are broken into large subgroups by tectonism, sea-level fall, strengthening of water-mass boundaries, or creation of bipolar distributions (Fig. 7).

These "dispersal-limiting" models have been widely applied to speciation in terrestrial clades (Brooks and McLennan 1991) as well as in benthic marine invertebrates and shallow-marine fishes (e.g., Palumbi 1994; Gold and Richardson 1998). In contrast, speciation models that permit low to moderate rates of genetic exchange include (1) isolation-by-distance models, where speciation occurs at the ends of a cline because of limited gene flow, and (2) isolation-by-depth/season models, where new taxa evolve within the ancestor's range or near its margins (Fig. 7). Pierrot-Bults and van der Spoel (1979) coined the term "depth parapatry" to describe speciation by changes in depth of reproduction in pelagic organisms. I would also add "seasonal parapatry," whereby populations become isolated by changing the season of reproduction (Fig. 7). Both "depth parapatry" and "seasonal parapatry" involve divergence between populations near the limits of the ancestral species' range. Sympatric models could also involve divergence in the place or season of reproduction in nascent species but with speciation occurring within the ancestor's range. Since both parapatric and sympatric models involve changes in the place of reproduction or its timing, I refer to them as "isolation-by-depth/season" models. Here, I discuss the types of evidence for these mechanisms of speciation.

Dispersal-Limited Models.—Some researchers have suggested that oceanic frontal systems formed at the intersection of surface water masses are as effective at subdividing oceanic populations and promoting their genetic isolation as might be the growth of tectonic barriers to exchange between ocean basins. Wei and Kennett (1988) proposed that the Tasman Front—an important boundary between subtropical waters off eastern Australia and tropical waters of the Coral Sea (Mulhern 1987)—played a crucial role in promoting speciation in globoconellid planktic foraminifera during the Pliocene. They proposed that the front effectively shut down genetic exchange between populations and made possible the radiation of taxa on either side of the water mass boundary. Endemic species are also known from modern marginal basins throughout the world (van der Spoel 1983), and there are numerous late Miocene examples from the central Paratethys Seaway (e.g., Rögl 1985). However, all these examples are open to reinterpretation given recent molecular genetic evidence that foraminifera frequently disperse across seemingly strong hydrographic boundaries. For example, de Vargas et al. (1999) showed that the North and South Atlantic subtropical gyres are inhabited by identical genotypes of *Orbulina universa*, the same that occurs in the waters off the Great Barrier Reef in the Pacific (Fig. 1). In another example, the foraminifer *Pullenatina obliquiloculata* and several other species repeatedly reinvaded the tropi-

cal Atlantic from refugia in the Indo-Pacific after being eliminated from the Atlantic during Pleistocene glacial cycles (Fig. 6). Apparently neither the equatorial upwelling system in the Atlantic nor the Isthmus of Panama has been a strong barrier to dispersal of pelagic species between subtropical populations or between different oceans.

Vicariant subdivision of large populations followed by speciation has been identified in many studies, particularly those in shallow-marine taxa isolated during major tectonic events such as the closure of the Central American Seaway (see numerous examples cited in Knowlton et al. 1993; Palumbi 1994; Collins 1996; Knowlton and Weight 1998; Lessios 1998). Vicariant processes are also cited in the evolution of the bathypelagic fish *Cyclothone*, following separation of the Atlantic and Pacific populations by growth of the Isthmus of Panama (Miya and Nishida 1997), and in isolation of eastern tropical Pacific populations of *Tridacna* from those in the western Pacific by changes in oceanic circulation patterns during the Pleistocene (Benzie and Williams 1997). Fleminger (1986) has suggested that speciation occurred in pelagic copepods after low sea-level stands of the Pleistocene isolated populations on either side of the Indonesian Seaway.

Isolation by Distance.—Geographic variation in morphology and ecology has long been a favorite source for speciation models (Endler 1977; Pierrot-Bults and van der Spoel 1979; Lazarus 1983; van der Spoel 1983). Limited dispersal between the ends of a cline could lead to speciation, particularly if intermediate morphotypes were eliminated (Pierrot-Bults and van der Spoel 1979). Genetic variation in Pacific sea urchins (Palumbi et al. 1997), *Tridacna* clams (Benzie and Williams 1997), deep-sea vent animals (Vrijenhoek 1997), and estuarine fishes (Gold and Richardson 1998) displays strong evidence for an isolation-by-distance model. In a similar vein, the pelagic euphausid *Meganyctiphanes norvegica* seems to have sufficiently low dispersal across the North Atlantic that populations in the northwest Atlantic have become genetically distinct from those in Norwegian fjords (Bucklin et al. 1997). In all these cases, nearby populations are more similar genetically than more geographically distant ones and there is also substantial population heterogeneity in genetic variants.

Clinal trends in morphology have been described from a wide variety of marine plankton (van Soest 1975; Lohmann and Malmgren 1983; van der Spoel 1983; Lohmann 1992). Some of these trends may well reflect a geographic succession of distinct species, since some of the trends have a "stepped" structure with sharp transitions in morphology across water-mass fronts (van Soest 1975; van der Spoel 1983). The latitudinal cline in morphology of the foraminifer *Globorotalia truncatulinoides* appears to be relatively continuous (Lohmann and Malmgren 1983; Lohmann 1992), but Healy-Williams et al. (1985) have shown that there are at least three distinct morphotypes in this cline. Given our experience with the subtle morphological differences between foraminifer genotypes (Huber et al. 1997; de Vargas et al. 1999), it seems possible that the cline in *G. truncatulinoides* may also reflect a series of species latitudinally arrayed across different water masses, as concluded by de Vargas et al. (in press).

Isolation by Depth.—Pierrot-Bults and van der Spoel (1979) and later Lazarus (1983) discussed examples of speciation along a depth gradient. As noted above, the oceans show strong gradients in temperature, light, nutrients, pressure, and salinity that could separate species and populations from one another. For example, many species of planktic foraminifera grow in the surface ocean and reproduce by sinking deeper into the water column (Orr 1967; Bé 1980; Bouvier-Soumagnac and Duplessy 1985; Lohmann 1992; Norris et al. 1996). The depth to which they sink could be an important means of establishing reproductive isolation. A species that uses the underside of the air/water interface to concentrate its gametes would be reproductively isolated from one that used the seasonal thermocline for this purpose. Given the large number of potential "surfaces" available in the upper ocean, it seems quite possible that speciation could occur through changes in reproductive depth habitat (Lazarus 1983; Norris 1992).

Speciation along a depth gradient could occur either in geographic sympatry or parapatry.

Several studies of planktic foraminifera have offered support for a depth parapatric, or depth sympatric, model. For example, Norris et al. (1993, 1996) observed, on the basis of stable isotope data, that the evolution of *Fohsella fohsi* in the middle Miocene involved a rapid shift in reproductive depth habitat. Remarkably, the change in reproductive ecology occurred in the middle of a long anagenetic morphocline, suggesting that the switch in depth of reproduction was decoupled from trends in morphological evolution. Kelly et al. (1996b, 1998) showed that the initial appearance of *Morozovella allisonensis* in the late Paleocene was associated with a shift toward deeper depth habitats relative to its ancestor, *Morozovella velascoensis*. Similar examples of closely related species that occupied different depth habitats have been described by Schneider and Kennett (1996), Berggren and Norris (1997), and Huber et al. (1997). In addition, the living variants of *G. truncatulinoides* also have different depth preferences, judging from isotopic and plankton tow data (Healy-Williams et al. 1985; Lohmann 1992). These examples from marine plankton are reminiscent of speciation in insects by host-switching, in which changes in the place of mating produces genetic isolation, even between species that can hybridize in laboratory culture (e.g., Tauber et al. 1977; Coyne 1992).

Isolation by Season.—Shifts in the timing of reproduction are a plausible, but generally untested, model for species formation in marine plankton. Many species show striking differences in seasonality of maximum population size (Fig. 8). Often, abundance peaks correlate with the timing of spring bloom, as in copepods and their predators (e.g., Bakun 1996; Gaard 1996; Gislason and Astthorsson 1996). Spring bloom shifts poleward in the subtropics with increased day length and surface ocean stratification (Honjo and Manganini 1993). Hence, populations utilizing the spring bloom may have different timing of reproduction and maturation along a latitudinal gradient. Among planktic foraminifera, there is a well-known seasonal succession of species that have distinct population maxima in the

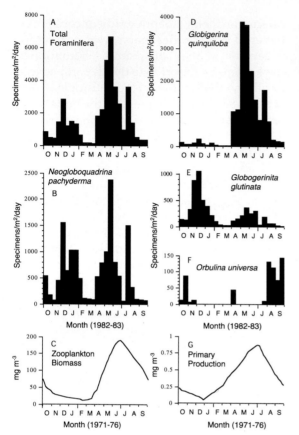

FIGURE 8. Seasonality of plankton populations (data from Reynolds and Thunell 1985; Thunell and Honjo 1987). Abundances of planktic foraminifera in sediment trap samples over a seasonal cycle showing tendency for a given morphospecies to have two (B, D, E) or three (F) peaks in abundance during the year. It is possible (but unproven) that the various populations of each morphospecies actually represent different genotypes, because one abundance peak tends to fall during the maximum in primary production (G) while the other is close to the minimum in production. Experience with genotypes of *Orbulina* (Figs. 5, 6) suggests that different genotypes are associated with different levels of primary production (de Vargas et al. 1999). In any event, a change in season of growth could effectively isolate one population from another within a species.

winter, the early and late spring, the summer, and the early fall (Deuser 1987; Sautter Reynolds and Thunell 1991; Thunell and Sautter Reynolds 1992). A modest shift in the timing of population growth owing to a shift in cues that trigger reproduction could produce isolation.

Indeed, very small shifts in reproductive timing could achieve isolation. Some species of planktic foraminifera are known to coordinate gametogenesis using the lunar cycle or

semilunar cycles (Bijma et al. 1990). The planktic foraminifera *Hastigerina pelagica* (Spindler et al. 1978; Spindler et al. 1979), *Globigerinoides sacculifer, Globigerinoides bulloides,* and *Orbulina universa* (Bijma et al. 1990) have lunar cycles of gametogenesis, and other species such as *Globigerinoides ruber* and *Globigerinella siphonifera* are known to have semilunar periodicity (Bijma et al. 1990). However, all these species show variation around a lunar or semilunar cycle, suggesting that reproduction could be shifted to a different period by selection or by genetic drift. *Hastigerina pelagica* apparently uses a dark–light cycle to coordinate gametogenesis since the periodicity can be altered by laboratory manipulation of day length (Spindler et al. 1979). However, *G. sacculifer, G. ruber,* and *O. universa* apparently use some other external environmental cue, possibly the magnetic field, to set their reproductive clock (Bijma et al. 1990). A genetic change that switches the response to these environmental variables could set the stage for speciation.

Certain examples of sympatric speciation could reflect changes in reproductive timing. Lazarus et al. (1995) describes the origins of the planktic foraminifer *G. truncatulinoides* in the south Pacific and concludes that it evolved sympatrically from an ancestral stock that includes *Globorotalia tosaensis* and *Globorotalia crassaforms*. All these species have similar geographic ranges and broadly similar depth ecology, but it is entirely possible that they diverged though small differences in depth habitat or seasonal displacement. Other examples include results of Norris et al. (1994), Schneider and Kennett (1996), and Pearson et al. (1997), in which small isotopic differences were observed between closely related foraminifer taxa. These isotopic data reflect ~1° to 2°C differences in calcification temperature and could plausibly be due to changes in either season or depth of growth. Given that seasonal changes in surface waters are typically in the range of several degrees or more in the Tropics and warm subtropics (Bottomley et al. 1990), it is quite reasonable to infer that divergence occurred by shifts in timing of gametogenesis.

Speciation Mechanisms

The mechanisms underlying "dispersal-limited" speciation in pelagic taxa are somewhat elusive considering evidence that hydrographic and tectonic barriers to dispersal seem to be remarkably leaky for many species. If tropical species of foraminifera can disperse from the Atlantic to the Pacific, why would a sea-level lowstand in the Indonesian archipelago (Fleminger 1986) create an insurmountable barrier for pelagic copepods? One explanation is that some of these vicariant events and clinal trends are not created by a shutdown in dispersal per se, but by divergence under strong selection in very distinct water masses. In essence, speciation is not really by allopatry, since some dispersal is maintained during speciation (see also Rice and Hostert 1993). Endler (1977) has shown how population divergence can occur along an environmental gradient even in the absence of strong barriers to gene flow. Gaard (1996) likewise suggests that strong temperature gradients across the Iceland-Faroes Front in the North Atlantic act to uncouple life-cycle events, including reproduction, in copepod populations. In species whose range spans a strong hydrographic front, the timing of reproduction may be offset sufficiently on both sides of the front to produce effective genetic isolation even in the face of continued dispersal. Changes in salinity and stratification of marginal seas might produce a similar discontinuity in the life cycles of formerly contiguous populations owing to shifts in the timing of reproduction. Hence, lowered sea level during the Pleistocene may have caused the hydrography of marginal basins in the western Pacific and the life histories of their resident populations to diverge enough from the open ocean to trigger speciation. The evolution of a diverse assemblage of endemic planktic foraminifera in the central Paratethys Seaway (e.g., Rögl 1985) may also have been produced by unusual oceanography in this marginal sea.

It is also conceivable that a species could experience sufficiently strong selection on either side of a hydrographic boundary to produce different average body size, growth rates, or skeletal shapes with consequent changes in mating recognition systems, particularly because these morphologic variables are known to be affected by changes in food supply, temperature, and predation intensity (Fleminger 1986; Palumbi 1994). Genetic divergence may be relatively easy to accomplish in species that spawn directly into the water through modification of the biochemistry involved in gamete recognition (Palumbi 1994, 1998). Simple mating cues, such as the pheromone system in rotifers (Snell and Hawkinson 1983), may be susceptible to changes in the activities of different genes as a function of water temperature or other hydrographic variables. Although there are few examples of environmentally triggered changes in mate recognition systems and development in marine plankton (but see Gaard 1996), such changes are well known from insects and other terrestrial taxa (e.g., Tauber et al. 1977, 1989; Coyne 1992; and numerous examples in Raff 1996). Hence, I think it is possible that although pelagic speciation may appear to involve allopatric or vicariant processes, it actually occurs in the face of sustained gene flow that is rendered ineffectual by changes in mating recognition cues or reproductive timing.

Relatively simple genetic switches may underlie changes in timing of reproduction in some taxa and contribute to species divergence. Population genetic studies of lacewing insects has shown that shifts in the timing of mating can be produced by as few as two genes (Tauber et al. 1977). One of the two lacewing species studied by Tauber et al. (1977) normally reproduces in the summer while the other does so in the spring, apparently in a genetically controlled response to changes in photoperiod. Since environmentally coordinated reproduction is common in marine plankton (Bijma et al. 1990), genetic switches similar to those in lacewings may be responsible for changes in reproductive timing and genetic isolation.

Finally, recent theoretical and empirical studies have lent increasing support for the widespread occurrence of sympatric speciation (see examples in Howard and Berlocher 1998). Recent theoretical treatments have suggested that sympatric speciation can result from selection for two divergent characteristics that exacts a penalty for intermediate morphotypes. For instance, strong intraspecific competition for a single food source could cause part of the population to evolve strategies to use a less sought-after food supply (Dieckmann and Doebell 1999). Selection would favor both new morphotypes because intermediates between them would be at a competitive disadvantage. Alternatively there might be two resources—say large and small prey—that cause selection to favor large and small predators while discouraging medium-sized predators (e.g., Kondrashov and Kondrashov 1999; Tregenza and Butlin 1999). These results agree with observations of speciation events in lake fishes that suggest that sympatry is common in aquatic ecosystems (Tregenza and Butlin 1999).

Patterns of Morphological Evolution: Anagenetic and Cladogenetic Trends

This review has mostly ignored the large literature of morphological analyses of anagenesis and cladogenesis in fossil plankton in large measure because these studies tend to be unable to resolve the geographic and ecological factors behind the morphological patterns. Still, there has been considerable effort to identify the processes of evolution through patterns of morphological change. Indeed, many studies have shown gradual, anagenetic trends in planktic foraminifera (Malmgren and Kennett 1981; Arnold 1983; Lohmann and Malmgren 1983; Malmgren et al. 1983; Malmgren and Berggren 1987; Hunter et al. 1988; Norris et al. 1996; Kucera and Malmgren 1998) as well as cladogenesis (Wei and Kennett 1988; Wei 1994; Lazarus et al. 1995; Malmgren et al. 1996). Similar studies have been done on morphological trends in radiolarians (Lazarus 1983, 1986) and diatoms (Sorhannus 1990a,b).

Comparisons of patterns of morphological and ecological data (principally species abundance, body size, and isotope-inferred depth habitats) suggest that morphological trends may not directly speak to the rate of speciation or even its mode (Fig. 9). Kucera and Malmgren (1998) argued that an anagenetic trend in Cretaceous planktic foraminifera probably reflects a gradual shift in relative proportions of morphotypes. These authors showed that

highly conical varieties of *Contusotruncana fornicata* evolved rapidly and gradually replaced low-conical forms, but that at any given time populations were normally distributed in terms of their shell conicity. Populations display a large increase in population variability about the time the first highly conical forms appear, suggesting that the mean morphological trend reflects an expansion of variance rather a concerted drift in morphology of a single population. In another example, the evolution of photosymbiosis in Paleocene foraminifera occurred in the middle of an anagenetic trend from *Praemurica uncinata* to *Morozovella angulata* (Kelly et al. 1996a; Norris 1996b). The appearance of a major ecological innovation in the middle of the morphological trend between *Praemurica* and *Morozovella* strongly suggests that one or more instances of speciation occurred within the anagenetic morphological series. The evolution of photosymbiosis also appears to have initiated the differentiation of *Praemurica* into the first representatives of *Acarinina* during the Paleocene (Quillévéré et al. in press). In all cases, it appears that an anagenetic pattern hides ecological subdivision of populations and cladogenetic speciation.

Norris et al. (1996) found that reproductive isolation was quite rapid in Miocene foraminifera even though morphological trends evolved very slowly (Fig. 9). The *Fohsella* clade displays a classic example of anagenetic evolution with a continual shift in average morphology from the ancestral form, *Fohsella praefohsi*, to the descendant form, *Fohsella fohsi* over a span of ~300–400 k.y. Population morphology retains normally distributed distributions with statistically similar variances throughout the time series, suggesting that there is no more than one taxon present at any given time. Yet isotopic data show a distinct bimodality in average $\delta^{18}O$ part way through the anagenetic trend, which suggests subdivision of the populations into those reproducing in thermocline waters from those reproducing in the surface ocean. Counts of the relative abundance of individuals with distinctive morphological features (in this case, a thickened band around the shell called a "keel") shows that keeled individuals gradually replace unkeeled

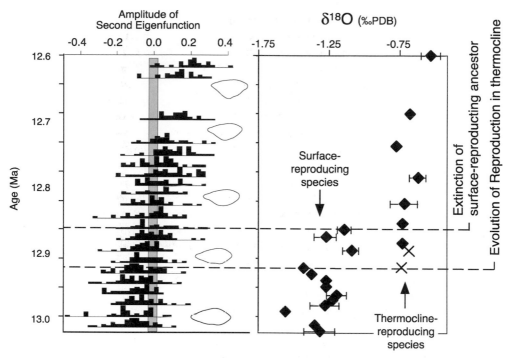

FIGURE 9. Ecological cladogenesis within an anagenetic morphological trend (after Norris et al. 1996). Left: frequency histograms showing gradual morphologic trend in shell outline in the planktic foraminifer genus *Fohsella* from the middle Miocene. Representative shell outlines shown for reference. Right: stable isotope analyses of the same specimens used in the morphological analysis showing abrupt divergence of thermocline-reproducing populations from their surface-reproducing ancestor. The ancestor apparently became extinct about 70 k.y. after the appearance of its descendant. The morphometric data suggest that no more than one population is present at any given time (but see Fig. 10).

individuals (Fig. 10). How are we to reconcile evidence for anagenesis and cladogenesis within the same populations?

It is possible to produce a reasonable facsimile of an anagenetic series by assuming that there really are two distinct species of fohsellids and that the ancestor is gradually replaced by the descendant (Fig. 10). For illustrative purposes, I assume that the morphological variability of the initial population reflects the unkeeled ancestor, *F. praefohsi*, whereas the variation in the final population represents the fully keeled descendant, *F. fohsi*. The gradual replacement of *F. praefohsi* by *F. fohsi* (as seen in the % keeled individuals) provides a measure of the change in relative proportions of the two taxa through time. I have sampled the two end-member populations (with replacement) at each step in the time series in their relative proportions given by the percentage-keeled data set.

The result is an apparent anagenetic trend that is set up by sampling different propor-

tions of the two end-member morphologies through the time series. The extreme end-member forms in both species are uncommon and are rarely sampled. Therefore, as the proportion of one species decreases, the likelihood that extreme morphologies of that species will be sampled falls. In cases where the two end-member species are morphologically similar, but not identical, the intermediate distributions representing roughly equal proportions of the two taxa may still be normally distributed. The result is a plausible anagenetic series that actually reflects cladogenesis and the gradual ecological replacement of the ancestor by its descendent.

As the morphological distinctiveness of the end-member forms increases, the intermediate distributions will become increasingly flat and the variance will increase. Eventually, very distinctive end-member taxa may yield bimodal distributions when both are sampled in roughly equal proportions or when the rare species is well sampled. The major giveaway

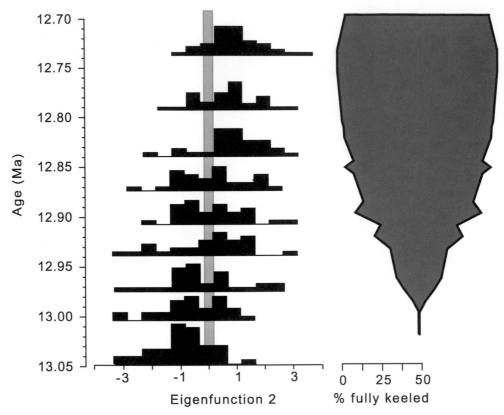

FIGURE 10. Simulated anagenetic trend (left) produced by random sampling (with replacement) of actual populations of fohsellids in the proportions suggested by the abundance of fully keeled individuals (right). The samples used to create the intermediate populations are shown at 12.73 and 13.05 Ma. Intermediate populations are mostly (though not entirely) normally distributed but tend to have higher variance and flatter distributions than those from which they were sampled. Experiments with different starting populations suggest that distributions become flatter as the end-member populations become more divergent, but bimodality is difficult to detect unless the end-members have little overlap or unless large numbers (hundreds) of individuals are analyzed in each sample.

that implicates mixing of the proportions of two distinct species in producing an apparent anagenetic trend is the increased variance of the intermediate distributions compared with either the ancestral or descendant populations. However, this increase in variance may not be obvious in fossil samples that include moderate numbers of the rare species for an extended interval of time. Convincing, but false, anagenetic trends also may be an artifact of the successive replacement of a series of intermediate taxa between the end-member species. For example, the *Fohsella* clade includes at least two additional, named taxa between *F. praefohsi* and the most extreme end-member form, *F. robusta*. The successive replacement of each of these species could produce a quite continuous-looking phyletic trend because the intermediate taxa have considerable morpho-

logical overlap with one another (e.g., Norris et al. 1996).

Another example of an apparently false anagenetic series is the evolution of *Globorotalia truncatulinoides* in the Plio-Pleistocene. Lohmann and Malmgren (1983) showed that *G. truncatulinoides* displays a morphological cline in the Atlantic between temperate waters and the Tropics. These authors and later Lohmann (1992) noted that down-core records from the South Atlantic show a gradual shift in mean morphology from a low-latitude morphotype to a midlatitude morphotype and proposed that the morphological trend reflects gradual environmental change over the past 700 k.y. Recently, de Vargas et al. (in press) demonstrated that the cline consists of a series of at least four distinct genotypes of *G. truncatulinoides* that are arrayed in different water masses between the Tropics and temperate waters south of the subtropical convergence. The genotypes recognized by de Vargas et al. broadly agree with a succession of morphotypes described by Healy-Williams et al. (1985). Hence, the anagenetic trend described by Lohmann (1992) appears to be produced by the gradual replacement of the more tropical genotype by a subtropical-gyre genotype.

Finally, perhaps one of the most famous examples of anagenesis—the "punctuated anagenetic" transition between *Globorotalia plesiotumida* and its descendant, *G. tumida* (Malmgren et al. 1983, 1984)—may actually not represent a case of anagenesis in any sense of the word! Recent stratigraphic analyses have demonstrated that *G. pleisotumida* persists well up in the range of its descendant (Chaisson and Leckie 1993; Chaisson and Pearson 1997; Norris 1998) and therefore cannot have given rise to *G. tumida* by wholesale replacement of the ancestral populations of *G. plesiotumida*. Apparently this most archetypal example of punctuated anagenesis actually represents cladogenesis accompanied by a rapid change in the proportions of the ancestral and descendant populations, like the example in Figure 9.

In sum, anagenetic trends in some groups of foraminifera may be explained by cladogenesis between morphologically similar taxa followed by the gradual replacement of one species by the other. Successive replacement of one closely related species after another need not reflect the evolution and extinction of each species, but could be a response to gradual changes in surface hydrography as inferred by Lohmann (1992) for the morphological trend in *G. truncatulinoides* and by Kucera and Malmgren (1998) for *C. contusa*. Likewise, the various fully keeled taxa in the Miocene *Fohsella* clade (*F. fohsi fohsi, F. fohsi lobata,* and *F. fohsi robusta*) are all thought to coexist for ~500 k.y. but could still replace one another locally to produce apparent evolutionary trends. I suspect that many gradual morphological trends may originate because of long-term shifts in surface-water hydrography that cause the regional, but not necessarily global, replacement of closely related species.

Gradual evolutionary trends undoubtedly represent remarkable, and complex, sets of evolutionary and ecological interactions. The origins of apparent phyletic trends by cladogenesis are ripe for modeling and for the development of new statistical tools. We particularly need models that assume the successive replacement of ancestral species and daughter species with different degrees of morphological overlap. Models that involve different numbers of taxa that successively replace one another would also be useful for examining changes in variance through morphological transitions. Empirical analyses will need to examine much larger numbers of specimens in any given population (hundreds of specimens, compared with the 30–40 specimens that are typically analyzed) to detect bimodality in anagenetic trends. I suspect that many anagenetic trends may be resolvable into cladogenetic splitting of closely related taxa but concede that not all cases of phyletic gradualism may be so easily explained.

Macroevolutionary Implications of Broad Dispersal in Pelagic Taxa

One of the paradoxes of evolution in the open ocean is that there is so much speciation and extinction, whereas one would expect the large population sizes and strong dispersal abilities of pelagic taxa to mute rates of turnover. For instance, many groups of planktic foraminifera and coccolithophorids show moderate levels of extinction (average duration of 5–10 m.y.) despite global distributions (Wei and Kennett 1986; Stanley et al. 1988; Collins 1989; Norris 1991, 1992, 1999). Furthermore, most extinction in planktic foraminifera and calcareous nannoplankton occurs by attrition in both the Atlantic and Pacific (e.g., Kitchell 1987), whereas one might expect the global forces that must cause extinction to affect many species simultaneously. Remarkably, extinctions rarely account for the loss of more than a few taxa in any single event (Wei and Kennett 1986; Kitchell 1987; Stanley et al. 1988).

Extinction may result from biological mechanisms (pathogens, parasitism, predation) rather than physical mechanisms (Emiliani 1982). The apparent genetic homogeneity of some pelagic species (e.g., *Orbulina* genotypes discussed in Darling et al. 1999 and de Vargas et al. 1999) suggest that they may not always have enough genetic diversity to resist a novel pathogen. A pathogen need not wipe out the entire species at once but could reduce its populations sizes sufficiently to cause a population bottleneck and speciation. Indeed, the relatively small effective population sizes observed in some copepod taxa (Bucklin and Wiebe 1998) suggest that they may be susceptible to population bottlenecks that drastically reduce genetic variation and increase susceptibility to extinction. Alternatively, species may be more susceptible to subtle changes in their habitats (perhaps owing to narrower tolerance limits than we generally appreciate) and so the habitats themselves may not be as large or as stable as is generally believed. Indeed, genetic evidence for numerous cryptic species in pelagic taxa is consistent with the idea that they are much more specialized than would be inferred from their zoogeographic distributions in traditional taxonomies.

Pelagic microfaunas are known for numerous examples of anagenetic morphological trends. Gradual changes in morphology suggest that there is complete genetic continuity between the ends of the morphoclines without the evolution of distinct, stable species. However, as noted above, the limited isotopic data available for some of these anagenetic transitions suggests that morphoclines may consist of a series of ecologically distinct species that differ in their depths of reproduction (e.g., Norris et al. 1993, 1996; Kelly et al. 1996b) or in the occurrence of photosymbiosis (e.g., Kelly et al. 1996a; Norris 1996a,b). Even morphological data, as in the data for *Fohsella, G. truncatulinoides,* and *C. contusa* described earlier, suggest that many of the classic cases of phyletic gradualism consist of a succession of subtly distinctive morphotypes that are good species in their own right. The presence of numerous "cryptic" species hidden within some modern morphospecies further suggests that clades may be much bushier than has been generally appreciated. A consequence of this hidden ecological and genetic diversity is that some anagenetic trends may actually contain numerous biological species.

Conclusions

Biogeographic, ecological, and genetic data suggest that morphological taxonomies have underestimated the number of pelagic species in a wide assortment of fish, zooplankton, and phytoplankton. Recent genetic studies also strongly suggest that some pelagic genotypes have essentially global distributions within large, relatively stable surface water masses. Other species may be broadly distributed but fail to maintain large populations in water masses that they are known to disperse into. Such cases strongly suggest that biogeographic ranges are determined more by where pelagic species can maintain viable populations than by dispersal. Therefore, speciation models that rely on limiting dispersal, such as vicariant and allopatric hypotheses, are probably not as satisfactory for explaining pelagic biodiversity as are sympatric and parapatric models. Speciation in the open ocean shares important similarities with that in widely dispersing insects, in which changes in season or location of mating are key to establishing reproductive isolation.

Major paleoceanographic events such as the closure of the Central American Seaway and the strengthening of hydrographic boundaries between water masses contribute to speciation by altering the vertical structure, seasonality, and ecology of the water masses in which viable reproductive populations can be maintained. For populations within marginal seas or those that span hydrographic fronts, the timing of reproduction or their mate recognition cues may be distorted from those of their conspecifics elsewhere. Paleoceanographic events that lead to extinction must be global events that destroy the species, or its habitat, before the species can repopulate by long-distance dispersal. The increasing recognition of numerous cryptic species within clades that have been traditionally regarded as low-diversity groups suggests that open-ocean taxa may have narrower tolerance limits and more circumscribed niches than has been generally appreciated.

Finally, the emerging recognition that many traditionally described species consists of clusters of closely related taxa suggests that at least some cases of phyletic gradualism will need to be reevaluated as the result of clado-genesis and successive replacement of sister species. Indeed, it may be that the oceanic microfossil record contains fewer instances of true gradual transitions and represents less of a challenge to cladogenetic models of speciation than has been believed.

Acknowledgments

The ideas in this paper have been derived principally from discussions with W. Chaisson, C. de Vargas, D. Kroon, D. C. Kelly, and D. Lazarus. I thank M. Hart and P. Pearson for insightful and constructive reviews. This work has been supported by grants from the Geology and Paleontology Division of the National Science Foundation.

Literature Cited

Antoine, D., J.-M. Andre, and A. Morel. 1996. Oceanic primary production 2. Estimation at global scale from satellite (coastal zone color scanner) chlorophyll. Global Biogeochemical Cycles 10:57–69.

Arnold, A. J. 1983. Phyletic evolution in the *Globorotalia crassaformis* (Galloway and Wissler) lineage: a preliminary report. Paleobiology 9:390–398.

Bakun, A. 1996. Patterns in the ocean. National Oceanic and Atmospheric Administration, Washington, D.C.

Bé, A. W. H. 1980. Gametogenic calcification in a spinose planktonic foraminifer, *Globigerinoides sacculifer* (Brady). Marine Micropaleontology 5:283–310.

Bell, T. H., Jr., A. B. Mays, and W. P. deWitt. 1974. Upper ocean stability: a compilation of density and Brunt-Väisälä frequency distribution for the upper 500 m of the World Ocean. Naval Research Laboratory, Washington, D.C.

Benzie, J. A., and S. T. Williams. 1997. Genetic structure of giant clam (*Tridacna maxima*) populations in the west Pacific is not consistent with dispersal by present-day ocean currents. Evolution 51:768–783.

Berggren, W. A., and R. D. Norris. 1997. Biostratigraphy, phylogeny and systematics of Paleocene trochospiral planktic foraminifera. Micropaleontology 43(Suppl. to No. 1):1–166.

Bijma, J., J. Erez, and C. Hemleben. 1990. Lunar and semi-lunar reproductive cycles in some spinose planktonic foraminifers. Journal of Foraminiferal Research 20:117–127.

Bottomley, M., C. K. Folland, J. Hsiung, R. E. Newell, and D. E. Parker. 1990. Global ocean surface temperature atlas. The Meteorological Office, Bracknell, England.

Bouvier-Soumagnac, Y., and J. C. Duplessy. 1985. Carbon and oxygen isotopic composition of planktonic foraminifera from laboratory culture, plankton tows and recent sediment: implications for the reconstruction of paleoclimatic conditions and of the global carbon cycle. Journal of Foraminiferal Research 15:302–320.

Briggs, J. C. 1994. Species diversity: land and sea compared. Systematic Biology 43:130–135.

Brooks, D. R., and D. A. McLennan. 1991. Phylogeny, ecology, and behavior. University of Chicago Press, Chicago.

Brown, C. W., and J. A. Yoder. 1994. Coccolithophorid blooms in the global ocean. Journal of Geophysical Research 99(C4): 747–782.

Bucklin, A. 1986. The genetic structure of zooplankton populations. Pp. 35–41 *in* Pierrot-Bults et al. 1986.

Bucklin, A., and P. H. Wiebe. 1998. Low microchondrial diversity and small effective population sizes of the copepods *Calanus finmarchicus* and *Nannocalanus minor*: possible impact of climatic variation during recent glaciation. Journal of Heredity 89:383–392.

Bucklin, A., T. C. LaJeunesse, E. Curry, J. Wallinga, and K. Garrison. 1996. Molecular diversity of the copepod: genetic evidence of species and population structure in the North Atlantic Ocean. Journal of Marine Research 54:285–310.

Bucklin, A., S. B. Smolenack, A. M. Bentley, and P. H. Wiebe. 1997. Gene flow patterns of the euphausiid, *Meganyctiphanes norvegica*, in the NW Atlantic based on mtDNA sequences for cytochrome *b* and cytochrome oxidase I. Journal of Plankton Research 19:1763–1781.

Chaisson, W. P., and R. M. Leckie. 1993. High-resolution Neogene planktonic foraminifer biostratigraphy of Site 806, Ontong Java Plateau (western equatorial Pacific). Pp. 137–178 *in* W. H. Berger, L. W. Kronke, L. A. Mayer, et al., eds. Proceedings of the Ocean Drilling Program, Scientific Results 130. College Station, Tex.

Chaisson, W. P., and P. N. Pearson. 1997. Planktonic foraminifer biostratigraphy at Site 925: middle Miocene–Pleistocene. Proceedings of the Ocean Drilling Program, Scientific Results 154:3–31. College Station, Tex.

Collins, L. S. 1989. Evolutionary rates of a rapid radiation: the Paleogene planktic foraminifera. Palaios 4:251–263.

Collins, T. 1996. Molecular comparisons of transisthmian species pairs: rates and patterns of evolution. Pp. 303–334 *in* J. B. C. Jackson, A. F. Budd, and A. G. Coates, eds. Evolution and environment in tropical America. University of Chicago Press, Chicago.

Coyne, J. A. 1992. Genetics and speciation. Nature 355:511–515.

Darling, K. F., C. M. Wade, D. Kroon, A. J. Leigh Brown, and J. Bijma. 1999. The diversity and distribution of modern planktic foraminiferal small subunit ribosomal RNA genotypes and their potential as tracers of present and past ocean circulations. Paleoceanography 14:3–12.

Darling, K. F., C. M. Wade, I. A. Stewart, D. Kroon, R. Dingle, and A. J. Leigh Brown. 2000. Molecular evidence for genetic mixing of Arctic and Antarctic subpolar populations of planktonic foraminifers. Nature 405:43–47.

de Vargas, C., R. Norris, L. Zaninetti, S. W. Gibb, and J. Pawlowski. 1999. Molecular evidence of cryptic speciation in planktonic foraminifers and their relation to oceanic provinces. Proceedings of the National Academy of Sciences USA 96: 2864–2868.

de Vargas, C., S. Renaud, H. Hilbrecht, and J. Pawlowski. In press. Pleistocene adaptive radiation in *Globorotalia truncatulinoides*: genetic, morphological and environmental evidence. Paleobiology 27(1).

DeLong, E. F., K. Y. Wu, D. D. Prizelin, and R. V. M. Jovine. 1994. High abundance of Archea in Antarctic marine picoplankton. Nature 371:695–697.

Deuser, W. G. 1987. Seasonal variations in isotopic composition and deep-water fluxes of the tests of perennially abundant planktonic foraminifera of the Sargasso Sea: results from sediment-trap collections and their paleoceanographic significance. Journal of Foraminiferal Research 17:14–27.

Dieckmann, U., and M. Doebell. 1999. On the origin of species by sympatric speciation. Nature 400:354–351.

Dortch, Q., and T. T. Packhard. 1989. Differences in biomass structure between oligotrophic and eutrophic marine ecosystems. Deep-Sea Research 36:223–240.

Emery, W. J., and J. Meincke. 1986. Global water masses: summary and review. Oceanologica Acta 9:383–391.

Emiliani, C. 1982. Extinctive evolution: extinctive and competitive evolution combine into a unified model of evolution. Journal of Theoretical Biology 97:13–33.

Endler, J. A. 1977. Geographic variation, speciation, and clines. Monographs in Population Biology Vol. 10. Princeton University Press, Princeton, N.J.

Etter, R. J., M. A. Rex, M. C. Chase, and J. M. Quattro. 1999. A genetic dimension to deep-sea biodiversity. Deep-Sea Research 46:1095–1099.

Fairbanks, R. G., and P. H. Wiebe. 1980. Foraminifera and chlorophyll maximum: vertical distribution, seasonal succession, and paleoceanographic significance. Science 209:1524–1526.

Fairbanks, R. G., M. Sverdlove, R. Free, P. H. Wiebe, and A. W. H. Bé. 1982. Vertical distribution and isotopic fractionation of living planktonic foraminifera from the Panama Basin. Nature 298:841–844.

Ferris, M. J., and B. Palenik. 1998. Niche adaptation in ocean cyanobacteria. Nature 396:226–228.

Fleminger, A. 1986. The Pleistocene equatorial barrier between the Indian and Pacific oceans and a likely cause for Wallace's line. Pp. 84–97 in Pierrot-Bults et al. 1986.

Fuhrman, J. A., and L. Campbell. 1998. Microbial microdiversity. Nature 393:410–411.

Gaard, E. 1996. Life cycle, abundance and transport of Calanus finmarhicus in Faroese waters. Ophelia 44:59–70.

Garfield, P. C., T. T. Packhard, G. E. Friederich, and L. A. Codispoli. 1983. A subsurface particle maximum layer and enhanced microbial activity in the secondary nitrate maximum of the northeastern tropical Pacific Ocean. Journal of Marine Research 41:747–768.

Gibbs, R. H. 1986. The stomioid fish genus Eustomias and the oceanic species concept. Pp. 98–103 in Pierrot-Bults et al. 1986.

Giovannoni, S. J., T. B. Britschgi, C. L. Moyer, and K. G. Field. 1990. Genetic diversity in Sargasso Sea bacterioplankton. Nature 345:60–63.

Gislason, A., and O. S. Astthorsson. 1996. Seasonal development of Calanus finmarchicus along an inshore–offshore gradient southwest of Iceland. Ophelia 44:71–84.

Gold, J. R., and L. R. Richardson. 1998. Mitochondrial DNA diversification and population structure in fishes from the Gulf of Mexico and western Atlantic. Journal of Heredity 89:404–414.

Hastenrath, S., and J. Merle. 1987. Annual cycle of subsurface thermal structure in the tropical Atlantic Ocean. Journal of Physical Oceanography 17:1518–1538.

Healy-Williams, N., R. Ehrlich, and D. F. Williams. 1985. Morphometric and stable isotopic evidence for subpopulations of Globorotalia truncatulinoides. Journal of Foraminiferal Research 15:242–253.

Hemleben, C., M. Spindler, and O. R. Anderson. 1989. Modern planktonic foraminifera. Springer, New York.

Honjo, S. 1996. Fluxes of particles to the interior of the open oceans. Pp. 91–154 in V. Ittekkot, P. Schäfer, S. Honjo, and P. J. Depetris, eds. Particle flux in the ocean. Wiley, New York.

Honjo, S., and S. J. Manganini. 1993. Annual biogenic particle fluxes to the interior of the North Atlantic Ocean: studies at 34°N 21°W and 48°N 21°W. Deep-Sea Research 40:587–607.

Honjo, S., J. Dymond, W. Prell, and V. Ittekkot. 1999. Monsoon-controlled export fluxes to the interior of the Arabian Sea. Deep-Sea Research II 46:1859–1902.

Howard, D. J., and S. H. Berlocher, eds. 1998. Endless forms: species and speciation. Oxford University Press, Oxford.

Huber, B. T., J. Bijma, and K. Darling. 1997. Cryptic speciation in the living planktonic foraminifer Globigerinoides siphonifera (d'Orbigny). Paleobiology 23:33–62.

Hunter, R. S. T., A. J. Arnold, and W. C. Parker. 1988. Evolution and homeomorphy in the development of the Paleocene Planorotalites psuedomenardii and the Miocene Globorotalia (Globorotalia) margaritae lineages. Micropaleontology 31:181–192.

Kelly, D. C., A. J. Arnold, and W. C. Parker. 1996a. Paedomorphosis and the origin of the Paleogene planktonic foraminiferal genus Morozovella. Paleobiology 22:266–281.

Kelly, D. C., T. J. Bralower, J. C. Zachos, I. Permoli Silva, and E. Thomas. 1996b. Rapid diversification of planktonic foraminifera in the topical Pacific (ODP Site 865) during the late Paleocene thermal maximum. Geology 24:423–426.

Kelly, D. C., T. J. Bralower, and J. C. Zachos. 1998. Evolutionary consequences of the latest Paleocene thermal maximum for tropical planktonic foraminifera. Palaeogeography, Palaeoclimatology, Palaeoecology 141:139–161.

Kitchell, J. F. 1987. The temporal distribution of evolutionary and migrational events in pelagic systems: episodic or continuous? Paleoceanography 2:437–487.

Knowlton, N. 1993. Sibling species in the sea. Annual Reviews of Ecology and Systematics 24:189–216.

———. 1997. Species of marine invertebrates: a comparison of the biological and phylogenetic species concepts. Pp. 199–219 in M. F. Claridge, H. A. Dawah, and M. R. Wilson, eds. Species: the units of biodiversity. Chapman and Hall, New York.

———. 2000. Molecular genetic analyses of species boundaries in the sea. Hydrobiologia 420:73–90.

Knowlton, N., and L. A. Weight. 1998. New dates and new rates for divergence across the Isthmus of Panama. Proceedings of the Royal Society of London B 265:2257–2263.

Knowlton, N., L. A. Weight, L. A. Solórzano, E. K. Mills, and E. Bermingham. 1993. Divergence in proteins, Mitochondrial DNA, and the reproductive compatibility across the Isthmus of Panama. Science 260:1629–1632.

Kondrashov, A. S., and F. A. Kondrashov. 1999. Interactions among quantitative traits in the course of sympatric speciation. Nature 400:351–354.

Kucera, M., and B. A. Malmgren. 1998. Differences between evolution of mean form and evolution of new morphotypes: an example from Late Cretaceous planktonic foraminifera. Paleobiology 24:49–63.

Lazarus, D. 1983. Speciation in pelagic protista and its study in the planktonic microfossil record: a review. Paleobiology 9: 327–341.

———. 1986. Tempo and mode of morphologic evolution near the origin of the radiolarian lineage Pterocanium prismatium. Paleobiology 12:175–189.

Lazarus, D., H. Hilbrecht, C. Pencer-Cervato, and H. Thierstein. 1995. Sympatric speciation and phyletic change in Globorotalia truncatulinoides. Paleobiology 21:28–51.

Lessios, H. A. 1998. The first stage of speciation as seen in organisms separated by the Isthmus of Panama. Pp. 186–201 in Howard and Berlocher 1998.

Lohmann, G. P. 1992. Increasing seasonal upwelling in the subtropical South Atlantic over the past 700,000 yrs: evidence from deep-living planktonic foraminifera. Marine Micropaleontology 19:1–12.

Lohmann, G. P., and B. A. Malmgren. 1983. Equatorward migration of Globorotalia truncatulinoides ecophenotypes through the late Pleistocene: gradual evolution or ocean change? Paleobiology 9:414–421.

Malmgren, B. A., and W. A. Berggren. 1987. Evolutionary changes in some late Neogene planktonic foraminifera lineages and their relationships to paleoceanographic changes. Paleoceanography 2:445–456.

Malmgren, B. A., and J. P. Kennett. 1981. Phyletic gradualism in a late Cenozoic planktonic foraminiferal lineage: DSDP 284, Southwest Pacific. Paleobiology 7:230–240.

Malmgren, B. A., W. A. Berggren, and G. P. Lohmann. 1983. Evidence for punctuated gradualism in the late Neogene *Globorotalia tumida* lineage of planktonic Foraminifera. Paleobiology 9:377–389.

———. 1984. Species formation through punctuated gradualism in plantonic foraminifera. Science 225:317–319

Malmgren, B. A., M. Kucera, and G. Ekman. 1996. Evolutionary changes in supplementary apertural characteristics of the late Neogene *Spheroidinella dehiscens* lineage (planktonic foraminifera). Palaios 11:96–110.

McGowan, J. A. 1971. Oceanic biogeography of the Pacific. Pp. 3–74 *in* B. M. Funnell and W. R. Riedel, eds. The micropalaeontology of the oceans. Cambridge University Press, Cambridge.

———. 1972. The nature of oceanic ecosystems. Pp. 9–28 *in* C. B. Miller, ed. The biology of the oceanic Pacific. Oregon State University Press, Corvallis.

———. 1986. The biogeography of pelagic ecosystems. Pp. 191–200 *in* Pierrot-Bults et al. 1986.

McGowan, J. A., and P. W. Walker. 1993. Pelagic diversity patterns. Pp. 203–214 *in* R. E. Ricklefs and D. Schluter, eds. Species diversity in ecological communities. University of Chicago Press, Chicago.

Michaels, A. F., and M. W. Silver. 1988. Primary production, sinking fluxes and the microbial food web. Deep-Sea Research 35:473–490.

Miya, M., and M. Nishida. 1997. Speciation in the open ocean. Nature 389:803–804.

Mooney-Seus, M. L., and G. S. Stone. 1997. The forgotten giants. Ocean Wildlife Campaign, Boston.

Moore, L. R., G. Rocap, and S. W. Chisholm. 1998. Physiology and molecular phylogeny of coexisting *Prochlorococcus* ecotypes. Nature 393:464–467.

Mulhern, P. J. 1987. The Tasman Front: a study using satellite infrared imagery. Journal of Physical Oceanography 17:1148–1155.

Norris, R. D. 1991. Biased extinction and evolutionary trends. Paleobiology 17:388–399.

———. 1992. Extinction selectivity and ecology in planktonic foraminifera. Palaeogeography, Palaeoclimatology, Palaeoecology 95:1–17.

———. 1996a. Symbiosis as an evolutionary innovation in the radiation of Paleocene planktic foraminifera. Paleobiology 22:461–480.

———. 1996b. Symbiosis as an evolutionary innovation in the radiation of planktic foraminifera. Paleontological Society Special Publication 8:291.

———. 1998. Neogene planktonic foraminifer biostratigraphy of Leg 159 sites; Eastern Equatorial Atlantic. Proceedings of the Ocean Drilling Program, Scientific Results 159:445–479. College Station, Tex.

———. 1999. Hydrographic and tectonic control of plankton distribution and evolution. Pp. 173–193 *in* F. Abrantes and A. Mix, eds. Reconstructing ocean history: a window into the future. Plenum, London.

Norris, R. D., and C. de Vargas. 2000. Evolution all at sea. Nature 405:23–24.

Norris, R. D., R. M. Corfield, and J. E. Cartlidge. 1993. Evolution of depth ecology in the planktic Foraminifera lineage *Globorotalia* (*Fohsella*). Geology 21:975–978.

———. 1994. Evolutionary ecology of *Globorotalia* (*Globoconella*) (planktic foraminifera). Marine Micropaleontology 23:121–145.

———. 1996. What is gradualism? Cryptic speciation in globorotaliid planktic foraminifera. Paleobiology 22:386–405.

Olsen, G. J. 1990. Variation among the masses. Nature 345:20.

Orr, W. N. 1967. Secondary calcification in the foraminiferal genus *Globorotalia*. Science 157:1554–1555.

Palumbi, S. R. 1998. Species formation and the evolution of gam-

ete recognition loci. Pp. 271–278 *in* Howard and Berlocher 1998.

———. 1994. Genetic divergence, reproductive isolation, and marine speciation. Annual Review of Ecology and Systematics 25:547–572.

Palumbi, S. R., G. Grabowsky, T. Duda, L. Geyer, and N. Tachino. 1997. Speciation and population genetic structure in tropical Pacific sea urchins. Evolution 51:1506.

Parsons, T. R., M. Takahashi, and B. Hargrave. 1984. Biological oceanographic processes. Pergamon, New York.

Pearson, P. N., N. J. Shackleton, and M. A. Hall. 1997. Stable isotopic evidence for the sympatric divergence of *Globigerinoides trilobus* and *Orbulina universa* (planktonic foraminifera). Journal of the Geological Society, London 154:295–302.

Pierrot-Bults, A. C., and S. van der Spoel. 1979. Speciation in macrozooplankton. Pp. 144–167 *in* S. van der Spoel and A. C. Pierrot-Bults, eds. Zoogeography and diversity of plankton. Halstead, New York.

Pierrot-Bults, A. C., S. van der Spoel, B. J. Zahuranec, and R. K. Johnson, eds. 1986. Pelagic Biogeography. UNESCO, Paris.

Quillévéré, F., R. D. Norris, I. Moussa, and W. A. Berggren. 2001. Role of photosymbiosis and biogeography in the diversification of early Paleogene acarininids (planktonic foraminifera). Paleobiology 27(2) (in press).

Raff, R. A. 1996. The shape of life. University of Chicago Press, Chicago.

Reynolds, L., and R. C. Thunell. 1985. Seasonal succession of planktonic foraminifera in the subpolar North Pacific. Journal of Foraminiferal Research 15:282–301.

Rice, W. R., and E. E. Hostert. 1993. Laboratory experiments on speciation: what have we learned in 40 years? Evolution 47:1637–1653.

Rögl, F. 1985. Late Oligocene and Miocene planktic foraminifera of the Central Paratethys. Pp. 315–328 *in* H. M. Bolli, J. B. Saunders and K. Perch-Nielsen, eds. Plankton stratigraphy. Cambridge University Press, Cambridge.

Sautter Reynolds, L. R., and R. C. Thunell. 1991. Seasonal variability in the $d^{18}O$ and $d^{13}C$ of planktonic foraminifera from an upwelling environment: sediment trap results from the San Pedro Basin, Southern California bight. Paleoceanography 6:307–334.

Schmitz, W. J., Jr., 1996. On the world ocean circulation, Vol. 1. Some global features: North Atlantic circulation. Woods Hole Oceanographic Institution Technical Report WHOI-96-03:1–141.

Schneider, C. E., and J. P. Kennett. 1996. Isotopic evidence for interspecies habitat differences during evolution of the Neogene planktonic foraminiferal clade *Globoconella*. Paleobiology 22:282–303.

Sieburth, J. M. 1986. Dominant microorganisms of the upper ocean: form and function, spatial distribution and photoregulation of biochemical processes. Pp. 157–186 *in* J. D. Burton, P. G. Brewer, and R. Chesselet, eds. Dynamic processes in the chemistry of the upper ocean. Plenum, London.

Silver, M. W., and A. L. Allredge. 1981. Bathypelagic marine snow: deep-sea algal and detrital community. Journal of Marine Research 39:501–530.

Snell, T. W., and C. A. Hawkinson. 1983. Behavioral reproductive isolation among populations of the rotifer *Branchionus plicatilis*. Evolution 37:1294–1305.

Sorhannus, U. 1990a. Punctuated morphological change in a Neogene diatom lineage: "local" evolution or migration? Historical Biology 3:241–247.

———. 1990b. Tempo and mode of morphological evolution in two Neogene diatom lineages. Pp. 329–370 *in* M. K. Hecht, B. Wallace, and R. J. MacIntyre, eds. Evolutionary biology. Plenum, London.

Spindler, M., O. R. Anderson, C. Hemleben, and A. W. H. Bé.

1978. Light and electron microscopic observations of gametogenesis in *Hastigerina pelagica* (Foraminifera). Journal of Protozoology 25:427–433.

Spindler, M., C. Hemleben, U. Bayer, A. W. H. Bé, and O. R. Anderson. 1979. Lunar periodicity in the planktonic foraminifer *Hastigerina pelagica*. Marine Ecology Progress Series 1:61–64.

Stanley, S. M., K. L. Wetmore, and J. P. Kennett. 1988. Macroevolutionary differences between two major clades of Neogene planktonic foraminifera. Paleobiology 14:235–249.

Tauber, C. A., and M. J. Tauber. 1989. Sympatric speciation in insects: perception and perspective. Pp. 307–344 *in* D. Otte and J. A. Endler, eds. Speciation and its consequences. Sinauer, Sunderland, Mass.

Tauber, C. A., M. J. Tauber, and J. R. Nechols. 1977. Two genes control seasonal isolation in sibling species. Science 197:592–593.

Thunell, R. C., and S. Honjo. 1987. Seasonal and interannual changes in planktonic foraminiferal production in the North Pacific. Nature 328:335–337.

Thunell, R., and R. L. Sautter Reynolds. 1992. Planktonic foraminiferal faunal and stable isotopic indices of upwelling: a sediment trap study in the San Pedro Basin, Southern California Bight. Pp. 77–91 *in* C. P. Summerhayes, W. L. Prell, and K. C. Emeis, eds. Upwelling systems: evolution since the Early Miocene. Geological Society of London, London.

Thunell, R. C., W. B. Curry, and S. Honjo. 1983. Seasonal variation in the flux of planktonic foraminifera: time series sediment trap results from the Panama Basin. Earth and Planetary Science Letters 654:44–55.

Tregenza, T., and R. K. Butlin. 1999. Speciation without isolation. Nature 400:311–312.

van der Spoel, S. 1983. Patterns in plankton distribution and the relation to speciation: the dawn of pelagic biogeography. Pp. 291–334 *in* R. W. Sims, J. H. Price, and P. E. S. Whalley, eds. Evolution, time and space: the emergence of the biosphere. Academic Press, London.

van Soest, R. W. M. 1975. Zoogeography and speciation in the Salpidae (Tunicata, Thaiacea). Beaufortia 23:181–215.

Vrijenhoek, R. C. 1997. Gene flow and genetic diversity in naturally fragmented metapopulations of deep-sea hydrothermal vent animals. Journal of Heredity 88:285–293.

Wefer, G. 1989. Particle flux in the Ocean: effects of episodic production. Pp. 139–153 *in* W. H. Berger, V. S. Smetacek, and G. Wefer, eds. Productivity of the ocean: present and past. Wiley, New York.

Wefer, G., and G. Fischer. 1993. Seasonal patterns of vertical particle flux in equatorial and coastal upwelling areas of the eastern Atlantic. Deep-Sea Research 40:1613–1645.

Wei, K.-Y. 1994. Allometric heterochrony in the Pliocene-Pleistocene planktic foraminiferal clade *Globoconella*. Paleobiology 20:66–84.

Wei, K.-Y., and J. P. Kennett. 1986. Taxonomic evolution of Neogene planktonic foraminifera and paleoceanographic relations. Paleoceanography 1:67–84.

———. 1988. Phyletic gradualism and punctuated equilibrium in the late Neogene planktonic foraminiferal clade *Globoconella*. Paleobiology 14:345–363.

Global climate change and North American mammalian evolution

John Alroy, Paul L. Koch, and James C. Zachos

Abstract.—We compare refined data sets for Atlantic benthic foraminiferal oxygen isotope ratios and for North American mammalian diversity, faunal turnover, and body mass distributions. Each data set spans the late Paleocene through Pleistocene and has temporal resolution of 1.0 m.y.; the mammal data are restricted to western North America. We use the isotope data to compute five separate time series: oxygen isotope ratios at the midpoint of each 1.0-m.y. bin; changes in these ratios across bins; absolute values of these changes (= isotopic volatility); standard deviations of multiple isotope measurements within each bin; and standard deviations that have been detrended and corrected for serial correlation. For the mammals, we compute 12 different variables: standing diversity at the start of each bin; per-lineage origination and extinction rates; total turnover; net diversification; the absolute value of net diversification (= diversification volatility); change in proportional representation of major orders, as measured by a simple index and by a G-statistic; and the mean, standard deviation, skewness, and kurtosis of body mass. Simple and liberal statistical analyses fail to show any consistent relationship between any two isotope and mammalian time series, other than some unavoidable correlations between a few untransformed, highly autocorrelated time series like the raw isotope and mean body mass curves. Standard methods of detrending and differencing remove these correlations. Some of the major climate shifts indicated by oxygen isotope records do correspond to major ecological and evolutionary transitions in the mammalian biota, but the nature of these correspondences is unpredictable, and several other such transitions occur at times of relatively little global climate change. We conclude that given currently available climate records, we cannot show that the impact of climate change on the broad patterns of mammalian evolution involves linear forcings; instead, we see only the relatively unpredictable effects of a few major events. Over the scale of the whole Cenozoic, intrinsic, biotic factors like logistic diversity dynamics and within-lineage evolutionary trends seem to be far more important.

John Alroy. National Center for Ecological Analysis and Synthesis, University of California, 735 State Street, Santa Barbara, California 93101. E-mail: alroy@nceas.ucsb.edu
Paul L. Koch and James C. Zachos. Department of Earth Sciences, University of California, 1156 High Street, Santa Cruz, California 95064. E-mail: pkoch@earthsci.ucsc.edu, jzachos@earthsci.ucsc.edu

Accepted: 22 June 2000

Introduction

Paleontologists since Cuvier and Lyell have sought to explain both sudden extinction events and long-term evolutionary trends by pointing to changes in global climate (Rudwick 1972). Two of the most obvious possible examples are the global extinction of dinosaurs and other vertebrates at the Cretaceous/ Tertiary (K/T) boundary, and the extinction of most large mammals in the Americas at the end of the Pleistocene. Both of these events can now be attributed to catastrophic perturbations that are unrelated, or only weakly related, to global climate change: the K/T mass extinction was probably due to the impact of a large bolide in the Yucatán peninsula (Alvarez et al. 1980; Hildebrand et al. 1991), and a consensus is forming that the end-Pleistocene

extinctions were caused largely, or possibly solely, by human impacts (MacPhee 1999).

Despite these developments, much of the literature going back to Matthew (1915) continues to argue for climate change as a major driver of North American mammalian evolution in the Tertiary. This literature is motivated by clearly climate-related geographical differences among Recent mammalian faunas. Very strong tropical-to-temperate zone gradients in species richness (Simpson 1964; Kaufman 1995) and differences in body mass distributions between wet and dry habitats (Legendre 1989) have to have evolved somehow, so it stands to reason that progressive climate change would cause these biotic variables to respond in a highly predictable way.

Many recent studies have worked within the climate/evolution paradigm. Rose (1981)

0094-8373/00/2604-0011/$1.00

quantified changes in alpha diversity through the Paleocene and early Eocene and attributed them directly to climate change. Webb (1977, 1984) and Barnosky (1989) attributed North American mammalian extinctions throughout the late Cenozoic to climate change. In a ground-breaking analysis, Stucky (1990) suggested that although continental diversity was largely static throughout the Cenozoic, climate change progressively decreased alpha diversity and increased beta diversity. Van Valkenburgh and Janis (1993) presented quantitative data that showed a clear decrease, not increase, in beta diversity—but still sought to explain both this pattern and the overall diversity trajectory in terms of climate change and orogeny. Janis (1993, 1997) and Janis and Wilhelm (1993) explicitly tied extinction rates, continental diversity levels, and changes in trophic structure and ecomorphology to global climate trends. Gunnell et al. (1995) emphasized climate as a driver of body mass evolution in Cenozoic mammals, and joined many other authors in arguing that mammalian body mass distributions are robust indicators of habitat and climate.

Because traditional timescales for the Cenozoic terrestrial fossil record have had poor (~3-m.y.) resolution, many other authors have attacked the climate/evolution problem by focusing on critical intervals of known climate change, and by working with highly resolved faunal data for localized stratigraphic sections. For example, Clyde and Gingerich (1998) were able to cite considerable evidence for abrupt climate change at the Paleocene/Eocene boundary both at the global scale (Zachos et al. 1993; papers in Aubry et al. 1998) and at regional scales (Koch et al. 1995; Wing et al. 1995; Fricke et al. 1998). Following Gingerich (1989), they showed that major faunal turnover events and ecological shifts in Wyoming were stratigraphically coincident with the boundary climate event. Thus, they were able to make a persuasive, albeit circumstantial, case for a causal connection between climate change and an important immigration pulse.

Meng and McKenna (1998) developed a similar argument for mammalian evolution across the Paleocene/Eocene and Eocene/Ol-

igocene boundaries in Asia, although evidence for regional climate change in Asia at either of these boundaries is currently lacking. In contrast, Alroy (1996, 1998d) and Prothero (Prothero and Heaton 1996; Prothero 1999) debated whether there was any substantial turnover among North American mammals at the Eocene/Oligocene boundary, but agreed that climate in general plays a minor role in pacing evolutionary change. Cerling et al. (1998) attributed late Miocene mammalian faunal turnover around the world to floral changes driven by a drop in CO_2 levels. Behrensmeyer et al. (1997) tested and rejected the hypothesis of Vrba (1985, 1992, 1995) that climate drove a major pulse of turnover in the Pliocene of Africa. Vrba (1992) tried to extend her theory to account for faunal turnover during the Great American Interchange, although these faunal changes seem already to be explained by direct effects of this large-scale migration between North and South America (Marshall et al. 1982).

Many other, similar studies could be cited. We are not opposed to the "critical interval" strategy that this literature employs, and in fact we employ it ourselves in part of our discussion. However, we will begin our analysis by trying to take a broader view. Our main goal will be to compare climate and fossil data with equal temporal resolution for a large part of the Cenozoic. Thanks to improvements in quantitative biochronological methods, we can now resolve the North American mammalian fossil record down to uniformly spaced, 1.0-m.y.-long intervals (Alroy 1992, 1994, 1996, 1998d, 2000). Likewise, greatly increased sampling of long deep-sea cores and improvements in correlation now make it possible to examine small-scale changes in climate, sometimes even at Milankovitch timescales, throughout much of the Cenozoic (e.g., Zachos et al. 1996, 1997). Earlier studies were only able to point to general, long-term trends and a small number of obvious, large-scale isotope value transitions (e.g., Miller et al. 1987), but the new isotope data even allow us to quantify the range of variation in climate within individual 1.0-m.y. bins.

Instead of looking for broad-brush, qualitative correspondences between our isotope and mammal data sets, we will begin by taking a rigidly quantitative approach based on basic time series analysis. Our strategy will be to look for any possible causal relationship between the data sets. Therefore, each and every mammalian time series will be cross-correlated with each and every isotope time series. Because of the large number of comparisons, we expect some correlations to be suggestively high just at random. Therefore, we will set a high bar in testing for statistical significance. We also will avoid overinterpreting cases in which two autocorrelated time series show a strong cross-correlation, because such correlations are expected of random walks regardless of whether there is any common causal factor (McKinney 1990).

Although this point about time series autocorrelation is widely understood in the statistical literature, we feel that it needs to be emphasized. There is a strong temptation to (say) look at a decreasing climate curve and an increasing diversity curve, note that each one has a strong trend, and then infer that deteriorating climate has spurred diversification. But any two curves with strong trends will show such apparent correlations. For example, the episodically increasing Cenozoic marine oxygen isotope curve would correlate strongly with data showing the longitude of North America during the same period of time, if only because the steady westward march of the continent happened to have occurred just while climate happened to be changing more or less unidirectionally. In general, then, the question isn't whether pairs of curves show trends that go in the same direction or opposite direction. Instead, we want to know whether the blow-by-blow, interval-by-interval variation in these curves is correlated. In other words, after taking the long-term trends and autocorrelation into account, are fine-scale correlations still visible (McKinney 1990)?

Our time series analysis should identify linear forcings of major aspects of mammalian evolution by climate, should they operate on the timescale of one or two m.y. or less. For example, noting the current latitudinal gradient in mammalian species richness (Simpson 1964; Kaufman 1995), and the fact that latitude is correlated with mean annual temperature, we might posit that time intervals with rapidly rising temperatures should be associated with increased speciation or decreased extinction rates. Yet climate may influence mammalian evolution in ways that are not strongly linear. Rather, the response to a climate change of a particular magnitude might depend on the recurrence time of such events, or even on whether it is the first such event in the Cenozoic. If mammal faunas respond to climate in ways that are strongly contingent on past major events, or if they respond only after extreme lags such as 5, 10, or 20 m.y., our time series approach may not capture the response.

To search for climatic forcing of this sort, we will conduct a second set of analyses somewhat analogous to the "critical interval" approach. We will identify all the time intervals where one or more of the mammalian data sets present clear outliers, and then test to see if any of the climatic parameters are also significantly outside the normal range of variation at the same time. If the majority of the mammalian "events" occurred at times when some aspect of climate is unusual, we would be able to conclude that climate exerts a nonlinear influence on mammalian evolution, with effects not being visible until some threshold of disturbance is crossed.

After running through all possible comparisons of the mammalian faunal data and isotopic data, we will conclude that not only are almost all of them insignificant, but the few significant linear correlations cannot be distinguished from statistical artifacts. We then will argue that the mammalian biota went through only a handful of major reorganizations during this interval, and although some kind of environmental change is thought to have occurred at some of these times, the biotic responses and types of environmental change vary in each case. Furthermore, reductionistic concerns about the relevance of continental data are misplaced because the mammalian data pertain only to a narrow, and climatically relatively uniform, region of the continent. Therefore, we will argue that no case can be made at this time for invoking long-term global climate change as a major and consistent driver of mammalian evolution.

Our analysis is unlikely to be the final word on this topic. As we discuss below, temperature estimates based on marine isotope data may have systematic errors and are not necessarily good estimates of climate in continental interiors even when they are accurate. Furthermore, temperature per se may not be as important to mammals as, say, precipitation. Nonetheless, we will suggest that the results are not due to lack of power and are likely to be replicated by future studies using these types of data. In making this argument, we will side with a relatively small number of authors who believe that most major patterns in mammalian evolution result from biotic factors like competition (Van Valen 1973; Alroy 1996, 1998d, 2000) and key innovation (Van Valen 1971; Hunter and Jernvall 1995; Jernvall et al. 1996), with the underlying biotic dynamics being reset periodically by major environmental perturbations.

Isotopes and Climate

Our climate proxy will be a composite curve of oxygen isotope values for benthic foraminifera from the Atlantic Ocean. This benthic oxygen isotope ($\delta^{18}O$) curve has been viewed as a proxy for "global climate" in studies of biotic response (e.g., Vrba 1995; Prothero 1999). It is worth briefly considering the controls on this record to evaluate the plausibility of this assumption.

The $\delta^{18}O$ of an organism is a function of growth temperature, as well as the $\delta^{18}O$ of the water from which it precipitates calcite (Epstein et al. 1951). The $\delta^{18}O$ of the world's oceans is relatively constant spatially, with subtle variations in the surface ocean induced by evaporation and freshwater runoff (Zachos et al. 1994). Yet the world's oceans experience fairly large temporal shifts in $\delta^{18}O$ owing to the preferential sequestration of ^{16}O-enriched water in glacial ice. Consequently, shifts in the $\delta^{18}O$ of benthic foraminifera at least partially record changes in the global volume of continental ice (Shackleton and Opdyke 1973). For example, in the Pleistocene it is estimated that two-thirds of the shift in benthic foraminiferal $\delta^{18}O$ values between glacial and interglacial

times results from ice volume changes (Schrag et al. 1996). For the mid-Miocene this proportion may be as high as five-sixths, and the Eocene/Oligocene shift may have resulted largely (50–100%) from the growth of Antarctic ice sheets (Zachos et al. 1993; Lear et al. 2000). However, in the warm Paleocene and early-to-middle Eocene the extent of continental glaciation was low (Zachos et al. 1994; Lear et al. 2000), so the impact of this parameter on short-term changes in benthic foraminiferal $\delta^{18}O$ values should be minimal. Shifts in $\delta^{18}O$ values in this time interval should correspond principally to changes in bottom-water temperature.

The temperature of ocean bottom water is determined at the sites of deep water formation, not at sites of sedimentary deposition. Today, deep water forms at high latitudes in the North Atlantic and off Antarctica, where temperatures are low and salinity increases seasonally as sea ice forms. This situation has existed since the late Miocene (Wright et al. 1992). Prior to the late Miocene, all evidence suggests that Antarctica had been the dominant source of global deep waters (Zachos et al. 1994). There has been speculation that deep-water formation may have taken place at low latitudes in some time periods: the mechanism would be the sinking of dense, warm brines generated in equatorial regions under extremely warm climates (e.g., during the Late Cretaceous or at the Eocene thermal maxima [Brass et al. 1982; Kennett and Stott 1991]). At present, however, there is little support for this view from either geochemistry or simulations of ocean circulation (Sloan et al. 1995). We suspect that deep-water formation has remained focused at polar regions for most, if not all, of the Cenozoic. Temperature variations in subpolar regions, and by extension the poles, not the Tropics or midlatitudes, are being recorded in the $\delta^{18}O$ of benthic foraminifera.

Thus, isotopic records from benthic organisms reflect a complex mixture of changes in both high-latitude temperature and ice volume, and are not recording "global" climate in any simple sense. Nonetheless, we believe there are several good reasons to compare our mammalian data sets with this record. Some reasons involve expediency, but the others have a more general rationale. The foremost practical issue is that only the benthic $\delta^{18}O$ record is both dense and complete enough to allow the types of tests presented here. Mg/Ca data may be a more direct proxy for ocean temperatures, but existing data sets are extremely sparse (Lear et al. 2000), and the robustness of this proxy remains to be fully tested. Records of climate on land for North America are incomplete. Isotopic records from surface-dwelling marine organisms in the middle latitudes of the Atlantic and Pacific are more complete, but still too patchy for our current purposes. Also, metabolic isotope effects associated with the presence of endosymbionts in the tissues of some planktonic foraminifera make it difficult to compare records among different taxa without careful, species-by-species analyses of paleoecology (Spero and Lea 1993; Norris 1996). More importantly, where comparisons between benthic $\delta^{18}O$ records and high-resolution records of continental North American climate are available, they do reveal synchronous trends—at least in the Paleogene (an issue we will return to in the Discussion). Finally, a number of authors who have linked climate change to mammalian evolution have used the marine record of climate change as their primary reference point.

Scale of Analysis

One might ask why we have restricted our analyses to a single, fairly large geographic region (western North America), and a single temporal scale of resolution (1.0 m.y.). There are three major reasons.

Implausibility of Opposed Local Effects.—A current trend in the literature is to claim that different geographic regions show fundamentally different macroevolutionary patterns (Miller 1998). However, splitting our mammalian data geographically would matter only if different subregions within western North America experienced not just different trends, but opposed ones. Only reversed climate gradients could cancel out biotic responses if the response functions are monotonic—otherwise, at least some part of our data set would still pick up the biotic response. That would occur, for example, if it

mattered to mammals that during the mid- and late Tertiary the Great Plains grew colder, drier, and more seasonal but the West Coast changed much less. Canceling out, as opposed to slight masking, only would be appreciable if (say) West Coast climate actually became warmer, wetter, and more equable during this period. Given that no major latitudinal changes and no joining or splitting of continents occurred near our study region during our study interval, mechanisms for generating these kinds of reversed intracontinental climate trends would have to be complex.

Lack of Finer-Scale Data.—Using substantially finer temporal bins is impossible because the temporal resolution of the mammalian data is just slightly less than 1.0 m.y. (Alroy 1996, 1998d, 2000), even though the method used to generate the time series is some three times more precise than conventional, subjective approaches (Alroy 1998c). Greater improvements in the mammalian timescale are highly unlikely given that the resolution of our ordination-based timescale is regulated by the underlying rate of faunal turnover, and that some 91% of all fossil mammal collections cannot be associated with independent geochronological age estimates (Alroy 1996, 1998d, 2000). Likewise, studying smaller geographic regions would create unavoidably lengthy temporal gaps. Even a study restricted merely to the Great Plains and Rocky Mountains would suffer from a major gap in the Uintan and Duchesnean (late Eocene), which can be filled only with data from Southern California (Alroy 1998d). Furthermore, even the two most fossiliferous states in the United States present highly episodic records: Nebraska has no substantial faunas older than about 38 Ma, and Wyoming yields an order of magnitude more data from the Paleocene and Eocene than from younger epochs.

Low Power of Coarser Analyses.— Expanding the geographic scale of coverage would add very little data, and therefore would provide a largely redundant data set. The already extreme geographic concentration of data is highlighted by the fact that 4787 (96%) of the lists fall in the western half of North America. Furthermore, 2966 (60%) can be placed in a much smaller rectangle centered within the

TABLE 1. Benthic foraminiferan $\delta^{18}O$ data for 1.0-m.y.-long Cenozoic sampling bins. n = number of samples falling in each bin; Midpoint = estimated $\delta^{18}O$ value at midpoint of each bin, based on linear regression of $\delta^{18}O$ on time for all values falling within the bin; Change = difference between current $\delta^{18}O$ midpoint value and preceding one; SD = standard deviation of all residuals around the regression line, i.e., a measure of variability independent of gradual secular trends within bins; Detrended SD = generalized difference of standard deviation (see text); N/A = not applicable because data are lacking. Volatility (absolute value of change) is omitted because these values differ from the changes only by their sign.

Bin (Ma)	n	Midpoint	Change	SD	Detrended SD
59.5	8	0.617	N/A	0.112	−0.047
58.5	19	0.414	−0.203	0.209	0.064
57.5	14	0.270	−0.144	0.188	0.005
56.5	13	0.113	−0.157	0.144	−0.032
55.5	95	−0.047	−0.160	0.467	0.307
54.5	34	−0.203	−0.156	0.197	−0.090
53.5	12	−0.228	−0.025	0.126	−0.057
52.5	7	−0.578	−0.350	0.172	0.016
51.5	4	−0.584	−0.006	0.146	−0.129
50.5	0	N/A	N/A	N/A	N/A
49.5	3	−0.600	N/A	0.226	0.098
48.5	6	−0.492	0.108	0.165	−0.034
47.5	14	−0.202	0.290	0.081	−0.095
46.5	13	0.075	0.277	0.163	0.019
45.5	6	0.197	0.122	0.164	−0.013
44.5	10	0.384	0.187	0.049	−0.129
43.5	31	0.223	−0.161	0.219	0.085
42.5	13	0.408	0.185	0.303	0.101
41.5	14	0.268	−0.140	0.343	0.107
40.5	17	0.584	0.316	0.123	−0.129
39.5	14	0.719	0.135	0.100	−0.067
38.5	15	0.850	0.131	0.091	−0.068
37.5	15	1.006	0.156	0.130	−0.027
36.5	27	1.112	0.106	0.227	0.054
35.5	38	1.179	0.067	0.166	−0.046
34.5	93	1.132	−0.047	0.161	−0.028
33.5	127	1.816	0.684	0.259	0.071
32.5	99	1.968	0.152	0.215	−0.012
31.5	114	1.850	−0.118	0.181	−0.030
30.5	67	1.969	0.119	0.213	0.014
29.5	31	2.115	0.146	0.214	0.002
28.5	48	2.038	−0.077	0.387	0.173
27.5	31	2.151	0.113	0.298	0.016
26.5	41	2.037	−0.114	0.385	0.137
25.5	139	1.198	−0.839	0.290	0.007
24.5	210	0.966	−0.232	0.260	0.013
23.5	199	1.355	0.389	0.295	0.059
22.5	99	1.117	−0.238	0.230	−0.021
21.5	9	1.408	0.291	0.192	−0.035
20.5	14	1.573	0.165	0.176	−0.037
19.5	17	1.414	−0.159	0.147	−0.060
18.5	19	1.298	−0.116	0.173	−0.024
17.5	23	1.343	0.045	0.163	−0.045
16.5	26	1.151	−0.192	0.163	−0.042
15.5	14	1.213	0.062	0.102	−0.104
14.5	7	1.209	−0.004	0.156	−0.027
13.5	84	1.621	0.412	0.192	−0.013
12.5	160	1.802	0.181	0.232	0.011
11.5	156	1.881	0.079	0.212	−0.025
10.5	63	1.982	0.101	0.165	−0.065
9.5	38	2.144	0.087	0.129	−0.090

TABLE 1. Continued.

Bin (Ma)	n	Midpoint	Change	SD	Detrended SD
8.5	186	2.057	0.075	0.179	−0.034
7.5	60	2.070	−0.074	0.199	−0.002
6.5	267	2.206	0.136	0.242	0.013
5.5	328	2.266	0.060	0.330	0.083
4.5	316	2.334	0.068	0.232	−0.050
3.5	304	2.457	0.123	0.252	0.007
2.5	493	2.898	0.441	0.327	0.073
1.5	513	3.121	0.223	0.361	0.077
0.5	528	3.433	0.312	0.492	0.194

Western Interior that spans less than 6% of the continental area. Meanwhile, analyzing the time series data with larger temporal bins (say, 2.0 m.y.) would reduce the number of data points so much that finding any robust cross-correlations between mammalian and isotopic data sets would be rendered almost impossible.

Data

Isotopic Data.—Isotopic data for benthic foraminifera were compiled from 17 DSDP cores in the Atlantic basin (Table 1, Fig. 1). Age models for each core were based on a combination of magnetostratigraphy and biostratigraphy, using dates for chron and biozone boundaries from Berggren et al. (1995). Where possible, correlations among cores at the 10^4 timescale were determined by matching cyclic patterns in carbon isotope values or sediment properties. Details of the construction and paleoclimatic implications of the benthic curve will be discussed in a future paper. The density of data generally decreases with increasing age (Fig. 2). This problem is so severe for the early Paleocene that we have had to restrict our analysis of isotopic data to the period from 60 Ma to the Recent.

Once placed in age models, $\delta^{18}O$ data from all cores were combined to produce a composite curve. The curve was divided into 1.0-m.y. bins to facilitate comparison with the similarly parsed mammal record. Four basic metrics were determined from these binned data (Table 1, Fig. 1A–D): midpoint $\delta^{18}O$ value, rate of change, volatility, and variability. To determine the midpoint $\delta^{18}O$ value of each bin, we regressed $\delta^{18}O$ on age for values falling in each bin, then calculated the $\delta^{18}O$ value for the

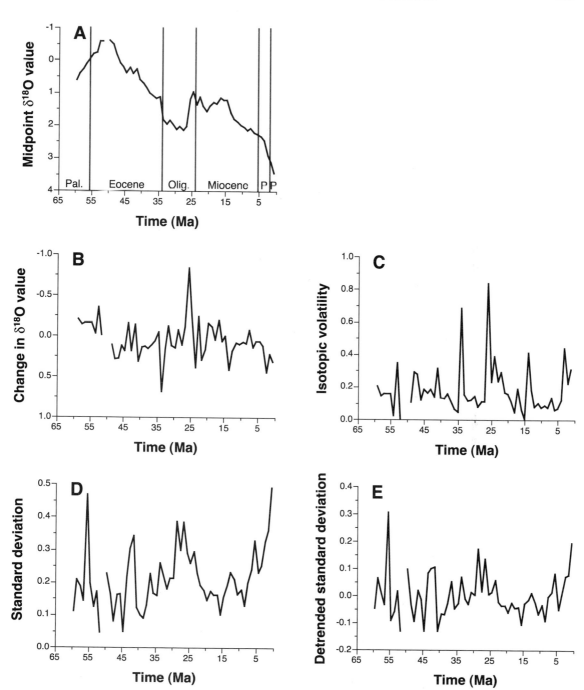

FIGURE 1. Cenozoic oxygen isotope records for Atlantic benthic foraminifera. Gap in each line at 50.5 Ma reflects missing data for that interval. Vertical gray lines show international epoch boundaries; epochs are listed at bottom. Abbreviations: Pal. = Paleocene; Olig. = Oligocene; P P = Pliocene and Pleistocene. A, Midpoint $\delta^{18}O$ value: estimated value at the midpoint of each bin based on a regression of $\delta^{18}O$ values against time within each bin. These data show a strong trend and strong autocorrelation. The y-axis is reversed because positive values are correlated with low temperatures. B, Change in value: difference between consecutive bins in midpoint $\delta^{18}O$ value. These data are first differences and show no trend and no autocorrelation. The y-axis is reversed as in the preceding panel; negative changes indicate warming/ice sheet melting and positive changes cooling/ice sheet growth. C, Isotopic volatility: absolute values of first differences. D, Standard deviation: variability of the residuals produced by regressing $\delta^{18}O$ values against time. These data show a weak trend and weak autocorrelation. E, Detrended standard deviation: generalized differences of the standard deviation data. These data show no trend and no autocorrelation.

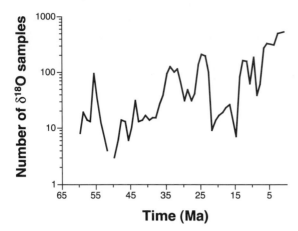

FIGURE 2. Number of oxygen isotope ratio determinations per 1.0-m.y.-long sampling bin.

bin's temporal midpoint from the regression equation (Fig. 1A). This method should provide a more robust measure of the average $\delta^{18}O$ of a time interval than simple averaging, especially in cases where data show directional, within-bin $\delta^{18}O$ shifts and the sampling is uneven through time. We measured the rate of change in $\delta^{18}O$ as the difference in midpoint $\delta^{18}O$ value between sequential bins (younger bin − older bin: Fig. 1B). We then took the absolute values of the changes in order to estimate isotopic volatility—that is, the magnitude of changes regardless of their direction (Fig. 1C). Finally, to estimate variability we determined the standard deviation of the regression residuals for each bin (Fig. 1D).

Both the midpoint and standard deviation time series show secular trends and autocorrelation. The midpoint data (Fig. 1A) are very strongly correlated with time ($r = -0.866; p < 0.001$), and even after removing this trend they still have a high serial correlation (i.e., the correlation between the variable and its own values lagged backwards by one bin: $r = +0.887; p < 0.001$). Cross-correlating a time series like this one with other autocorrelated variables would be meaningless, so detrending and differencing is necessary. The generalized differencing method discussed by McKinney (1990) would be appropriate, but owing to the strong serial correlation it turns out that the first differences we already have obtained (i.e., rate of change: Fig. 1B) and generalized differences (not shown) are almost

identical for this variable. Therefore, we treat the differenced midpoint $\delta^{18}O$ value data set (Fig. 1B) as our main standard for evaluating the Cenozoic trend in average $\delta^{18}O$ values. This data set exhibits neither a significant correlation with time ($r = +0.227$; n.s.) nor a significant serial correlation ($r = +0.141$; n.s.). The volatility data set (Fig. 1C) is just a simple transform of the unproblematic differenced data, so it was not subjected to further manipulations.

By contrast, the standard deviation data (Fig. 1D) present more typical features of time series: a weak but significant correlation with time ($r = -0.300; p < 0.05$), and a similarly weak and significant serial correlation ($r = +0.391; p < 0.01$). A correction is needed, but taking first differences would remove too much of the signal in this data set. Therefore, here we did apply the much less draconian generalized differencing method (McKinney 1990): the data were regressed against time, residuals were taken, the correlation of the residuals with themselves at a lag of one time interval was computed, each of the lagged values was multiplied by this correlation, and finally the lagged values were subtracted from the data. The resulting, corrected curve (Fig. 1E) resembles the raw data fairly closely because of the weak correlations, but differs in lacking a visible secular trend (correlation against time: $r = -0.033$; n.s.) and having less dramatic medium-duration swings (serial correlation: $r = 0.000$; n.s.).

While we are confident that the values for midpoint $\delta^{18}O$ and rate of change are adequately captured for all bins with more than 10 or 15 values, we are less sanguine about our estimator of variability. We explored the sensitivity of our estimator to the dramatic temporal variation in sample size (Fig. 2) by analyzing data from the 0.5-Ma bin. This bin not only had the most data but also showed the highest variability, no doubt in response to Pleistocene glacial cycles. We randomly sampled the regression residuals from this bin and generated new sequences of residuals with samples sizes of 500, 300, 150, 75, 50, 25, 10, and 5. We generated 20 randomized sequences for sample sizes from 500 to 25, and 40 sequences for sample sizes of 10 and 5. We recalculated standard deviations for each randomly generated sequence of residuals, then determined the mean and standard deviation of these 20 or 40 sequences at each sample size. For bins with sample sizes great than 50, one standard deviation for the randomized replicates was less than 0.05‰. For samples sizes of 25, one standard deviation rose to 0.07‰, and for sample sizes of 10 or less, the deviation was 0.1‰. Thus, our interpretations of the relationship between variability and mammalian evolution must be tempered by the fact that our confidence in the variability values is low for many bins. Fortunately, the density of $\delta^{18}O$ samples is high around many (but not all) of the key intervals of mammalian evolution discussed below.

Mammalian Faunal and Body Mass Data.—The primary data used here to document North American mammalian evolution are 4978 lists of species occurring in faunal samples and 23,125 measurements of individual first lower molar teeth (Alroy 2000). The lists form the basis of our diversity curves and turnover rate estimates, and the tooth measurements are used as proxies for body mass. The data have figured in a series of earlier publications (Alroy 1992, 1996, 1998a,b,c,d, 1999a,b; Wing et al. 1995). Both data sets are restricted to terrestrial, nonflying mammals; are derived from a set of 2828 publications; and are taxonomically standardized using the companion database of synonymies and genus-species combinations included in the North American fossil mammal systematics database (NAFMSD: http://www.nceas.ucsb.edu/~alroy/nafmsd.html).

The set of faunal lists includes 30,951 taxonomic occurrences and now extends back to the Early Cretaceous, documenting age ranges for 1241 genera and 3243 species. The lists can be viewed on the World Wide Web at the North American mammalian paleofaunal database (http://www.nceas.ucsb.edu/~alroy/nampfd.html). All of the Cretaceous and Cenozoic lists were used in a maximum likelihood appearance event ordination analysis (Alroy 2000), which establishes age ranges for genera and species. For reasons outlined by Alroy (1996, 1998d), 191 Cenozoic localities from outside of the west (i.e., Mexico, south-

western Canada, and the western United States) were excluded from the diversity analyses.

The expanded body mass data set now includes 3398 population samples of 1969 different species. A current version is available at the NAFMSD. Although body mass estimates are available for only 61% of the species, these species tend to be common and long-ranging. The estimates were derived from the lower first molar measurements using standard allometric equations that are cited and discussed elsewhere (Alroy 1998b, 1999a,b). Most of these equations depend on the logarithm of the length-times-width of the first lower molar, but following Damuth (1990) the log of length is used for ungulates, and second lower molar measurements were used for proboscideans. The regression-based estimates may have substantial errors, and indeed some families and genera may exhibit systematic departures from the relationship seen for their respective orders. However, occasional estimates that are off by a factor of two, three, or even ten are unlikely to have much of an effect on distributions spanning nearly seven orders of magnitude.

Biotic Time Series Statistics

Methods used to transform the faunal lists into age ranges and diversity curves are described elsewhere (Alroy 1996, 1998d, 2000). In particular, Alroy 2000 serves as a companion to this paper. It details some of the new methods used to prepare the data, which expand and refine earlier techniques. There have been improvements in the multivariate ordination method used to establish the mammalian age-range data; the interpolation method used to calibrate these age ranges to numerical time; and the random subsampling methods used to standardize the resulting diversity and turnover-rate data. Furthermore, we use the new, much more methodologically sound equations of Foote (1999) to translate raw age-range data into proper turnover rates. In total we consider five turnover metrics in addition to the standardized diversity curve: instantaneous origination and extinction rates; net diversification (the difference of origination and extinction); diversification

volatility (the absolute value of net diversification); and total turnover (the sum of origination and extinction).

Additionally, we computed two statistics intended to measure the rate of change of proportional species richness of major orders. In
other words, these "proportional volatility"
statistics measure the rate with which orders
replace each other. The first statistic, a simple
index, is correlated with turnover rates. The
second, a *G*-statistic (i.e., likelihood ratio),
measures how much the dominance of different orders changes in excess of what one
would expect at random given the observed
overall origination and extinction rates in each
bin, plus the average turnover rates through
the whole time series for each order. Full equations and time series data are given by Alroy
(2000).

For the body mass data, we relied upon four
standard univariate statistics (the mean, standard deviation, skewness, and kurtosis) computed across all of the species that ranged
anywhere into each 1.0-m.y. bin. Alroy (2000)
discusses our reasons for believing that these
slightly time- and space-averaged data are biologically meaningful, and for rejecting the
much more common use of "cenogram" statistics (Legendre 1989).

Time Series Analyses

Diversity and Turnover.—While the diversity
curve (Fig. 3) reflects patterns seen in earlier
analyses (Alroy 1996, 1998d, 1999a), several
features are subtly different. In this study
there is less of a consistent difference between
early Eocene and mid-to-late Eocene diversity,
resulting in a more unpredictable overall Eocene trend. Likewise, the recovery from the
latest Miocene extinction is sharper, putting
late Pleistocene levels close the Miocene average.

Also of interest is overall variation within
the turnover rate curves (Fig. 4A–E; see also
Alroy 2000: Table 4). There is a general decline
in turnover rates, especially after the Paleocene; spikes in origination are higher than
spikes in extinction, especially during the Paleocene; and very few discrete pulses of turnover are visible in any of the curves. Nonetheless, in a later section we discuss how some of

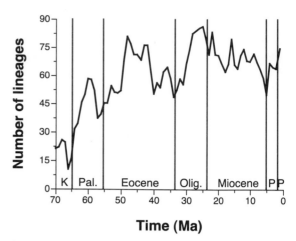

Time (Ma)

FIGURE 3. Sampling-standardized diversity curve for
North American Cenozoic mammals. Curve is based on
subsampling lists that total occurrences-squared quotas
shown in Alroy (2000: Fig. 4). The method allows for
gradual changes through time in alpha diversity.

the outlying points in these curves may relate
to episodes of global climatic change.

Standard deviations within our 60-m.y.-
long study interval are comparable for origination (0.100), extinction (0.110), and net diversification (0.124). Because of mathematical
constraints, values are lower for diversification volatility (0.077) and higher for total turnover (0.169). These sampling-standardized
data do exhibit comparable variability for
origination and extinction, but fully 36% of the
variation in the origination rates could be removed by correlating this variable against log
standing diversity for the relatively monoto-
nous 60-m.y. interval ($r = -0.602$, $t = 5.737$, $p
< 0.001$). In other words, more than a third of
the variance can be explained by diversity-dependent models that are based on entirely intrinsic, biological interactions like competition
(Alroy 1998d).

Meanwhile, regressing extinction rates
against log diversity ($r = -0.224$; $t = 1.749$; $p
< 0.10$) would reduce the variance by a mere
5%. Although unimpressive, this relationship
also could be attributed to intrinsic factors.
The weak correlation is due to the fact that extinction rates fall off during the Cenozoic as
diversity rises. Thus, it may be a side effect of
the Paleocene/Eocene replacement of volatile,
archaic groups such as multituberculates with
less volatile "modern" groups such as rodents
(Alroy 2000: Fig. 7). A decrease in average ex-

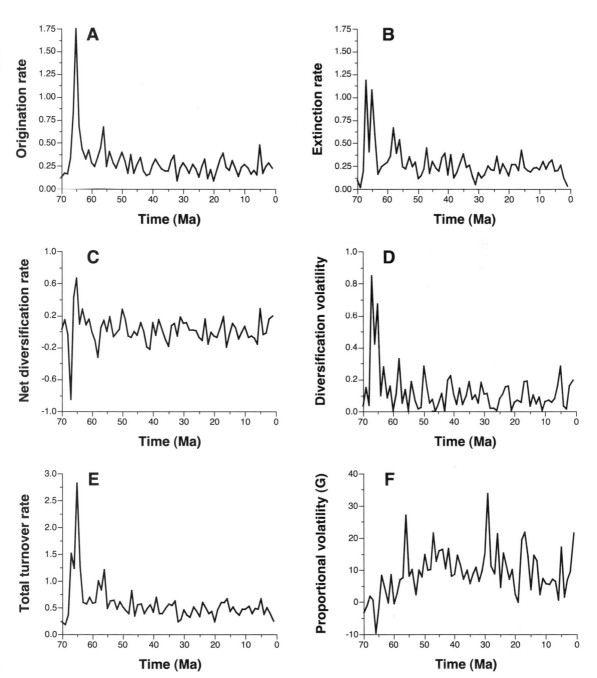

FIGURE 4. Cenozoic trends in turnover rates and the proportional volatility of ordinal diversity. The statistics are discussed by Alroy (2000). A, Origination. B, Extinction. C, Net diversification. D, Diversification volatility. E, Total turnover. F, Proportional volatility G-statistic.

tinction rates that results from the loss of volatile taxonomic groups is well known for Phanerozoic marine invertebrates, and it has been held up as a clear demonstration that major features of biodiversity dynamics can be governed by intrinsic, biological factors (Gilinsky 1994).

In light of the fact that so much of the turnover pattern can be explained by intrinsic dynamics, it comes as no great surprise that cross-correlations of the five taxonomic-turnover time series against the isotope data fail to recover compelling evidence for a causal interrelationship (Table 2). Before discussing

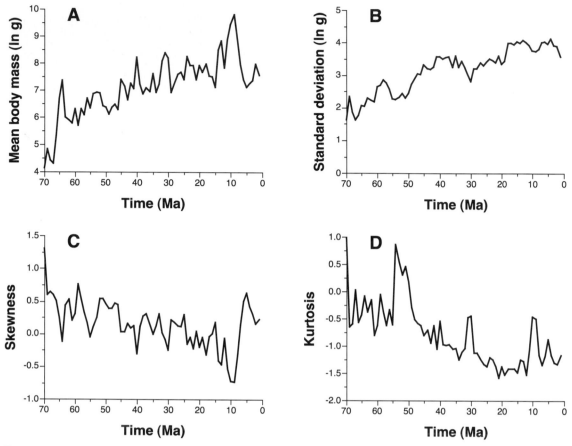

FIGURE 5. Cenozoic trends in body mass distributions. Each point describes the distribution of body mass estimates for all species ranging into that 1.0-m.y.-long sampling bin. A, Mean. B, Standard deviation. C, Skewness. D, Kurtosis.

these results in detail, we need to set some ground rules for evaluating nominally "significant" results:

1. Many of the variables are more-or-less normal but do include notable outliers (which we will discuss in detail in our evaluation of "critical intervals"). Because of this, we will be careful to make sure that a correlation holds up even after removing one or two outlying data points.

2. Because of the presence of outliers and slight skewness in many variables, we will explore both Pearson's parametric and Spearman's nonparametric correlation techniques. These results typically will agree, but when they do not we will take this as evidence that the correlations are not very robust.

3. We will be careful not to overinterpret correlations with low significance values, because our reporting of a large number of cor-

relations makes it relatively likely that a few "significant" values will crop up at random.

The results are completely unambiguous. Several weak correlations involving the midpoint $\delta^{18}O$ data crop up, but these consistently reflect coincidental trends in each time series and are not replicated by the correlations for change in midpoint $\delta^{18}O$. Meanwhile, just one correlation is significant for any comparison between $\delta^{18}O$ data and the remaining four measures of mammalian turnover. This involves a weak correlation between detrended $\delta^{18}O$ SD and origination. However, the relationship is seen only when using Pearson's product-moment coefficient (r_P); it disappears once we switch to the nonparametric Spearman's rank-order correlation coefficient (r_S).

Despite the lack of even a close call here, the question still arises whether some real correlations might be masked by errors in the data

TABLE 2. Time series analyses contrasting benthic foraminiferal oxygen isotope data with mammalian biotic data. Pearson's product-moment correlation coefficients are reported above; Spearman's rank-order correlations are reported below. The latter are favored because they lessen potential problems with outliers and non-normal distributions of variables. "Volatility" is just the absolute value of the change in $\delta^{18}O$ values. Sample size is 60 data points for serial correlation and correlation against time; 59 for correlations with midpoint and standard deviation isotope data, which omit the 50.5-Ma bin because of lack of data; and 57 for correlations with changes in $\delta^{18}O$ and volatility, which additionally omit the 59.5-bin (because of differencing) and 49.5-Ma bin (because of lack of data). * = $p <$ 0.05; ** = $p <$ 0.01; *** = < 0.001.

Variable	Time	Serial	Midpoint	Change	Volatility	SD	Detrended SD
			Pearson's product-moment correlations				
Log richness	−0.433**	0.792***	0.294*	−0.032	0.165	0.149	−0.002
Origination	0.317*	0.141	- 0.275*	0.063	0.012	0.109	0.275*
Extinction	0.359**	0.247	−0.360**	0.122	−0.003	−0.136	−0.021
Net diversification	−0.061	0.040	0.097	−0.059	0.012	0.208	0.239
Diversification volatility	0.015	0.019	−0.051	−0.023	−0.103	0.122	0.138
Total turnover	0.420***	0.282*	−0.397***	0.166	0.005	−0.024	0.149
Proportional volatility (index)	0.477***	0.244	−0.444***	0.145	0.191	0.083	0.208
Proportional volatility (G)	0.051	0.086	−0.045	0.030	0.170	0.341**	0.379**
Mass (mean)	−0.712***	0.762***	0.611***	−0.110	−0.041	0.024	−0.143
Mass (standard deviation)	−0.865***	0.939***	0.724***	−0.284*	−0.050	0.111	−0.081
Mass (skewness)	0.430***	0.702***	−0.311*	0.090	0.035	0.129	0.224
Mass (kurtosis)	0.686***	0.812***	−0.639***	0.169	−0.165	−0.237	−0.021
			Spearman's rank-order correlations				
Log richness	−0.370**	0.780***	0.220	−0.112	0.188	0.211	0.135
Origination	0.319*	0.097	−0.263*	0.053	0.120	−0.075	0.039
Extinction	0.284*	0.136	−0.272*	0.120	0.005	0.248	−0.154
Net diversification	−0.011	0.065	0.056	−0.051	0.030	0.191	0.156
Diversification volatility	−0.043	−0.019	−0.009	0.047	0.006	0.111	0.141
Total turnover	0.371**	0.149	−0.354**	0.145	0.052	−0.225	−0.112
Proportional volatility (index)	0.467***	0.190	−0.416***	0.054	0.184	−0.115	−0.121
Proportional volatility (G)	0.098	0.127	−0.150	−0.151	0.199	0.155	0.147
Mass (mean)	−0.745***	0.722***	0.655***	−0.141	−0.054	0.124	−0.070
Mass (standard deviation)	−0.865***	0.904***	0.673***	−0.273*	−0.158	0.105	−0.080
Mass (skewness)	0.393***	0.690***	−0.278*	0.036	0.043	0.119	0.272*
Mass (kurtosis)	0.708***	0.829***	−0.535***	0.110	−0.100	−0.230	−0.038

or other factors. Perhaps these difficulties make immediate causal relationships between climate and evolution hard to see in side-by-side comparisons of values. Three such factors might be important: (1) miscalibration of one time series or another may mean that we are comparing bins of nominally identical, but actually different, ages; (2) backwards-smearing of extinction events and forwards-smearing of origination events may mean (for example) that climate events in a bin 1.0 m.y. after an apparent series of extinctions may actually have caused those extinctions; and (3) causal interactions may be mediated by intermediate factors like evolutionary changes in vegetation that may delay responses on the million-year timescale. For all of these reasons, it is of interest to examine correlations where we lag the isotope data by one sampling bin either

backward or forward relative to the faunal turnover data.

The lagged correlations show no greater biological significance (Table 3). At first glance, it seems as if there is a series of weak but perhaps important correlations involving change in $\delta^{18}O$. Most of these involve extinction or total turnover. However, these correlations are highly suspect for two reasons: (1) none of the correlations are also seen in the unlagged data, and (2) the extinction correlation is seen either with backwards or forwards lagging, which makes no sense in causal terms—how could an isotopic shift 1 m.y. in the future be just as important as one 1 m.y. in the past? Instead, the pattern of correlations suggests a persistent problem with artifactual cross-correlations between time series that have temporal trends: all four of the biotic variables

TABLE 3. Lagged cross-correlations among key time series. Lags are of 1.0 m.y. (i.e., one sampling bin) in either direction. Abbreviations as in Table 6. Forward lags are considered more meaningful because if climate change drives turnover, then climate change in the past that is lagged forward one interval may be relevant to the current interval. In contrast, climate change in the future that is lagged backward cannot be.

Variable	Backward change	Backward volatility	Backward detrended SD	Forward change	Forward volatility	Forward detrended SD
			Pearson's product-moment correlations			
Log richness	−0.056	0.193	−0.090	−0.032	0.143	0.085
Origination	0.246	−0.057	0.051	−0.046	−0.060	−0.106
Extinction	0.188	−0.005	−0.098	0.262*	−0.083	−0.072
Net diversification	0.038	−0.042	0.127	−0.279*	0.025	−0.022
Diversification volatility	−0.189	−0.021	−0.178	0.040	0.035	0.077
Total turnover	0.268*	−0.037	−0.031	0.139	−0.087	−0.109
Proportional volatility (index)	0.308*	−0.118	−0.107	0.186	−0.104	0.072
Proportional volatility (G)	0.110	−0.090	−0.068	−0.109	−0.003	0.115
Mass (mean)	−0.056	−0.070	−0.073	−0.155	−0.019	−0.190
Mass (standard deviation)	−0.288*	0.040	−0.046	−0.272*	−0.003	−0.135
Mass (skewness)	−0.045	0.146	0.143	0.082	0.115	0.143
Mass (kurtosis)	0.157	−0.080	−0.031	0.137	−0.139	−0.145
			Spearman's rank-order correlations			
Log richness	−0.080	0.178	0.047	−0.117	0.198	0.197
Origination	0.248	0.031	0.020	0.005	−0.081	−0.159
Extinction	0.283*	0.004	−0.111	0.373**	−0.088	−0.133
Net diversification	−0.014	0.063	0.101	−0.232	−0.049	0.006
Diversification volatility	−0.208	−0.028	−0.194	0.016	0.122	0.042
Total turnover	0.340**	0.012	−0.071	0.241	−0.163	−0.156
Proportional volatility (index)	0.315*	0.018	−0.090	0.250*	−0.065	0.123
Proportional volatility (G)	0.093	0.073	−0.012	−0.132	0.163	0.204
Mass (mean)	−0.074	−0.108	−0.017	−0.175	−0.051	−0.096
Mass (standard deviation)	−0.270*	−0.081	−0.080	−0.248	−0.157	−0.099
Mass (skewness)	−0.062	0.175	0.167	0.015	0.037	0.171
Mass (kurtosis)	0.096	−0.039	0.005	0.077	−0.092	−0.081

show moderately to very strong trends (Table 2), and even though the +0.227 correlation of the change in $\delta^{18}O$ values with time was insignificant at the $p < 0.05$ level, it was barely so. Indeed, after detrending the isotopic variable every single one of the suggestive cross-correlations drops far below the level of significance.

Even if one wanted to interpret the few consistent correlations, it would be unwise because we have subjected the data to a large number of analyses without correcting for multiple comparisons. Just at random, we should expect several comparisons to be significant at the $p < 0.05$, and possibly even $p < 0.01$, level.

For example, a standard Bonferroni correction for these multiple comparisons would work as follows: we have performed backward, forward, and non-lagged cross-correlations for three key isotope variables matched with five distinct turnover rate metrics, so if we want to examine all of our tests one by one

then we really should multiply our alpha levels by $3 \times 3 \times 5 = 45$. Therefore, an apparent $p < 0.001$ result actually is equivalent to $p < 0.045$, and an apparent $p < 0.05$ or even $p < 0.01$ outcome is simply not significant. As it happens, none of the Pearson's correlations in Table 3 are "significant" at $p < 0.01$.

The lack of any strong results here argues against the "turnover pulse" scenario outlined by Vrba (1985, 1992, 1995), which has been criticized for other reasons in previous studies (Alroy 1996, 1998d; Behrensemeyer et al. 1997). This theory specifically predicts that both origination and extinction rates should rise at the same time as, or no later than 1.0 m.y. after, any major positive excursion in isotope values. That is, global cooling/ice volume buildup should strongly encourage both speciation and extinction. No such pattern is seen in our data.

In summary, we find no evidence for linear abiotic forcings of turnover rates throughout the Cenozoic. Quite to the contrary, this and earlier studies (Stucky 1990; Alroy 1996, 1998d) have shown that a strong linear *biotic* forcing accounts for a substantial amount of the variation in origination rates. Thus, the data clearly would have been capable of providing support for a strong causal mechanism if one had existed.

Relative Diversity of Major Orders.—We examined two measures of the rate of change in species richness within mammalian orders: the proportional volatility index and G-statistic (Alroy 2000: Fig. 7). Both show insignificant autocorrelation but only the index shows a relatively strong trend (Table 2). As discussed by Alroy (2000), the two statistics differ substantially in that the simple proportional volatility index is strongly correlated with taxonomic turnover, whereas the proportional volatility G-statistic is not. For this reason, the G-statistic curve seems to be far more informative and our discussion will focus on it.

The G-statistic time series (Fig. 4F) suggests four distinct patterns. First, despite the lack of a general correlation with the turnover rate data, there still are instances in which high turnover at the species level results in high turnover at the ordinal level. For example, the Paleocene/Eocene transition shows unusual turnover at both levels.

Second, the rapid and near-total replacement of archaic groups like the multituberculates and "condylarths" by the modern orders may largely be a side effect of generally high Paleogene turnover rates. Some key steps in this transition occur at times of little proportional volatility: multituberculates decline steadily throughout the Paleocene, well before the modern groups establish themselves; and "condylarths" lose their foothold during a stretch of several million years in the early Eocene, well after the Paleocene/Eocene boundary and well before the massive mid-Eocene radiation of artiodactyls.

Third, there is a weak trend of increasing proportional volatility, but it is not visible for the 60 data points in the study interval (Table 2; G vs. time: $r = 0.051$; $t = 0.388$; n.s.). The low points before 60 Ma cannot be explained as a side effect of unusual turnover rates, not only because secular variation in turnover rates was accounted for in computing the G-

statistic (Alroy 2000), but because rates were actually higher, not lower, in the latest Cretaceous and early Eocene. Still, the persistently high G-values suggest the possibility of nonrandom replacement between pairs of ecologically similar major groups, not just the stochastic loss of volatile groups. Testing for direct competitive replacement would be the subject of another, much more complex analysis.

Fourth and most obviously, a handful of data points stand out from the rest (Fig. 4). The second-highest marks the Clarkforkian/Wasatchian (Paleocene/Eocene) boundary at ~55 Ma, which has been the focus of considerable paleoecological interest (e.g., Clyde and Gingerich 1998). Surprisingly, there is no high point at the K/T boundary, even though high turnover rates and massive changes in ordinal composition are obvious and well known (e.g., Alroy 1999a). Several other high points are scattered throughout the Oligocene and early Miocene, an interval where poor sampling (Alroy 2000: Fig. 4) often complicates any straightforward biological interpretations of turnover.

Perhaps the most interesting part of the time series is a stretch in the mid-Eocene that includes four of the five highest G-values (46.5, 44.5, 43.5, and 41.5 Ma, i.e., Bridgerian and Uintan). This interval corresponds to the radiation of artiodactyls (and to some degree perissodactyls), matched by the terminal decline of primates (and to some degree "condylarths"). At 47 Ma, the two modern ungulate orders account for 13.9% of standing diversity, and the primates are near their all-time peak at 12.1%; by 44 Ma, these figures are 35.6% and 5.1%. By 37 Ma, primates have gone temporarily extinct in North America.

In light of the obvious evidence here for a major ecological transition, it is almost disappointing to report our failure to find any meaningful correlation between the proportional volatility data and the isotope data (Table 2). (1) As expected, the proportional volatility index does correlate significantly (if not very strongly) with the midpoint $\delta^{18}O$ curve, because these two time series both show secular trends. Therefore, this correlation must be put aside as being a potential statistical artifact. (2) None of the correlations involving the proportional volatility statistic appear when using both the Pearson's coefficient and the Spearman's coefficient. The collapse of a correlation involving the $\delta^{18}O$ detrended standard deviation and the G-statistic is particularly shocking ($r_P = +0.379$; $r_S = +0.147$). (3) The lagged correlations (Table 3) are every bit as uninteresting, with correlations involving the change in $\delta^{18}O$ variable and the proportional volatility index disappearing after detrending the former (as discussed above).

Body Mass Distributions.—The moment statistics depict strong patterns of change in the overall body mass distribution (Fig. 5). All of these individual statistics show strong serial correlation and strong trends through time (Table 2), making them unsuitable for being cross-correlated with the autocorrelated isotope time series (i.e., midpoint value and standard deviation), as discussed in the Introduction. Therefore, after describing some major features of the body mass data, we will focus our discussion on comparisons with the remaining three isotope data sets that lack strong autocorrelation.

Mean body mass increases abruptly at the Cretaceous/Tertiary boundary (Fig. 5A). As discussed elsewhere (Alroy 1999a), some of this shift is attributable to differential immigration of relatively large species such as basal ungulates, but most of it seems to be due to in situ evolution. The mean continues to increase throughout most of the Cenozoic, but slowly. Most of the trajectory is attributable to strong evolutionary trends operating within medium- and large-sized lineages (Alroy 1998b), which could be explained by any number of biotic mechanisms: biomechanical or physiological optimality, escape from competition with diverse small-sized taxa, an evolutionary arms race between predators and prey, or high speciation and/or low extinction rates of larger species resulting from the scaling of demographic factors.

A relatively rapid drop in mean body mass at around 7 Ma is too large to be a sampling artifact. One might think that the Miocene data up to this point differentially sample large species. However, the remarkably stable standard deviation values—which vary only

between 3.73 and 4.13 units between 18 and 1 Ma—suggest that any size-related sampling biases are essentially constant throughout this entire interval. The pattern more likely reflects bona fide differential extinction of large mammals in the latest Miocene (Alroy 1999b).

Like the somewhat noisier curve for the mean, the standard deviation curve increases steadily throughout the Cenozoic. The data do show a major downward excursion between about 56 and 49 Ma (Fig. 5B). The possible ecological significance of this drop is discussed below.

We employ skewness as a measure of whether faunas are dominated by small species (positive skewness) or large species (negative skewness). The skewness curve shows an erratic drift toward negative values, meaning that initially there is a long, thin tail of larger species, but that by the Miocene large mammals are dominant (Fig. 5C). The trend is reversed in the latest Miocene. Brown and Nicoletto (1991) and Marquet and Cofre (1999), who respectively studied Recent regional (biome-scale) distributions in North and South America, also found positive skews (although their values are higher than ours). Confirmation of their results is important because Recent distributions suffer from a clear historical bias: they are influenced by a catastrophic extinction of megafaunal species that was almost certainly anthropogenic, had no clear antecedent in the Tertiary record, and had major ramifications for local, regional, and continental-scale body mass distributions (Alroy 1999b). Marquet and Cofre (1999) also argued for the importance of historical factors in governing these distributions. Thus, the current data suggest that even if positive skewness in the Recent biota is exaggerated, the pattern still reflects a remarkable late Miocene reversal in the overall Cenozoic trend that dominated before about 7 Ma.

The most important observation for our purposes is that skewness did generally drop through the Tertiary. The overall trend makes sense given a dual-equilibrium pattern: the distribution around the small-size equilibrium point was constant and well filled from the very start, but only a few rapidly evolving lineages were able to approach the large-size equilibrium point prior to the late Tertiary (Alroy 1998b). Filling of the upper size range resulted in the distribution's becoming more evenly balanced, with lower (and eventually even negative) skewness.

The last statistic is kurtosis, a measure of whether a distribution has more than the expected number of species in the mid-size range (positive, or leptokurtic, values) or fewer, in which case a mid-sized gap may be visible (negative, or platykurtic, values). Kurtosis shows an even more interesting pattern than skewness (Fig. 5D). Values are slightly negative in the Late Cretaceous and early Paleocene. They rise quickly around 54 Ma and drift downwards until finally dropping to nearly -1 by 45 Ma, where they largely remain throughout the rest of the Cenozoic. Hence, a large gap in the body mass distribution opens in the mid-Eocene and remains unfilled (Alroy 1998b: Fig. 1). Alroy (2000: Fig. 9D) showed that exactly the same bimodal pattern is seen in a majority of diverse quarry-scale samples younger than 45 Ma.

The kurtosis pattern has three important implications. First, the Paleocene/Eocene immigration pulse (Gingerich 1989; Clyde and Gingerich 1998) registers not just in taxonomic turnover data, but in body mass data: a drop in the standard deviation and rise in kurtosis indicate the rapid filling of the middle size range by small ungulates, carnivores, and primates. Second, the primate/ungulate transition, which has such a remarkable impact on the proportional diversity of major orders, is also expressed as a substantial increase in the standard deviation and decrease in kurtosis during the Bridgerian (50.3–46.2 Ma [Alroy 2000]), and arguably continuing into the Uintan (46.2–42.0 Ma). Hence, the mid-Eocene ecological transition easily ranks as one of the most important of the entire Tertiary. Third, the fact that kurtosis hovers around -1, evidencing no strong trend after 45 Ma—despite episodic drops in mean temperature and increases in seasonality—suggests that gaps in the mammalian body mass distribution cannot be linked to climate change in a simple way. If a link is indicated here, it is in the form of a nonlinear threshold effect whereby the

middle size range becomes emptied once climate deteriorates past a certain point.

The lack of any simple relationship between climate and body mass distributions is supported by the correlation analyses. We find a series of significant but potentially artifactual cross-correlations involving midpoint isotope values and the assorted body mass measures (Table 2). All of these variables have strong secular trends, so the correlations are inevitable and, therefore, uninformative.

For the remaining 16 comparisons we find exactly one case of significance at the $p < 0.05$ level that is suggested by both the Pearson's and Spearman's coefficients. Of course, 16 comparisons frequently should yield one such case just at random. The relevant correlation involves standard deviation of body mass and rate of change of $\delta^{18}O$. Unfortunately, we find that removing a single outlier changes the result. The change in $\delta^{18}O$ of -0.350 units going into the 52.5-Ma bin is the second highest in the time series; this bin's body mass standard deviation of 2.327 is the fourth highest. Without this point, the correlations are insignificant ($r_P = -0.232$; n.s.; $r_S = -0.236$; n.s.). The lagged correlations (Table 3) are even less impressive. Apart from a similar correlation including this pair of variables, which again disappears once the isotopic data are detrended, not one of them approaches the $p < 0.05$ level of significance.

Critical Intervals

Our exhaustive comparisons of isotopic and biotic records provide substantial evidence that mammal evolution and extinction are not linearly forced by global climate change. The few cases of apparent cross-correlation are all attributable either to the presence of outlying data points or to the presence of strong, misleading autocorrelation in the time series. Our results should not be misinterpreted as merely showing that the data are noisy. We do see robust patterns in the mammalian fossil record, but most of them can be attributed to biological processes such as ecological release in the wake of the K/T mass extinction (Alroy 1999a), equilibrial diversity dynamics later on in the Cenozoic (Alroy 1998d), and evolutionary trends within individual taxonomic line-

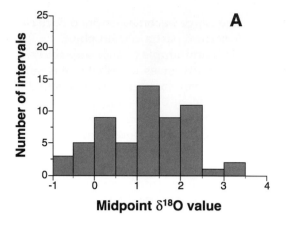

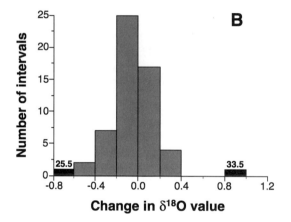

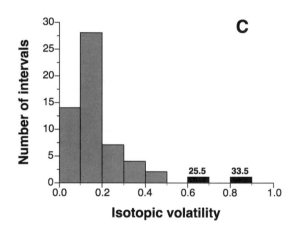

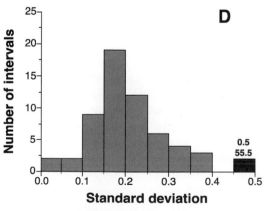

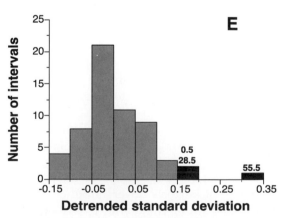

FIGURE 6. Frequency distributions of statistics describing benthic isotope records. Black boxes and numbers indicate the five critical intervals of climate change discussed in the text. A, Estimated midpoint $\delta^{18}O$ value for bin based on linear regression of values against time. B, Change in midpoint $\delta^{18}O$ value from preceding bin to current bin. C, Isotopic volatility (absolute value of change). D, Standard deviation of residual $\delta^{18}O$ values (variability). E, Generalized differences of standard deviations (detrended variability).

ages (Alroy 1998b), all of which are better explained by intrinsic factors like ecological competition than by extrinsic factors like mean temperature and precipitation. Similarly, the highly averaged isotopic records used in our analysis do show major features that have been interpreted in the past as evidence for global climatic change, such as the Paleocene/Eocene boundary event, the Eocene thermal maximum, warming/ice volume decrease in the early Miocene, and shifts toward cooler climates and/or greater ice volume at the Eocene/Oligocene boundary, in the late Miocene, and in the Plio-Pleistocene.

Even though our analyses failed to uncover strong evidence of linear forcings, they do suggest an intriguing pattern: a handful of cases in which a correlation results from one or two highly anomalous data points. Perhaps, then, climate does have an influence on mammalian evolution—but only in the form of occasionally and unpredictably producing unique, one-time-only biotic responses. To test this hypothesis more rigorously, here we examine the state of the climate system during critical intervals documented in the biotic records.

This exercise is substantively different from the analyses undertaken by Alroy (1998d, 1999b) and Prothero (1999). In those studies, the authors first subjectively identified intervals of rapid change in the climate record and then examined mammalian responses during those intervals. In some sense, this is equivalent to hypothesizing linear forcing; the exercise only makes sense if one expects every major climate event to instigate some sort of biotic response. The test conducted here depends on a less restrictive view of the impact of climate on mammalian evolution: we merely consider whether the climate system was in an atypical state during time periods for which we have objective evidence of major changes in the biota.

There are six such biotic events in our records, most of which correspond to generally recognized boundaries in the traditional North American land mammal age system (Woodburne 1987; Alroy 1998a, 2000). They are the Tiffanian/Clarkforkian boundary (~57 Ma), the Clarkforkian/Wasatchian (i.e.,

Paleocene/Eocene) boundary (~56 to 55 Ma), the early Uintan (~45 to 44 Ma), the mid-Oligocene diversification (~31 to 28 Ma), the latest Miocene and possibly earliest Pliocene (~7 to 4 Ma), and the Pleistocene/Recent boundary (1 to 0 Ma).

The massive faunal reorganization associated with the late Pleistocene extinction was a dramatic and highly selective catastrophe that focused almost exclusively on large mammals (Alroy 1999b). Because we consider human–faunal interaction the most likely cause of this event, we preclude it from further discussion here. Our data for the 0.99 m.y. immediately preceding the extinction do suggest a net diversification with high volatility and other subtle shifts, like a decrease in the standard deviation of body mass. However, these statistical wobbles are most likely attributable to residual sampling artifacts created by the late Pleistocene's truly exceptional coverage—in terms of geography, taphonomy, and sheer quantity of data. For the other five events, we will briefly describe the biotic data that define each one, and then turn to an examination of the globe's climatic state at the time.

Before proceeding, we will note that several other events may merit attention at a future date. Currently, however, we are not confident that they are anything more than statistical anomalies. A typical case is the subtle faunal transition around the very end of the Eocene (Chadronian/Orellan boundary: 33.9 Ma). Diversity drops slightly in the wake of elevated extinction rates for the 34.5-Ma bin, and yet proportional volatility is truly unremarkable and shifts in the body mass distribution are small and seemingly aimless. The vast majority of intervals are marked by similar patterns, with no concordance among statistical measures and data points that all fall comfortably within two or even 1.5 standard deviations of the mean. The few exceptions—especially for body mass distributions—seem to relate to residual sampling artifacts, with some periods failing to sample either large or (more commonly) small mammals (e.g., data for 32–29 Ma).

Our climatic analysis is simple. In Figures 6–8, we present histograms respectively summarizing the benthic isotope, taxonomic turn-

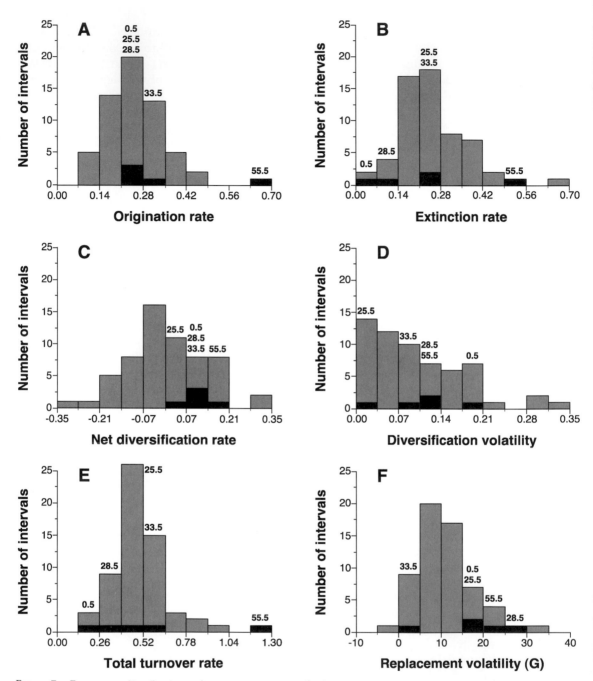

FIGURE 7.   Frequency distributions of turnover rates over the last 60 Ma. All rates are instantaneous and apply to a single 1.0-m.y.-long sampling interval. Black boxes and numbers indicate critical climate change intervals. A, Instantaneous rates of origination. B, Instantaneous rates of extinction. C, Net diversification rates. D, Diversification volatility. E, Total turnover rates. F, Proportional volatility G-statistic.

over, and body mass distributions. The body mass data have been transformed into first differences (changes between consecutive bins) in order to remove strong autocorrelation (see Table 2). If biotic events in the mam-

mal record are associated with unusual climates, we would expect them to occur in, or immediately after, intervals with climatic parameter values (Fig. 6) more than two standard deviations from typical Cenozoic values.

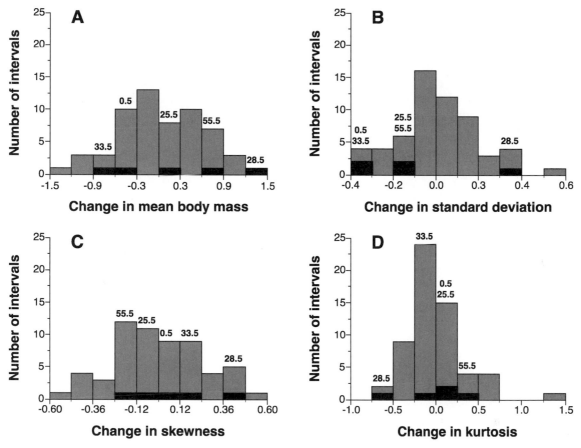

FIGURE 8. Frequency distributions of changes in body mass moment statistics over the last 60 Ma. Black boxes and numbers indicate critical climate change intervals. A, Changes in the mean. B, Changes in the standard deviation. C, Changes in skewness. D, Changes in kurtosis.

These limits are −0.409 to +0.509 for change in midpoint $\delta^{18}O$ value, zero (the lowest possible value) to +0.478 for isotopic volatility, +0.025 to +0.395 for standard deviation of isotope value residuals, and −0.161 to +0.164 for detrended isotopic standard deviation. By this standard, the only significant climate events are in the following five bins: 55.5 Ma (high standard deviation and detrended standard deviation), 33.5 Ma (rise in midpoint $\delta^{18}O$, high isotopic volatility), 28.5 Ma (high detrended standard deviation), 25.5 Ma (drop in midpoint $\delta^{18}O$, high isotopic volatility), and 0.5 Ma (high standard deviation and detrended standard deviation). As we will argue, just two of these episodes also correspond to one of the five intervals of major biotic change—even though we defined the intervals rather broadly. However, one other biotic event is associated with a longer run of unusual climatic

values, perhaps indicating a nonlinear response to gradual climatic forcing.

1. The Tiffanian/Clarkforkian (Ti/Cf) transition, which essentially falls in the 57.5-Ma bin, is marked by a sudden drop in diversity from ~52 to 38 species (sampling-standardized data; see Fig. 3). The extinction rate (0.666) is the most extreme in our entire study interval. As a result, the volatility and total turnover rate values are the highest and second-highest in the data set (0.327, 1.004), and the net diversification value is the lowest (−0.327). Sampling intensity is uniform throughout this interval, so this drop cannot be attributed to variation in sampling per se.

Because the Ti/Cf boundary occurs so near the start of our isotopic time series, it is difficult to evaluate the role of climate change in this event. The benthic marine $\delta^{18}O$ record reached an early Cenozoic maximum at 60

Ma, then declined steadily for 10 m.y. (Table 1). This trend has been interpreted as a sign of gradual global warming (Miller et al. 1987; Zachos et al. 1994). While the Ti/Cf boundary occurs very near the inflection point in the benthic curve (i.e., very close to the point where temperatures begin to rise), it is not clear that the two precisely coincide and neither the change in midpoint $\delta^{18}O$ value nor the measure of $\delta^{18}O$ variability is at all unusual near the boundary.

Rose (1981) argued that Tiffanian as well as Clarkforkian alpha diversity was aberrantly low. If this were true, it would make more sense in light of evidence that a shift to cooler climates occurred in the early Paleocene, before the Tiffanian. Maas et al. (1995) concurred, arguing that regional diversity also was essentially constant during the middle and late Paleocene (i.e., Tiffanian–Clarkforkian). Gunnell et al. (1995) thought that Clarkforkian faunas evidenced cooler, drier habitats than later, Wasatchian faunas, and went on to argue that Clarkforkian diversity may be underrepresented because of sampling biases that exclude many small species.

A simple relationship between climate and diversity would predict low diversity in the cooler Tiffanian and steadily rising diversity afterwards. In our data, though, diversity crashes suddenly after the Tiffanian—which raises reasonable doubts about the reliability of both our biotic and our climatic data sets. However, our marine isotope data are strongly supported by leaf-margin analyses (Wing et al. 1995; Wilf 1997, 2000), which show clear evidence for warming between the Tiffanian and Clarkforkian in the same northern Rocky Mountains region as the mammal faunas studied by Rose (1981), Gunnell et al. (1995), and Maas et al. (1995).

As for the diversity data: (1) a drop going into the Clarkforkian is present in all earlier versions of our data set (Alroy 1996, 1998d), even including one that was not corrected for sampling effects (Wing et al. 1995); (2) several high-quality Clarkforkian faunas such as those of Rose (1981) are included in our analyses; (3) there is no evidence for undersampling of small species in the form of any large, sudden shift in any of the body mass moment

statistics between the 58.5- and 57.5-Ma bins; and (4) likewise, proportional volatility is low throughout the late Tiffanian and Clarkforkian, and small groups like the Insectivora do not experience a major drop in relative diversity—at 57 Ma they actually constitute 12.9% of the fauna, which is a bit higher than their Cenozoic mean of 9.9%.

If our data are meaningful, then how could a gradual warming trend have resulted in a sudden, dramatic diversity crash? One possible scenario is that warming may have led to the northward shift of a mammal community that heretofore was restricted to the very southern margin of North America. Perhaps this subtropical community was depauperate because of the small geographic area of the biome it occupied prior to the Clarkforkian (most of Central America would not emerge for another 55 m.y.). We find this argument ad hoc. Furthermore, it has been argued that southern limits of Recent mammals are set in large part not by climatic conditions, but by biotic interactions (Kaufman 1995). If this speculation is correct, then Tiffanian holdover species might have persisted in situ even as a set of more warm-adapted species immigrated from the south—so the addition of even a depauperate subtropical fauna could have increased, not decreased, standing diversity. In summary, we believe that the Clarkforkian diversity crash is a legitimate biological mystery that remains to be solved.

2. The Clarkforkian/Wasatchian (Cf/Wa) boundary is marked by a major ordinal replacement event: true primates, hyaenodontid creodonts, artiodactyls, and perissodactyls all make their first appearance in North America (Gingerich 1989; Clyde and Gingerich 1998). The proportional volatility G-statistic (Fig. 4F) is one of two truly extreme values in the data set and shows clearly that this event goes beyond the limits of background turnover. In addition, the body mass data show that the boundary is close to a large drop in the standard deviation (−0.301: 54.5-Ma bin) and the largest increase in kurtosis (+1.478: 53.5-Ma bin) in the entire record (Fig. 5B,D). These statistics record the filling of the middle body size range by members of each of the new groups.

The Cf/Wa boundary coincides with the marine bin (55.5 Ma) that has the second highest standard deviation (0.467, which is more than two standard deviations above the mean) and by far and away the highest detrended standard deviation (+0.307). This portion of the Paleogene is well sampled in the marine record, so it is very unlikely that the high value is an artifact of insufficient sampling. Rather, it reflects the extremely unusual $\delta^{18}O$ values associated with a warm climatic excursion at 55 Ma (Kennett and Stott 1991). Our million-year binning is too coarse to illustrate the event, but the climatic transient is large enough to increase observed variability for the entire million-year block. Highly resolved biostratigraphic, chronostratigraphic, and paleoclimatic analyses from the western United States show that changes in the relative diversity of orders and in the body size distribution are precisely coincident with the climatic transient (Koch et al. 1992; Clyde and Gingerich 1998). However, even at this coarse temporal scale the relationship between climatic change and faunal response is obvious.

The proximate links between climate change and turnover at the Paleocene/Eocene boundary remain obscure. The warming at this boundary is extremely rapid (<30,000 years [Bains et al. 1999; Norris and Röhl 1999]), and some of the mammalian first appearances are tightly coupled to the warming (Koch et al. 1992). As a consequence, many workers have focused on immigration, rather than in situ evolution, as a primary factor in this turnover (Gingerich 1989; Krause and Maas 1990; Beard 1998; Clyde and Gingerich 1998). Under this scenario, the appearance event at the Paleocene/Eocene boundary in the Holarctic continents is thought to reflect rapid faunal homogenization via high-latitude land bridges that is facilitated by extreme high-latitude warming, though the ultimate source of these immigrants is debated (Gingerich 1989; Koch et al. 1992; Beard 1998; Hooker 1998).

Despite the addition of these potential immigrant taxa, the biotic transition at the Paleocene/Eocene boundary in North America is actually dominated by the products of in situ cladogenesis immediately following the climatic transient. Sampling-standardized data show that primates, artiodactyls, and perissodactyls (the three new orders) together constitute 14.3% of the fauna by 55 Ma and 18.6% by 54 Ma, and raw data show that at least 121 and 117 species existed at these two times. Thus, four or five founding species in these groups quickly gave rise to dozens of others. If cladogenesis in these groups had proceeded at background rates instead, the immigration events would have appeared to be completely unremarkable.

3. The early Uintan, and to some extent the preceding Bridgerian, is marked by yet another major ordinal shift, as primates and "condylarths" decline while artiodactyls and perissodactyls increase. The clearest evidence here is a series of elevated proportional volatility G-values throughout the mid-Eocene, with particularly high values for the 46.5-, 44.5-, 43.5-, and 42.5-Ma bins. With respect to patterns in body size, the key 44.5 early Uintan bin includes the third-largest increase in the mean (+1.102), the fifth-largest standard deviation increase (+0.307), and the fifth-biggest drop in skewness (−0.407), all of which indicate a trade-off between medium- and large-sized mammals. Turnover rates per se are not remarkable, but the data set's second-highest net diversification value is in the 49.5-Ma bin (i.e., earliest Bridgerian), which puts diversity at the levels it maintained throughout most of the Eocene. Many of these quantitative measures track the loss of primates, which creates a gap in the middle of the body size distribution. The unusual pattern of taxonomic turnover and body mass change around 45–44 Ma qualifies as a bona fide ecological transition, one with few or no parallels elsewhere in the record. Whether the transition corresponded to a single, abrupt extinction event or a gradual replacement lasting several million years is not yet determinable.

At first inspection, the isotopic standard deviation and detrended standard deviation values for this interval seem to be absolutely typical. The rate of change is one standard deviation higher than the mean for the two blocks between 48.5 and 46.5 Ma, and is positive, but much closer to the mean in the temporal blocks bracketing this interval. In addition to these slightly higher than normal rates of $\delta^{18}O$ increase, this portion of the record is conspicuous for being the first long interval after the earliest Paleocene (immediately prior to our first isotopic records at 60 Ma) in which $\delta^{18}O$ values steadily increase. In other words, the time period from ~49.5 to 44.5 Ma represents the first sustained, significant pulse of global cooling experienced by mammals in the Cenozoic. While no single 1.0-m.y.-long block was highly unusual, the aggregate effect of climate change in this interval was to present the presumably warm/wet adapted faunas of the latest Paleocene–early Eocene with a cooler world. A plausible argument could be made that this trend included an increase in seasonality, a decrease in precipitation and temperature, or both.

In the Recent, primates and mammalian arboreal folivores/frugivores are most diverse in tropical areas with low seasonality and high mean annual precipitation and temperature, whereas tropical ungulates dominate in areas with greater seasonality and lower precipitation (Gunnell et al. 1995; Gagnon 1997). The fact that mammalian data for just a few 1.0-m.y. intervals signal such a major transition suggests that if environmental factors were important, the biotic response involved a nonlinear threshold effect: perhaps a critical amount of environmental deterioration resulted in primates no longer being able to prosper and artiodactyls and perissodactyls suddenly being able to radiate.

4. The mid-Oligocene is a period of substantial diversification following a long interval, going back arguably to the end of the mid-Eocene upheaval at about 41 Ma, that witnessed exceptionally low regional diversity (Alroy 1996). Although visually dramatic, the mid-Oligocene diversification is hard to pin down to a particular bin. Instead, modest but consistent net diversification in each bin from 31 to 28 Ma boosts diversity by some 49%. Apart from this run, late Oligocene origination, extinction, volatility, and total-turnover rates are not very remarkable. However, the 28.5-Ma bin does have the highest proportional volatility G-statistic in the study interval (33.964). Body mass data record nothing of great interest, with all of the small shifts dur-ing the late Oligocene being rendered dubious by persistently low counts of measured species. Thus, the shifts may reflect variable reporting of small mammals like insectivorans and small rodents.

There is strong evidence for significant climate change in this interval. The 28.5-Ma bin records the third-highest detrended standard deviation value (+0.173), and the standard deviation seems high in general throughout the late Oligocene (Fig. 1D, E). The single largest decrease in the midpoint $\delta^{18}O$ value separates the 26.5- and 25.5-Ma bins. Unfortunately, poor faunal correlations make it hard to say which of these isotopic shifts, if either, correlates with the mammalian diversification. In any event, the implied melting of ice sheets and/or global warming event that is indicated by these data marks the beginning of an isotope plateau that persists until perhaps the mid-Miocene, at which point isotope values again begin to climb steadily.

Here, as across the Paleocene/Eocene boundary, there is evidence for both an ameliorating climate and an increase in diversity. However, the mechanism is profoundly different. This time there is no evidence for the immigration of important new taxonomic groups, and shifts in ordinal dominance are short lived: a relative increase in rodent and carnivoran diversity at the expense of insectivorans, perissodactyls, and arguably artiodoctyls, all of which is reversed within a few million years. These shifts could conceivably relate to the monographic bias that is suggested by the body mass data. Another source of worry is the fact that the late Oligocene data include extensive faunal lists from the West Coast (i.e., the John Day area) in addition to the northern Great Plains, whereas the early Oligocene data are heavily concentrated in the Great Plains region. Despite these concerns about data quality, we do believe that the easily visible, half-again increase in regional diversity between the early and late Oligocene is real. Understanding the exact nature of this biotic response will require obtaining much better paleoecological and taxonomic data than are currently available in the published literature.

5. The last conspicuous set of mammalian

events occurred in the late Miocene and Pliocene, from ~7 to 4 Ma. Here it is hard to tell whether one or more distinct events are involved. A very important shift in body mass distributions occurs going into the 6.5-Ma bin, with the time series' third-largest drop in the mean (−0.955) and biggest increase in skewness (+0.488); the preceding bin witnesses another large drop in the mean (−0.898) and the second-largest decrease in kurtosis (−0.661). Some of these patterns may reflect sampling biases—small mammal faunas are unusually poor during the preceding several million years. However, the drop in the mean and increase in skewness both mark long-term transitions between the first and second halves of the Neogene, and are therefore likely to be biologically meaningful.

A second pattern involves high turnover very close to the Miocene/Pliocene boundary, which may or may not be independent. The last two Miocene bins (6.5 and 5.5 Ma) witness a pair of high extinction values (0.284, 0.316), and diversity crashes quite noticeably from 64 species at 7 Ma to 50 at 5 Ma. Immediately afterwards there is a sharp recovery to 67 species, marked by the data set's second-highest origination value (+0.476) and a high proportional volatility G-statistic (17.336). Although another spike in diversity at the very end of the time series is visible, this may be attributable to a "pull of the Recent" or some other residual bias.

Essentially, the data suggest that something fundamental sets off the large mammal-rich faunas of the late Miocene from the Pliocene and Pleistocene. The exact timing and nature of this transition remain to be documented in more detail.

Here the marine isotope records are truly unexceptional. Values for rate of change are quite low before 3 Ma, and both versions of the standard deviation values are also completely typical. The only marine "event" is a rapid increase in midpoint $\delta^{18}O$ at 13.5 Ma, well before diversity starts to decline. After this event, the entire late Miocene was marked by steadily increasing $\delta^{18}O$ values.

Given available data from benthic foraminifera, there is no reason to suppose that this was an interval of sudden change in the phys-

ical environment: $\delta^{18}O$ values do rise throughout the late Miocene as they did during the mid-to-late Eocene, but in general more slowly. Unlike the latter situation, a large fraction of the late Miocene rise almost certainly reflects an ice volume increase in addition to cooling. Interestingly, the late Miocene did witness the global expansion of C4 plants in low-to-middle terrestrial ecosystems. Cerling et al. (1997) have argued that this expansion was linked to a decline in pCO_2 levels below some critical threshold, but several lines of geochemical and paleobotanical data have failed to support their conjecture (van der Burgh et al. 1993; Palmer et al. 1998; Pagani et al. 1999). Whatever its source, however, the expansion of C4 plants may have had a negative impact on mammal faunas. C4 plants have lower protein levels, are less digestible, and are richer in phytoliths than C3 plants (see discussions in Koch 1998 and Stanley 1998). Several authors have suggested that one or all of these factors may have contributed to extinctions and faunal change in the late Miocene (Cerling et al. 1998; Koch 1998; Stanley 1998), although no rigorous tests of these hypotheses have been conducted.

In summary, we found an association between unusual climatic states and unusual biotic states at two times: across the Paleocene/Eocene boundary and in the mid-Oligocene. One other critical interval in the biotic record (early Uintan) occurs during a time of prolonged and somewhat more rapid climate deterioration, but there is no clear link to any truly severe climate event. The two remaining critical biotic intervals (Tiffanian/Clarkforkian; late Miocene/Pliocene) show no clear association with unusual climate states at all.

Discussion

Our results are clear in demonstrating that major changes in the mammalian biota, as measured by origination and extinction rates, proportional taxonomic turnover, and body size, are not linearly forced by the changes in global climate recorded in benthic isotope data. We see evidence for an association between climate and only two of the five critical transitions in the mammalian biota: the Paleocene/Eocene reorganization documented by

Gingerich (1989) and Clyde and Gingerich (1998), and the mid-Oligocene diversification that we illustrated above. One other important episode of mammalian turnover—also in the Eocene—appears to occur near the beginning of a long-term, unidirectional shift in climate. These three intervals show no consistent pattern. One involves diversification without marked immigration (mid-Oligocene), another both diversification and immigration (Paleocene–Eocene), and the third shifts in body mass distributions and taxonomic composition that don't relate to long-term changes in taxonomic diversity (early Uintan).

Most importantly, these isolated events fail to match up with our isotope records in any simple way. The presumed warming events (Paleocene/Eocene; mid-Oligocene) match with increases in diversity, but the basic mechanisms are fundamentally different: immigration and an attendant pulse of cladogenesis in the first case, simple cladogenesis in the second. The presumed mid-Eocene cooling event had a very different impact—a permanent and profound shift in the biota's ecomorphological character (Uintan).

Despite the fact that both the isotope and mammal data sets are robust, we acknowledge that future research might show more subtle connections between climate and evolution than we have been able to prove. The major reason for concern is that our data do not directly address the key climate variables that structure terrestrial ecosystems. To begin with, there is a reasonable concern that marine isotope data may do a poor job of recording mean annual temperature trends in continental interiors, or even in ocean basins (Lear et al. 2000). The decoupling of the marine isotope record from continental climate change was likely to have been more severe in the Neogene because of the growing impact of continental ice sheets on the isotope records. Thus, events like the mid-Oligocene diversification could in theory be tied to continental climatic changes that are not reflected, or are only very weakly reflected, in the marine data.

We believe, however, that the use of marine data as a proxy is fairly well justified, especially in the Paleogene—which contains most of the critical biotic intervals. Several studies have shown that similar isotope signals are seen in both continental and marine records for narrowly defined intervals of Cenozoic time (e.g., around the Paleocene/Eocene boundary [Fricke et al. 1998; Bao et al. 1999]). Furthermore, the existing terrestrial climate proxy data (Wolfe 1994; Wing et al. 1995; Bao et al. 1999; Wilf 2000) seem to show a reasonable match to the marine records, with key climate shifts like the late Eocene deterioration turning up in both arenas.

Another concern is our lack of a marine record that in any way serves as a quantitative proxy for mean annual precipitation, which might be very important in structuring mammalian communities. Increasingly sophisticated analyses of leaf physiognomic data (Wilf et al. 1998) should eventually make it possible to get at this issue. For the moment, we will note merely that available paleobotanical data seem to point to a crude correlation between mean annual temperature and mean annual precipitation: early Cenozoic climates were warm and wet, late Cenozoic climates colder and drier. Seasonality of temperature and precipitation also may be captured in part by mean temperature records. Therefore, we suspect that a good precipitation record would not change our results substantially, but we strongly encourage this kind of future research.

The crudeness of using isotope data as our sole climate indicator is not the only potential shortcoming of our analysis: we would be the first to admit that our mammalian time series do not necessarily capture every important aspect of paleoecological change. For example, we have not been able to examine changes in alpha (local-scale) diversity, beta (among-region) diversity, and ecomorphological diversity. Stucky (1990) suggested that a major drop in alpha diversity took place somewhere close to the Eocene/Oligocene boundary, which also marks a major global climate shift. Van Valkenburgh and Janis (1993) made a reasonable case for gradual decreases in beta diversity through time. Although there are some concerns that their approach might be influenced by spatial variation in sampling intensity and other effects, it is likely that a more detailed study would yield a statistically informative beta diversity time series. Janis and Wilhelm (1993) showed quantitatively that locomotor adaptations in ungulates and carnivores have evolved greatly during the Tertiary, with ungulates quickly adopting their modern cursorial strategy but carnivores lagging far behind. Hunter and Jernvall (1995) and Jernvall et al. (1996) showed that dental morphology evolved quickly in the early Paleogene and remained largely static through the middle and late Cenozoic, which might relate to increasing dietary specialization as mammalian lineages abandoned generalized omnivory in favor of herbivory or carnivory. All of these patterns could be influenced by climate trends.

Additional ecological indicators could and should be analyzed in future studies. However, we have reason to believe that few surprises would result. Stucky (1990) did not attempt to illustrate variation in alpha diversity within the Paleocene–Eocene, or the Oligocene–Neogene. Preliminary data of the same kind suggest that very little such variation exists, and therefore that the Eocene/Oligocene drop in alpha diversity is another singular paleoecological transition of the same kind as the mid-Eocene changeover of ordinal groups and opening of a gap in the body size distribution. Therefore, a general correlation with long-term climate change is unlikely to be demonstrated. We are similarly skeptical about being able to demonstrate fine-scale variation in beta diversity; most sampling throughout the Tertiary is restricted to the Western Interior of the United states, an area hardly large enough to encompass significant amounts of biogeographic turnover.

Finally, ecomorphological shifts such as increasing dietary and locomotor specialization are of great interest, but we suspect that our analysis of proportional volatility has captured the key ecomorphological transition that is to be found in the western North American record: the loss of the arboreal frugivore/folivore guild and greatly increased richness of the terrestrial large herbivore guild in the middle Eocene. This interpretation is consistent with the data of Janis and Wilhelm (1993), Hunter and Jernvall (1995), and Jernvall et al. (1996), all of which suggest rapid ecomor-

phological transitions during the Eocene. Thus, the mid-Eocene diversification of ungulates was probably related to their cursorial locomotor adaptations (Janis and Wilhelm 1993), well-developed hypocones (Hunter and Jernvall 1995), and selenodont and lophodont dentitions (Jernvall et al. 1996). The climatic deterioration following the Eocene thermal maximum probably played a part in this diversification, but the actual dynamics may have been rooted in evolutionary and ecological interactions among component taxa.

Conclusion

The idea that organisms ride passively on an evolutionary vehicle driven by global climate change has pervaded the paleontological literature for two centuries. Nothing in paleobiology seems to be much harder than testing such a hypothesis rigorously. Taxonomic richness is notoriously hard to quantify, with sampling biases potentially obscuring any real trends. Turnover rates turn out to be conceptually slippery, with most commonly used metrics having serious methodological shortcomings (Foote 2000). Ecomorphological trends cannot be studied objectively without obtaining exhaustive, hard-won morphometric data sets. Temporal correlations in the terrestrial realm are difficult and traditional timescales are so coarse as to be almost useless. Finally, despite decades of work collecting isotopic data we are only now beginning to understand the intimate details of climate trends throughout the entire Cenozoic.

In light of these problems, it is not really surprising that our efforts have yielded no strong evidence in support of the traditional climate-evolution scenario. Still, though, we feel that we have done everything imaginable to make our analysis rigorous and quantitative, and to remove every important statistical and conceptual bias from our representation of the data. Even the fact that we have worked mostly at one spatial scale is simply a necessary consequence of (1) the dense geographic concentration of the data and (2) the difficulty of computing precise statistics, such as moments of body mass distributions, using local assemblage data (Alroy 2000).

More importantly, this and other studies

(Alroy 1996, 1998b,d) have shown that much of mammalian evolution can be explained by intrinsic dynamics: species richness tracks an equilibrium throughout most of the Cenozoic; extinction events are rare; major orders maintain roughly proportionate diversity levels for tens of millions of years; and body mass distributions slowly move to equilibrium points. All of these features suggest the importance of biological factors like competition, escape from predation, and perhaps biomechanical and physiological optimality over and above the more unpredictable influences of climate. We therefore would be very surprised to see our results completely overturned by a new analysis comparing terrestrial vertebrate data with quantitative climate indicators, even if the climate data were derived solely from continental records. Thus, the key question raised by our analyses is not whether data quality problems may have caused us to overlook some hidden climatic effect, but rather why paleobiologists have traditionally been so willing to embrace climate-driven evolutionary scenarios.

Acknowledgments

We thank M. Foote, D. Fox, and J. Hunter for comments on the manuscript, and M. Pagani for assistance in compiling and interpreting the benthic marine isotope records. This work was conducted while J. A. was a postdoctoral associate at the National Center for Ecological Analysis and Synthesis, a center funded by the National Science Foundation (grant DEB-94-21535); the University of California, Santa Barbara; the California Resources Agency; and the California Environmental Protection Agency.

Literature Cited

Alroy, J. 1992. Conjunction among taxonomic distributions and the Miocene mammalian biochronology of the Great Plains. Paleobiology 18:326–343.
———. 1994. Appearance event ordination: a new biochronologic method. Paleobiology 20:191–207.
———. 1996. Constant extinction, constrained diversification, and uncoordinated stasis in North American mammals. Palaeogeography, Palaeoclimatology, Palaeoecology 127:285–311.
———. 1998a. Diachrony of mammalian appearance events: implications for biochronology. Geology 26:23–27.
———. 1998b. Cope's rule and the dynamics of body mass evolution in North American mammals. Science 280:731–734.
———. 1998c. Diachrony of mammalian appearance events: implications for biochronology—Reply. Geology 26:956–958.
———. 1998d. Equilibrial diversity dynamics in North American mammals. Pp. 232–287 in M. L. McKinney and J. Drake, eds. Biodiversity dynamics: turnover of populations, taxa and communities. Columbia University Press, New York.
———. 1999a. The fossil record of North American mammals: evidence for a Paleocene evolutionary radiation. Systematic Biology 48:107–118.
———. 1999b. Putting North America's end-Pleistocene megafaunal extinction in context: large scale analyses of spatial patterns, extinction rates, and size distributions. Pp. 105–143 in R. D. E. MacPhee, ed. Extinctions in near time: causes, contexts, and consequences. Plenum, New York.
———. 2000. New methods for quantifying macroevolutionary patterns and processes. Paleobiology 26:707–733.
Alvarez, L. W., W. Alvarez, F. Asaro, and H. V. Michel. 1980. Extraterrestrial cause of the Cretaceous–Tertiary extinction. Science 208:1095–1108.
Aubry, M.-P., S. G. Lucas, and W. A. Berggren, eds. 1998. Late Paleocene–early Eocene climatic and biotic events in the marine and terrestrial records. Columbia University Press, New York.
Bains, S., R. M. Corfield, and R. D. Norris. 1999. Mechanisms of climate warming at the end of the Paleocene. Science 285:724–727.
Bao, H. M., P. L. Koch, and D. Rumble. 1999. Paleocene–Eocene climatic variation in western North America: evidence from the delta O-18 of pedogenic hematite. Geological Society of America Bulletin 111:1405–1415.
Barnosky, A. D. 1989. The late Pleistocene event as a paradigm for widespread mammal extinction. Pp. 235–254 in S. K. Donovan, ed. Mass extinctions: processes and evidence. Bellhaven, London.
Beard, K. C. 1998. East of Eden: Asia as an important center of taxonomic origination in mammalian evolution. Bulletin of the Carnegie Museum of Natural History 34:5–39.
Behrensmeyer, A. K., N. E. Todd, R. Potts, and G. E. McBrinn. 1997. Late Pliocene faunal turnover in the Turkana Basin, Kenya and Ethiopia. Science 278:1589–1594.
Berggren, W. A., D. V. Kent, C. C. Swisher III, and M.-P. Aubry. 1995. A revised Cenozoic geochronology and chronostratigraphy. SEPM Special Publication 54:129–212.
Brass, G. W., J. R. Southam, and W. H. Peterson. 1982. Warm saline bottom water in the ancient ocean. Nature 296:620–623.
Brown, J. H., and P. F. Nicoletto. 1991. Spatial scaling of species composition—body masses of North American land mammals. American Naturalist 138:1478–1512.
Cerling, T. E., J. M. Harris, B. J. MacFadden, M. G. Leakey, J. Quade, V. Eisenmann, and J. R. Ehleringer. 1997. Global vegetation change through the Miocene/Pliocene boundary. Nature 389:153–158.
Cerling, T. E., J. R. Ehleringer, and J. M. Harris. 1998. Carbon dioxide starvation, the development of C-4 ecosystems, and mammalian evolution. Philosophical Transactions of the Royal Society of London B 353:159–170.
Clyde, W. C., and P. D. Gingerich. 1998. Mammalian community response to the latest Paleocene thermal maximum: an isotaphonomic study in the northern Bighorn Basin, Wyoming. Geology 26:1011–1014.
Damuth, J. 1990. Problems in estimating body masses of archaic ungulates using dental measurements. Pp. 229–253 in J. Damuth and B. J. MacFadden, eds. Body size in mammalian paleobiology. Cambridge University Press, Cambridge.
Epstein, S., R. Buchsbaum, H. A. Lowenstam, and H. C. Urey. 1951. Carbonate-water isotopic temperature scale. Geological Society of America Bulletin 62:417–425.
Foote, M. 1999. Morphological diversity in the evolutionary ra-

diation of Paleozoic and post-Paleozoic crinoids. Paleobiology 25:1–115.

———. 2000. Origination and extinction components of taxonomic diversity: general problems. Paleobiology 26:578–605.

Fricke, H. C., W. C. Clyde, J. R. O'Neil, and P. D. Gingerich. 1998. Evidence for rapid climate change in North America during the latest Paleocene thermal maximum: oxygen isotope compositions of biogenic phosphate from the Bighorn Basin (Wyoming). Earth and Planetary Science Letters 160:193–208.

Gagnon, M. 1997. Ecological diversity and community ecology in the Fayum sequence (Egypt). Journal of Human Evolution 32:133–160.

Gilinsky, N. L. 1994. Volatility and the Phanerozoic decline in background extinction intensity. Paleobiology 20:445–458.

Gingerich, P. D. 1989. New earliest Wasatchian mammalian fauna from the Eocene of northwestern Wyoming: composition and diversity in a rarely sampled high-floodplain assemblage. University of Michigan Papers on Paleontology 28:1–97.

Gunnell, G. F., M. E. Morgan, M. C. Maas, and P. D. Gingerich. 1995. Comparative paleoecology of Paleogene and Neogene mammalian faunas: trophic structure and composition. Palaeogeography, Palaeoclimatology, Palaeoecology 115:265–286.

Hildebrand, A. R., G. T. Penfield, D. Kring, M. Pilkington, A. Camargo Z., S. B. Jacobsen, and W. Boynton. 1991. Chicxulub crater: a possible Cretaceous/Tertiary boundary impact crater on the Yucatan peninsula, Mexico. Geology 19:867–871.

Hooker, J. J. 1998. Mammalian faunal change across the Paleocene–Eocene transition in Europe. Pp. 428–450 in M.-P. Aubry, S. G. Lucas, and W. A. Berggren, eds. Late Paleocene–early Eocene climatic and biotic events in the marine and terrestrial records. Columbia University Press, New York.

Hunter, J. P., and J. Jernvall. 1995. The hypocone as a key innovation in mammalian evolution. Proceedings of the National Academy of Sciences USA 92:10718–10722.

Janis, C. M. 1993. Tertiary mammal evolution in the context of changing climates, vegetation, and tectonic events. Annual Review of Ecology and Systematics 24:467–500.

———. 1997. Ungulate teeth, diets, and climatic changes at the Eocene/Oligocene boundary. Zoology: Analysis of Complex Systems 100:203–220.

Janis, C. M., and P. B. Wilhelm. 1993. Were there mammalian pursuit predators in the Tertiary? Dances with wolf avatars. Journal of Mammalian Evolution 1:103–125.

Jernvall, J., J. P. Hunter, and M. Fortelius. 1996. Molar tooth diversity, disparity, and ecology in Cenozoic ungulate radiations. Science 274:1489–1492.

Kaufman, D. M. 1995. Diversity of New World mammals: universality of the latitudinal gradients of species and bauplans. Journal of Mammalogy 76:322–334.

Kennett, J. P., and L. D. Stott. 1991. Abrupt deep-sea warming, paleoceanographic changes and benthic extinctions at the end of the Paleocene. Nature 353:225–229.

Koch, P. L. 1998. Isotopic reconstruction of past continental environments. Annual Review of Earth and Planetary Sciences 26:573–613.

Koch, P. L., J. C. Zachos, and P. D. Gingerich. 1992. Correlation between isotope records in marine and continental carbon reservoirs near the Paleocene-Eocene boundary. Nature 358:319–322.

Koch, P. L., J. C. Zachos, and D. L. Dettman. 1995. Stable-isotope stratigraphy and paleoclimatology of the Paleogene Bighorn Basin (Wyoming, USA). Palaeogeography Palaeoclimatology Palaeoecology 115:61–89.

Krause, D. W., and M. C. Maas. 1990. The biogeographic origins of late Paleocene–early Eocene mammalian immigrants to the Western Interior of North America. Pp. 71–105 in T. M. Bown and K. D. Rose, eds. Dawn of the Age of Mammals in the northern part of the Rocky Mountain Interior, North America. Geological Society of America Special Paper 243.

Lear, C. H., H. Elderfield, and P. A. Wilson. 2000. Cenozoic deep-sea temperatures and global ice volumes from Mg/Ca in benthic foraminiferal calcite. Science 287:269–272.

Legendre, S. 1989. Les communautés de mammifères du Paléogène (Éocène supérieur et Oligocène) d'Europe occidentale: structures, milieux et évolution. Münchner Geowissenschaftliche Abhandlungen, Reihe A, Geologie und Paläontologie 16:1–110.

Maas, M. C., M. R. L. Anthony, P. D. Gingerich, G. F. Gunnell, and D. W. Krause. 1995. Mammalian generic diversity and turnover in the Late Paleocene and Early Eocene of the Bighorn and Crazy Mountains Basins, Wyoming and Montana (USA). Palaeogeography, Palaeoclimatology, Palaeoecology 115:181–207.

MacPhee, R. D. E., ed. 1999. Extinctions in near time: causes, contexts, and consequences. Plenum, New York.

Marquet, P. A., and H. Cofre. 1999. Large temporal and spatial scales in the structure of mammalian assemblages in South America: a macroecological approach. Oikos 85:299–309.

Marshall, L. G., S. D. Webb, J. J. Sepkoski Jr., and D. M. Raup. 1982. Mammalian evolution and the great American interchange. Science 215:1351–1357.

Matthew, W. D. 1915. Climate and evolution. Annals of the New York Academy of Sciences 24:171–318.

McKinney, M. L. 1990. Classifying and analysing evolutionary trends. Pp. 28–58 in K. J. McNamara, ed. Evolutionary trends. University of Arizona Press, Tucson.

Meng, J., and M. C. McKenna. 1998. Faunal turnovers of Palaeogene mammals from the Mongolian plateau. Nature 394:364–367.

Miller, A. I. 1998. Biotic transitions in global marine diversity. Science 281:1157–1160.

Miller, K. G., R. G. Fairbanks, and G. S. Mountain. 1987. Tertiary oxygen isotope synthesis, sea level history, and continental margin erosion. Paleoceanography 2:1–19.

Norris, R. D. 1996. Symbiosis as an evolutionary innovation in the radiation of Paleocene planktic foraminifera. Paleobiology 22:461–480.

Norris, R. D., and U. Röhl. 1999. Carbon cycling and chronology of climate warming during the Paleocene/Eocene transition. Nature 401:775–778.

Pagani, M., K. H. Freeman, and M. A. Arthur. 1999. Late Miocene atmospheric CO_2 concentrations and the expansion of C4 grasses. Science 285:876–879.

Palmer, M. R., P. N. Pearson, and S. J. Cobb. 1998. Reconstructing past ocean pH-depth profiles. Science 282:1468–1471.

Prothero, D. R. 1999. Does climatic change drive mammalian evolution? GSA Today 9:1–7.

Prothero, D. R., and T. H. Heaton. 1996. Faunal stability during the Early Oligocene climatic crash. Palaeogeography, Palaeoclimatology, Palaeoecology 127:257–283.

Rose, K. D. 1981. The Clarkforkian land-mammal age and mammalian faunal composition across the Paleocene–Eocene boundary. University of Michigan Papers on Paleontology 26:1–196.

Rudwick, M. J. S. 1972. The meaning of fossils. Macdonald, London.

Schrag, D. P., G. Hampt, and D. W. Murray. 1996. Pore fluid constraints on the temperature and oxygen isotope composition of the glacial ocean. Science 272:1930–1932.

Shackleton, N. J., and N. D. Opdyke. 1973. Oxygen isotope and paleomagnetic stratigraphy of equatorial Pacific core V28–238: oxygen isotope temperatures and ice volumes on a 10^5 to 10^6 year scale. Quaternary Research 3:39–55.

Simpson, G. G. 1964. Species density of North American Recent mammals. Evolution 15:413–446.

Sloan, L. C., J. C. G. Walker, and T. C. Moore Jr. 1995. Possible role of oceanic heat transport in early Eocene climate. Paleoceanography 10:347–356.

Spero, H. J., and D. W. Lea. 1993. Intraspecific stable isotope variability in the planktic foraminifer *Globigerinoides sacculifer*: results from laboratory experiments. Marine Micropaleontology 22:193–232.

Stanley, S. M. 1998. Macroevolution: pattern and process. Johns Hopkins University Press, Baltimore.

Stucky, R. K. 1990. Evolution of land mammal diversity in North America during the Cenozoic. Pp. 375–430 *in* H. H. Genoways, ed. Current mammalogy, Vol. 2. Plenum, New York.

van der Burgh, J., H. Visscher, D. L. Dilcher, and W. M. Kurschner. 1993. Paleoatmospheric signatures in Neogene fossil leaves. Science 260:1788–1790.

Van Valen, L. 1971. Adaptive zones and the orders of mammals. Evolution 25:420–428.

———. 1973. A new evolutionary law. Evolutionary Theory 1:1–30.

Van Valkenburgh, B., and C. M. Janis. 1993. Historical diversity patterns in North American large herbivores and carnivores. Pp. 330–340 *in* R. E. Ricklefs and D. Schluter, eds. Species diversity in ecological communities: historical and geographical perspectives. University of Chicago Press, Chicago.

Vrba, E. S. 1985. Environment and evolution: alternative causes of the temporal distribution of evolutionary events. South African Journal of Science 81:229–236.

———. 1992. Mammals as a key to evolutionary theory. Journal of Mammalogy 73:1–28.

———. 1995. The fossil record of African antelopes (Mammalia, Bovidae) in relation to human evolution and paleoclimate. Pp. 385–424 *in* E. S. Vrba, G. H. Denton, T. C. Partridge and L. H. Burckle, eds. Paleoclimate and evolution with emphasis on human origins. Yale University Press, New Haven, Conn.

Webb, S. D. 1977. A history of savannah vertebrates in the New World, Part I. North America. Annual Review of Ecology and Systematics 8:355–380.

———. 1984. Ten million years of mammal extinction in North America. Pp. 189–210 *in* P. S. Martin and R. G. Klein, eds. Quaternary extinctions. University of Arizona Press, Tucson.

Wilf, P. 1997. When are leaves good thermometers? A new case for leaf margin analysis. Paleobiology 23:373–390.

———. 2000. Late Paleocene–early Eocene climate changes in southwestern Wyoming: paleobotanical analysis. Geological Society of America Bulletin 112:292–307.

Wilf, P., S. L. Wing, D. R. Greenwood, and C. L. Greenwood. 1998. Using fossil leaves as paleoprecipitation indicators: an Eocene example. Geology 26:203–206.

Wing, S. L., J. Alroy, and L. J. Hickey. 1995. Plant and mammal diversity in the Paleocene to Early Eocene of the Bighorn Basin. Palaeogeography, Palaeoclimatology, Palaeoecology 115:117–155.

Wolfe, J. A. 1994. Tertiary climatic changes at middle latitudes of western North America. Palaeogeography, Palaeoclimatology, Palaeoecology 108:195–205.

Woodburne, M. O. 1987. Cenozoic mammals of North America: geochronology and biostratigraphy. University of California Press, Berkeley.

Wright, J. D., K. G. Miller, and R. G. Fairbanks. 1992. Early and middle Miocene stable isotopes; implications for deepwater circulation and climate. Paleoceanography 7:357–389.

Zachos, J. C., K. C. Lohmann, J. C. G. Walker, and S. W. Wise. 1993. Abrupt climate change and transient climates during the Paleogene: a marine perspective. Journal of Geology 101:191–213.

Zachos, J. C., L. D. Stott, and K. C. Lohmann. 1994. Evolution of early Cenozoic marine temperatures. Paleoceanography 9:353–387.

Zachos, J. C., T. M. Quinn, and K. A. Salamy. 1996. High-resolution (10^4 years) deep-sea foraminiferal stable isotope records of Eocene–Oligocene climate transition. Paleoceanography 11:251–266.

Zachos, J. C., B. P. Flower, and H. Paul. 1997. Orbitally paced climate oscillations across the Oligocene/Miocene boundary. Nature 388:567–570.

Modeling fossil plant form-function relationships: a critique

Karl J. Niklas

Abstract.—Attempts to model form-function relationships for fossil plants rely on the facts that the physiological and structural requirements for plant growth, survival, and reproductive success are remarkably similar for the majority of extant and extinct species regardless of phyletic affiliation and that most of these requirements can be quantified by means of comparatively simple mathematical expressions drawn directly from the physical and engineering sciences. Owing in part to the advent and rapid expansion of computer technologies, the number of fossil plant form-function models has burgeoned in the last two decades and encompasses every level of biological organization ranging from molecular self-assembly to ecological and evolutionary dynamics. This recent and expansive interest in modeling fossil plant form-function relationships is discussed in the context of the general philosophy of modeling past biological systems and how the reliability of models can be examined (i.e., direct experimental manipulation or observation of the system being modeled). This philosophy is illustrated and methods of validating models are critiqued in terms of four models drawn from the author's work (the quantification of wind-induced stem bending stresses, wind pollination efficiency of early Paleozoic ovulate reproductive structures, population dynamics and species extinction in monotypic and "mixed" communities, and the adaptive radiation of early vascular land plants). The assumptions and logical (mathematical) consequences (predictions) of each model are broadly outlined, and, in each case, the model is shown to be overly simplistic despite its ability to predict the general or particular behavior or operation of the system modeled. Nonetheless, these four models, which illustrate some of pros and cons of modeling fossil form-function relationships, are argued to be pedagogically useful because, like all models, they expose the internal logical consistency of our basic assumptions about how organic form and function interrelate.

Karl J. Niklas. Department of Plant Biology, Cornell University, Ithaca, New York 14853. E-mail: KJN2@cornell.edu

Accepted: 3 December 1999

Model. *A standard for imitation or comparison; a pattern; an exemplar; also, a representation, generally in miniature.*

> The New Century Dictionary

"What are the roots that clutch, what branches grow Out of this stony rubbish? Son of man, You cannot say or guess, for you know only A heap of broken images. . . ."

> T. S. Eliot

Introduction

The benefits of modeling physical or biological systems are generally acknowledged. Some systems are unavailable for direct observation or experimentation because they are too big, small, expensive, rare, or fragile to observe or manipulate directly. Modeling these systems to analyze their behavior is a convenient approach to coping with these limitations. Likewise, some systems that are subject to direct observation may be so complex or to direct observation may be so complex or

imperfectly known that modeling them provides the only tractable method to explore the details of their operation. Here, models can be used to generate hypotheses that can be empirically examined subsequently. Finally, models are an important part of our pedagogic armamentarium. They provide a method to communicate and summarize complex ideas or operations easily yet accurately.

Modeling the relationship between animal form and function has a long and distinguished career (e.g., Richter 1928; Thompson 1942; Ostrom 1964; Rudwick 1964; Raup and Michelson 1965; Clarkson 1966; Stanley 1970). In comparison, modeling has only recently been used to explore the quantitative relationships between fossil plant form and function. Nonetheless, within the last few decades, virtually every level of biological organization has been explored for fossil plants, ranging from the level of molecular self-assembly (e.g., Hemsley et al. 1996, 1998; Hemsley 1998) to

0094-8373/00/2604-0012/$1.00

community structure, ecological interactions, and evolutionary dynamics (Niklas 1982, 1986a, 1994a, 1997a,b,c, 1999; Beerling et al. 1992, 1993a,b), including aspects of ontogeny and development (Niklas 1976, 1977, 1979, 1993; Niklas and Chaloner 1976; Niklas and Phillips 1976; Stein 1993, 1998; Hotton and Stein 1994; Meyer-Berthaud and Stein 1995), water transport and hydraulic design (Niklas 1984, 1985a; Niklas and Banks 1985; Roth et al. 1994, 1995, 1998; Empacher et al. 1995; Mosbrugger and Roth 1996; Roth and Mosbrugger 1996; see also Bower 1930), reproduction and propagule dispersal (Phillips 1979; Niklas 1981a,b, 1983, 1985b), taphonomic phenomena (Rex and Chaloner 1983), and paleoclimatology (Wagner et al. 1999).

Although the reasons for this exuberant interest in modeling fossil plants are obvious, the reasons that fossil plant models are late coming are not obvious, especially since modeling how organic form and function interrelate is arguably easier for plants than for animals. Unlike animals, the vast majority of plant species perform the same tasks to grow, survive, and reproduce—plants use sunlight to manufacture their living substance, exchange atmospheric gases with their external environment, absorb and transport water and minerals, cope with the mechanical stresses induced by externally applied mechanical forces, and they capture or disperse spores, seeds, fruits, or other similar structures to reproduce or colonize new sites. Thus, unlike the case for animals, the metabolic operations and requirements of plants, both past and present, are nearly identical in their broad outline (Gates 1965; Nobel 1983; Taiz and Zeiger 1991; Salisbury and Ross 1992). Likewise, most plants are sedentary and all lack neurological and muscular systems, and thus they grow, reproduce, and die in much the same location in which they began their existence. Plants, therefore, can be modeled more as "structures" than as "mechanisms" (Niklas 1978, 1986a,b,c, 1990, 1992, 1998; Niklas and Kerchner 1984; Speck and Vogellehner 1988; Speck et al. 1990a,b).

For these reasons, the botanist can turn directly to the techniques and concepts formulated by the physical and engineering sciences to model virtually any plant function requiring energy or mass transport (e.g., photosynthesis, respiration, water and mineral absorption) or involving solid and fluid mechanics (e.g., heat dissipation, mechanical stability, passive or active dispersal of spores or propagules, and the bulk transport or movement of water or air). Analytical geometry can be used to evaluate the capacity of a plant to intercept sunlight (Niklas 1988a,b); equations such as Fick's law can be used to model the passive diffusion of carbon dioxide, oxygen, or any other substance through its tissues (Niklas 1994b); the transport of water through plant vascular systems can be evaluated by means of the Hagen-Poiseuille equation (Zimmermann 1983; Niklas 1992); and the long-distance passive dispersal of reproductive structures by wind or water can be treated with the aid of Stokes' law or comparatively simple physical or mathematical models (e.g., Vogel 1981; Phillips 1979; Okubo and Levin 1989; Habgood et al. 1998). On a macroscopic scale, the ability of any plant to cope with the effects of gravity or drag induced bending moments and stresses can be evaluated with the aid of standard engineering techniques and concepts (Niklas 1986a,b,c,d; Speck and Vogellehner 1988; Speck et al. 1990a,b), whereas ecological interactions among conspecifics or among members of different plant species can be modeled often on the basis of simple allometric "rules" that appear to hold true through across the full spectrum of plant size and all clades (Niklas 1994b,c,d; West et al. 1999).

The botanist is also at advantage because the spectrum of plant bodyplans is narrower than that of animals. Whereas the bodyplans of the algae are extremely diverse (e.g., unicellular, colonial, filamentous, thalloid, and treelike), the land plant species share the same basic bodyplan predicated on one or more cylindrical organs (stems or stem-like analogues) bearing flattened organs (leaves or leaf-like analogues) anchored to a substrate by a variety of means (e.g., rhizoids and multicellular scales) of which roots are the most familiar. Regardless of their apparent complexity, all plant bodyplans can be modeled using one or more simple geometric forms (e.g., terete cylinders and oblate or prolate spheroids) whose shape and size can be adjusted independently or allometrically.

Modeling any fossil form-functional relationship, however, has its pitfalls. Whereas the fundamental tenet of modeling is that the first principles of physics, chemistry, engineering, and mathematics cannot be violated, it is also true that organisms typically obviate (or at least obscure) the effects of some of these well-known principles by virtue of growth, reproduction, or unique adaptations, aspects of which are sometimes poorly preserved or entirely obscured in the fossil record. Fick's law for passive diffusion always holds true, just as Euler's buckling formula accurately predicts how tall a tree can grow before it elastically deflects under its own weight. But Fick's law is irrelevant for species that actively transport carbon dioxide in the form of bicarbonates (as do some aquatic plants), whereas Euler's buckling formula gives different estimates of the tree height for woods differing in their lignin content and thus stiffness. Even the most credulous may reasonably doubt that the fossil record retains an unbroken image of a plant's active transport system or of tissue lignin content.

The objective of this paper is to review some of the issues raised by modeling past plant life, first, by discussing the philosophy of modeling in general, and second, by using four examples of form-function models to discuss practical aspects of the approach. Since the emphasis will be on the pitfalls of modeling form-function relationships, I have used models of my own creation.

Types, Uses, and "Fit"

What constitutes a model and how can a model be tested? These issues are not easily resolved because models come in many forms, are put to different uses, and must survive the challenge of different kinds of tests to gain credibility.

Although there are many models, there are only three general types: iconic models, analogue models, and mathematical models. Iconic models are physical conceptions of reality (e.g., scale models or drawings of fossil plants). Analogue models are schematic representations of dynamic processes or opera-tions (e.g., photosynthesis rendered in chemical notation as $CO_2 + H_2O \xrightarrow{light} (CH_2O)_n + O_2$). Mathematical models are the most abstract of the three general types because they express reality in terms of mathematical operations (e.g., the Nernst equation). Importantly, each type of model can be used as a descriptive, behavioral, or decision-making tool. A descriptive model represents the operational relationship, order, or sequencing among the components of a real system (e.g., a reconstruction of a fossil plant used to show organographic or anatomical relationships). Behavioral models are used to predict the response of a real system to perturbation (e.g., a scale model of plant placed in a wind tunnel to estimate wind-induced drag forces and stresses). Decision-making models are used to identify which among available alternative responses of a system is the most likely or favorable according to a priori criteria (e.g., cladistic algorithms and statistical inference models). Complex models can be constructed using two or more types of models in combination as descriptive, behavioral, and decision-making tools.

Regardless of its type or intended use, however, every model is a conceived image of reality. Every modeler, therefore, must deal with the ongoing tension between the ideal (conceived image) and the real (observed data). This does not imply that all useful models are complex. A model's complexity depends as much on the objective of modeling as it does on the structure, object, operation, or process being modeled. When put to some simple use, a model can successfully mimic a structure or process with a minimum number of assumptions or stipulations. For example, to evaluate convective cooling, a cow can be modeled as a cylinder (provided its surface area with respect to volume is scaled properly). Indeed, some very useful models bear little resemblance to the actual appearance of the object or system being represented. But it is always true that a model is only as useful as the extent to which its behavior or operation faithfully accords with that of reality. A cylindrical cow is a useful model in terms of understanding the physics of heat dissipation, but, in terms

Table 1. Underlying assumptions for each of four fossil plant form-function models. For further details, see text.

	Assumptions
Model 1	a. Stems are terete cylinders or tapered cones
	b. Plants grow in crowded communities, and thus
	c. Wind speed profile modeled as a "square" function
	d. Accurate estimates of stem projected areas
	e. No elastic stem flexure; leaves sometimes neglected
Model 2	a. Ovulate species are obligative anemophilous species
	b. Pollen grains are small, spherical, and do not clump
	c. Ovules exposed to wind
	d. Ovule morphology corresponds to that of when ovule is receptive
	e. Non-turbulent airflow directly upwind of ovules
Model 3	a. Plants consist of three "compartments" (i.e., stem, leaf, reproduction)
	b. Energy allocation to three "compartments" obeys allometric rules
	c. Dispersal radius is proportional to height of "parent"
	d. Juvenile mortality due to insufficient light
	e. Mature plants die randomly
Model 4	a. Relative fitness determined by performing one or more of four tasks
	(light interception, mechanical stability, dispersal, water conservation)
	b. Morphology approximated by five variables per axis
	(probability of branching, rotation and bifurcation angles, length, and diameter)
	c. Morphological transformations non-saltational
	d. Relative fitness either increases or remains the same
	e. Phenomena such as frequency dependency, phenotypic variance,
	etc. are entirely neglected

of locomotion, a cylinder rolls whereas a cow walks.

The logic that a model's behavior must "fit" that of its corresponding reality and that this "fit" must be rigorously validated is remorseless. It is especially irksome when the reality being modeled occurred in the past, because, by their very nature, past events, processes, or systems are incompletely known and not subject to direct observation or experimental manipulation. Direct observation or experimental manipulation, however, are among the very few ways we gain any real confidence in our models. Hence, an unavoidable problem confronts those of use who try to model the past: our conceived images of the past are broken images that we try to mend with the aid of models that can rarely if ever be validated to complete satisfaction.

Form-Function Models: Four Examples

The previous discussion of model types, uses, and validation provides a context in which to critique four fossil plant form-function models, all drawn from my research. Two of these models have been published extensively (models 1 and 2); the others are presented here for the first time. For this reason, the assumptions underlying each model are

listed in Table 1. What follows is an exposé of each model's logical (mathematical) consequences (predictions).

Model 1.—The first model purports to calculate the maximum bending stresses that stems located anywhere within the vascular plant bodyplan experience due to wind-induced drag forces. The model was designed to explore the hypothesis that the evolution of land plant branching patterns was adaptive to the increase in wind speed that would be expected to attend the evolutionary increase in vascular plant height (see Niklas 1998).

The mathematics of the model rests on the fact that tensile or compressive bending stresses σ invariably reach their maximum intensities at the surface of any stem, and that these stresses are quantified by the formula $\sigma_i = d_i M_i / 2 I_i$, where the subscript i refers to any portion (element) of the stem, d_i is diameter, M_i is the bending moment, and I_i is the axial second moment of area. Since $I = \pi d^4 / 64$ for any stem with a terete cross-section, it follows that $\sigma_i = 32 M_i / \pi d_i$. In this formula, the wind-induced bending moment (M_i) is the drag force (D_i) multiplied by the distance over which this force is exerted (ℓ_i); i.e., $M_i = \Sigma_{j=1}^{i} D_i \ell_i$. As expected from first principles, the bending moment increases from the tip ($\ell = 0$) to the base of any stem ($\ell = h$, where

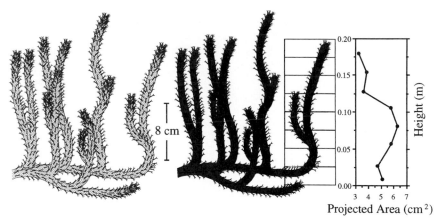

FIGURE 1. Reconstruction (left), stem projected areas (middle), and plot of stem projected area as a function of height above ground (right) in intervals of 2.5 cm for the Middle Devonian genus *Protolepidodendron*. Reconstruction adopted from Taylor and Taylor 1993 (as provided by D. A. Eggert).

h is plant height). Since the drag force at any location along the length of a stem is given by the formula $D_i = 0.5 \rho u_i^2 S_i C_D$ (where ρ is the density of air, u_i is the local wind speed, S_i is the surface area projected toward the wind by stem element i, and C_D is the drag coefficient), it mathematically follows that $M_i = \Sigma_{j=1}^{i} (0.5 \rho u_i^2 S_i C_D) \ell_i$. Thus, the model reduces to the simple formula $\sigma_i = 32 \Sigma_{j=1}^{i} (0.5 \rho u_i^2 S_i C_D) \ell_i / \pi d_i^3$.

These mathematics show that the maximum stresses generated by any wind drag force acting along the length of any stem or entire plant can be calculated provided that the wind speed, stem diameter, and projected surface area are known for different locations along the length of a plant. Stem diameter and projected surface areas can be measured directly from well-preserved compression fossils, or they can be estimated from fossil plant reconstructions scanned into a computer equipped with software to measure them. (Naturally, different protocols incur different measurement errors—an issue that will be addressed later.) Likewise, provided that each plant is assumed to live in a community composed of individuals some of which are of comparable height, the vertical wind speed profile can be calculated from the formula $u_i = u_{max}[1 - (l_i / h)]^2$, where u_{max} is the maximum wind speed measured at plant height h. This formula gives the typical 'square' wind speed profile, which has been empirically measured for dense vegetation.

For purposes of illustration only, this model is illustrated for three fossil lycopods (the Middle Devonian genus *Protolepidodendron*, the Pennsylvanian species *Lepidodendron scleroticum*, and the Triassic species *Pleuromeia longicaulis*, for which stem diameters and projected surface areas were measured from reconstructions; see Figs. 1–3) and for one extant dicot species (*Prunus serotina*, for which stem surface areas were measured empirically). Assuming a 20 m/second wind at the top of each of these four taxa, the stem-stress model predicts the highest stresses for *P. serotina*, whereas the stresses estimated for the much taller *L. scleroticum* are comparable to those of *P. serotina* only for terminal branches and at ~50% total plant height (Fig. 4). The lowest stresses are predicted for *P. longicaulis*, which had a thicker stem but was of comparable height to the geologically older *Protolepidodendron*. Among the inferences that could be drawn from this very simple and preliminary comparison is that, even if neighboring plants shielded one another and so reduced each other's drag, the terminal branches of *L. scleroticum* and *P. serotina* trees are expected to be more prone to mechanical failure than older portions of these plants. This expectation is consistent with the observations that large *P. serotina* trees often shed many small branches in wind storms and that large numbers of disarticulated *Lepidodendron* "twigs" and small branches are found in the fossil record. Clearly, shedding terminal branches is functionally adaptive because their absence reduces the drag (and thus the

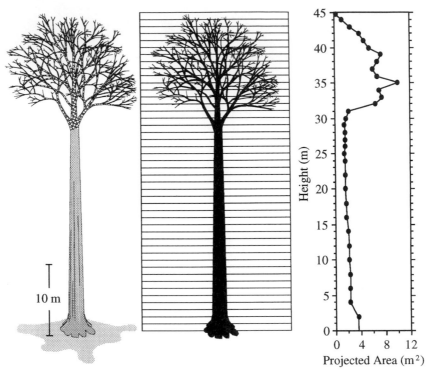

FIGURE 2.   Reconstruction (left), stem projected areas (middle), and plot of stem projected area as a function of height above ground in intervals of 1 m (right) for the Pennsylvanian coal swamp species *Lepidodendron scleroticum*. Reconstruction adopted from DiMichele 1981.

bending moment and stresses) exerted on older portions of the plant body. In this way, a tree can sacrifice juvenile and less "costly" stems in mechanical deference to older and more metabolically costly branches or trunk.

In this regard, it is interesting that, unlike those of *P. serotina*, the trunks of arborescent lycopods such as *L. scleroticum* relied on large amounts of periderm and on a comparatively slender core of wood for mechanical support. The periderm of living plants has a much lower yield stress compared to wood, but the model indicates that the bending stresses along most of the *L. scleroticum* trunk are much lower in comparison to those of the extant *P. serotina* tree (Fig. 4). Taken at face value, the model indicates that arborescent lycopods may have been better adapted at sustaining high wind loads than some modern-day tree species. Finally, because thick stems reduce the magnitude of bending stresses regardless of their tissue construction and because unbranched stems incur less drag than their branched counterparts, the disproportionate-

ly thick and unbranched stem of *P. longicaulis* (in comparison to that of geologically older *Protolepidodendron*) appears to be especially conducive to sustaining high winds.

It is evident, nonetheless, that my model is based on reasonably criticized assumptions (see Table 1). For example, with the exception of *P. serotina*, the model relies on calculations based on stem diameters and projected surface areas estimated from fossil reconstructions scanned into a computer. Arguably, these reconstructions convey the "typical" (stereotypical) appearance of each fossil taxon (and so perhaps make morphometric comparisons among different species more meaningful), yet over the years, the "image" of even "well-known" fossil taxa can change, often dramatically. Likewise, the assumption that wind speeds can be estimated using an equation that assumes a densely packed community is questionable. In some cases, we know nothing about the community structure of fossil plants, and different wind speed profiles based on other assumptions about community

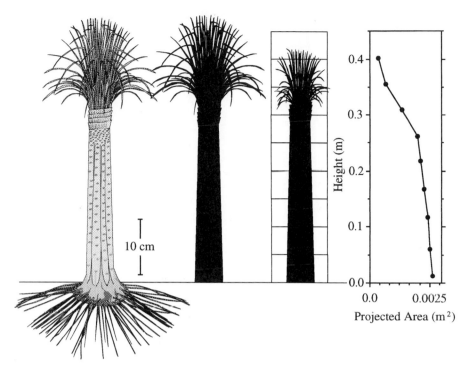

FIGURE 3. Reconstruction (left), stem projected areas with unflexed (and wind-flexed leaves, assuming a tissue composition similar to that of a modern day grass) (middle), and plot of stem projected area as a function of height above ground in intervals of 5 cm (right) for the Triassic species *Pleuromeia longicaulis*. Reconstruction adopted from Retallack 1975.

physiognomy give different estimates of bending stresses. The model also entirely neglects wind-induced elastic stem or leaf flexure, which can reduce drag substantially. Finally, what does it matter if calculations predict that twigs or larger branches are more

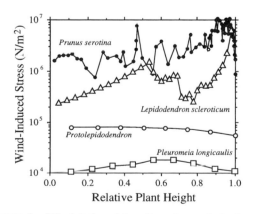

FIGURE 4. Wind-induced bending stress plotted as a function of relative plant height (i.e., distance along the length of stem normalized with respect to total stem height) for *Protolepidodendron*, *Lepidodendron scleroticum*, *Pleuromeia longicaulis* (see Figs. 1–3), and *Prunus serotina*. For further details, see text.

likely to break than the trunk, when plants can mechanically fail or die by wind-throw as a result of uprooting? Wind-throw depends on a number of phenomena, some of which are known for some fossil plants (e.g., root morphology) and some of which are rarely known with confidence (e.g., the degree of soil water-saturation). In summary, this model is biophysically realistic, but its assumptions can be easily thrown into doubt.

Model 2.—The second model purports to estimate the efficiency with which some early Paleozoic reproductive organs captured wind-borne pollen. The original version of this model consisted of miniature (iconic) models of Devonian and Carboniferous ovules, which were placed in a wind tunnel to evaluate the hypothesis that streamlined morphologies possessing well-formed integuments and micropyles were more efficient at capturing pollen than were geologically contemporaneous ovules bearing branch-like appendages surrounding a funnel-shaped nucellar extension that served as a functional an-

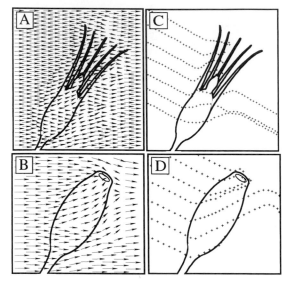

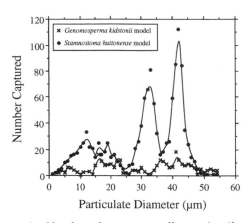

FIGURE 6. Number of spores or pollen grains (from a diverse number of extant species) captured by iconic models of *Genomosperma kidstonii* and *Stamnostoma huttonense* ovules (see Fig. 7) plotted as a function of spore or pollen grain diameter. Equivalent numbers of spores or pollen grains from various extant species were released 40 cm upwind of each model positioned downwind in a small wind tunnel. Data were collected over many "release-capture" trials. Solid lines are running averages of data.

FIGURE 5. Computer simulations of airflow patterns (A, B) and seven pollen grain trajectories (C, D) for two ovules differing in morphology. Direction of airflow in each diagram is from left to right. Length of arrows in A and B proportional to local wind speed.

alogue to the micropyle (i.e., a salpinx; see Niklas 1981a,b, 1983, 1985b). The behavior of these scale models was used to design a computer (mathematical) model capable of predicting both the airflow patterns generated by and around reproductive organs and the trajectories of airborne particulates in these airflow patterns (Fig. 5). This mathematical model is predicated on the terminal settling velocity, u_t, of a small sphere (used to mimic the behavior of a spore or pollen grain), which is given by the formula $u_t = [2r^2g(\rho - \rho_o)]/9\mu$, where r is the radius of the sphere, g is the gravitational acceleration, ρ is the density of the spore or pollen grain, and ρ_o and μ are, respectively, the density and the dynamic viscosity of air. This formula is derived by balancing the weight and buoyancy of the particle, and is based, in turn, on Stokes' law for the drag force, D, acting on very small spheres (i.e., $D = 6\pi\mu r u$). For large spheres, the simple formula for terminal settling velocity does not hold true, but approximate solutions can be used to estimate drag forces using the drag coefficient for large spherical objects, e.g., $C_D = (24/Re) + [6/(1 + Re^{0.5})] + 0.4$, where Re is the Reynolds number.

The predictions of the mathematical model

and the behavior of iconic models of ovules placed in a wind tunnel support the hypothesis that ovules possessing well-formed integuments and micropyles (e.g., *Stamnostoma huttonense*) are more efficient at wind pollination than ovules bearing branch-like appendages variously configured around a salpinx (e.g., *Genomosperma kidstonii* and *G. latens*). Aerodynamically, streamlined ovules "direct" the trajectories of airborne spores or pollen toward the micropyle, whereas those with appendages tend to trap airborne particulates before they reach the salpinx (Fig. 5). More interestingly, streamlined ovules capture more particulates across a wide range of particle size classes than do ovules with appendages, and they can preferentially capture spores or pollen of specific size classes (Fig. 6). Clearly, if this behavior holds true for early Paleozoic seed plants, then the model indicates that species with streamlined ovules had a reproductive advantage over those producing ovules with appendages and a salpinx.

When originally advanced, this explanation for the evolutionary rise to dominance of species with streamlined reproductive organs was vigorously criticized (Rothwell and Taylor 1982). Although the model has now achieved sufficient respectability to be mentioned in two subsequently published paleo-botany textbooks (Stewart and Rothwell 1993; Taylor and Taylor 1993), the validity of some of its assumptions is still very much in doubt (see Table 1). For example, the model assumes that the fossil morphologies upon which icon-ic models were based are those of ovules re-ceptive to pollination, whereas many of these fossil morphologies are aborted and mature reproductive structures. The model further as-sumes that pollen size and geometry accord with those of small spheres (i.e., that Stokes' law or its approximate solutions for drag hold true). However, the meiospores of many an-cient wind-pollinated species are large and deviate from the geometry of a sphere, such that the airborne trajectories of real meios-pores may have been very different from those simulated by the model. Finally, even though the airflow patterns around models of ovulate organs can be precisely determined, in the ab-sence of the aerodynamic context created by the rest of the plant body, these patterns may be irrelevant (e.g., the model assumes that these ancient ovulate morphologies were ex-posed to the wind, rather than shielded by leaves or other reproductively "inert" struc-tures).

Subsequent paleobotanical discoveries have shed light on a number of these issues. For ex-ample, recent reconstructions of the Late De-vonian seed plants *Elkinsia polymorpha* and *Moresnetia zalesskyi* depict leafless branches bearing terminal clusters of cupulate ovules at their tips. These are very similar in appear-ance to those anticipated by the model be-cause the absence of leaves capable of trap-ping pollen grains is particularly conducive to wind pollination. In addition, simple (unpub-lished) elastic models of the slender branching systems of *E. polymorpha* and *M. zalesskyi* in-dicate that wind-induced stem flexure (a fea-ture neglected in Model 1) may have permit-ted ovules to "search" air currents, thereby raising the possibility that meiospores could be captured by direct inertial collision, much as grass flowers arranged in loose panicles do today (Fig. 7). However, these suppositions and inferences are based on imperfectly known fossil remains, and, under any circum-stances, it is evident that pollination efficiency

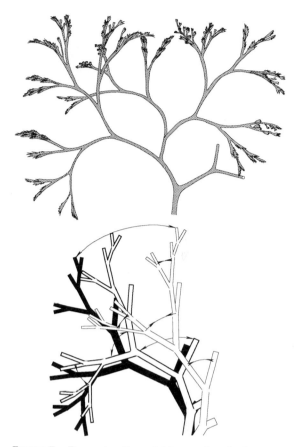

FIGURE 7.—Reconstruction of *Moresnetia zalesskyi* (top; taken from Fairon-Demaret and Scheckler 1987) and simulation of elastic stem flexure for a portion of the re-construction (bottom) based on the assumption that cu-pule-bearing axes have the mechanical properties of grass stems. Direction of arrows between adjoining axes indicate direction of wind-induced stem flexure. See text for further details.

is only one of many factors that contribute to the relative fitness of species. In summary, this model has many weaknesses, yet it appears reasonable in certain respects.

Model 3.—This unpublished model, which is being developed in collaboration with Brian J. Enquist (University of New Mexico), attempts to mimic the behavior of plant populations in an effort to determine the influence of energy and biomass allocation patterns on the surviv-al of individuals or fossil species. Despite the large number of random and nonrandom var-iables known to influence plant population dynamics, the mathematical model is based on five simple assumptions: (1) the total pho-tosynthetic energy gained by leaves per individual per unit time (day, week, etc.) is differentially allocated for future growth among three 'compartments' (leaf, stem, and reproduction) according to allometric rules selected to accord with those of a real or hypothetical species; (2) reproductive biomass is partitioned (per unit time) into N-number of propagules according to some arbitrarily specified or empirically determined allometric relationship; (3) propagules are dispersed randomly to a maximum distance proportional to the height of each parent "plant" (which is proportional to stem biomass); (4) a "juvenile" shaded by a more mature individual receives less light, and thus grows more slowly or dies according to the attenuation of light through the "canopy" above it; and (5) mature "plants" die randomly according to a specified stochastic process (see Table 1).

Starting with an initial "random seeding" of a site, this model mimics much of the behavior of real plant populations (Fig. 8). Simulated plants grow in size and reproduce; individuals that germinate under larger plants grow more slowly than unshaded juveniles, and some ultimately die in the understory; those that do survive reach a maximum height, increase in stem girth, height, and canopy coverage, continue to contribute juveniles to the population, and eventually die, only to make room for the next cadre of juveniles, which reiterate the process. As these biological operations are mathematically simulated over many cycles of "birth and death," typical sigmoid population-growth curves are achieved. However, since the parameters of these curves differ depending on the allometric rules assigned to each species, the model predicts that some species competing for light and space in a mixed community dwindle in number and eventually become locally extinct (Fig. 9).

This model is undeniably overly simplistic, and the phenomena it predicts are nothing more that the a priori consequences of the allometric relationships attributed to each species. From a pedagogical perspective, however, this is the model's strength as well as its weakness, because it shows that the behavior of very complex systems is potentially reducible to a very few, simple operations whose bi-

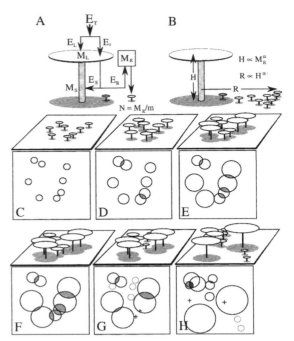

FIGURE 8. Mathematical model for energy E and biomass M allocation patterns to leaf, stem, and reproductive compartments (subscripts L, S, and R, respectively) of the plant body used to simulate population dynamics for monotypic or mixed communities of plants (see Fig. 6). A, Total energy (E_T) from sunlight allocated to leaf energy and the rest of the plant body (E_L and E_R, respectively). Energy to "leaf" compartment (E_L) used to construct leaf biomass (M_L); energy allocated to the rest of the plant E_R is partitioned into energy for the stem and reproduction (E_S and E_R), which is used to construct stem biomass (M_S) and reproductive biomass (M_R). Reproductive biomass is used to construct N number of propagules each with biomass m. B, Dispersal radius R for propagules is proportional to total plant height (H), which is proportional to reproductive biomass (M_R), both of which scale according to specified or empirically determined allometric rules (with scaling exponents α and α', respectively). C–H, Diagrammatic renditions of the "growth" of a population after an initial random seeding of the site (C). Overlapping of "canopies" shown as shaded arcs of circles; "germination" of new propagules indicated by thin-lined circles; death of individuals indicated by +.

ological reality can be experimentally tested on the basis of observing living plants. Likewise, the model permits us to immediately explore the logical (mathematical) consequences of its assumptions about how real populations behave. In summary, the model can be used to construct hypotheses and design experiments dealing with real plants, regardless of whether its initial assumptions are proven to be correct.

Model 4.—The last model to be critiqued has

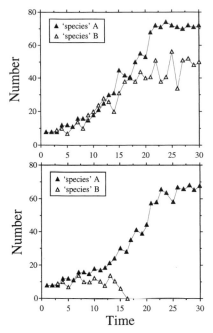

FIGURE 9. Simulations of population "growth" for two hypothetical species predicted by a simple model (see Fig. 5). Top graph: Number of individuals in monotypic populations of "species" A and "species" B differing in the allometric rules governing energy and biomass allocation to leaf, stem, and reproductive compartments of the plant body plotted as a function of time (in arbitrary units). Bottom graph: Numbers of individuals in a mixed population composed of "species" A and "species" B plotted as a function of time (in arbitrary units) assuming an equivalent initial number of "juveniles" of both "species." This simulation results in the local "extinction" of "species" B after 16 cycles of "birth and death" for individuals competing for light and space.

been published elsewhere (see Niklas and Kerchner 1984; Niklas 1994a, 1997a,b, 1999). It attempts to evaluate the hypothesis that the major morphological transitions seen in the early evolution of vascular land plants were the consequence of natural selection acting on the performance of one or more of four functional obligations: light interception, mechanical stability, long-distance dispersal of reproductive structures, and conservation of water. At its core, the model consists of two components, a "morphospace" containing a large number (i.e., on the order of 10^9) of mathematically conceivable leafless vascular plant sporophytes, and a total of 15 "fitness landscapes" each of which quantifies the performance of one or more of the four tasks for each variant in the morphospace. Adaptive "walks" through the morphospace begin at

the same location, the generalized sporophyte morphology of *Cooksonia*, one of the earliest bona fide vascular land plants. The algorithm driving each walk locates one or more neighboring morphological variants with a higher relative fitness than that of the *Cooksonia* analogue and proceeds through each of the 15 "fitness landscapes" locating morphological variants with progressively higher relative fitness surrounded by other less fit ones. Fitness landscapes can be shifted arbitrarily during the progress of a "walk" to mimic the effects of changes in the focus of natural selection (e.g., a "walk" can begin on the "landscape" for water conservation and end on the landscape for maximizing light interception), or a "walk" can be barred from entering certain portions of the morphospace in an effort to evaluate the consequences of developmental constraints on morphological evolution.

The model makes a number of predictions, some of which may appear controversial or pedantic. For example, the model predicts that only a small number of hypothetical morphologies can maximize the performance of a single functional obligation, whereas the number of morphologies that can optimize the performance of two or more tasks equally well increases in proportion to the number of tasks performed simultaneously (Fig. 10). Since the difference in the relative fitness of hypothetical morphologies decreases as the number of tasks performed increases, an inference drawn from the model is that morphological diversification is fostered whenever natural selection acts on the performance of two or more tasks rather than one. This prediction echoes the well-known engineering maxim that there are many options for the appearance of a "machine" or "structure" required to perform many as opposed to one task, but these manifold design options are far less efficient at performing each task because of conflicting design requirements. Another prediction of the model is that natural selection likely acted on the obligation to conserve water during the initial phase of early land plant evolution but expanded to focus to the performance of all four tasks once mechanisms had evolved to conserve water. This conclusion comes from the fact that only those "adaptive walks" pass-

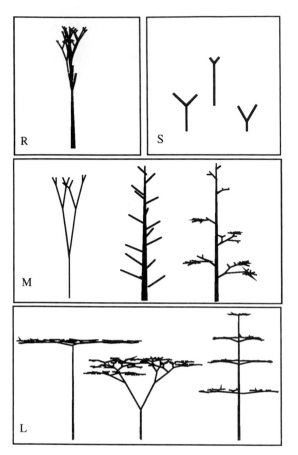

FIGURE 10. Appearance of simulated early vascular land plant sporophytes terminating "adaptive walks" on "fitness landscapes" quantified by maximizing reproductive efficiency (R), conservation of water (S), mechanical stability (M), and light interception (L).

ing through shifting "landscapes" starting with the water conservation and ending with the performance of all four tasks obtain sequences of hypothetical morphologies that agree reasonably well with the major morphological trends observed in the fossil record of early vascular land plants (Fig. 11).

That evolution on land required adaptations to acquire and conserve water is not particularly controversial. Likewise, we have no compelling reason to argue against the logic that a machine or a structure can, in theory, maximize the performance of a single task but can only optimize the performance of two or more tasks. Nonetheless, the model for early land plant evolution is based on a number of assumptions or simplifications, some of which are biologically unrealistic (Table 1). For example, it assumes that the relative fitness of morphological variants can be quantified solely on the basis of performing one or more of four tasks, thereby neglecting many other factors that influence relative fitness (e.g., fecundity, survival of progeny, and the intensity of herbivory). The model further assumes that no variation exists in the branching pattern of each hypothetical morphology, whereas adaptive evolution requires some degree of phenotypic variation in populations. The model also neglects the role of frequency dependency and other phenomena that are known to alter relative fitness. Finally, as modeled here, adaptive "walks" preclude saltational evolution—abrupt morphological transformations are excluded. None of these assumptions is realistic in terms of how real plants evolve, yet the model mimics many well-known trends observed for land plant evolution, drawing attention to the fact that a model may appear to "fit" reality, but, perhaps, for all the wrong reasons.

Models and Past Realities

Each of the four fossil plant form-function models discussed in the previous section was selected not as a "straw man" but to show that the conclusions drawn from each axiomatically follow from the assumptions used to frame reality. In this sense, each model was used as a "predictive tool" not to reveal how reality works, but to demonstrate the logical (mathematical) consequences of its underlying assumptions about how reality works. This can be especially useful when the quantitative evaluation of a form-function relationship necessitates numerous reiterative mathematical operations, which typically have hidden or emergent properties.

Any model can transcend this pedagogic role and become a predictive tool, but only if the reality it purports to mimic is amenable to experimental manipulation or direct observation. Models of living systems are easily tested in this way, but models of past events or operations are far less tractable. Clearly, any model of a past system can be rejected if it fails to accord with the fossil record. But historical events cannot be manipulated directly or observed always in great detail, and so con-

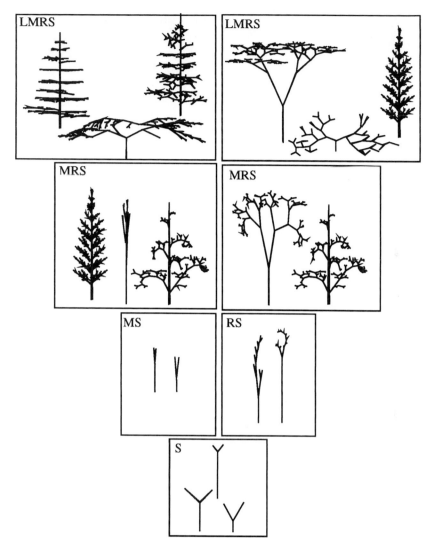

FIGURE 11.   Appearance of simulated early vascular land plant sporophytes terminating different stages in two "adaptive walks" passing through shifting "fitness landscapes" quantified first by maximizing the conservation of water (S) and subsequently quantified by optimizing two, three, and finally all of four functional obligations: reproductive efficiency (R), mechanical stability (M), and light interception (L). Groupings of letters shown in upper left of each diagram denote the two or more tasks each sporophyte is capable of optimizing. The two "walks" (series of diagrams shown on left and right) proceed from bottom to top.

fidence in a model for a fossil form-function relationship can be gained only if the model successfully predicts or accords with *future* discoveries made from the fossil record, or if the model successfully predicts the behavior of a modern-day analogue (which can be experimentally manipulated or directly observed).

Some of my models have been scrutinized in this way. For example, the model for the aerodynamic behavior of ovulate structures has been "tested" by placing the reproductive structures of living plants in wind tunnels and comparing observed pollen grain trajectories with those simulated mathematically (e.g., Niklas and Buchmann 1987). Likewise, the model for early vascular land plant evolution successfully predicted the morphology of the "average" *Salicornia europaea* plant in monotypic populations differing in density (Ellison and Niklas 1988). However, validating a model on the basis of predicting past events or the extent to which it complies with the behavior of a living species has its pitfalls. Modern-day

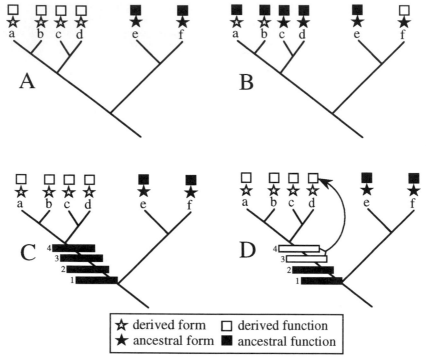

FIGURE 12. Illustration of the criteria of historical congruity (A, B) and biophysical exclusion (C, D) used to test form–function adaptationist hypotheses in terms of a hypothetical cladogram composed of six species (a–f). The derived and ancestral forms are denoted as ☆ and ★, respectively; the efficiencies of the derived and ancestral forms are depicted by □ and ■, respectively. A, The criterion of historical concordance is complied with and supports the adaptationist form-function hypothesis because the evolutionary appearance of form ☆ accords with function □ for species a–d and the appearance of form ★ accords with function ■ for species e and f. B, The criterion of historical concordance is violated because species a and b with form ☆ are less efficient than species c and d with form ★ and because species f has form ★ yet functions as well as species a and b with form ☆. C, The criterion of biophysical exclusion supports the form-function hypothesis because a biophysical model shows that other derived features (closed rectangles 1–4) shared by species a–d cannot account for function □. D, The criterion of biophysical exclusion is violated because two other derived features (open rectangles 3 and 4) shared by species a–d account for function □. (Adapted from Niklas 1997c.)

analogues for past form-function relationships may be unavailable, imprecise in some significant but unappreciated way, or, in the worse case, incorrectly identified. The real danger is that a model may mimic the behavior of a fossil system for all the wrong reasons.

Any model can be tinkered with endlessly to account for discrepancies in its correspondence with reality. Undue pessimism about modeling the past, however, is unjustified. First, it is always necessary to balance hardnosed skepticism with sober-minded pragmatism; second, in the absence of direct experimentation, a model can still be modified or rejected as our knowledge about past reality advances; and, third, even a bad model can be intellectually useful. Logic quickly demonstrates that no model can ever offer secure

knowledge about past or present reality even after exhaustive testing. Consequently, the extent to which a model is scrutinized must always be gauged in terms of how it is used. Some models require no formal testing because the inferences we draw from them are uncontroversial; other more controversial models require more intensive testing before their implications are accepted as valid. In the absence of experimental data or direct observation, controversial models can be "tested" against the two criteria of historical concordance and biophysical exclusion (Fig. 12). That is, it is necessary to show that the evolutionary appearance of a structure or anatomy ("form") believed to confer a selective advantage is congruent with the performance of a particular biological task ("function"). This can be evaluated with the aid of phylogenetic (cladistic) hypotheses (models), which evaluate whether historical discordance between form and function exists. But, since correlation does not imply causation and since shared derived features other than those postulated to confer a selective advantage characterize lineages, it is also necessary to show precisely which among the various features shared among taxa is responsible for the efficient performance of function. This criterion of biophysical exclusion requires a biomechanical model that predicts precise quantitative relationships between form and function. In this way, adaptationist hypotheses are constrained and thus indirectly tested. Finally, as noted, every model, even a bad one, is pedagogically useful because its assumptions can be quickly shown to be misleading or incomplete. Testing our preconceptions of how past realities may have worked may be the greatest benefit of modeling.

Literature Cited

Beerling, D. J., W. G. Chaloner, B. Huntley, J. A. Pearson, M. J. Tooley, and F. I. Woodward. 1992. Variations in the stomatal density of *Salix herbacea* L. under the changing atmospheric carbon dioxide concentrations of late and post-glacial time. Philosophical Transactions of the Royal Society of London B 336:215–224.

Beerling, D. J., W. G. Chaloner, B. Huntley, J. A. Pearson, and M. J. Tooley. 1993a. Stomatal density responds to glacial cycle of environmental change. Proceedings of the Royal Society of London B 251:133–138.

Beerling, D. J., D. P. Mattey, and W. G. Chaloner. 1993b. Shifts in the delta carbon-13 composition of *Salix herbacea* L. leaves in response to spatial and temporal gradients of atmospheric carbon dioxide concentration. Proceedings of the Royal Society of London B 253:53–60.

Bower, F. O. 1930. Size and form in plants, with special reference to conducting tracts. Macmillan, London.

Clarkson, E. N. K. 1966. Schizochroal eyes and vision in some Silurian acastid trilobites. Palaeontology 9:1–29.

DiMichele, W. A. 1981. Arborescent lycopods of Pennsylvanian age coals: *Lepidodendron*, with a description of a new species. Palaeontographica 175B:85–125.

Ellison, A. M., and K. J. Niklas. 1988. Branching patterns of *Salicornia europaea* (Chenopodiaceae) at different successional stages: a comparison of theoretical and real plants. American Journal of Botany 75:501–512.

Empacher, N., V. Mosbrugger, A. Roth, M. Wolf, and A. Wunderlin. 1995. Qualitative mathematical discussion of different evolutionary states in water transport systems of plants. Journal of Biological Physics 21:241–264.

Fairon-Demaret, M., and S. E. Scheckler. 1987. Typification and redescription of *Moresnetia zalesskyi* Stockmans, 1948, an early seed plant from the Upper Famennian of Belgium. Bulletin de l'Institut Royal des Sciences Naturelles de Belgique, Sciences de la Terre 57:183–199.

Gates, D. M. 1965. Energy, plants, and ecology. Ecology 46:1–16.

Habgood, K. S., A. R. Hemsley, and B. A. Thomas. 1998. Modelling of the dispersal of *Lepidocarpon* based on experiments using reconstructions. Review of Palaeobotany and Palynology 102:101–114.

Hemsley, A. R. 1998. Nonlinear variation in simulated complex pattern development. Journal of Theoretical Biology 192:73–79.

Hemsley, A. R., P. D. Jenkins, M. E. Collinson, and B. Vincent. 1996. Experimental modelling of exine self-assembly. Botanical Journal of the Linnean Society 121:177–187.

Hemsley, A. R., B. Vincent, M. E. Collinson, and P. C. Griffiths. 1998. Simulated self-assembly of spore exines. Annals of Botany 82:105–109.

Hotton, C. L., and W. E. Stein. 1994. An ontogenetic model for the Mississippian seed plant family Calamopityaceae. International Journal of Plant Science 155:119–142.

Meyer-Berthaud, B., and W. E. Stein. 1995. A reinvestigation of *Stenomyelon* from the Late Tournaisan of Scotland. International Journal of Plant Science 156:863–895.

Mosbrugger, V., and A. Roth. 1996. Biomechanics in fossil plant biology. Review of Palaeobotany and Palynology 90:195–207.

Niklas, K. J. 1976. Morphological and ontogenetic reconstruction of *Parka decipiens* Fleming and *Pachytheca* Hooker from the Lower Old Red Sandstone, Scotland. Transactions of the Royal Society of Edinburgh 69:483–499.

———. 1977. Ontogenetic constructions of some fossil plants. Review of Palaeobotany and Palynology 23:337–357.

———. 1978. Branching patterns and mechanical design in Paleozoic plants: a theoretic assessment. Annals of Botany 42:33–39.

———. 1979. Simulations of apical developmental sequences in bryophytes. Annals of Botany 44:339–352.

———. 1981a. Simulated wind pollination and airflow around ovules of some early seed plants. Science 211:275–277.

———. 1981b. Airflow patterns around some early seed plant ovules and cupules: implications concerning efficiency in wind pollination. American Journal of Botany 68:635–650.

———. 1982. Computer simulations of early land plant branching morphologies: canalization of patterns during evolution? Paleobiology 8:196–210.

———. 1983. The influence of Paleozoic ovule and cupule morphologies on wind pollination. Evolution 37:968–986.

———. 1984. Size-related changes in the primary xylem anatomy of some early tracheophytes. Paleobiology 10:487–506.

———. 1985a. The evolution of tracheid diameter in early vascular plants and its implications on the hydraulic conductance of the primary xylem strand. Evolution 39:1110–1122.

———. 1985b. The aerodynamics of wind pollination. Botanical Review 51:328–386.

———. 1986a. Computer-simulated plant evolution. Scientific American 254:78–86.

———. 1986b. Evolution of plant shape: design constraints. Trends in Ecology and Evolution 1:67–72.

———. 1986c. Computer simulations of branching-patterns and their implications on the evolution of plants. Lectures on Mathematics in the Life Sciences 18:1–50.

———. 1988a. Biophysical limitations on plant form and evolution. Pp. 185–220 *in* L. D. Gottlieb and S. K. Jain, eds. Plant evolutionary biology. Chapman and Hall, London.

———. 1988b. The role of phyllotactic pattern as a "Developmental Constraint" on the interception of light by leaf surfaces. Evolution 42:1–16.

———. 1990. Biomechanics of *Psilotum nudum* and some early Paleozoic vascular sporophytes. American Journal of Botany 72:590–606.

———. 1992. Plant biomechanics: an engineering approach to plant form and function. University of Chicago Press, Chicago.

———. 1993. Ontogenetic-response models and the evolution of plant size. Evolutionary Trends in Plants 7:43–49.

———. 1994a. Morphological evolution through complex domains of fitness. Proceedings of the National Academy of Sciences USA 91:6772–6779.

———. 1994b. Plant allometry: the scaling of form and process. University of Chicago Press, Chicago.

———. 1994c. The scaling of plant and animal body mass, length, and diameter. Evolution 48:44–54.

———. 1994d. Predicting the height of fossil plant remains: an allometric approach to an old problem. American Journal of Botany 81:1235–1242.

———. 1997a. Adaptive walks through fitness landscapes for early vascular land plants. American Journal of Botany 84:16–25.

———. 1997b. Effects of hypothetical developmental barriers and abrupt environmental changes on adaptive walks in a computer-generated domain for early vascular land plants. Paleobiology 23:63–76.

———. 1997c. The evolutionary biology of plants. University of Chicago Press, Chicago.

———. 1998. The influence of gravity and wind on land plant evolution. Review of Palaeobotany and Palynology 102:1–14.

———. 1999. Evolutionary walks through a land plant morphospace. Journal of Experimental Botany 50:39–52.

Niklas, K. J., and H. P. Banks. 1985. Evidence for xylem constrictions in the primary vasculature of *Psilophyton dawsonii*, an Emsian trimerophyte. American Journal of Botany 72:674–685.

Niklas, K. J., and S. L. Buchmann. 1987. The aerodynamics of pollen capture in two sympatric *Ephedra* species. Evolution 41:104–123.

Niklas, K. J., and W. G. Chaloner. 1976. Simulations of the ontogeny of *Spongiophyton*, a Devonian plant. Annals of Botany 40:1–11.

Niklas, K. J., and V. Kerchner. 1984. Mechanical photosynthetic constraints on the evolution of plant shape. Paleobiology 10:79–101.

Niklas, K. J., and T. L. Phillips. 1976. Morphology of *Protosalvinia* from the Upper Devonian of Ohio and Kentucky. American Journal of Botany 63:9–29.

Nobel, P. S. 1983. Biophysical plant physiology and ecology. W. H. Freeman, New York.

Okubo, A., and S. A. Levin. 1989. A theoretical framework for data analysis of wind dispersal of seeds and pollen. Ecology 70:329–338.

Ostrom, J. H. 1964. A functional analysis of the jaw mechanics in the dinosaur *Triceratops*. Postilla (Peabody Museum of Natural History, Yale University) 88:1–35.

Phillips, T. L. 1979. Reproduction of heterosporous arborescent lycopods in the Mississippian-Pennsylvanian of Euramerica. Review of Palaeobotany and Palynology 27:239–289.

Raup, D. M., and A. Michelson. 1965. Theoretical morphology of the coiled shell. Science 147:1294–1295.

Retallack, G. 1975. The life and times of a Triassic lycopod. Alcheringa 1:3–29.

Rex, G. M., and W. G. Chaloner. 1983. The experimental formation of plant compression fossils. Palaeontology 26:231–252.

Richter, R. 1928. Psychische Reaktionen fossiler Tiere. Palaeobiologica 1:226–244.

Roth, A., and V. Mosbrugger. 1996. Numerical studies of water conduction in land plants: evolution of early stele types. Paleobiology 22:411–421.

Roth, A., V. Mosbrugger, and H. J. Neugebauer. 1994. Efficiency and evolution of water transport systems in higher plants: a modelling approach: II. Stelar evolution. Philosophical Transactions of the Royal Society of London B 345:153–162.

Roth, A., V. Mosbrugger, G. Belz, and H. J. Neugebauer. 1995. Hydrodynamic modelling study of angiosperm leaf venation types. Botanica Acta 108:121–126.

Roth, A., V. Mosbrugger, and A. Wunderlin. 1998. Computer simulations as a tool for understanding the evolution of water transport systems in land plants: a review and new data. Review of Palaeobotany and Palynology 102:79–99.

Rothwell, G. W., and T. N. Taylor. 1982. Early seed plant wind pollination studies: a commentary. Taxon 31:308–309.

Rudwick, M. J. S. 1964. The inference of function from structure in fossils. British Journal of Philosophy and Science 15:27–40.

Salisbury, F. B., and C. W. Ross. 1992. Plant physiology. Wadsworth, Belmont, Calif.

Speck, T., and D. Vogellehner. 1988. Biophysical examinations concerning the mechanical stability of various stele types and the kind of stabilizing system in early vascular land plants. Palaeontographica, Abteilung B 210:91–126.

Speck, T., H.-C. Spatz, and D. Vogellehner. 1990a. Contributions to the biomechanics of plants I. Stabilities of plant stems with strengthening elements of different cross-sections against weight and wind forces. Botanica Acta 103:111–112.

———. 1990b. Contributions to the biomechanics of plants II. Stability against local buckling in hollow plant stems. Botanica Acta 103:123–130.

Stanley, S. M. 1970. Relation of shell form to life habits in the Bivalvia (Mollusca). Geological Society of America Memoir 125.

Stein, W. E. 1993. Modeling the evolution of stelar architecture in vascular plants. International Journal of Plant Sciences 154:229–263.

———. 1998. Developmental logic: establishing a relationship between developmental process and phylogenetic pattern in primitive vascular plants. Review of Palaeobotany and Palynology 102:15–42.

Stewart, W. N., and G. W. Rothwell. 1993. Paleobotany and the evolution of plants. Cambridge University Press, Cambridge.

Taiz, L., and E. Zeiger. 1991. Plant physiology. Benjamin-Cummings, Redwood City, Calif.

Taylor, T. N., and E. L. Taylor. 1993. The biology and evolution of fossil plants. Prentice Hall, Englewood Cliffs, N. J.

Thompson, D'A. W. 1942. On growth and form. Cambridge University Press, Cambridge.

Vogel, S. 1981. Life in moving fluids. Willard Grant, Boston.

Wagner, F., S. J. P. Bohncke, D. L. Dilcher, W. M. Kürschner, B. van Geel, and H. Visscher. 1999. Century-scale shifts in Early Holocene atmospheric CO_2 concentration. Science 284:1971–1973.

West, G. B., J. H. Brown, and B. J. Enquist. 1999. The fourth dimension of life: fractal geometry and allometric scaling of organisms. Science 284:1677–1679.

Zimmermann, M. H. 1983. Xylem structure and the ascent of sap. Springer, Berlin.

Invention by evolution: functional analysis in paleobiology

Roy E. Plotnick and Tomasz K. Baumiller

Abstract.—Functional analysis of fossils is and should remain a key component of paleobiological research. Despite recently expressed doubts, conceptual and methodological developments over the past 25 years indicate that robust and testable claims about function can be produced. Functional statements can be made in at least three different hierarchical contexts, corresponding to the degree of structural information available, the position in the phylogenetic hierarchy, and the degree of anatomical specificity. The paradigm approach, which dominated thinking about function in the 1960s and 1970s, has been supplanted with a methodology based on biomechanics. Paleobiomechanics does not assume optimality in organismal design, but determines whether structures were *capable* of carrying out a given function. The paradigm approach can best be viewed as a way of generating, rather than testing, functional hypotheses. Hypotheses about function can also be developed and supported by well-corroborated phylogenetic arguments. Additional functional evidence can be derived from studies of trace fossils and of taphonomy. New computer techniques, including "Artificial Life" studies, have the potential for producing far more detailed ideas about function and mode of life than have been previously possible. Functional analysis remains the basis for studies of the history of adaptation. It is also an essential component of many paleoecological and paleoenvironmental studies.

Roy E. Plotnick. Department of Earth and Environmental Sciences, University of Illinois at Chicago, 845 West Taylor, Chicago, Illinois 60607. E-mail: plotnick@uic.edu

Tomasz K. Baumiller. Department of Geological Sciences and Museum of Paleontology, University of Michigan, 1109 Geddes Road, Ann Arbor, Michigan 48109-1079. E-mail: tomaszB@umich.edu

Accepted: 1 May 2000

Introduction

"A science of form is now being forged within evolutionary theory. It studies adaptation by quantitative methods, using the organism-machine analogy as a guide; it seeks to reduce complex form to fewer generating factors and causal influences" (Gould 1970: p. 77).

"The flowering of functional morphology has yielded a panoply of elegant individual examples and few principles beyond the unenlightening conclusion that animals work well . . . . I, at least, once harbored the naive belief that a simple enumeration of more and more cases would yield new principles for the study of form. But Newtonian procedures yield Newtonian answers, and who doubts that animals tend to be well designed?" (Gould 1980: p. 101).

"[W]e have placed unwarranted faith in our understanding of the relationship between form and function . . . . Of particular concern . . . are the nature and accuracy of predictions of function from morphology in fossil taxa" (Lauder 1995: p.2).

Thirty years have passed since Gould expressed his enthusiasm for human engineered devices as guides to the study of fossil forms and 20 years since his more pessimistic appraisal of functional morphology. His later skepticism apparently stemmed from a concern that although functional morphology is good at determining functional design of particular organisms, the significance of these studies to understanding evolution is obscure. Lauder (1995) went further, asserting that the structure-function linkage is so weak that inferences of function from morphology in fossils are themselves questionable.

These concerns make clear that we must distinguish between the methodology of functional morphology and its goals, in order to properly assess its importance in paleontology. In this we follow Fisher (1985), who distinguished functional *analysis* from functional morphology. Functional morphology, which focuses on the "nature, evolution, and historical consequences of adaptation" (Fisher 1985: p. 121), provides the historical context for the essentially ahistorical results of functional analysis.

This paper will focus on methods and approaches to functional analysis. We will discuss some recent methodological developments, provide an example of a taphonomic

  0094-8373/00/2604-0013/$1.00

approach to function, critique the paradigm method, and argue that although biomechanics (or paleobiomechanics) has proven to be the best approach to the testing of functional hypotheses, it can be usefully supplemented by other methods. We will then briefly discuss the broader implications and uses of functional analysis, including the insights it can provide into the dynamics of the evolutionary process.

We will not be concerned here with recent developments in other aspects of the "science of form" (Gould 1970), such as allometric studies or theoretical morphology (McGhee 1998). Instead, we will focus on approaches to the determination of both the "function" and "biological role" (sensu Bock and von Wahlert 1965) of the preserved structures of fossil organisms and on the evaluation of their evolutionary and ecological significance.

The Form-Function Relationship in Fossils

Form and Function.—In a highly influential paper, Bock and von Wahlert (1965) attempted to clarify the terminology surrounding functional morphology. Their key definitions are

1. *Feature*: any part of an organism, including morphological, behavioral, and physiological; the structures of the organism are its morphological features.
2. *Form*: the appearance, configuration, composition, shape, etc. of a feature.
3. *Function*: what a feature does or how it works; includes chemical and physical properties arising from its form. A given feature can have multiple functions.
4. *Faculty*: the combination of a given form and a particular function; this is the "form-function complex." Faculty is defined as "what the feature is capable of doing in the life of the organism"(p. 277).
5. *Biological role*: how the organism uses the faculty during its lifetime, in the context of its environment. The same faculty can have multiple biological roles. Bock and von Wahlert stressed that the biological role cannot be predicted with certainty from the study of form and function and must be directly observed. They considered this a particular problem for

fossil organisms. Biological role generally corresponds to the concept of "life habit."

In this context the shape of a bird's wing is part of its form, the production of lift is one of its functions, the use of the wing for flight is a faculty, and the use of flight to escape predators is a biological role. Similarly, the arrangement of bones in a skull is a form, the forces the skull can exert are a function, the use of these forces to bite is a faculty, and the biting of a prey animal is a biological role. This sequence also corresponds to the degree of certainty available to a paleontologist in a functional analysis; i.e., we can be quite certain about the form, model or test the function, make reasonable hypotheses about faculties, and speculate about biological roles (e.g., Witmer and Rose 1991).

These definitions can provide a very useful framework for functional interpretation of fossil organisms. Nevertheless, the Bock and von Wahlert definition of function, referring to little more than the physical and chemical properties of structure, is perhaps unnecessarily narrow. In common usage, the term function encompasses their concepts of function and of faculty. This paper will generally follow the usual practice and use the term in this broader sense. In specific cases, the distinction between the two concepts will be made explicit.

Resolving Lauder's Dilemma: Is Function Predictable from Structure?—In two recent articles Lauder (1995, 1996) expressed marked skepticism about the ability to decipher function from structure. These doubts are not based on an assumption that a relationship between form and function does not exist or cannot be deciphered, but that the required structural data to do so are rarely, if ever, available for fossil taxa. In particular, this implies that the most widely used fossil data in functional analysis, skeletal morphology, is of little direct use in interpreting fossil function. We would argue that the situation is not quite so dire as Lauder suggests, mostly because his concerns are relevant only to a very specific subset of structure–function problems.

First, Lauder (1995) uses a concept of function different from that of Bock and von Wahlert (1965). Lauder defines function as the mechanical or physical role that a structure plays in the organism; that is, how a phenotypic feature is used. This definition seems closer to Bock and von Wahlert's concept of a faculty. Implicitly, Lauder's view of function appears much narrower than this, referring to species- (or even population-) level statements about precise patterns of structural kinematics. Since these kinematics are dependent on neuromuscular features that cannot be observed in fossils, Lauder inescapably rejects the possibility of unequivocally predicting function from form in fossils. Functional studies would thus be restricted to direct observations in living organisms (Savazzi 1999).

Second, Lauder's admonition about the weakness of the link between morphology and function is scale-dependent. At the lower histological level and an upper "general level of behavior and ecology," he accepts that a much tighter correlation between structure and function can be demonstrated. For example, at the histological level, cross-sectional area of muscle is a good predictor of muscle force, and at the more general level, accurate predictions of habitat or diet can be made from analyses of structure. His caution is restricted to what he refers to as the intermediate level of generality, that is, where the neuromuscular system interacts with the skeletal system to generate patterns of movement. At that level, predictions about patterns of movement require understanding the interactions between the skeleton, muscles, and the nervous system. For example, a number of osteoglossomorph fishes have the same evolutionary novelty, the tongue-bite. Although these taxa have basically the same musculoskeletal systems, they have markedly different feeding kinematics due to differences in their nervous systems. Differences in kinematics of feeding in these fishes cannot be predicted from their morphology alone; information on their nervous system is required. More generally, given that a particular morphology is consistent with a wide repertoire of possible movements (i.e., functions in Lauder's view), and that the structure of the nervous system is unknown in fossils, function at this level is unrecoverable from fossil organisms.

This is undeniably true, but whereas Lauder sees this as a lack of fit between structure and function, we view it somewhat differently. The critical issue is the level of desired precision of the functional analysis. In the case of osteoglossomorph fish, the significant question is, What is the functional significance of the presence/absence of a tongue-bite? As pointed out by Lauder (1995: p. 7): the "tongue-bite is a significant evolutionary novelty that, if present in an extinct taxon, would provoke functional speculation and hypotheses as to its role in the feeding mechanism." For a paleontologist, therefore, the problem would be identifying functions that correlate with the presence of a tongue-bite. If it could be shown that fishes with the tongue-bite exhibit characteristic feeding behaviors, then regardless of the range of those behaviors, a fit between structure and function would have been established at this level of analysis.

Has such a fit been established for Lauder's example? Since "one could be reasonably confident given these results, that a fossil taxon possessing a tongue-bite morphology used these teeth to manipulate and puncture prey" (Lauder 1995: p. 8), the presence of a tongue-bite indicates a limited and characteristic repertoire of functions. Lauder makes the glass appear half empty by stressing the imprecision inherent in inferring function from structure because a structure can have a wide repertoire of functions. We see the glass as half full, in that when a unique morphology is observed, it is predictably associated with such a repertoire. We would thus reinterpret Lauder's claims as a lesson about the precision of the link between structure and function. The precision with which function can be inferred will depend upon the amount of structural information available, with the degree of precision of the functional statements increasing as we progressively add more information about muscles, the nervous system, etc. Functional statements are thus hierarchical; e.g., a general statement based on the skeleton alone includes a variety of more precise statements possible if direct information on musculature was also available.

Hierarchical approaches to biological systems have received a great deal of attention over the past several decades (e.g., Jacob 1977;

Eldredge 1985; Allen and Hoekstra 1992; Valentine and May 1996). As pointed out by Medawar (1974, cited in Valentine and May 1996; see also Jacob 1977), as one descends the ranks of a hierarchy, the smaller becomes the scope but the greater the complexity of the possible phenomena. For example, "bone" can be considered one rank of a form-function hierarchy. Although bone is a complex tissue, the variety of bone morphology pales in comparison to the diversity of structural elements that can be made from it. Functional statements about bone, such as its compressive and tensile strength, are of much more general nature than those about particular bones. A hierarchy of anatomical specificity and corresponding functional statements can be identified; e.g., bone—vertebral bone—thoracic vertebrae—first thoracic vertebra. In the same way, a hierarchy of functional statements can be based on the amount of information available; a statement based only on hard-part anatomy is more inclusive of possible faculties than one that includes additional anatomical information.

Another form-function hierarchy parallels the phylogenetic one; functional statements become more specific as one descends the Linnean hierarchy. In the same sense that a paleontologist might look at the diversity of families or genera, rather than species, functional studies tend to focus on general attributes characteristic of large taxonomic groups; e.g., pterosaurs (Padian 1991), stromatoporoids (LaBarbera and Boyajian 1991), or eurypterids (Plotnick 1985). A notable exception is Fisher (1977), who elegantly examined function in a single species of horseshoe crab. In addition, especially among invertebrate paleontologists, these functional analyses are general statements of life habits or function rather than specific statements of kinematics or behavior (e.g., Baumiller 1990; Labandeira 1997); i.e., they are at Lauder's upper hierarchical level. At these levels of analysis we believe the form-function relationship to be reliable.

In sum, functional statements can be made in at least three different hierarchical contexts, corresponding to the degree of structural information available, the position in the phylogenetic hierarchy, and the degree of anatomical specificity. The situation thus is not quite so dire as Lauder suggests, mostly because his concerns are relevant only to a restricted category of structure-function problems. These concerns are, however, valid cautions against overestimating the precision of functional interpretations.

Functional Analysis of Fossil Organisms

Phylogenetic Approach.—The interpretation of function in fossils has primarily followed a variety of inductive, comparative approaches (Savazzi 1999). Principal among these has been the comparison of homologous structures in fossil and living organisms; i.e., the function of a structure in an extinct organism is inferred to be similar to that of the homologous structure in a living relative. For example, on the basis of their phylogenetic relationship with modern *Limulus*, it has been suggested that eurypterids swam on their backs, an idea rejected by Plotnick (1985) on both anatomical and hydrodynamic grounds. Cowen (1979) considered this approach the most reliable of those available.

In recent years, more explicit methods for inferring function in fossils using the homology approach have been formulated (e.g., Lauder 1990; Weishampel 1995; Witmer 1995). These methods rely on the use of phylogenies, and they treat functions as traits and character optimization as a criterion for assessing the distribution of these traits among taxa. The basic premise is that genealogy can serve as a guide for reconstructing the unknown traits of organisms. Since functional characters can be treated as any other organismal attribute, they, just like structures, may have synapomorphic, apomorphic, or plesiomorphic distributions.

In principle, the phylogenetic approach allows functional inferences to be made purely by optimizing the functional characters on the cladogram; no knowledge of the distribution of the functionally relevant morphological traits is necessary. By keeping function and form separate, one may then use correlates of form and function or other relevant information as independent tests for reconstructing function. In the simplest case, given an independently corroborated phylogenetic hypothesis for three taxa, if taxa A and C share a known function and phylogenetically bracket taxon B, such that A represents the outgroup and C the sister taxon to B, we may infer that B shares the function of A and C. This inference can be further corroborated if a tight linkage exists between a given structure and the function in taxa A and C and if the structure is also found in taxon B.

It is clear that the phylogenetic methods offer little for structures and functions that are uniquely derived in fossils, since they base their inferences on homologous structures and functions. The chief danger of the argument from homology is that a highly specific function is assigned on the basis of a very general homology; i.e., homologous structures often perform very different functions in even closely related organisms. The more detailed the described function, and the more distantly related the taxa are, the more likely that the homologous structure-function relationship will break down (Lauder 1995). Also, for many paleontologically interesting questions, such as about flight in *Archaeopteryx* or pterosaurs, functionally bracketing the fossil taxa is not possible.

Another danger is that the wrong homology may be used. For example, Jacobs and Landman (1993) strongly questioned the common use of *Nautilus* as a model for the life habits of ammonoids. Instead, they pointed to phylogenetic evidence for a coleoid-ammonoid relationship and suggested that the biology of ammonoids be interpreted on that basis. Purnell (1999) described similar problems with the interpretation of conodont elements.

Analogy and the Paradigm Approach.—If suitable homologies are not available, the tendency has been to argue for function based on analogy, usually biological. Radinsky (1987) termed this the "form-function correlation approach." It assumes that a close relation exists between form and function, so that the latter can be predicted from the former. For example, as discussed by Radinsky (1987), since extant animals with long legs are usually fast runners, it is reasonable to assume that extinct animals with long legs, whether or not they are related to modern forms, also ran fast (note that "running fast" is a faculty). This approach also underlies Stanley's (1970) classic

analysis of the relationship between bivalve shell form and life habit and Labandeira's (1997) interpretation of insect feeding mechanisms based on mouthpart morphology.

When biological analogues are not available, mechanical ones have often been used. For example, Cowen (1975) argued for a "flapping valve" in richtofeniacean brachiopods, based on an analogy with a single-valved pump (cf. Grant 1975). Similarly, Myhrvold and Currie (1997), using analogy with whips, suggested that sauropod tails cracked and were used in communication.

The identification of a functional analogue for a structure in a fossil is a hypothesis that must be tested. By far the most influential conceptual approach to the functional morphology of extinct invertebrates, the "paradigm" approach of Rudwick (1964) has been suggested as a way to carry out such a test. The paradigm method was extensively described in older reviews of the field (e.g., Raup 1972; Gould and Lewontin 1979) and remains prominent in more recent articles and textbooks (Hickman 1988; Lauder 1995; Prothero 1998; Moon 1999; Paul 1999).

Rudwick's (1964) original statement of the concept was that a paradigm is "the structure that would be capable of fulfilling the function with the maximum efficiency attainable under the limitations imposed by the nature of the materials" (p. 36). Function is used here in the broader definition; i.e., as essentially synonymous with faculty. This approach involves several steps:

1. A function is suggested for a morphologic feature, perhaps based on analogy with a living organism or with a mechanical device.

2. From a knowledge of engineering and of the nature of the biological materials involved, a *paradigm* is developed for the performance of this function. The paradigm is thus a model (in Rudwick's term, a "structural prediction") of the optimum structure for the performance of the function.

3. The paradigm is compared with the observed structure. The degree of correspondence between the two acts as a test of the paradigm as a functional hypothesis. The expectation is that if the paradigm is valid, and no

other constraints hold, the paradigm and the structure will closely agree.

4. Each alternative function for a given feature generates its own paradigm. The function whose paradigm most closely matches the observed structure would have been "fulfilled most effectively" by the structure.

An often overlooked point is that Rudwick (1964), in his discussion of the paradigm approach, indicated that the comparison between paradigm and structure shows whether the structure would be *capable* of performing the function, but "cannot however establish in fact that it did fulfil that function" (p. 38).

The paradigm approach has been both strongly criticized (Grant 1972, 1975; Lauder 1995) and defended (Cowen 1975; Paul 1975, 1999; DeMar 1976; Fisher 1985). The key objections can be summarized as follows:

1. The paradigm method assumes that natural selection produces an optimal structure for a particular function. This assumption is suggested to be invalid since other factors, including developmental constraints and phylogenetic history, can exert comparable control over morphology (Seilacher 1970; Grant 1972; Signor 1982; Seilacher and LaBarbera 1995). Implicitly, this concept is contained in the "limitations imposed by the nature of the materials" of the original formulation of the paradigm concept.

2. There may be multiple possible structural optima; i.e., alternative equally (or nearly equally) valid paradigms could exist for a single function (Signor 1982).

3. Competing functional requirements ("trade-offs") may produce suboptimal structures for the functions considered independently (a point conceded by Rudwick [1964]).

4. The comparison between the paradigm and the observed structure is essentially visual and qualitative; the "test" is therefore subjective (Signor 1982).

5. Paradigms rely too heavily on analogies to mechanical devices, they thus tend to overlook important biological factors such as physiology (Grant 1972, 1975).

Grant concluded that the paradigm method, as stated by Rudwick, is a "point of view,

an approach . . . it is not a complete methodology . . . " Signor (1982) was far more critical, suggesting that it should be used only if other methods are not available.

Despite these criticisms, the paradigm method still has proponents. Even Grant (1972), in a generally critical article, labeled it "a watershed in the conceptual methodology of invertebrate paleontology" (p. 236). Paul (1999) asserted that it was a simple step-by-step approach that allows the rejection of inappropriate hypotheses and makes it possible to compare competing hypotheses. We will argue below that the value of the paradigm method is as a source of testable hypotheses for function, not as the test itself. Before we do so, we want to correct two claims about the method that we believe to be misconceptions. We call these the *Fallacy of the Perfect Engineer* and the *Fallacy of the Mechanical Analogy*.

The Fallacy of the Perfect Engineer.—François Jacob, in his generally overlooked essay "Evolution and Tinkering" (Jacob 1977), pointed out several differences between the process of natural selection and actions of an engineer. One of these is that "the objects produced by the engineer, at least by a good engineer, approach the level of perfection made possible by the technology of the time. In contrast, evolution is far from perfection" (p. 1161). As discussed above, the "imperfection" of morphology produced by biological evolution is a key part of much of the criticism of the paradigm method (e.g., Gould and Lewontin 1979). Later on, in comparing the action of evolution to that of a "tinkerer," rather than an engineer, Jacob stated, "Unlike engineers, tinkerers who tackle the same problem are likely to end up with different solutions. This also applies to evolution . . . " (p. 1164). On this premise, Jacob made a strong case for the importance of contingency in evolution. Again, there is the explicit assumption that a trained engineer will wind up with a perfect, optimal product. This idea, that an engineering approach will unerringly produce the optimal form to solve a given functional problem, has been fundamental to the paradigm method. We argue that this concept, which we term "the fallacy of the perfect engineer," is both false and misleading.

We base our argument on a reading of the nontechnical literature of engineering, especially the popular works of civil engineer Henry Petroski (1985, 1993, 1996). Petroski has extensively analyzed the engineering design process, used in producing such mundane items as paper clips and can openers and such spectacular items as bridges. One of his key observations (Petroski 1993) is that even for engineers, form does not follow function. Instead, form follows *failure;* i.e., engineering design advances by recognizing the limitations of existing products. New forms develop as an attempt to overcome these failures. Implicit in this is the concept that very few, if any, human-designed objects are optimal for their task. There is always room for improvement.

In addition, human-engineered objects are subject to many of the same kinds of constraints and influences that Seilacher (1970) recognized for biologically evolved forms. For instance, there is clear evidence for a form of phylogenetic constraint in engineering design. The design maxim known as MAYA, "most advanced yet acceptable" (Petroski 1996), indicates that new designs cannot be too radically different from existing forms, or they won't be adopted. Many details of the first iron bridges closely resembled those of their wooden predecessors, even though this was not required by the nature of the materials (Petroski 1996). In addition, even "ideal" engineering objects, such as the standard paper clip (Petroski 1993), have identifiable shortcomings. These shortcomings result from such factors as limitations imposed by the nature of the materials, competing functional requirements, or simply design mistakes (Dennett 1998). And of course, as argued by Gould and Lewontin (1979), many features of human-made structures are not "adaptive" but inescapable side effects of how the structure must be built. The products of engineering design cannot be considered as unerringly optimal, for the same reasons that organic design cannot. They thus should not be used as tests of functional hypotheses for extinct or living organisms.

Interestingly, as pointed out by Vogel (1998), many of those who have previously recognized the failings of human design have pointed to nature as providing examples of design excellence! Manned flight provides an excellent example of where a too slavish attempt to copy nature, by the construction of ornithopters, led to a technological dead end.

In summary, the concept that the engineering design process leads to optimality whereas evolution does not is incorrect. In fact, there may be more similarities between the two than has been generally accepted. In both, historical legacies, material constraints, costs in production of different structures under different conditions, and competing functional requirements mean that the concept of global optima is less useful than that of optimization—the climb to local peaks on an ever changing landscape. Nevertheless, the similarities between the engineered and the evolved cannot be carried too far; although the processes may be similar, we will argue in the next section that the failures of organism-machine comparisons stem from the use of too strict an analogy between the *results* of natural and human design.

The Engineered and the Evolved: The Fallacy of the Mechanical Analogy.—Implicit in the paradigm method and in other discussions of functional interpretation (e.g., Gould 1970; Cowen 1975; Frazetta 1975; Hickman 1988) is the use of analogy; i.e., the comparison of the observed structure with "simple machines, architecture, industrial design . . . and other man-made systems designed for efficient and cost-effective function" (Hickman 1988: p. 782). Recent examples include the comparison of sauropod tails with bullwhips by Myhrvold and Currie (1997) and the "ammonites as Cartesian divers" hypothesis of Seilacher and LaBarbera (1995; cf. Jacobs 1996).

The difficulty with this approach was cogently stated by Wainwright (1988: p. 8): "Man-made buildings are large, dry, rectangular, rigid, and static. In comparison, plants and animals are small, damp, cylindrical, flexible and dynamic." As discussed by Vogel (1998) and Dennett (1998), the technology of nature and human technology have far more differences than similarities; these differences spring from both the nature of the materials and the design process. They include the following:

1. Unlike many manufactured artifacts, there are very few corners or right angles in nature; organisms tend to favor round surfaces and cylindrical shapes.

2. Units of engineered structures tend be homogeneous, whereas biological units are internally variable (i.e., individual steel beams have the same physical properties throughout, but individual bones or crab sclerites have regions with different composition and organization).

3. Metallic materials are absent in organisms.

4. Very few organisms roll, and the wheel and axle are essentially absent in the living world (LaBarbera 1983).

5. Human artifacts are designed to be stiff and are consequently often brittle; organismal design favors strength over stiffness and thus produces toughness.

6. In most complicated mechanical devices, each separate part usually performs one or two discrete functions; multiple functions for each part are rare. For example, in a computer printer the paper feeder, the drum unit, and the output tray each perform a separate and single role. In contrast, in biological systems the same feature can perform multiple functions (e.g., the jaw) and many functions are performed by the joint action of many structures.

Obviously, the list can go on; the reader is referred to Vogel (1998) for a far more complete rendition. The essential point is simply that most machines make poor analogues to living organisms. The use of engineering structures as analogues to biological systems is fraught with difficulties and must be used with extreme caution.

Paleobiomechanical Approach.—Our discussion of the machine-organism analogy may sound pessimistic; but by becoming cognizant of the very real differences between machines and organisms, we can focus on their similarities. These similarities, as pointed out by Vogel, come from "inescapable physical rules and environmental circumstances" (p. 292). It is the relationship between these physical rules and organisms that is the foundation of the paleobiomechanical approach to functional analysis.

At its most fundamental, biomechanics examines the interrelationships between biological structures and physical processes (cf. definition in Rayner and Wootton 1991). The assumption is that such factors as the strength of biological materials, the kinetics of linked mechanisms, fluid drag and lift, and diffusion all have directly observable and measurable consequences on both the possible faculties and the biological roles of morphological features (Wainwright et al. 1976; LaBarbera 1990; Vogel 1994). Biomechanics thus allows us to quantify the functional properties of biological structures and thus test their effects on faculties and biological roles.

Paleobiomechanics, therefore, is simply the uniformitarian extension of this; the consequences of physical processes existed to the same extent in the past as they do today (Alexander 1989). As a result, the principles of physics that describe bird, bat, and airplane flight can be used to understand pterosaur flight (Padian 1991). The dynamics of waves along rocky coasts were the same in the Devonian as today and thus had the same influence on morphology (Denny 1995). The biomechanics of extinct organisms is thus one of the only areas within paleontology amenable to direct experimental investigation (taphonomy is another).

Paleobiomechanics also does not require the existence of a living homologue or living or machine analogue (Radinsky 1987; Witmer and Rose 1991), although one can be suggested as a starting point of the analysis. Instead, principles of physics and engineering are directly applied to the observed structure to infer its function and faculty; as we will discuss below, this is directly comparable to the practice of *reverse engineering*.

We can summarize the paleobiomechanical approach as follows; it is clearly derived from the paradigm method, but does not rely on the flawed assumptions of that approach:

1. A possible faculty (*not* a biological role) for a structure is proposed. This proposal is a hypothesis that could be derived from homology or analogy. For example, it is straight-

forward to hypothesize that the forelimbs of pterosaurs and *Archaeopteryx* were used for flight (Padian 1991; Rayner 1991).

2. The hypothesized faculty is then used to make a prediction of function (sensu Bock and von Wahlert 1965) and of form. If, for example, the wing of *Archaeopteryx* was used for flight, then it should have been capable of generating sufficient lift to support the weight of the animal. It should also have a form consistent with the production of that lift (Rayner 1991); e.g., a cambered wing produces more lift than one with a symmetrical cross-section (Vogel 1998). Similarly, the aerodynamics of flapping flight predict a large discrepancy in muscle mass between downstroke and upstroke muscles (Greenewalt 1975).

3. A model, either physical or computer-based (see below), is produced that allows the experimental determination of the structure's function (sensu Bock and von Wahlert 1965) and a test of whether the observed structure is *capable* of carrying out the hypothesized faculty. In the case of *Archaeopteryx*, one could place a model of the wing in a wind tunnel and measure the amount of lift produced. If the measured lift proves sufficient to support the weight of the animal in air under a reasonable set of conditions, the wing's hypothesized faculty has not been rejected. Note that the question is not whether the wing of *Archaeopteryx* is optimally designed, but whether the actual wing of the animal could produce sufficient lift to overcome its weight and body drag.

4. If direct experimental tests are not possible, or in addition to experiments, predicted aspects of form are compared with the observed form. Is the wing of *Archaeopteryx* cambered? Does the skeleton reflect greater downstroke muscle mass? Again, the goal is not seeing whether the form is the optimal for a particular function (lift generation), but instead whether the form has characteristics that are associated with performance of the function. A structure may indeed be optimal, but this is a hypothesis to be tested rather than an assumption of the approach.

A superb example of the paleobiomechanical approach to function and faculty is the study of the skull and jaw of *Diatryma* by Witmer and Rose (1991). They began by pointing out that there are no modern avian analogues to the *Diatryma*, so that the form-function correlation approach (Radinsky 1987) could not be used. Instead, they utilized beam theory to predict what design features a bird skull should have to maximize biting forces and decided that "*Diatryma* exhibits virtually all of the predicted features" (p. 103). They concluded that the jaw apparatus of *Diatryma* was capable of exerting tremendous bite forces. Commendably, they clearly distinguish the biological role of the jaw (e.g., herbivory or carnivory) from its function and faculty ("Whatever *Diatryma* ate, it could bite hard" [p. 117]). After analyzing food availability and considering the requirements of jaw form imposed by eating vegetation versus meat and bones, Witmer and Rose then interpreted the bird as a carnivore. The forces generated by the mechanism are functions; these functions are appropriate for crushing certain objects in biting (faculty); this allowed Witmer and Rose to evaluate the biological role of *Diatryma* jaws by considering objects that could actually be crushed.

In another example, Plotnick and Baumiller (1988) examined two alternative hypotheses for the faculty of the wide flat telson of pterygotid eurypterids. Each of these hypotheses made testable predictions about the morphology and function of the telson and of the rest of the animal. The first hypothesis was that the telson actively flapped and acted to propel the animal, similar to the caudal fins of cetaceans. This hypothesis of faculty predicts morphologic features such as large condyles for flexibility, large muscle insertions, and a high aspect ratio (width to length) for the telson. None of these features are found in pterygotids. The second hypothesis of faculty was that the telson was used to steer the animal, that it acted as a rudder. The functional properties of a rudder require numerous characteristics and these can be used to make morphologic and allometric predictions. All of these predictions are consistent with the observed features of pterygotids and their telsons. Further, a comparison of alternative telson designs showed that the observed morphology produced greater steering forces than the alternatives, suggesting that it approaches an optimal design.

In a study of crinoid functional morphology, Baumiller (1992) examined the hypothesis that lift on the crinoid crown was sufficient to maintain the position of the crown above the substrate; i.e., that the faculty of the crinoid crown was to act as a kite, with the stem acting as the string that tethered it to the substrate. His experiments allowed him to estimate the lift that the crowns of two representative crinoid genera may have experienced. These results, combined with analyses of the other forces (weight, drag) acting on the crinoid led Baumiller to conclude that the kite hypothesis was untenable at the level of function, and therefore of faculty and role.

As a final example, in a pair of studies Boyajian and LaBarbera (Boyajian and LaBarbera 1987; LaBarbera and Boyajian 1991) explored alternative hypotheses for the systematics and paleoecology of stromatoporoids, specifically the role of the astrorhizae. Boyajian and LaBarbera (1987) examined whether the astrorhizae represented an excurrent canal system for a filter feeder, similar to those in living sclerosponges. Using scale models, they studied flow patterns through astrorhizae and concluded they were indeed able to function as excurrent canals. They compared alternative hypotheses for astrorhizae, based on alternative predictions made for the distributions of canal diameters (LaBarbera and Boyajian 1991). Their results allowed them to reject several hypotheses, but were consistent with the interpretation of the astrorhizae as sponge-like mass transport systems. Their conclusion of a close relationship between stromatoporoids and sponges suggests how functional analysis could be used as a tool for phylogeny reconstruction.

Functional Analysis As Reverse Engineering.— One noticeable similarity between functional analysis, particularly paleobiomechanics, and a human design process is to *reverse engineering* (Petroski 1996; Dennett 1998). In reverse engineering, a product is disassembled, frequently by a business competitor, to determine how it works and how it might be duplicated. This is perhaps most common today

for software, where executable code is reverse engineered to obtain the original programming. Reverse engineering can be distinguished from "forward" engineering, which creates the object.

Cognitive scientist D. C. Dennett (1999: p. 256) pointed out that "in spite of the difference in the design processes, reverse engineering is just as applicable a methodology to systems designed by Nature, as to systems designed by engineers." If reasons for suboptimality and historical contingency are recognized, the techniques of reverse engineering should lead to a sound understanding of the design of organisms. In fact, Dennett (1999: p. 256) even went on to redefine biology as the "reverse engineering of natural systems." For living things, evolution is the forward design process.

A key similarity between the reverse engineering of extant organisms and that of engineered systems is that the "purpose" of the reverse-engineered entity is already known; the goal is to uncover the specific way it performs this purpose. We already know that a computer printer produces printed output and that a bird flies in order to find food. On the other hand, implicit in the *methodology* of reverse engineering is the determination of how a particular part of the device performs its role; e.g., How much current does a particular circuit carry? How much lift does a particular airplane wing design generate? How much bending can a particular strut withstand? The purpose of a reverse-engineered object thus corresponds to the biological role of Bock and von Wahlert (1965), whereas the properties uncovered by reverse engineering correspond to their definitions of function and faculty.

Paleontologists, in contrast, are in a position similar to those who try to uncover the often forgotten uses of obsolete tools and utensils (Petroski 1993). From familiarity with similar objects (i.e., either through homology or analogy), a purpose can be surmised. We then "reverse engineer" the fossil, to see if it could carry out the surmised purpose. The critical point here is that we do not ask Was it the best structure to do the assumed task? but Could it have been at least minimally capable of carrying out this task? Physical rules may be inescapable, but they are not dictatorial; multiple alternatives could exist to perform the same function.

Real Animals in a Virtual World: Computers and A-life.—No survey of any recent development in science is complete without a discussion of the role played by computers. Computers allow the detailed examination of far more complex systems than are generally amenable to standard experimental methods. Four areas can be identified in which functional analysis has benefited or could benefit from the new technology: kinematics of complex skeletal systems, finite element analysis of stress and strain, computational fluid mechanics, and artificial life.

Vertebrate kinematics are usually studied by manipulation of bones or models of bones. This is often impractical because there may be many separate elements, their pattern of connections could be complex, or the bones themselves might be very large (Stevens and Parrish 1999). As a result, investigators have begun to use software similar to that used by engineers to model these systems. One example is the previously cited work of Myhrvold and Currie (1997) on sauropod tails. Morphologic information on sauropod tail vertebrae was input into a physics-based simulation program. The flexibility and possible velocity of motion of the tail along its length were modeled, leading to the conclusion that the tip may have moved at supersonic speeds. Another example is the work of Stevens and Parrish (1999), who examined the flexibility and posture of the sauropod neck. They decided that the neck was markedly less flexible than previously suggested.

Finite element analysis is an important and widely used technique among engineers and physicists (Huebner et al. 1995; Gershenfeld 1999). It is a method for finding approximate solutions for the values of variables, such as stress, within a complexly shaped surface or volume. It does this by dividing (discretizing) the region into contiguous pieces or elements and solving the relevant equations (usually a polynomial) within each element separately. The results from all elements are then assembled to produce a set of simultaneous equations that describe the behavior of the desired property for the entire region. The equations are then solved for a specific set of boundary conditions (Huebner et al. 1995). Finite element methods have the potential for studying far more complex structural situations than are usually amenable to direct experimental analysis.

Recent applications of finite element methods in functional analysis include Philippi and Nachtigall (1996), Daniel et al. (1997), and Kesel et al. (1998). Kesel et al. analyzed the distribution of material in the wings of dragonflies and flies and examined the function of wing veins for stiffening. Philippi and Nachtigall investigated the distribution of forces in the test of regular echinoids under different loadings and rejected the pneu hypothesis for test shape.

A specifically paleontological application was Daniel et al. (1997). They used finite element analysis to examine the distribution of pressure stresses in ammonoid septa. They tested the idea that greater septal complexity allowed greater resistance to hydrostatic pressure, so that sutural complexity correlated positively with greater depth during life. Their results suggested that highly complex sutures actually lead to diminished resistance to hydrostatic pressures.

The use of computer models that incorporate the basic principles of fluid dynamics, such as the Navier-Stokes equations (Vogel 1994), is also possible. These computational fluid dynamics (CFD) models often rely on the finite element approach discussed above (Huebner et al. 1995). These models make it possible to analyze the fluid flow around complex objects and have become extensively used in the aircraft industry, replacing more traditional physical modeling approaches (Petroski 1996). We are aware of no studies applying these models to functional analysis, but they clearly have great promise.

One example of a computer study of fossil fluid mechanics is that of Knight (1996). He developed a computer model that incorporated the equations for lift and drag. These were then used to study how eurypterids may have swum. He suggested that lift was the primary mechanism for eurypterid swimming.

An especially exciting recent development

in computer science, with direct implications for paleontology and the functional analysis of fossils, is "artificial life," often called AL or A-life (Plotnick 1997; Dennett 1998). Ray (1994: p. 179), one of its leading developers, stated that "Artificial Life (AL) is the enterprise of understanding biology by constructing biological phenomena out of artificial components, rather than breaking natural life forms down into their component parts." The basic approach of A-life is to create entities that possess properties and operate under rules similar to those of biological entities and systems. For example, an "individual" in an A-life system, which is in reality a string of computer code, can reproduce (the code duplicates) and mutate (the code can change during replication). Individuals may also be allowed to mate (code pieces are exchanged), to find resources, and to die. A group of similar individuals (i.e., a "species") can thus be subject to natural selection. Even given a simple set of rules and properties, highly complex behaviors can emerge. It is this development of emergent system properties that is characteristic of A-life; Dennett (1998: p. 256) described it as a form of "bottom up reverse engineering."

Two A-life projects particularly relevant to functional analysis are Karl Sims's "creatures" (Sims 1994) and the "artificial fish" of Terzopoulos and colleagues (Terzopoulos et al. 1994, 1996). Sims's creatures are morphologically simple virtual organisms, made up of rectangular blocks of various sizes. They are supplied with basic control systems and occupy a virtual environment with realistic physical laws (e.g., gravity, fluid mechanics). The codes for these organisms, which describe their form and their behavior, can reproduce and mutate. Natural selection is introduced by allowing only those forms that are best able to perform a task in the environment, such as swimming, to survive. Although morphologically crude, the resulting creatures exhibit a wide range of plausible behaviors, including undulatory swimming and sculling. These behaviors were not introduced, but arise as a consequence of the interaction of the evolving virtual life forms with their environment.

The goal of Terzopoulos and his colleagues is to produce visually realistic virtual organisms that are self-animating; that is, they are given a set of rules and behaviors and then act autonomously, rather than being directed by a programmer. To this end, they designed artificial fishes, whose external morphology is based on living examples and whose body movements are controlled by relatively realistic representation of the skeletal and neuromuscular systems. These virtual fish combine simple yet realistic algorithms for biomechanics (including fluid mechanics), perception, and learning (Grzeszczuk and Terzopoulos 1995). Movements of their bodies allow the fish to move in their environment; those movement patterns that lead to faster movements are kept, whereas other movement patterns are discarded. As a result, the artificial fish "learn" how to swim. The resulting movement patterns closely resemble those seen in biological fish. A similar learning pattern results in fish that pursue "prey."

The methods developed by Terzopoulos and Sims have tremendous potential for studying function in fossils. In particular, they may lead to at least a partial solution to Lauder's (1995) concerns about kinematics of fossil forms. We envision, for example, a virtual fish based on the anatomy of such organisms as Silurian ostracoderms. A virtual experiment is then conducted in which the artificial ostracoderm fish learns to swim. If properly designed, a set of virtual experiments could determine if there are uniquely predicted kinematics for ostracoderm swimming or if there is a range of equally likely alternatives.

Ichnological and Taphonomic Approaches.—Other types of paleontological data, besides morphology, are relevant to reconstructing fossil function and life habits. Although the inability to test function in extinct organisms directly is sometimes viewed as a limitation, it provides for paleontologists an opportunity to find different means of extracting functional information from the fossil record (Savazzi 1999). In particular, data from trace fossils and taphonomy have great potential for developing and testing functional and life-habit hypotheses.

An extensive body of literature on ichnofossil characterization and interpretation exists

Skeletal morphology (rupture point)	absent	absent	absent	absent	present
Function (stalk-shedding)	no	no	no	yes	yes
Life habit	sessile	sessile	sessile	?	crawling/swimming

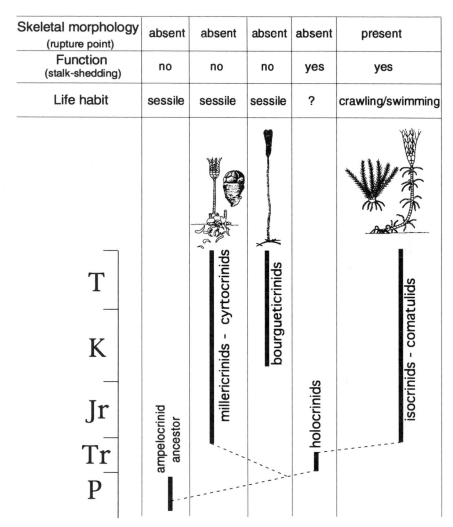

FIGURE 1. A, Distribution of the stalk-shedding function and life habit among extant crinoids. B, Stratigraphic ranges and inferred relationships of the five extant crinoid groups and the Triassic holocrinids (modified from Simms 1999). The position of the bourgueticrinids is controversial; they are thought to be either a neotenous offshoot of the comatulids (Simms 1988) or a subgroup of the millecrinids (Gislén 1938; Roux 1978).

(e.g., Bromley 1990) and we will not review it in detail here. Ichnofossils have provided information on modes and speeds of locomotion in groups such as arthropods (Briggs et al. 1991) and vertebrates (Alexander 1989). Ichnofossils have also been used for the interpretation of life habit (e.g., suspension versus deposit feeding) or behavior of extinct organisms (Seilacher 1964; Bromley 1990). Obviously, the use of trace fossils in functional interpretation requires that the trace maker be correctly identified.

Ichnofossils often act as tests of hypotheses of kinematics based on limb morphology. One

especially exciting study is that of Gatesy et al. (1999), who analyzed a suite of footprints of Triassic theropods. Combining detailed examination of the tracks, studies of modern bird locomotion, and computer graphics, they produced a detailed reconstruction of the foot movements of these dinosaurs.

Taphonomy is another source of paleontological data that can provide critical insights into function, either directly or by providing critical information on soft-tissue morphology. Although biostratinomic processes are often viewed as filters that remove information (Behrensmeyer and Kidwell 1985), decay and fossilization can also leave signatures of life habit or function that would otherwise be unavailable. For example, the preserved postmortem gape of extinct lamellibranchs provides information relevant to reconstructing soft-tissue distribution (muscles and ligament) in these organisms. This information, in turn, constrains hypotheses of function.

Taphonomic information used in reconstructing soft tissues and function in extinct taxa may even be used to overturn skeletal proxies. One example is the analysis of stalk function in fossil crinoids. Two of the five extant groups of crinoids, comatulids and isocrinids, are capable of freeing themselves from the substrate and crawling or swimming (Fig. 1) (Messing et al. 1988). The functional consequences of a free-living versus fully sessile life habit may have profound ecological and evolutionary implications (Meyer and Macurda 1977), so it is important to determine the distribution and origin of the free-living habit within the post-Paleozoic clade.

In all extant crinoids the juveniles are cemented by a holdfast to the substrate. The free-living habit thus requires that the animal be capable of shedding its stalk at some stage of life. In comatulids, the stalk is shed in early ontogeny and is never regenerated. In isocrinids, the stalk continues to grow throughout life, but as new elements are added in the proximal portion, the older, distal end of the stalk is shed. Stalk shedding in isocrinids occurs at specialized rupture points that are spaced at regular intervals along the length of the stalk. These rupture points possess a characteristic skeletal morphology and specialized ligamentary organization that are not found at the non-shedding articulations.

Holocrinus is the first crinoid to appear following the Permo-Triassic extinction and the sister taxon of the entire isocrinid-comatulid clade (Simms and Sevastopulo 1993; Hagdorn 1995). This genus lacks skeletally differentiated articulations. Since ligament organization cannot be studied directly in fossils, the skeletal data suggest that *Holocrinus* was incapable of shedding its stalk and that this function is derived in the comatulid-isocrinid clade.

However, stalk shedding also produces a characteristic non-random pattern in the shed stalk segments (Baumiller and Ausich 1992; Baumiller et al. 1995). This pattern can be used as an independent taphonomic test of the shedding function. An analysis of *Holocrinus* stalk segments revealed just such a pattern (Baumiller and Hagdorn 1995). Thus *Holocrinus* was capable of stalk shedding, despite the failure of skeletal morphology to reflect this specialization.

The original functional hypothesis for *Holocrinus* was based on a correlation between morphology (articulation type) and function (shedding ability) in extant taxa and was overturned using taphonomic criteria. This example thus appears to support Lauder's claims about a lack of fit between structure and function. In contrast, we believe it conveys a different and instructive message; that basing claims of a linkage between structure and function solely on correlation can lead to faulty conclusions not only about a specific situation but also about the general fit between function and structure. For example, since almost all airplanes have wheels, using only correlation we might construe a link between wheels and flight. Seeing a wheel-less seaplane could then lead to the conclusion that the link between structure and function was weak. Obviously, this conclusion would be incorrect because the wrong structure-function couplet was initially selected. That is also why we are skeptical of studies concluding a general lack of structure-function fit from examinations only of congruence between structure and function without exploring the biomechanical linkage between function and structure (see Lauder 1995, his Case Study 2). In linking function and structure it is critical to develop and test biomechanically how a morphological feature affects function (Lauder 1991). The goal of functional morphology should be not simply to find correlations between structure and function, but to find causal explanations for them.

The Roles of Functional Morphology in Paleobiology

In the previous sections we have discussed some of the available methods for reconstructing function in fossils and for getting at the link between structure and function. In this section we will show that functional interpretations remain at the core of many areas of paleobiological research and that their uses go beyond demonstrating good design and current utility. Paleobiologists attempt to interpret patterns in the history of life in functional and ecological terms. Implicitly, function is causally connected with the origin of phenotypes, evolutionary trends, evolutionary properties of taxa, and long-term changes in the structure and dynamics of the biosphere. In addition, the association of functions and life habits with morphology, at whatever hierarchical level, is an essential part of paleoecology, including the distribution of organisms among environments and the distribution of paleocommunities. As pointed out by Boucot (1990) in his extensive review, functional analysis is also the best available method for the study of the evolution of behavior in the fossil record. The underlying assumption of these approaches is that functional attributes of individuals, populations, species, and clades can have ecological and evolutionary consequences. A few examples will suffice to illustrate the role that functional interpretations play in paleobiological research.

Jacobs et al. (1994) used a well-known biomechanical principle and experiments to explain the pattern of distribution of different ammonite morphs among facies. They showed that more spherical, less compressed morphs experience a lower total drag under conditions when drag is dominated by frictional forces (small Reynolds number) than do compressed, streamlined morphs. The opposite is true under conditions when drag is dominated by dynamic forces (large Reynolds number). Since overcoming drag is energetically expensive for active swimmers, it was hypothesized that environments with different current energies should be characterized by the presence of the least drag-inducing ammonite morphologies. The patterns observed in different facies of the Western Interior Cretaceous Seaway were consistent with this prediction.

In a seminal paper, Bambach (1983) introduced the concept of the "guild" to paleontology, where it has become commonly used

in paleocommunity studies (e.g., Watkins 1994). Species that belong to the same guild exploit environmental resources in a similar way. Species in a paleocommunity are assigned to a guild on the basis of their phylogenetic class (e.g., Trilobita), their feeding type and food source, and their life habit or life position. As indicated by Bambach, recognition of food source, feeding type, life positions, and life habits for a particular species relies heavily on functional analysis (see also Bambach 1994).

Paleobiologists also continue to offer ecological descriptions of evolutionary trends based on functional claims. Vermeij (1977) examined the interactions between predators and prey in what he termed the Mesozoic marine revolution. He documented in detail an increase over time in the frequency of gastropod shell designs resistant to crushing, including changes in shell coiling and the abundance and elaboration of ornamentation, concurrent with an increase in diversity of durophagous predators. The identification of shell crushers and of designs resistant to shell crushing is based on functional arguments.

Thayer (1979) attributed the changes in the structure of benthic communities during the Phanerozoic to the diversification of deposit feeders. The diversification led to increased disturbance of sediment and a consequent replacement of immobile suspension feeders living on soft substrates by mobile taxa and immobile hard-surface dwellers.

Labandeira (1997) examined the evolutionary history of insect mouthparts. Using cluster analysis, he identified 34 distinctive mouthpart classes among modern insects. Many of these classes are polyphyletic. Each mouthpart class is associated with a characteristic feeding strategy, such as piercing and sucking. By examining the history of insect mouthparts, Labandeira was able to identify five phases in hexapod evolution, which he associated with increased partitioning of food resources. The correlation between preserved morphology and function is essential to his analyses.

Taxon longevities have also been explored using functional arguments. Baumiller (1992) used filtration theory, experiments, and energy budgets to claim that filter morphology placed constraints on distributions of passive filter feeders among environments of different current energies. This claim was confirmed by documenting patterns of distribution of fossil crinoids with different filter morphotypes among facies. More widely distributed, and thus "eurytopic," morphotypes were predicted to be less prone to extinction than narrowly distributed, and thus "stenotopic," morphotypes. Stratigraphic ranges of crinoids categorized by filter morphotype confirmed this prediction (Baumiller 1993; Kammer et al. 1998).

What the above examples illustrate, and what we would like to emphasize, is that the question of function permeates paleobiological research. However, functional arguments are not always well constrained by rigorous functional analyses, nor is the link between functional attributes and their evolutionary consequences always made clear. Often implicit is the notion that functional traits confer some performance advantage, and that natural selection is the mechanism for the origin and proliferation of particular phenotypes. But because natural selection is not the only mechanism of evolutionary change and because a variety of processes of sorting and selection can affect the distribution of traits at different levels of the biological hierarchy, an assumption of natural selection as the cause requires further testing (Lewontin 1978; Gould and Lewontin 1979; also see Rose and Lauder 1996a and Vermeij 1996). When inferences on function are well constrained, adaptive scenarios are plausible and become good starting points for deeper evolutionary analyses into the origin, maintenance, or evolutionary consequences of traits. Tests with varying degrees of rigor can be applied to such scenarios. Tests might include data from phylogenetic analyses to supply information on the history of transformation of traits and their independent or nonindependent origins (convergence vs. homology), as well as data from paleoecology to provide relevant information on selective regimes. Alternate scenarios deemed more plausible can replace them and, likewise, be tested.

Conclusions

Despite the concerns expressed over the past two decades, we are convinced that functional analysis is capable of producing robust and testable statements about function and life habits in fossil organisms. These functional statements should not be more detailed than is allowed by the amount of preserved information and should be made in their appropriate hierarchical contexts. These hierarchies are based on the anatomical detail of the analysis, the amount of structural information available, and the phylogenetic level of the group studied. A study of the function of the septa of nautiloids has quite different data requirements and range of generality than one of forelimb movements in *Anomalocaris*.

There is no single source of information or methodology sufficient on its own for reconstructing function in fossils. Data and methods derived from biomechanics, phylogenetics, ichnology, and taphonomy all can play important roles. We do not believe that functional morphology would benefit at this time from rigid methodological standardization, and we encourage the use of new, nonstandard methods and data. We are especially excited about the potential for new methods derived from studies of artificial life and artificial intelligence to produce far more detailed functional reconstructions than have been previously available.

For paleobiologists, functional attributes of taxa continue to be a source of explanatory hypotheses about ecological and evolutionary patterns and trends. Their appeal is in part based on the theoretical underpinning that Darwin provided. But in using functional explanations one is not constrained solely to invoking natural selection: functional explanations may play a role in scenarios invoking sorting or selection operating at a variety of levels (Vermeij 1996) or even those that do not invoke selection at all.

The focus of this journal, as given by its title, is paleo*biology*. To paleobiologists, fossils are not simply organic constituents of rocks; they are remains of once living organisms. We are denied our neontological colleagues' ability to make the direct observations that are some of the chief joys and interests of natural history, to see our organisms swim, fly, walk, mate, and eat.

We strongly believe that the attempt to breathe life back into extinct animals, to attempt to visualize a once living world, is "scientifically, as well as spiritually, uplifting and rewarding" (Eldredge 1979: p. 195).

Acknowledgments

G. Lauder and an anonymous reviewer are thanked for their insightful comments which greatly improved the manuscript. L. Ivany and P. Kaplan kindly read over the manuscript and made many useful comments. We also thank the editors, S. Wing and D. Erwin, for their incredible patience and forbearance. Finally, we would like to gratefully acknowledge three individuals who shaped and inspired our research into the functional morphology of living and extinct animals: S. Vogel, M. LaBarbera, and D. C. Fisher. Partial support was provided by the National Science Foundation (grant EAR-97601 to T. K. B.).

Literature Cited

Alexander, R. M. 1989. Mechanics of fossil vertebrates. Journal of the Geological Society, London 146:41–52.

Allen, T. F. H., and T. W. Hoekstra. 1992. Toward a unified ecology. Columbia University Press, New York.

Bambach, R. K. 1983. Ecospace utilization and guilds in marine communities through the Phanerozoic. Pp. 719–746 in M. J. S. Tevesz and P. L. McCall, eds. Biotic interactions in fossil and Recent benthic communities. Plenum, New York.

———. 1994. Seafood through time: changes in biomass, energetics, and productivity in the marine ecosystem. Paleobiology 19:372–397.

Baumiller, T. K. 1990. Physical modeling of the batocrinid anal tube: functional analysis and multiple hypothesis-testing. Lethaia 23:399–408.

———. 1992. Importance of hydrodynamic lift to crinoid autecology, or, could crinoids function as kites? Journal of Paleontology 66:658–665.

———. 1993. Survivorship analysis of Paleozoic Crinoidea: effect of filter morphology on evolutionary rates. Paleobiology 19:304–321.

Baumiller, T. K., and W. I. Ausich. 1992. The broken-stick model as a null hypothesis for crinoid stalk taphonomy and as a guide to the distribution of connective tissue in fossils. Paleobiology 18:288–298.

Baumiller, T. K., and H. Hagdorn. 1995. Taphonomy as a guide to functional morphology of Holocrinus, the first post-Paleozoic crinoid. Lethaia 28:221–228.

Baumiller, T. K., G. Llewellyn, C. G. Messing, and W. I. Ausich. 1995. Taphonomy of isocrinid stalks: influence of decay and autotomy. Palaios 10:87–95.

Behrensmeyer, A. K., and S. M. Kidwell. 1985. Taphonomy's contributions to paleobiology. Paleobiology 11:105–119.

Bock, W. J., and G. von Wahlert. 1965. Adaptation and the form-function complex. Evolution 19:269–299.

Boucot, A. J. 1990. Evolutionary paleobiology of behavior and coevolution. Elsevier, Amsterdam.

Boyajian, G. E., and M. LaBarbera. 1987. Biomechanical analysis of passive flow of stromatoporoids—morphological, paleoecological, and systematic implications. Lethaia 20:223–229.

Briggs, D. E. G., J. E. Dalingwater, and P. A. Selden. 1991. Biomechanics of locomotion in fossil arthropods. Pp. 37–56 in Rayner and Wootton 1991.

Bromley, R. G. 1990. Trace fossils. Unwin Hyman, London.

Coddington, J. A. 1988. Cladistic tests of adaptational hypotheses. Cladistics 2.53–67.

Cowen, R. 1975. 'Flapping valves' in brachiopods. Lethaia 8:23–29.

———. 1979. Functional morphology. Pp. 487–489 in R. Fairbridge and D. Jablonski, eds. Encyclopedia of paleontology. Dowden, Hutchinson, and Ross, Stroudsburg, Penn.

Daniel, T. L., B. S. Helmuth, W. B. Saunders, and P. D. Ward. 1997. Septal complexity in ammonoid cephalopods increased mechanical risk and limited depth. Paleobiology 23:470–481.

DeMar, R. 1976. Functional morphological models: evolutionary and non-evolutionary. Fieldiana (Geology) 33:333–354.

Dennett, D. C. 1998. Brainchildren: essays on designing minds. MIT Press, Cambridge.

Denny, M. 1995. Predicting physical disturbance—mechanistic approaches to the study of survivorship on wave-swept shores. Ecological Monographs 65:371–418.

Eldredge, N. 1979. Cladism and common sense. Pp. 165–198 in J. Cracraft and N. Eldredge, eds. Phylogenetic analysis and paleontology. Columbia University Press, New York.

———. 1985. Unfinished synthesis. Oxford University Press, New York.

Fisher, D. C. 1977. Functional morphology of spines in the Pennsylvanian horseshoe crab Euproops danae. Paleobiology 3:175–195.

———. 1985. Evolutionary morphology: beyond the analogous, the anecdotal, and the ad hoc. Paleobiology 11:120–138.

Frazetta, T. H. 1975. Complex adaptations in evolving populations. Sinauer, Sunderland, Mass.

Gatesy, S. M., K. M. Middleton, F. A. Jenkins Jr., and N. H. Shubin. 1999. Three-dimensional preservation of foot movements in Triassic theropod dinosaurs. Nature 399:141–144.

Gershenfeld, N. 1999. The nature of mathematical modeling. Cambridge University Press, Cambridge.

Gislén, T. 1938. A revision of the recent Bathycrinidae. Acta Universitatis Lundensis 34:1–30.

Gould, S. J. 1970. Evolutionary paleontology and science of form. Earth Science Reviews 6:77–119.

———. 1980. The promise of paleobiology as a nomothetic, evolutionary discipline. Paleobiology 6:96–118.

Gould, S. J., and R. C. Lewontin. 1979. The spandrels of San Marco and the Panglossian paradigm: a critique of the adaptationist programme. Proceedings of the Royal Society of London B 205:581–598.

Grant, R. E. 1972. The lophophore and feeding mechanism of the Productidina (Brachiopoda) Journal of Paleontology 46:213–249.

———. 1975. Methods and conclusions in functional analysis: a reply. Lethaia 8:31–34.

Greenewalt, C. J. 1975. The flight of birds. Transactions of the American Philosophical Society, new series, 65(4):1–67.

Grzeszczuk, R., and D. Terzopoulos. 1995. Automated learning of muscle-actuated locomotion through control abstraction. SIGGRAPH (Conference 1995). Computer graphics proceedings, annual conference series, pp. 63–70. Special Interest Group on Computer Graphics, Association for Computing Machinery, New York.

Hagdorn, H. 1995. Triassic crinoids. Zentralblatt für Geologie und Paläontologie, Teil II:1–22.

Hickman, C. 1988. Analysis of form and function in fossils. American Zoologist 28:775–783.

Holland, N. D., J. C. Grimmer, and K. Wiegmann. 1991. The structure of the sea lily Calamocrinus diomedae, with special reference to the articulations, skeletal microstructure, symbiotic bacteria, axial organs, and stalk tissues (Crinoidea, Millericrinida). Zoomorphology 110:115–132.

Huebner, K. H., E. A. Thornton, and T. G. Byrom. 1995. The finite element method for engineers. Wiley-Interscience, New Work.

Jacob, F. 1977. Evolution and tinkering. Science 196:1161–1167.

Jacobs, D. K. 1996. Chambered cephalopod shells, buoyancy, structure and decoupling: history and red herrings. Palaios 11:610–614.

Jacobs, D. K., and N. H. Landman. 1993. Nautilus—a poor model for the function and behavior of ammonoids. Lethaia 26:101–111.

Jacobs, D. K., N. H. Landman and J. A. Chamberlain Jr. 1994. Ammonite shell shape covaries with facies and hydrodynamics: iterative evolution as a response to changes in basinal environment. Geology 22:905–908.

Kammer, T. W., T. K. Baumiller, and W. I. Ausich. 1998. Evolutionary significance of differential species longevity in Osagean–Meramecian (Mississippian) crinoid clades. Paleobiology 24:155–176.

Kesel, A. B., U. Philippi, and W. Nachtigall. 1998. Biomechanical aspects of the insect wing: an analysis using the finite element method. Computers in Biology and Medicine 28:423–437.

Knight, G. J. 1996. Making rocks swim. In J. E. Repetski, ed. Sixth North American paleontological convention, Abstracts of papers. Paleontological Society Special Publication 8:214.

Labandeira, C. C. 1997. Insect mouthparts: ascertaining the paleobiology of insect feeding strategies. Annual Review of Ecology and Systematics 28:153–193.

LaBarbera, M. 1983. Why the wheels won't go. American Naturalist 121:395–408.

———. 1990. Principles of design of fluid transport systems in zoology. Science 249:992–1000.

LaBarbera, M., and G. E. Boyajian. 1991. The function of astrorhizae in stromatoporoids—quantitative tests. Paleobiology 17:121–132.

Lauder, G. V. 1990. Functional morphology and systematics: studying functional patterns in an historical context. Annual Review of Ecology and Systematics 21:317–340.

———. 1991. Biomechanics and evolution: integrating physical and historical biology in the study of complex systems. Pp. 1–19 in Rayner and Wootton 1991.

———. 1995. On the inference of function from structure. Pp. 1–18 in Thomason 1995.

———. 1996. The argument from design. Pp. 55–91 in Rose and Lauder 1996b.

Lewontin, R. C. 1978. Adaptation. Scientific American 239:156–169.

McGhee, G. 1998. Theoretical morphology: the concept and its applications. Columbia University Press, New York.

Medawar, P. 1974. A geometric model of reduction and emergence. Pp. 57–63 in F. J. Ayala and T. Dobzhansky, eds. Studies in the philosophy of biology. University of California Press, Berkeley and Los Angeles.

Messing, C. G., M. C. RoseSmyth, S. R. Mailer, and J. E. Miller. 1988. Relocation movement in a stalked crinoid (Echinodermata). Bulletin of Marine Science 42:480–487.

Meyer, D. L., and D. B. Macurda. 1977. Adaptive radiation of comatulid crinoids. Paleobiology 3:74–82.

Moon, B. R. 1999. Testing an inference of function from struc-

ture: snake vertebrae do the twist. Journal of Morphology 241:217–225.

Myhrvold, N. P., and P. J. Currie. 1997. Supersonic sauropods? Tail dynamics in the diplodocids. Paleobiology 23:393–409.

Padian, K. 1991. Pterosaurs: were they functional birds or functional bats? Pp. 145–160 in Rayner and Wootton 1991.

Paul, C. R. C. 1975. A reappraisal of the paradigm method of functional analysis in fossils. Lethaia 8:15–21.

———. 1999. The paradigm method. Pp. 25–28 in E. Savazzi, ed. Functional morphology of the invertebrate skeleton. Wiley, Chichester, England.

Petroski, H. 1985. To engineer is human. St. Martins, New York.

———. 1993. The evolution of useful things. Knopf, New York.

———. 1996. Invention by design. Harvard University Press, Cambridge.

Philippi, U., and W. Nachtigall. 1996. Functional morphology of regular echinoid tests (Echinodermata, Echinoida): a finite element study. Zoomorphology 116:35–50.

Plotnick, R. 1985. Lift-based mechanisms for swimming in eurypterids and portunid crabs. Transactions of the Royal Society of Edinburgh 76:325–337.

———. 1997. Wonderful interactions: the Digital Burgess conference. American Paleontologist 5:2–4.

Plotnick, R., and T. Baumiller. 1988. The pterygotid telson as a biological rudder. Lethaia 21:13–27.

Prothero, D. 1998. Bringing fossils to life. WCB/McGraw Hill, Boston.

Radinsky, L. B. 1987. The evolution of vertebrate design. University of Chicago Press, Chicago.

Raup, D. 1972. Approaches to morphologic analysis. Pp. 28–45 in T. J. M. Schopf, ed. Models in paleobiology. Freeman, Cooper, San Francisco.

Ray, T. 1994. An evolutionary approach to synthetic biology: Zen and the art of creating life. Artificial Life Journal 1:179–209.

Rayner, J. M. V. 1991. Avian flight evolution and the problem of Archaeopteryx. Pp. 183–212 in Rayner and Wootton 1991.

Rayner, J. M. V., and R. J. Wootton, eds. 1991. Biomechanics in evolution. Society for Experimental Biology Seminar Series 36. Cambridge University Press, Cambridge.

Rose M. R., and G. V. Lauder. 1996a. Post-spandrel adaptationism. Pp. 1–8 in Rose and Lauder 1996b.

———, eds. 1996b. Adaptation. Academic Press, San Diego.

Roux, M. 1978. Ontogenèse, variabilité et évolution morphofonctionnelle du pédoncule et du calice chez les Millericrinida (Echinodermes, Crinoïdes). Geobios 11:213–241.

Rudwick, M. J. S. 1964. The inference of function from structure in fossils. British Journal for the Philosophy of Science 15:27–40.

Savazzi, E. 1999. Introduction to functional morphology. Pp. 3–14 in E. Savazzi, ed. Functional morphology of the invertebrate skeleton. Wiley, Chichester, England.

Seilacher, A. 1964. Biogenic sedimentary structures. Pp. 293–316 in J. Imbrie and N. D. Newell, eds. Approaches to paleoecology. Wiley, New York.

———. 1970. Arbeitskonzept zur Konstruktions-Morphologie. Lethaia 3:393–396.

Seilacher, A., and M. LaBarbera. 1995. Ammonites as Cartesian divers. Palaios 10:493–506.

Signor, P. 1982. A critical re-evaluation of the paradigm method of constructional inference. Neues Jahrbuch für Geologie und Paläontologie, Abhandlungen 164:59–63.

Simms, M. J. 1988. The phylogeny of post-Paleozoic crinoids. Pp. 269–284 in C. R. C. Paul and A. B. Smith, eds. Echinoderm phylogeny and evolutionary biology. Clarendon, Oxford.

———. 1999. Systematics, phylogeny and evolutionary history. Pp. 31–40 in H. Hess, W. I. Ausich, C. E. Brett, and M. J. Simms, eds. Fossil crinoids. Cambridge University Press, Cambridge.

Simms, M. J., and G. D. Sevastopulo. 1993. The origin of articulate crinoids. Palaeontology 36:91–109.

Sims, K. 1994. Evolving virtual creatures. SIGGRAPH (Conference 1994). Computer graphics proceedings, annual conference series, pp. 15–22. Special Interest Group on Computer Graphics, Association for Computing Machinery, New York.

Stanley, S. M. 1970. Relation of shell form to life habits in the Bivalvia (Mollusca). Geological Society of America Memoir 125.

Stevens, K. A., and J. M. Parrish. 1999. Neck posture and feeding habits of two Jurassic sauropod dinosaurs. Science 284:798–800.

Terzopoulos, D., X. Tu, and R. Grzeszczuk. 1994. Artificial fishes: autonomous locomotion, perception, behavior, and learning in a simulated physical world. Artificial Life 1:327–351.

Terzopoulos, D., T. Rabie, and R. Grzeszczuk. 1997. Perception and learning in artificial animals. Pp. 1–8 in C. G. Langton and K. Shimohara, eds. Artificial life V: proceedings of the fifth international workshop on the synthesis and simulation of living systems, Nara-shi, Japan, 1996. MIT Press, Cambridge.

Thayer, C. W. 1979. Biological bulldozers and the evolution of marine benthic communities. Science 203:458–461.

Thomason, J., ed. 1995. Functional morphology in vertebrate paleontology. Cambridge University Press, Cambridge.

Valentine, J. M., and C. M. May. 1996. Hierarchies in biology and paleontology. Paleobiology 22.23–33.

Vermeij, G. J. 1977. The Mesozoic marine revolution: evidence from snails, predators, and grazers. Paleobiology 3:245–258.

———. 1996. Adaptations of clades: resistance and response. Pp. 363–380 in Rose and Lauder 1996b.

Vogel, S. 1994. Life in moving fluids: the physical biology of flow. Princeton University Press, Princeton, N.J.

———. 1998. Cats' paws and catapults. Norton, New York.

Wainwright, S. 1988. Axis and circumference: the cylindrical shape of plants and animals. Harvard University Press, Cambridge.

Wainwright, S., W. Biggs, J. Currey, and M. Gosline. 1976. Mechanical design in organisms. Edward Arnold, London.

Watkins, R. 1994. Evolution of Silurian pentamerid communities in Wisconsin. Palaios 9:488–499.

Weishampel, D. B. 1995. Fossils, function, and phylogeny. Pp. 34–54 in Thomason 1995.

Witmer, L. M. 1995. The extant phylogenetic bracket and the importance of reconstructing soft tissues in fossils. Pp. 19–33 in Thomason 1995.

Witmer, L. M., and K. D. Rose. 1991. Biomechanics of the jaw apparatus of the gigantic Eocene bird Diatryma: implications for diet and mode of life. Paleobiology 17:95–120.

Fossils, genes, and the origin of novelty

Neil H. Shubin and Charles R. Marshall

Abstract. — The origin of evolutionary novelty involves changes across the biological hierarchy: from genes and cells to whole organisms and ecosystems. Understanding the mechanisms behind the establishment of new designs involves integrating scientific disciplines that use different data and, often, different means of testing hypotheses. Discoveries from both paleontology and developmental genetics have shed new light on the origin of morphological novelties. The genes that play a major role in establishing the primary axes of the body and appendages, and that regulate the expression of the genes that are responsible for initiating the making of structures such as eyes, or hearts, are highly conserved between phyla. This implies that it is not new genes, per se, that underlie much of morphological innovation, but that it is changes in when and where these and other genes are expressed that constitute the underlying mechanistic basis of morphological innovation. Gene duplication is also a source of developmental innovation, but it is possible that it is not the increased number of genes (and their subsequent divergence) that is most important in the evolution of new morphologies; rather it may be the duplication of their regulatory regions that provides the raw material for morphological novelty. Bridging the gap between microevolution and macroevolution will involve understanding the mechanisms behind the production of morphological variation. It appears that relatively few genetic changes may be responsible for most of the observed phenotypic differences between species, at least in some instances. In addition, advances in our understanding of the mechanistic basis of animal development offer the opportunity to deepen our insight into the nature of the Cambrian explosion. With the advent of whole-genome sequencing, we should see accelerated progress in understanding the relationship between the genotype, phenotype, and environment: post-genomics paleontology promises to be most exciting.

Neil H. Shubin.* Department of Biology, University of Pennsylvania, Philadelphia, Pennsylvania 19104-6018

Charles R. Marshall. Department of Earth and Planetary Sciences, and Organismic and Evolutionary Biology, 20 Oxford Street, Harvard University, Cambridge, Massachusetts 02138. E-mail: marshall@eps. harvard.edu

*Present address: Department of Organismal Biology and Anatomy, The University of Chicago, 1027 East Fifty-seventh Street, Chicago, Illinois 60637. E-mail: nshubin@uchicago.edu

Accepted: 10 July 2000

"There may be nothing new under the sun, but permutation of the old within complex systems can do wonders" (Gould 1977).

Introduction

The publication of Gould's *Ontogeny and Phylogeny* in 1977 was a landmark event in the intellectual history of paleobiology because it outlined a synthesis of two fields that, until 1977, had grown far apart. Indeed, the 1970s were a pivotal time in this field as the papers of Britten and Davidson, as well as Gould and Valentine, began to suggest that developmental biology, evolution and paleontology could be unified to understand the evolution of form. In the late nineteenth century, embryology and paleontology had a common purpose—understanding the mechanisms behind morphological transformations. Notions such

as recapitulation, for example, served to link studies of morphological change over both developmental and paleontological timescales. By the early twentieth century, paleontology and embryology diverged as each field began to use different methods to test hypotheses and collect data. Paleontology, with its reliance on comparative methods and the consilience of evidence, differed from the new embryology that became increasingly reliant on experimental manipulation to test hypotheses (Gould 1977; Maienschein 1991). By placing a new emphasis on the links between the evolution of regulatory genes and evolutionary change, among paleontologists Gould (1977) foresaw the renaissance of integrated studies of development and evolution. Today, this perspective provides the means to link evolutionary histories across vastly different scales:

0094-8373/00/2604-0014/$1.00

from genes and cells to organs and ecosystems.

Recent discoveries have crystallized interest around one of the classic themes of paleobiology: the origin of evolutionary novelties. How are new organs and bodyplans assembled? The origin of new designs involves genetic, developmental, functional, and environmental modifications. Understanding the links between changes at each of these levels involves integrating fields that use different data and, often, very different means of testing hypotheses.

In the last 25 years, the study of evolutionary novelty has become an integrative discipline involving the analysis of genes, development, and fossils (Müller and Wagner 1991; Raff 1997; Shubin et al. 1997; Duboule and Wilkins 1998; Knoll and Carroll 1999; Marshall et al. 1999; Valentine et al. 1999). Untangling the web of ecological, developmental, and genetic interactions is difficult. A key question is which changes occur first: new structures, new environments, new behaviors, or new developmental capacities? Determining the relative importance and relationship between these events during the inception of a novelty is a challenge because developmental and paleontological data often differ in type and quality. Morphological data is often richer in diversity than genetic data and, if it includes fossils, contains information on absolute time and paleoenvironment. Data from the function of regulatory genes, while often derived from a small number of model systems, reveals direct evidence of patterning mechanisms. Accordingly, our ability to infer the evolution of genetic systems in Deep Time is limited by the quality of phylogenetic hypotheses, the quality of fossils, and the phylogenetic position of the organisms used in genetic analyses. Indeed, while we constantly decry the poor quality of the fossil record, we could equally decry the gaps in our knowledge of comparative developmental genetics.

Perhaps the most fundamental problem in the study of evolutionary novelty is delineating just what is "new" in evolution. This problem is not as trivial as it seems, because organisms often use similar genes, processes, or structures in different ways. Jacob (1977) cap-

tured this notion in his principle, bricolage ("tinkering"). Jacob noted that evolution does not build new things from scratch but, during each generation, molds new features by the modification or change of existing characters, whether they are genes, developmental processes, or enzymes. This metaphor defines an important role for history in the origin of evolutionary novelty. The importance of history is abundantly clear when we try to understand the role of regulatory genes in the origin of new organs and bodyplans. Many of the genes, proteins, and cell types that are involved in producing the great diversity of organisms are deeply ancient. Gerhart and Kirschner (1997: p. 234) describe much of cellular evolution as occurring by "reshuffling the deck of existing proteins, linking them in new combinations, and putting their expression under the control of new contingencies." A large part of understanding the origin of novelties, then, comes down to deciphering the history of the modification of existing genetic pathways and developmental processes.

In the last 25 yr, new data from genetics have dealt some profound surprises to the evolutionary biology community. Perhaps the most striking discovery is the extent to which major patterning genes and regulatory interactions are deeply conserved across vast expanses of time and phylogeny. Taxa from cnidarians through deuterostomes share homologous regulatory genes (Finnerty and Martindale 1998). Indeed, in many cases, the developmental role of these homologous genes is also conserved in creatures with different bodyplans. Strikingly, many homologous genes appear to perform the same function in structures that share functional similarities but lack a common evolutionary history. However, homologous structures need not share a common genetic basis (Wray and Abouheif 1998). These contrasting generalities make understanding the mechanisms behind the conservatism and lability of genes and morphology one of the central puzzles in the field.

The fossil record adds a critical dimension to an understanding of morphological innovation. The appearance of morphological novelties is irregular in rate, in phylogenetic pattern, and in the manner in which taxonomic diversification relates to genetic change. Certain geological intervals contain the first appearance of major novelties—for example, the Cambrian for metazoan phyla (e.g., see Bengtson 1994; Knoll and Carroll 1999; Smith 1999), the Devonian for gnathostome fishes (Carroll 1988; Zhu et al. 1999), and the Triassic for diapsid vertebrates (Carroll 1988). These times appear to be special because the characters and taxa that arise often correspond to fundamental ecological transitions. However, the phylogenetic analysis of fossils and genes often suggests an incongruity between the patterns of genetic, morphological, and taxonomic diversification. To cite just one example, scleractinian corals unequivocally appear for the first time in the fossil record in the Triassic, while molecular-clock data suggest crown-group scleractinians were already present deep in the Paleozoic (Romano and Palumbi 1996). Changes at any one of these levels need not be synchronous or even related mechanistically to changes at another level.

Together, new discoveries from genetics and the fossil record have the potential to answer fundamental questions about the origin of homologous structures. In so doing, this synthetic approach reinvigorates one of the classic problems in paleobiology: the relationships between microevolutionary and macroevolutionary change in the origin of morphological variation.

Deep Homology: Trivial or Profound?

The notion of deep homology derives from the observation that certain aspects of development are highly conserved across vast time frames and over great phylogenetic expanses. Insights into this evolutionary conservatism are gained from comparisons of arthropods and vertebrates where common genes are frequently involved in the development of similar structures. Examples include eye formation (*Pax6/eyeless*), heart differentiation (*tinman/Nkx 2.5-DMEF2*), head and cranial nervous system development (e.g., *spalt/Xsal-1*), anterior-posterior patterning (Hox genes), and head development (*otd/otx*).

As more taxa become added to comparative genetic studies, these genetic similarities become even more striking. A prime example of this deep conservation lies in the diversity of uses of the transcription factor, *distal-less* (Panganiban et al. 1997). *distal-less* is essential for outgrowth in insect appendages and is present in the distal component of the developing limb. Comparative analyses of *distal-less* expression reveal fundamental similarities among diverse creatures and organs: *distal-less* is expressed in the distal portion of tube feet of echinoderms, the siphons of tunicates, the parapodia of annelids, and the lobopodia of onychophorans. Indeed, one of the homologues of *distal-less*, a member of the Dlx gene family, is expressed in the distal portion of the limb buds of vertebrates. What common theme underlies the structure or function of each of these organs? In each case, an outgrowth of the body wall utilizes a homologous transcription factor, expressed in the same compartment of the organ.

Paleontology informs and is informed by these examples of deep genetic homology. No working biologist would argue that tunicate siphons and chicken wings are homologous as appendages. Accordingly, this example points to a fundamental dissociation of the histories of genes, their functions, and organs (for a general review see, Wray and Abouheif 1998). These different histories take several forms. Homologous structures can be patterned by different genetic interactions; there are numerous cases where the contribution of individual genes differs between homologous structures. In limbs, for example, the gene *Wnt3a* is essential in the formation of the apical ectodermal ridge (AER) of a chick, but is not even expressed in this component of the mouse limb bud (see discussion in Tabin et al. 1999). Both mouse limbs and chicken wings are clearly homologous as the appendages of tetrapods, yet they differ in the role that *Wnt3a* plays in the development of the proximodistal axis. Another example lies in the observation that similar genes can be involved in the development of nonhomologous structures. For example, large parts of signal transduction pathways are conserved across vastly different cell types in distantly related organisms. In addition, ligands and their receptors tend to be highly conserved for functional reasons, and their presence in diverse cells does not imply any special homology between cell types, only a homology of parts of the signal transduction cascade. The Notch signaling pathway, for example, is involved in the development of diverse structures in *Drosophila* (e.g., wings and bristles); homologous genes in vertebrates are involved in everything from T-lymphocytes to feathers (see further discussion in Wray and Abouheif 1998).

Animal Appendages

Deep homologies have the potential to tell us about both developmental mechanisms and evolutionary history. Animal appendages serve as an excellent example of the degree to which genetic processes can be conserved. Homologous genes, either orthologues or paralogues, are involved in patterning the three ordinate axes of the wings of *Drosophila* and chicks (Shubin et al. 1997; Tabin et al. 1999). These similarities extend to more than gene sequence, structure or expression. Experimental manipulation of several of these genetic pathways reveals functional conservation as well. Appendages are patterned along three axes: anteroposterior (a/p), proximodistal (p/d), and dorsoventral (d/v). Similarities between each of these axes is discussed in turn below:

The Anteroposterior Axis.—The wing disk of *Drosophila* is divided into two compartments, anterior and posterior. The posterior portion of the disk expresses the gene *hedgehog* (Lee et al. 1992; Tabata et al. 1992). This gene encodes a signal that induces cells running along the border between the compartments to produce another protein encoded by the gene *decapentaplegic* (*dpp*) (Basler and Struhl 1994). *dpp*, in turn, plays a role in specifying fates to cells lying in the anterior and posterior compartments. Misexpression of either *hedgehog* or *dpp* in the anterior part of the disk yields mirror image duplications of limb structures (Basler and Struhl 1994).

In vertebrate limbs, the main signal involved with anteroposterior patterning is *Sonic hedgehog* (*Shh*). There are several similarities to the situation seen in the wing of *Drosophila*: (1) *Shh* is a homologue of *hedgehog* (Krauss et al. 1993; Riddle et al. 1993); (2) *Shh* is localized

in the posterior portion of the vertebrate limb bud (Riddle et al. 1993); and (3) experimental manipulations of *Shh* produce in the vertebrate limb similar effects to those resulting from manipulation of *hedgehog* in the wing of *Drosophila* (Riddle et al. 1993; Chang et al. 1994). In addition, a homologue of *Drosophila*'s *dpp*, *BMP-2* is expressed in the limb bud in response to *Shh*. Indeed, the parallels extend to the responses to experimental manipulations: misexpression of *Shh* anteriorly causes mirror-image duplications analogous to those caused by *hedgehog* misexpression in flies (Riddle et al. 1993).

The Proximodistal Axis.—In the wing disk of *Drosophila*, the proximodistal axis is originally organized by a group of cells running the length of the dorsoventral (d/v) border: the "wing margin." The transcription factor *apterous* specifies dorsal-specific cell fate and is expressed within the dorsal compartment of the disk (Diaz-Benjumea and Cohen 1993). *apterous* controls the expression of the secreted protein, *fringe*. The interface between cells that express *fringe* and those cells that do not becomes the wing margin (Irvine and Weischaus 1994).

Proximodistal outgrowth of vertebrate limbs appears to be established by a very similar set of genes and circuits. Signals that emanate from the apical ectodermal ridge (AER) are involved with the outgrowth of the limb. The AER is analogous to the wing margin in that it extends along the d/v border of the distal limb. A vertebrate homologue of *fringe*, *Radical-fringe*, is expressed in the dorsal half of the limb ectoderm prior to formation of the AER. The AER forms at the border between cells expressing *Radical-fringe* and cells not expressing *Radical-fringe* (Rodriguez-Esteban et al. 1997).

Other patterning factors are also shared between *Drosophila* and chicks. The homologues *Meis1/2* in chicks and *Hth* in *Drosophila* are both factors that are initially expressed in the proximal compartment of the limb (Mercader et al. 1999). These factors perform analogous functions within the cells of chick and *Drosophila* limbs. In both cases they promote the transport of another factor (*Pbx1* in chicks, *exd* in *Drosophila*) into the nucleus. Indeed, ectopic expression of these factors causes analogous morphological defects: in both cases distal-to-proximal transformations are induced. Mercader et al. (1999) propose that this system of genes reflects a shared strategy to allow outgrowth of appendages in the common ancestor of chicks and *Drosophila*.

The Dorsoventral Axis.—Genes specifying d/v polarity in both *Drosophila* and chicks are paralogues of one another and often perform analogous functions. In *Drosophila*, the gene *wingless*, a member of the *Wnt* family, is expressed in the ventral compartment of the wing. This gene is necessary for the proper d/v patterning of the wing (Couso et al. 1993; Williams et al. 1993). The expression of *apterous* defines the dorsal compartment and specifies dorsal cell fates (Diaz-Benjumea and Cohen 1993). In chicks, the early expression of a Wnt family member is also required for d/v patterning. This gene, *Wnt7a*, is expressed in the dorsal ectoderm and is both necessary and sufficient for dorsal patterning (Parr and McMahon 1995). *Wnt7a* acts by inducing the expression of the gene *Lmx-1* in the limb mesoderm (Riddle et al. 1995; Vogel et al. 1995). Like *apterous*, *Lmx-1* is a member of the LIM-homeodomain family of transcription factors. *Lmx-1* defines a dorsal compartment; being expressed early throughout the dorsal half of the limb bud, its expression is sufficient to convey dorsal cell fate. In both of these counts, the expression and function of *Lmx-1* is similar to *apterous* in *Drosophila*.

Implications of Shared Genes

What are the implications of these extensive similarities between the wings of *Drosophila* and those of chicks? The phylogenetic definition of homology is very specific. Mouse forelimbs and bird wings are homologous as the limbs of tetrapods because they share common ancestors that possessed corresponding structures. Indeed, homologues of the humerus, radius, and ulna in the skeleton of chicks and mice are seen in the fins of many Devonian sarcopterygian fish (Shubin 1995). However, bird wings are clearly analogous to the wings of *Drosophila* because the putative common ancestor of these taxa did not appear to have limbs (basal deuterostomes, such as echi-noderms, and basal chordates, such as urochordates and cephalochordates, do not possess appendages and are presumed not to have lost them). Indeed, the wings of *Drosophila* differ both developmentally and structurally from those of a chick: *Drosophila* appendages develop from imaginal disks, not limb buds. Acknowledging that it is virtually impossible to claim that the two appendages can be homologous as limbs, what can be made of the genetic similarities?

The answer to this question comes from understanding the manner in which developmental systems evolve. Indeed, the ways that development evolves provide the means for linking histories (or homologies) between different levels of the biological hierarchy. This reasoning is similar to that used to interpret the differences between gene trees and species trees. Cladograms of taxa are based on the notion that divergence follows cladogenesis (speciation). Different taxa diverge from one another through the patterns of divergence that occur at the time of, or between, speciation events: the divergence of different lineages is the result of the evolution of barriers to gene flow. Divergence between orthologous genes also result from common descent and genetic isolation. On the other hand, gene trees are based on the notion that the primary means of producing new members of gene families is not the result of the genetic isolation related to speciation events. Different members of the globin gene family or Hox gene family are produced by events of gene duplication and later divergence, divergence that starts within the same gene pool. The fact that speciation and gene duplication need not be causally linked allows us to differentiate orthologous from paralogous genes and to differentiate between species and gene trees. If descent with modification at the level of clades can be attributed to speciation and divergence, how can we characterize descent with modification at the level of developmental processes and organs?

One important mode of regulatory evolution involves the cooption of primitive genes or regulatory pathways in the development of new organs. This cooption need not involve the replacement of one role for another, but the elaboration of roles of each gene. Cooption has clearly played a role in the evolution of Hox genes, for example. One major role of Hox genes is to regulate the development of structures at specific positions along the anteroposterior axis. In vertebrates, they are involved in a variety of functions, from the specification of regional identities along the vertebral column to the development of appendages, visceral mesoderm, and genitalia. The functional roles of these genes are very general, and they are involved in the development of structures that develop at different times and in different places during development. For example, vertebrate pectoral and pelvic appendages develop at different sites along the body axis. Often, the development of these two sets of structures can be delayed by hours, days, or even months, as is the case with many amphibians. Despite these temporal and spatial differences, similar members of the Hox gene family are expressed during the development of pelvic and pectoral appendages.

The phylogenetic result of cooption and divergence is that homology at one structural level need not correlate to homology at another. A simple example lies in the relationship between the different types of vertebrate appendages. If one were to construct an "organ tree" for the history of life (as an analogue of a "gene tree"), one relies on the notion of cooption as the primary means of descent and later divergence. In vertebrates, if one constructs an "organ tree" for appendages, one would find homology between these serially homologous organs: forelimbs, hindlimbs, dorsal fins, anal fins, caudal fins, etc. This limb tree would differ (slightly) from the tree of vertebrate clades, because some clades contain numerous unpaired appendages, others have only paired appendages, and still others have no appendages at all. Indeed, if cooption, divergence, and loss are major events in the history of regulatory sequences and organs, one would expect that different organs would share similar genetic regulatory circuits.

The phylogenetic distribution of genes, circuits, and organs sets up several possibilities for cooption during the evolution of appendages. The simplest conclusion is that the genes and circuits that pattern animal appendages are more primitive than the appendages themselves (Shubin et al. 1997; Tabin et al. 1999). Two pieces of evidence suggest that the gene circuits are more primitive than the appendages they pattern: they are seen in other taxa that lack appendages, and in the development of many organs. The key detail in their phylogenetic distribution is that they are not seen in toto in any other organ. For example, portions of the anteroposterior module are involved in patterning of a variety of metazoans and a variety of different metazoan organs. The similarities in dorsoventral patterning, and their linkage to proximodistal circuits are most specific to body outgrowths. Several scenarios can account for these similarities. The first scenario is one of simple convergent evolution at both the morphological and genetic levels. This hypothesis holds that the systems that specify the three major axes were coopted independently during arthropod and vertebrate evolution. This convergent evolution could have happened piecemeal or in toto. Different components of the limb patterning program could have been recruited independently at different stages of metazoan evolution, or the whole set of circuits that pattern all three axes could have been coopted independently ensuite. The prediction of scenarios of convergent evolution is that the appendages of cladistically intermediate groups will lack these genetic circuits in their limbs. The other hypothesis posits a single origin of the patterning modules, modules that underlie the development of other organs in metazoans. What these other organs might have been before the evolution of limbs can only be surmised by understanding the function of the shared elements in arthropods and vertebrates. In the limb example above, the modules regulate the development of organs that are patterned along secondary axes (Shubin et al. 1997). These developmental axes need not be limbs, or even appendages. Any patterned secondary axis—the branchial arches, for example—could serve as a genetic intermediate (Shubin et al 1997; Tabin et al. 1999).

Each hypothesis of cooption leads to specific predictions that are testable by both paleontological and genetic data. One interpretation of the deep homologies of limb-patterning genes is that intermediates between chordates and arthropods will have secondary axes that are patterned by similar sets of genes. Dorsoventral patterning is particularly interesting in this regard. Parts of the d/v patterning system are highly conserved (*fringe*/ *radical fringe*) while others (*wg*/*wnt7a*) are more labile. For example, *wingless* homologues are expressed in different compartments in *Drosophila* and chicks: *wg* is ventral and *wnt7a* is dorsal. This situation implies either that the appendage patterning system was coopted piecemeal, or that the system was coopted in the common ancestor of chicks and *Drosophila*, but that the function of the *wingless* homologues has diverged between the two lineages.

Untangling the different histories of clades, organs, and genes involves developing phylogenies at each of these levels. The recognition of a process of cooption is a post-hoc event, usually entailing a good phylogeny of the taxa involved, a fossil record that includes key stem taxa, and a comparative knowledge of gene structure and expression. Two types of comparative genetic data are important: comparisons across different taxa and comparisons of genes across different organs in the same animal.

Leigh van Valen (1982) described homology as the "continuity of information" during phylogeny. One implication of the new genetic approaches is that—in a genetic sense—this continuity of information may reside in any number of different genes, genetic interactions, or epigenetic processes. Importantly, the organs or tissues specified by these mechanisms need not be phylogenetically homologous to one another. Indeed, evolution may utilize a very small genetic repertoire of genes, circuits, and cellular events during morphogenesis. If this is the case, diverse organisms are mosaics of similar genes and organs that have been redeployed in different functional contexts during the evolution of gene regulatory interactions.

The Function of Universal Genes

Many of us are first introduced to the relationship between the genotype and phenotype through the description of Mendelian genetics. Particular alleles are associated with specific phenotypes, such as smooth or wrinkled textures of peas, or red or white eyes. In this tradition, textbooks largely concentrate on subtle differences in morphologies that are manifest at the end of the organism's development. Thus, it is perhaps easy to see how nonmolecular biologists come to associate specific genes with specific morphologies, and tend to associate a gene's function directly with a specific phenotype. Obviously in the case of eye color, the relevant alleles will be associated with the production or transport of the appropriate pigment to the eye. However, in the case of the surface textures of peas, the relevant allele is more likely to belong to a gene, or several genes, that play a role in the cell surface or adhesion properties of the pea's cells. Indeed, it is quite possible that it would be difficult to determine from the function of the gene what the effect of altering it would be. In fact, most individual genes, even developmentally important genes, do rather mundane things: bind to specific DNA sequences (e.g., transcription factors); play a role in the passing of signals between cells; regulate rates of cell proliferation of death; or affect cell surface properties, such as its adhesion properties, or general shape. Understanding the 'local' function of individual genes turns out to be largely uninformative; what matters is how genes work together in networks, or circuits, to produce specific morphologies.

Genes studied by comparative developmental biologists usually play a role in the early stages of ontogeny, and much of the literature on the developmental basis of morphological innovation has centered on conserved genes with widespread phylogenetic distributions, such as the Hox genes, or those associated with determining the axes in limb development (see above). Of course part of the reason of the emphasis on conserved genes is that they are most easily identified. Rising above the level of the simple function of single conserved genes (e.g., binding DNA, functioning in a signaling pathway), one can ask, what do these conserved genes do?

Some of these genes are involved in laying out the coordinates of the developing body. As discussed above, a whole suite of conserved genes is involved in establishing the proximodistal, dorsoventral, and anteroposterior axes of the limbs. Other genes are involved in establishing the dorsoventral polarity in the body. In the case of the Hox genes, many are involved in laying and interpreting the "Cartesian coordinates" of the body, effectively dividing the embryo into specific regions, which later will grow specific structures (e.g., wings or halteres). The Hox genes also play a role in the cell proliferation itself. Thus the genes that currently dominate discussions of morphogenesis do not themselves make structures, but demarcate regions and groups of cells for specific future fates. Many of these early genes are referred to as patterning genes, and many are not themselves functionally tied to specific structures. As a consequence, these genes are general in their functions and in their phylogenetic distribution. While many of these genes are highly conserved, others, such as *bicoid* (central to the anteroposterior patterning in *Drosophila*), have seen very rapid evolution. This evolutionary pattern makes is difficult to make simple generalizations between gene function and evolutionary rates. As the *bicoid* example clearly shows, genes with roles in early development and with general functions can evolve rapidly.

Evolutionary Innovation: Changes in Regulation

Mechanistically (in terms of gene function), it is interesting to ask where morphological innovation comes from. It is tempting to deduce from the general statement that the genotype produces the phenotype that morphological innovation stems from new genes. However, the fact that the same key developmental genes are found over huge taxonomic distances undermines this assumption: very different organisms seem to be made from the same genes, or at least if not exactly the same genes, certainly genes from the same or similar gene families. Thus, it is becoming apparent that major evolutionary changes may not be due to changes in the number or structure of genes per se, but may be due to changes in their regulation (e.g., see Carroll 1995, 2000). Indeed, changes in the spatial pattern and timing of gene activity play an important role in generating variation at both small and large

phylogenetic scales (Averof and Patel 1997; Stern 1998). For example, small changes in the activity of the *Drosophila* homeobox gene, *Ubx*, correlate with differences between species. Indeed, large-scale distinctions between higher taxonomic groups of crustacean arthropods correlate with patterns of *Ubx* and *Abd-A* expression patterns, presumably directed by changes in their regulation (Averof and Patel 1997).

Evolutionary Innovation: Gene Duplication

While changes in gene regulation are a key component to the genetic basis of morphological innovation, another important means of developmental evolution lies in the duplication and divergence of regulatory genes (Holland 1998). The expansion of the Hox gene family during metazoan history has involved successive phases of increases in size of the cluster through gene duplication and later divergence (e.g., Holland 1998; de Rosa et al. 1999; Finnerty and Martindale 1999). One might expect that the duplication of regulatory genes would be directly correlated with the development of increasingly complex organisms. However, the correlation between duplication and morphological change does not appear to be simple, although there is certainly a relationship. Indeed, the Hox gene family has been a persistent feature of animal evolution, and individual duplications do not appear to be directly synchronous with either morphological change or taxonomic diversification. New data on this issue have come from several fronts. Analyses of metazoan phylogeny support the division of protostomes into two major clades, the Lophotrochozoa and Ecdysozoa (Garey and Schmidt-Rhaesa 1998). Finnerty and Martindale (1998), in a comparative study of Hox gene sequences in basal metazoans (i.e., cnidarians), argue that initial expansion of the Hox gene family preceded the origin of major phyla. The presence of nine Hox genes appears to be a general character of bilaterians and has preceded major morphological change within the Bilateria, but may well have been associated with the origin of the bilaterians. Furthermore, one main implication of the new phylogeny is that there has been at least one major example of a reduction in the numbers of Hox genes. Nematodes, formerly thought be primitive in having a relatively small cluster, are now considered to be derived in the loss of members of the cluster (de Rosa et al. 1999).

This work in no way discounts the importance of gene duplications in the origin of evolutionary novelties, long seen as a major route to evolutionary innovation (e.g., Haldane 1932; Muller 1935; Ohno 1970). In fact, the Hox gene themselves seem to owe their existence to the duplication of small set of genes, the protoHox cluster (Brooke et al. 1998; Holland 1998). Gene duplications are associated with divergence in that they occasionally occur in stem lineages that lie at the base of major clades (expansion along the Bilaterian stem, cluster duplication along the vertebrate stem). This distribution suggests that gene duplications are one important way to generate raw material for evolution and, as a consequence, may precede rather than be synchronous with other types of evolutionary change. Over time, some lineages may be more evolvable if they are able to produce more types of selectable variation. In this regard, the importance of duplications may not reside in the duplication of individual Hox genes alone. Duplication of the Hox cluster, for example, involves Hox genes and the associated *cis*-regulatory elements that direct their expression. Adding new *cis*-regulatory elements may, indeed, be the important part of this process. There are several ways that this type of duplication can affect variation. First, it can offer redundancy by multiplying regulatory units— increasing numbers of *cis*-regulatory elements can offer more potential for variation. Second, duplication can lead to greater regulatory fragmentation: the variation of different regions of the body can now be under the control of different regulatory elements. As both regulatory genes and their associated regulatory elements proliferate, organisms can be broken down into independently varying modules. For example, the fact that forelimbs and the axial skeleton can evolve independently in many taxa may be the result of this proliferating regulatory fragmentation. The proliferation and diversification of *cis*-regulatory elements may promote fine-scaled selectable variation within populations.

It might seem paradoxical, but this regulatory fragmentation can, under the right circumstances, also enable the correlated evolution of structures, as described below.

Evolutionary Innovation: The Importance of Correlated Evolution of Structures

The correlated evolution of structures is an important way that novel functions and morphologies can evolve. The origin of the snake bodyplan is an excellent example of the evolutionary utility of cooption and correlated evolution. The evolutionary correlation between body elongation and limb reduction has long been associated on functional grounds. Indeed, the correlated evolution of limb reduction and body elongation is so common as to be the rule in vertebrate evolution. This pattern of evolution is seen in urodele and caecilian amphibians, lizards, snakes, and amphisbaenians. Gans (1975) argues that body elongation increases the efficiency of locomotion in crevasses and tuft grasses. In his view, limb reduction is only a secondary correlate of selection for an elongate body. However, there would be clear selective reasons to maintain a genetic correlation between the patterning of the limbs and the body axis.

Recent studies have suggested a possible molecular mechanism for the correlated evolution of the axial skeleton and limbs (Cohn and Tickle 1999). Hox genes are involved with the regional specification of the identity and position of axial structures in vertebrates (Burke et al. 1995). In addition, the position of forelimbs and hindlimbs along the body axis is influenced by these same genes (Cohn and Tickle 1996).

The boundaries between different vertebral types correlate with the expression of Hox genes (Burke et al. 1995). Each region of the axial skeleton has a distinctive pattern of expression of different members of the Hox gene family. Boundaries between cervical and thoracic vertebrae and thoracic and lumbar vertebrae correlate precisely to borders of Hox gene expression. For example, the expression of Hox C6 and Hox C8 are correlated with the development of vertebrae with a thoracic identity in most tetrapods. In chicks and mice, the expression of these genes is localized in the thoracic region. Furthermore, the anterior expression boundary of these genes correlates to the position of the forelimb and shoulder.

Pythons have over 300 vertebrae, no fore-limbs, and highly rudimentary hindlimbs that contain homologues of the pelvis and femur. The morphology of the vertebrae also is unique, with ribs (a feature of thoracic verte-brate) extending anteriorly to articulate with all vertebrae but the atlas. All of the vertebrae, then, appear to have a thoracic identity.

This unique morphological bodyplan of snakes is correlated with a novel pattern of Hox gene expression along the body axis (Cohn and Tickle 1999). Hox C6 and Hox C8 are expressed in the entire body wall anterior to the hindlimbs. This simple shift is correlat-ed both with the expansion of thoracic iden-tities and with the loss of the forelimbs. A ge-netic correlation between vertebral and limb evolution may provide a basis for the morpho-logical transformations seen in snake phylog-eny. The successive loss of forelimbs, loss of the regionalization of the vertebrae, and loss of hindlimbs may be accounted for by the pro-gressive expansion of Hox domains along the snake body axis (Cohn and Tickle 1999). Cohn and Tickle (1999) speculate that the dramatic reduction in the hindlimbs of pythons is re-lated to changes of Hox gene expression along the body axis. This correlation is suggested by the observation that the reduced hindlimb of pythons differs from other amniotes in not ex-pressing Sonic hedgehog, in lacking an apical ectodermal ridge, and in retaining polarizing potential across the entire limb bud. All of these effects may be due to changes in Hox ex-pression within the limb. If this hypothesis is true, changes in the limb and in the axial skel-eton are due to changes in the regulation of the same Hox genes.

The integrated evolution of the snake bod-yplan can be viewed as a consequence of the genetic events that happened early in the his-tory of metazoans. In metazoans, the role of Hox genes in specifying regional identities of the body axis is more primitive than its role in the appendages (Shubin et al. 1997; Cohn and Coates 1999). Indeed, the involvement of Hox genes in the specification of anterior-posterior identities along the body axis is a possible bi-laterian synapomorphy (Knoll and Carroll

1999; Peterson and Davidson 2000). The role of the 5' Hox paralogues in patterning the ap-pendages is a shared-derived feature of ver-tebrates (Shubin et al. 1997; Cohn and Coates 1999). At least three different scenarios have been proposed to account for how Hox genes became coopted into limbs. Laufer and Tabin (1993) and Shubin et al. (1997) propose that overlapping patterns of Hox expression first evolved to pattern the body axis; later coop-tion of Hox genes into limbs happened in one or both sets of limbs simultaneously. Cohn and Coates (1999) argue against these views and propose that localized patterns of 5' Hox expression first evolved to pattern the meso-derm of the gut. The end result of both of these similarities is the same: 5' Hox genes are involved in anterior-posterior patterning of paraxial, lateral plate, and limb mesoderm. These linkages provide the raw material for the types of variation necessary to evolve new bodyplans.

The involvement of Hox genes in the devel-opment of two different structures does not guarantee that they will evolve in concert. Limbs can often vary independently of the ax-ial skeleton, guts, and genitalia—these link-ages are not necessary constraints on verte-brate evolution. The evolution of new regula-tory interactions, through modification of cis-regulatory elements, can break or anneal genetic links between structures.

Why No Post-Cambrian Origins of New Phyla?

Despite our general ignorance of how, in de-tail, most morphological features are actually built, our level of understanding of the key el-ements needed to establish the basic bodyplan has reached a point where we can begin to speculate on the origin of different bodyplans, and thus on why we see such a unique restric-tion of bodyplan originations in the Neopro-terozoic.

Perhaps the most important aspect of the diversity of animal bodyplans is our difficulty in identifying homologies between the animal phyla. Certainly the lack of obvious synapo-morphies among the adult bodyplans of the phyla has made it difficult to construct reli-able morphological phylogenies of the phyla, and most analyses are heavily dependent on developmental characters, etc. There are two end-member explanations for the observed disparity between the phyla; either now-disparate structures are homologous, but have undergone sufficient morphological divergence to obliterate their plesiomorphic similarity, or most structures and bodyplans are not homologous, but were autapomorphically developed within each phylum.

Davidson, Peterson, and Cameron (Davidson et al. 1995; Peterson et al. 1997, 2000; Peterson and Davidson 2000) have developed a hypothesis based on developmental considerations that supports the latter. They argue that the organisms that belong to the stem-group bilaterians (that is, triploblastic animals that predate the divergence of the modern bilaterian animal phyla) were small organisms that developed through stereotypical and limited cell cleavages (Davidson's type-I embryogenesis). According to their evolutionary scenario, a common ancestor of the bilaterians then developed the capacity to proliferate cell masses without having to specify immediately their fates. These are termed set-aside cells (set aside in the sense that they are set aside, or freed, from the immediate constraints of the developing animal). They argue that the Hox genes were deployed at this stage in evolution to pattern the proliferating set-aside cells, laying down the coordinates for later gene expression that ultimately direct different regions of the set-aside cell mass to develop legs, or wings, mouthparts, etc. These workers argue that different lineages of primitive bilaterians recruited different sets of genes to the regions demarcated by the patterning genes, and that the different configurations of organs that result from this post-patterning developmental evolution produced the different adult animal bodyplans that we now recognize as the separate phyla. In this sense patterning of the adult bodies is homologous between the bilaterian phyla, but the organization and structures built by the adults are not (in the same way that capacity for language is plesiomorphic for Europeans and Japanese, even though the words for individual objects are not homologous but represent separate autapomorphic features of English and Japanese).

The hypothesis is not without controversy, and key questions arise. Is the prediction that the first bilaterians were small type-I developing organisms true (that is to say, is it true that indirect development is plesiomorphic for the bilaterians)? What were the set-aside cells doing at their time of origin, before post-patterning gene batteries were hooked up the patterning genes?

However, if the hypothesis is true, it suggests that the disparity between the phyla is real and explains the difficulties in morphological phylogenies of the phyla. It may also explain why there was just one distinct phase of phylum level innovations, for it suggests that the phyla originated during the time when preexisting sets of morphogenetic genes were hooked together in different ways, under the direction of genes that patterned the set-aside cells. Presumably animals that had committed to a specific configuration of post-patterning development (and thus committed to developing specific adult morphologies, including the deployment of sense organs, limbs, jaws, etc.) out-competed relatively undifferentiated lineages that had yet to adequately differentiate their set-aside cells. Through the competitive elimination of the lineages with uncommitted set-aside cells, the capacity for new innovation at the phylum level was lost. An analogous event happened with the origin of pleurodire and cryptodire turtles, where the origin of the shell set up the selective pressures and morphological opportunity for the development of neck retraction, ultimately leading to these two forms of neck retraction (Marshall 1995). But, with the extinction of the plesiomorphic groups that were unable to retract their necks, the evolutionary opportunity for a third neck retraction mechanism was closed.

It is important to remember that the differences in genetic architecture that today distinguish different taxa need not relate to the original patterns of variation that existed when they diverged from their common ancestor (Budd 1999). This perspective provides a different approach to problems like the Cambrian explosion. The Cambrian explosion is as much about the evolution of stable complexes of characters and the origin of new patterns of selectable variation as it is about the origin of new taxa. The observation that few phyla have appeared since the Cambrian is as much a statement about the manner in which variation has changed since that time as it is about the phylogeny of groups. When we make the observation that no new phyla have arisen, we are basically saying that certain types of characters no longer vary in the same manner as they did in the Cambrian. Characteristics of symmetry, number of eyes, position of the nervous system, and cleavage pattern appear to be more fundamental in several senses. First, they appear at deep nodes in cladograms, and are properties of very inclusive groups. Second, these characters originally were fixed early in the fossil record and are relatively stable in the members of each group. Bilaterians, for example, all are bilaterally symmetrical and have been since at least the Neoproterozoic: modifications of the bodyplan of Bilaterians do not involve major changes in symmetry.

One interesting aspect of the Cambrian explosion is that the events that might be responsible for it may, indeed, be general properties of evolving developmental systems. The recognition of bodyplan-level characters is, in part, a post-hoc procedure. Characters gain this status only when they are relatively stable in the face of large-scale diversification. It is also true that there is a hierarchy of characters that exists outside of phylogeny. Organisms must have cells before they have tissues, patterns of cleavage before embryos, patterns of symmetry and segmentation before bodyplans. It is the interplay within this hierarchy of structural necessity, combined with historical contingency, that results in the recognition of bodyplan-level, or phylotypic, characters.

One result of this interplay is that the patterns of intraspecific variation in modern phyla are qualitatively different from those that must have existed in the taxa that lived during Neoproterozoic and probably Early Cambrian times.

A promising area of research lies in understanding the mechanisms by which genetic var-

iation becomes entrained. Many of the characteristics that evolved during the origin of phyla are no longer able to change. The reason for this is that selectable variation is absent: either the characters are invariant or mutants that carry this variation are sterile or lethal. How does this entrainment evolve? A large part of understanding this issue may involve understanding how regulatory networks are, themselves, assembled over evolutionary time. The evolution of new morphologies may involve the evolution of new pleiotropic interactions. The elaboration of pleiotropic interactions could, in turn, entrain (and limit) variation. An evolutionary way out of this dead end could involve duplications, thereby promoting phenotypic variation through the evolution of new enhancers (cis-regulatory) and other gene regulatory elements.

Microevolution and Macroevolution Revisited

New discoveries from genetics and paleontology are likely to return us to one of the most fruitful themes to have graced the pages of *Paleobiology* in the past 25 years: the relationship between microevolution and macroevolution (Erwin 2000).

The disconnect between rates of genetic and morphological change is as vexing a problem for population geneticists as it is for paleontologists. For evolutionary geneticists this problem has been the source of numerous debates about the origin of adaptations. One of the key discussions has centered on the number of genes and degree of effects that underlie adaptive change. These range from classical Fisherian views of gradual change in genes with small effects to more recent views that defend the plausibility of genes of larger effect in the origin of adaptation. For paleontologists, faced with problems like the Cambrian explosion, an understanding of the relationship between genotypic and phenotypic change promises to provide an understanding of the mechanisms behind evolutionary radiations.

The study of the developmental mechanisms by which variation is produced is one way to bridge the gap between studies of microevolution and macroevolution. Natural selection can only act on the heritable variation present in a population at any given time. Currently, much of developmental evolutionary biology has been concerned with macro-level comparisons. These comparisons show that the gross differences between bodyplans may be correlated with major changes in the action and/or number of a few regulatory genes. Does this observation inevitably lead to the conclusion that much of large-scale evolution is due to macromutation? At first glance, it does—these studies seem to lead to a reaffirmation of Goldschmidt's ideas of macromutation.

In all the examples above, and in fact in most of the genetic analyses of the developmental basis for morphological change, individual features (such as halteres or wings in *Drosophila*) and just a few genes are considered. An important question is, how many changes does it take to account for *all* the differences between species? Does the evolution of animal and plant form typically result, for instance, from the action of many genes of small effect or from a few genes of large effect? Historically, evolutionists have favored the view that the evolution of animal and plant form typically results from the action of many genes of small effect, and traditional quantitative genetic theory largely rests upon the so-called infinitesimal assumption, i.e., the assumption that phenotypic change involves many factors of very small effect each.

Recent quantitative trait locus (QTL) analyses have, however, called this view into question. QTL analysis allows the genetic dissection of the morphological differences between pairs of crossable taxa. By producing hybrids who carry random combinations of chromosome regions from two taxa (where the species identity of the chromosomal regions is inferred from already-known molecular markers), and by scoring the mean phenotype of each genotype, one can map, count, and estimate the effects of genes underlying the trait studied (Tanksley 1993). This is very expensive and time consuming work, but the economic importance of crop plants has provided the resources to carry out the work.

QTL analyses routinely reveal that morphological differences involve a modest number of chromosome regions of substantial effect. Recent studies by Doebley and colleagues (Doebley and Stec 1991; Doebley and Wang 1997), for instance, have shown that morphological differences between maize and its ancestor, teosinte, may involve as few as five relatively small chromosomal regions. One of the major differences between teosinte and maize, the difference in lateral branching pattern, appears to reflect the action of a single gene, *teosinte branched-1* (*tb1*) (Doebley and Wang 1997).

It is possible that the relative few "loci" involved may be a function of the history of strong artificial selection in crops. For example, the evolution of maize from teosinte involved concentrated and relatively unidirectional selection pressures that may not be common without human intervention. Thus, under less-directional and weaker selection pressures, many more changes may accumulate and be necessary for the observed differences between closely related species.

But the unexpectedly small number of loci involved in the maize-teosinte transition may not be unusual. A growing body of evidence suggests that "natural adaptations" may also involve a modest number of genetic factors (Orr and Irving 1997; Bradshaw et al. 1998; Jones 1998). It is beginning to appear that, while many genes of small effect may contribute to the morphological differences between species, just a few genetic differences might be responsible for many of the observed differences (Doebley and Wang 1997).

In contrast, virtually all theoretical analyses of the number of genetic differences responsible for the observed phenotypic difference between species assume that these differences reflect either the cumulative effects of many infinitesimal contributions (quantitative genetics) or many changes at single loci (population genetics), assumptions that the empirical data from the QTL analyses seem to contradict.

However, recent theoretical advances now draw the empirical and theoretical predictions into closer accord. The advances (Orr 1998) take advantage of Fisher's (1930) idealization of evolution, under which organisms are viewed as comprising n independent characters, with fitness falling off from the optimum at the same rate for all traits. Fisher used this model to calculate the probability that mutations of a given phenotypic size would be favorable, showing that, while small mutations have a good chance of being advantageous, larger ones suffer a rapidly decreasing probability. Fisher thus arrived at his now famous conclusion that small mutations are the stuff of adaptation. It was only much later realized, however, that Fisher had neglected an important aspect of adaptation—the probability of fixation. To contribute to adaptation, mutations not only must be favorable but also must escape random loss when rare. Noting that the probability of such escape increases with the size of the phenotypic effect of a mutation, Kimura (1983) argued that mutations of *intermediate* size are the most likely to underlie adaptation.

However, Orr (1998) has recently argued that Kimura's analysis also is incomplete, for following some change in the environment, evolution toward a new morphological optimum might involve many evolutionary steps (substitutions), not one, as Kimura tacitly assumed. Thus, the distribution of greatest biological interest concerns phenotypic effects among factors fixed when summing over an *entire* "adaptive walk" to an optimum This distribution approximates the evolutionary path revealed by QTL analysis. Specifically, Orr (1998) showed that the distribution is approximately exponential, unlike those of Fisher and Kimura. It appears the result is robust to changes in the distribution of mutational effects, as long as small mutations are more common than large. And perhaps most interestingly, this work shows that the expected size of the largest genetic factor fixed during adaptation is quite large. In short, the theoretical work provides population-genetic support for the search for genes of large effect underlying morphological change.

Post-genomics Paleontology

The fiftieth anniversary issue of Paleobiology will be published in a post-genomics age. Knowledge of genome structure, gene expression, and genome diversity is increasing exponentially, following dramatic technological progress. New methods for obtaining gene sequences, analyzing gene expression, and understanding gene function are generating new data faster than it can be analyzed. Genome projects are underway—or completed—in an astounding variety of organisms: microbes (*Archaeoglobis, Helicobacter, Borrelia, Methanococcus, Mycobacterium, Mycoplasma, Theromotoga, Treponema, Plasmodium, Trypanosomes*); vertebrates (*Rattus, Mus, Homo, Fugu, and Danio*); invertebrates (*Drosophila, Caenorabditis*); and plants (potato, rice, tomato, *Arabidopsis, Maize*). These new data sets are going to have a profound impact on paleontology (Banfield and Marshall 2000). Our understanding of genetic evolution will segue from gene-by-gene analyses of diversity to genome-by-genome comparisons.

The most obvious impact of these new data will be on phylogenetic systematics. Technological advances are likely to redefine our concept of molecular characters in phylogenetic analyses. Current molecular phylogenies are done on a gene-by-gene basis: substitutions are currently the bread and butter of molecular phylogenies. Future phylogenies will be based on entire genomes. These approaches will require a quantum leap in the computational methods used to align and analyze sequence data. As with any technological advance, there is no guarantee that phylogenies will be necessarily more robust with the input of this vast amount of new data. Our only guarantee is that phylogenies will be more exhaustive and sample a greater variety of characters.

Some of the most exciting technological advances relate to understanding the genetic basis of morphological adaptation. Right now, we have to settle for studies of gene expression and gene function that are performed on single genes or small groups of genes. For example, much of the comparative data on Hox gene evolution has been based on studies of the sequence, expression and function of orthologous or paralogous genes. New studies provide only a glimpse of what is to come in the next few years; rapid technological advances are going to provide a whole genome approach to this problem. DNA microarray technology is but one example of what the future holds. These techniques allow gene expression to be analyzed across the entire genome. Recent experimental studies on yeast genes provide an exciting example of the utility of this technique (Ferea et al. 1999). Ferea et al. (1999) integrated a study of experimental evolution with an analysis of whole-genome expression. Populations of yeast were raised in a new selective environment, one that was glucose-limited. As expected, the populations adapted to this new environment quickly, in 500 generations or less. What was not expected, however, was the nature of the genetic changes involved in this adaptation. DNA microarray analysis revealed that this simple laboratory-based adaptation involved changes in the expression of literally hundreds of genes over a very small number of generations of adaptation. This example points to an important role of changes in gene regulation in producing metabolic changes. While Ferea et al. (1999) do not know the nature of these regulatory mutations, the role of gene regulation in producing these adaptations is supported by coordinated changes among genes in several replicate populations of yeast that were adapting to the same selective conditions.

Finally, we emphasize that fundamental new insights and discoveries may await us, advances that may change some of our most cherished views about the relationship between genotype, phenotype, and the environment. For example, Rutherford and Lindquist (1998) recently proposed a mechanism that offers the opportunity for rapid, environmentally triggered evolutionary change. Their proposal stems from their work with the heat shock protein Hsp90, which, among its various functions, helps maintain the three-dimensional configuration of key proteins in cell-to-cell signaling pathways used in development. This homeostatic function of Hsp90 may mask significant variation in the proteins of these pathways from the action of natural selection. Rutherford and Lindquist proposed that at times of stress this variation might be revealed, allowing for rapid rates of evolutionary change across much of the phenotype at exactly the time when it might be most needed—at time of environmental stress.

Conclusions

1. Genes that play a major role in establishing the primary axes of the body and appendages, and that regulate the expression of the genes that initiate the making of structures such as eyes or hearts are highly conserved between phyla.

2. This implies that it is not new genes, per se, that underlie morphological innovation, but that it is changes in when and where these and other genes are expressed that constitute the underlying mechanistic basis of morphological innovation. Gene duplication is also a source of developmental innovation, but it is possible that it is not the increased number of genes (and their subsequent divergence) that is important in the evolution of new morphologies; rather it may be the increased number of regulatory regions that results from gene duplication which provides the raw material for morphological innovation.

3. One of the major ways that regulatory interactions can evolve is by the modification of *cis*-regulatory elements (enhancers and promoters). Changes in these elements can bring about major new patterns of selectable variation. In addition, this type of evolutionary change can either make or break the genetic linkages between different structures.

4. Advances in our understanding of the mechanistic basis of animal development offers opportunity to understand the roles of development and developmental canalization in the origin of the animal phyla, deepening our insight into the nature of the Cambrian explosion.

5. Bridging the gap between microevolution and macroevolution will involve understanding the mechanisms behind the production of morphological variation. In some instances it is becoming apparent that relatively few genetic changes may be responsible for most of the observed phenotypic differences between species.

6. The last two and a half decades have seen enormous advances in our understanding of the relationship between genes and evolutionary novelty. However, especially with the number of whole-genome sequences coming on line we might anticipate a deepening of our

understanding of the relationship between the genotype, phenotype, and environment that overshadow the spectacular advances already made. Post-genomics paleontology promises to be an exciting time.

Acknowledgments

We thank D. Erwin and P. Holland for valuable reviews and S. Carroll for helpful discussion. This work was partially supported by National Aeronautics and Space Administration grant NCC2-1053 (to C. R. M.) and National Science Foundation grant EAR-980600 (to N. H. S.).

Literature Cited

Averof, M., and N. Patel. 1997. Crustacean appendage evolution associated with changes in Hox gene expression. Nature 388: 682–686.

Banfield, J. F., and C. R. Marshall. 2000. Genomics and the geosciences. Science 287:605–606.

Basler, D., and G. Struhl. 1994. Compartment boundaries and the control of Drosophila limb pattern by hedgehog protein. Nature 368:208–214.

Bengtson, S., ed. 1994. Early life on Earth (Nobel Symposium No.84). Columbia University Press, New York.

Bradshaw, H. D., K. Otto, B. Frewen, J. K. McKay, and D. Schemske. 1998. Quantitative trait loci affecting differences in floral morphology between two species of monkeyflower (Mimulus). Genetics 149:367–382.

Brooke, N. M., J. Garciafernandez, and P. W. H. Holland. 1998. The paraHox gene cluster is an evolutionary sister of the Hox gene cluster. Nature 392:920–922.

Budd, G. E. 1999. Does evolution in body patterning genes drive morphological change—or vice versa? BioEssays 21:326–332.

Burke, A. C., C. E. Nelson, B. A. Morgan, and C. Tabin. 1995. Hox genes and the evolution of vertebrate axial morphology. Development 121:333–346.

Carroll, R. L. 1988. Vertebrate paleontology and evolution. W. H. Freeman, New York.

Carroll, S. B. 1995. Homeotic genes and the evolution of arthropods and chordates. Nature 376:479–485.

———. 2000. Endless forms: the evolution of gene regulation and morphological diversity. Cell 101:577–580.

Chang, D. T., A. Lopez, D. P. Vonkessler, C. Chiang, B. K. Simandl, R. B. Zhao, M. F. Seldin, J. F. Fallon, and P. A. Beachy. 1994. Products, genetic linkage and limb patterning activity of a murine hedgehog gene. Development 120:3339–3353.

Cohn, M. J., and M. Coates. 1999. Vertebrate axial and appendicular patterning: the early development of paired appendages. American Zoologist 39:676–685.

Cohn, M. J., and C. Tickle. 1996. Limbs—a model for pattern formation within the vertebrate body plan. Trends in Genetics 12:253–257.

———. 1999. Developmental basis of limblessness and axial patterning in snakes. Nature 399:474–479.

Couso, J. P., M. Bate, and A. A. Martinez-Arias. 1993. Wingless-dependent polar coordinate system in Drosophila imaginal discs. Science 259:484–489.

Davidson, E. H., K. J. Peterson, and R. A. Cameron. 1995. Origin of adult bilaterian body plans: evolution of developmental regulatory mechanisms. Science 270:1319–1325.

de Rosa, R., J. Grenier, T. Andreeva, C. Cook, A. Adoutte, M. Akam, S. B. Carroll, and G. Balavoine. 1999. Hox genes in brachiopods and priapulids and protostome evolution. Nature 399:772–776.

Diaz-Benjumea, F., and S. M. Cohen. 1993. Interaction between dorsal and ventral cells in the imaginal disc directs wing development in Drosophila. Cell 75:741–752.

Doebley, J., and A. Stec. 1991. Teosinte branched1 and the origin of maize: evidence for epistasis and the evolution of dominance. Genetics 129:285–295.

Doebley, J., and R.-L. Wang. 1997. Genetics and the evolution of plant form: an example from maize. Cold Spring Harbor Symposia on Quantitative Biology 22:361–367.

Duboule, D., and A. S. Wilkins. 1998. The evolution of 'bricolage'. Trends in Genetics 14:54–59.

Erwin, D. H. 2000. Macroevolution is more than repeated rounds of microevolution. Evolution and Development 2:78–84.

Ferea, T. L., D. Botstein, P. O. Brown, and R. F. Rosenzweig. 1999. Systematic changes in gene expression patterns following adaptive evolution in yeast. Proceedings of the National Academy of Sciences USA 96:9721–9726.

Finnerty, J. R., and M. Q. Martindale. 1998. The evolution of the Hox cluster: insights form outgroups. Current Opinion in Genetics and Development 8:681–687.

———. 1999. Ancient origins of axial patterning genes: Hox genes and ParaHox genes in the Cnidaria. Evolution and Development 1:16–23.

Fisher, R. A. 1930. The genetical theory of natural selection. Oxford University Press, Oxford.

Gans, C. 1975. Tetrapod limblessness: evolution and functional corollaries. American Zoologist 15:455–467.

Garey, J. R., and A. Schmidt-Rhaesa. 1998. The essential role of "minor" phyla in molecular studies of animal evolution. American Zoologist 38:907–917.

Gerhart, J., and M. Kirschner. 1997. Cells, embryos and evolution. Blackwell Science, Malden Mass.

Gould, S. J. 1977. Ontogeny and phylogeny. Harvard University Press, Cambridge.

Haldane, J. B. S. 1932. The causes of evolution. Longmans and Green, London.

Holland, P. W. H. 1998. Major transitions in animal evolution: a developmental genetic perspective. American Zoologist 38: 829–842.

Irvine, K., and E. Weischaus. 1994. fringe, a boundary-specific signaling molecule, mediates interactions between dorsal and ventral cells during Drosophila wing development. Cell 79: 595–606.

Jacob, F. 1977. Evolution and tinkering. Science 196:1161–1166.

Jones, C. D. 1998. The genetic basis of Drosophila sechellia's resistance to a host plant toxin. Genetics 149:1899–1908.

Kimura, M. 1983. The neutral theory of molecular evolution. Cambridge University Press, Cambridge.

Knoll, A. H., and S. B. Carroll. 1999. Early animal evolution: emerging views from comparative biology and geology. Science 284:2129–2137.

Krauss, S., J. P. Concordet, and P. W. Ingham. 1993. A functionally conserved homolog of the Drosophila segment polarity gene hh is expressed in tissues with polarizing activity in zebrafish embryos. Cell 75:1431–1444.

Laufer, E., and C. Tabin. 1993. Hox genes and serial homology. Nature 361:692–693.

Lee, J. J., D. P. von Kessler, S. Parks, and P. A. Beachy. 1992. Secretion and localized transcription suggest a role in positional signaling for products of the segmentation gene hedgehog. Cell 71:33–50.

Maienschein, J. 1991. The origins of Entwicklungsmechanik. Developmental Biology 7:43–61.

Marshall, C. R. 1995. Darwinism in an age of molecular revolution. Pp. 1–30 in C. R. Marshall and J. W. Schopf, eds. Evolution and the molecular revolution. Jones and Bartlett, Sudbury, Mass.

Marshall, C. R., H. A. Orr, and N. H. Patel. 1999. Morphological innovation and developmental genetics. Proceedings of the National Academy of Sciences USA 96:9995–9996.

Mercader, N., E. Leonardo, N. Azpiazu, A. Serrano, G. Morata, A. C. Martinez, and M. Torres. 1999. Conserved regulation of proximodistal limb axis development by *Meis/Hth*. Nature 402:425–429.

Müller, G., and G. P. Wagner. 1991. Novelty in evolution: restructuring the concept. Annual Reviews of Ecology and Systematics 22:229–256.

Muller, H. J. 1935. The origination of chromatin deficiencies as minute deletions subject to insertion elsewhere. Genetics 17:237–252.

Ohno, S. 1970. Evolution by gene duplication. Springer, Berlin.

Orr, H. A. 1997. The population genetics of adaptation—the distribution of factors fixed during adaptive evolution. Evolution. 52:935–949.

Orr, H. A., and S. Irving. 1998. The genetics of adaptation—the genetic basis of resistance to wasp parasitism in *Drosophila melanogaster*. Evolution 51:1877–1885.

Panganiban, G., S. M. Irvine, C. Lowe, H. Roehl, L. Corley, B. Sherbon, J. K. Grenier, J. F. Fallon, J. Kimble, M. Walker, G. A. Wray, B. J. Swalla, M. Q. Martindale, and S. B. Carroll. 1997. The origin and evolution of animal appendages. Proceedings of the National Academy of Sciences USA 94:5162–5166.

Parr, B. A., and A. P. McMahon. 1995. Dorsalizing signal *Wnt-7a* required for normal polarity of D-V and A-P axes of mouse limb. Nature 374:350–353.

Peterson, K. J., and E. H. Davidson. 2000. Regulatory evolution and the origin of bilaterians. Proceedings of the National Academy of Sciences USA 97:4430–4433.

Peterson K. J., R. A. Cameron, and E. H. Davidson. 1997. Set-aside cells in maximal indirect development: evolutionary and developmental significance. BioEssays 19:623–631.

———. 2000. Bilaterian origins: significance of new experimental observations. Developmental Biology 219:1–7.

Raff, R. 1997. The shape of life. University of Chicago Press, Chicago.

Riddle, R. D., R. L. Johnson, E. Laufer, and C. Tabin. 1993. *Sonichedgehog* mediates the polarizing activity of the ZPA. Cell 75:1401–1416.

Riddle R. D., M. Ensini, C. Nelson, T. Tsuchida, T. M. Jessell, and C. Tabin. 1995. Induction of the LIM homeobox gene *Lmx-1* by *Wnt-7a* establishes dorsoventral pattern in the vertebrate limb. Cell 83:631–640.

Rodriguez-Esteban, C., J. W. R. Schwabe, J. Delapena, B. Foys, B. Eshelman, J. C. I. Belmonte. 1997. *Radical fringe* positions the apical ectodermal ridge at the dorsoventral boundary of the vertebrate limb. Nature 386:360–366.

Romano, S. L., and S. R. Palumbi. 1996. Evolution of scleractinian corals inferred from molecular systematics. Science 271:640–642.

Rutherford, S. L., and S. Lindquist. 1998. Hsp90 as a capacitor for morphological evolution. Nature 396:336–342.

Shubin, N. H. 1995. The evolution of paired fins and the origin of tetrapod limbs: phylogenetic and transformational approaches. Evolutionary Theory 28:39–86.

Shubin, N. H., C. Tabin, and S. B. Carroll. 1997. Fossils, genes and the evolution of animal limbs. Nature 388:639–648.

Smith, A. B. 1999. Dating the origin of metazoan body plans. Evolution and Development 1:138–142.

Stern, D. 1998. A role of *Ultrabithorax* in morphological differences between species. Nature 396:463–466.

Tabata, T., S. Eaton, and T. B. Kornberg. 1992. The *Drosophila hedgehog* gene is expressed specifically in posterior compartment cells and is a target of *engrailed* regulation. Genes and Development 6:2635–2645.

Tabin C. J., S. B. Carroll, and G. Panganiban. 1999. Out on a limb: parallels in vertebrate and invertebrate limb patterning and the origin of appendages. American Zoologist 39:650–663.

Tanksley, S. D. 1993. Mapping polygenes. Annual Reviews of Genetics 27:205–233.

Valentine, J. W., D. Jablonski, and D. H. Erwin 1999. Fossils, molecules and embryos: new perspectives on the Cambrian explosion. Development 126:851–859.

Van Valen, L. M. 1982. Homology and causes. Journal of Morphology 173:305–312.

Vogel, A., C. Rodriguez, W. Warnken, and J. C. I. Belmonte. 1995. Dorsal cell fate specified by chick *Lmx1* during vertebrate limb development. Nature 378:716–720.

Williams, J. A., S. W. Paddock, and S. B. Carroll. 1993. Pattern formation in a secondary field: a hierarchy of regulatory genes subdivides the developing *Drosophila* wing disc into discrete sub-regions. Development 117:571–584.

Wray, G. A., and E. Abouheif. 1998. When is homology not homology. Current Opinion in Genetics and Development 8:675–680.

Zhu, M., X. B. Yu, and P. Janvier. 1999. A primitive fossil fish sheds light on the origin of bony fishes. Nature 397:607–610.

Phylogenetic analyses and the fossil record: tests and inferences, hypotheses and models

Peter J. Wagner

Abstract.—Tree-based paleobiological studies use inferred phylogenies as models to test hypotheses about macroevolution and the quality of the fossil record. Such studies raise two concerns. The first is how model trees might bias results. The second is testing hypotheses about parameters that affect tree inference.

Bias introduced by model trees is explored for tree-based assessments of the quality of the fossil record. Several nuisance parameters affect tree-based metrics, including consistency of sampling probability, rates of speciation/extinction, patterns of speciation, applied taxonomic philosophy, and assumed taxonomy. The first two factors affect probabilistic assessments of sampling, but also can be tested and accommodated in sophisticated probability tests. However, the final three parameters (and the assumption of a correct phylogeny) do not affect probabilistic assessments.

Often paleobiologists wish to test hypotheses such as rates of character change or rates of preservation. Assumptions about such parameters are necessary in simple phylogenetic methods, even if the assumptions are that rates are homogeneous or that sampling is irrelevant. Likelihood tests that evaluate phylogenies in light of stratigraphic data and/or alternative hypotheses of character evolution can reduce assumptions about unknowns by testing numerous unknowns simultaneously. Such tests have received numerous criticisms, largely based in philosophy. However, such criticisms are based on incorrect depictions of the logical structures of parsimony and likelihood, misunderstandings about when arguments are probabilistic (as opposed to Boolean), overly restrictive concepts of when data can test a hypothesis, and simply incorrect definitions of some terms.

Likelihood methods can test multiparameter hypotheses about phylogeny and character evolution (i.e., rates, independence, etc.). The best hypothesis positing a single rate of independent character change (with no variation among character states) is determined for each topology. Hypotheses about rate variation among characters or across phylogeny, character independence, and different patterns of state evolution then are examined until one finds the simplest (i.e., fewest varying parameters) hypothesis that cannot be rejected given knowledge of a more complicated hypothesis. This is repeated for alternative topologies. An example is presented using hyaenids. Two trees are contrasted, one of which requires the minimum necessary steps and the other of which requires at least seven additional steps. Given either tree, likelihood rejects fewer than three general rates of character change and also rejects the hypothesis of independence among the characters. However, hypotheses of changes in rates across the tree do not add substantially to the tree likelihood. The likelihoods of the trees given stratigraphic data also are determined. Both morphologic and stratigraphic data suggest that the multiparameter hypothesis including the parsimony tree is significantly less likely than the multiparameter hypothesis including a different tree.

Peter J. Wagner. Department of Geology, Field Museum of Natural History, 1400 Lake Shore Drive, Chicago, Illinois 60605. E-mail: pwagner@fmnh.org

Accepted: 16 May 2000

Introduction

Numerical taxonomy introduced a battery of techniques for repeatably inferring phylogenetic relationships (see Sneath 1995 and Edwards 1996 for reviews). These developments accompanied a growing awareness that phylogenies are primary explanations for character distributions among taxa (Raup and Gould 1974; Gould and Lewontin 1979; Felsenstein 1985). Workers also realized that phylogenies offer implicit hypotheses about origination times (Fisher 1982; Paul 1982) and the relative diversities of clades (Raup et al. 1973; Sanderson and Donoghue 1994). These progressions led to a plethora of "tree-based" research programs, which use inferred phylogenies as models when testing hypotheses about macroevolution (see Harvey and Pagel 1991) and other issues, such as the quality of the fossil record (e.g., Smith 1988).

The main focus of this paper is how paleobiologists should use phylogenies when testing hypotheses. Tree-based studies implicitly assume either that an inferred phylogeny is

0094-8373/00/2604-0015/$1.00

correct or that inference error does not bias results toward favoring one set of hypotheses over others (Harvey and Pagel 1991: p. 121). However, several simulation studies show that incorrect phylogenetic inferences present biased depictions of evolution (Kuhner and Felsenstein 1994; Mooers et al. 1995; Huelsenbeck and Kirkpatrick 1996) and the fossil record (Wagner 2000a). The first part of this paper examines the latter issue in detail. In particular, it evaluates the claim that tree-based assessments of the fossil record might make fewer assumptions than do probabilistic assessments of sampling. The paper then explores an alternative to classic tree-based methodology, which tests multiparameter hypotheses about both phylogeny and character evolution. Because debate exists concerning the goals and utilities of quantitative and nonquantitative methods for testing historical hypotheses, the paper first reviews the logical structures of alternative methods as well as the logical structures of the arguments supporting and opposing those different approaches.

Terms and Definitions

My use of "model" in this paper follows the definition give by Edwards (1992: p. 3): "that part of the description which is not at present in question, and may be regarded as given. . . ." My use of "hypothesis" also follows the definition given by Edwards (1992: pp. 3–4): "parameters or entities being in question, and the subject of investigation." This distinction becomes important when there are numerous "nuisance parameters," which are unknowns other than the test hypotheses that affect predictions (Edwards 1992: p. 109) (philosophers of science label these "auxiliary assumptions" [T. Grantham personal communication 2000]). If factors other than phylogenetic topology affect the predictions made by a hypothesized tree, then the identification of these nuisance parameters is necessary. Once identified, one can either assume some condition for nuisance parameters (i.e., modeling) or make those parameters components of a multiparameter hypothesis. A primary theme of this paper will be making hypotheses of as many parameters as possible.

This paper also makes a distinction between inferences and tests. "Inferences" here refer to best explanations given data and some set of assumptions. "Tests" here refer to methods that can reject explanations given data and some set of assumptions. A main theme of this paper is that we should prefer methods that test over methods that infer. The identification of nuisance parameters becomes important here because the primary reason for identifying such parameters (and for devising tests for multiparameter hypotheses) is to increase our confidence that a hypothesis is not rejected because of incorrect assumptions about nuisance parameters.

Inferred Phylogenies As Models for Assessing the Fossil Record

Assessments of Sampling—Do Tree-Based Methods Require Fewer Assumptions Than Do Probabilistic Methods?—A basic issue in paleobiology concerns the quality of a taxon's fossil record. One basic approach involves looking at the distribution of gaps in the fossil record. An assumed phylogeny implies such gaps if either (1) two sister taxa have different first appearances or (2) a gap exists between the last appearance of an inferred ancestor and the first appearance of its inferred descendant (Fisher 1982; Paul 1982) (Fig. 1). These implied gaps, termed range extensions by A. Smith (1988) (= ghost lineages/taxa [Norell 1992, 1993]), represent either unsampled representatives of known morphotypes or unsampled morphotypes that workers presumably would recognize as distinct taxa were they sampled. Stratigraphic debt (Fisher 1988) (= sum of minimum implied gaps or SMIG of Benton and Storrs [1994]) is the sum of these gaps.

Workers also have devised numerous probabilistic methods for testing hypotheses about taxon durations (e.g., Paul 1985; Springer and Lilje 1988; Strauss and Sadler 1989; Marshall 1990, 1994, 1997; Solow 1996; Solow and Smith 1997; Wagner 2000c) and rates of sampling (e.g., Foote 1996, 1997; Foote and Raup 1996; Foote and Sepkoski 1999; Foote et al. 1999; Holland this volume). These methods derive

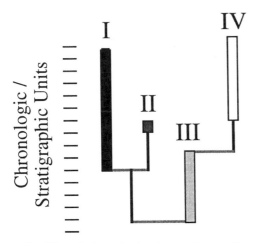

FIGURE 1. How phylogenies imply gaps in sampling. If two taxa are sister taxa (i.e., derivatives of a common ancestor) but have different first appearances (i.e., I and II or [I+II] and [III+IV]), then unsampled members of the clade (thin black lines) must span the durations separating the taxa. Similarly, if two taxa are ancestor and descendant (e.g., III and IV) but a gap separates the two species, then another unsampled portion must be inferred. Stratigraphic debt is the sum of these durations (Fisher 1988). In this example, the tree implies ten units of stratigraphic debt. Stratigraphic debt depends on exact phylogenetic inferences. For example, if III and IV are considered sister taxa instead of ancestor-descendant, the tree implies an additional five units of stratigraphic. See papers by Paul (1982, 1985), Smith (1988), Fisher (1991, 1994) and Norell (1992, 1993) for additional discussion.

the probability of observing stratigraphic data (either distributions of finds and absences, or distributions of stratigraphic ranges) given alternative hypotheses of sampling and/or true durations. Many of these methods (especially the older ones) require simplifying assumptions about nuisance parameters such as consistency of sampling over time or extinction rates (see Foote 1997; Solow and Smith 1997; Weiss and Marshall 1999).

Benton (1995) justifies evaluating the fossil record with cladograms by noting that cladogram construction is independent of sampling order in the fossil record. Several workers (e.g., Benton and Storrs 1994; Smith and Littlewood 1994; Wills 1998, 1999) use this argument to justify various metrics that standardize stratigraphic debt. Such metrics use fractions in which stratigraphic debt is part of or the entire numerator. In a series of papers, Benton and Hitchin (1997) (see also Benton

1995, 1998; Hitchin and Benton 1997) use debt metrics to test whether the sampling of particular groups is significantly better or worse than sampling of other groups.

The justification for tree-based assessments of the fossil record might be unsound for two reasons. First, sampling order and phylogeny reconstruction might not be independent. Sampling intensity obviously affects the former (Paul 1982). Simulation studies indicate that sampling also affects accuracy of cladograms (Huelsenbeck 1991; Graybeal 1998; Wagner 2000a). This is important because stratigraphic debt reflects actual gaps only if the phylogeny is correct (contra Norell 1996). Second, comparing implied gaps might make assumptions about sampling. Another class of tree-based metrics examines the "consistency" between inferred branching order and observed appearance order (Gauthier et al. 1988; Cloutier 1991; Norell and Novacek 1992a,b; Huelsenbeck 1994). Simulations by Wagner and Sidor (2000) show that stratigraphic consistency metrics reflect numerous nuisance parameters other than simple sampling intensity (R, the probability of sampling a taxon per stated interval). These parameters include (1) variation in R over time; (2) cladogenesis intensity (λ) and extinction intensity (μ); (3) speciation patterns; (4) the status of examined taxa (i.e., paraphyletic [Fig. 2A], stem-based monophyletic [Fig. 2B], or apomorphy-based monophyletic [Fig. 2C]); and (5) accurate cladograms. Attributing differences in consistency indices among groups to differences in R alone therefore requires assuming that all nuisance parameters are equivalent in each group.

The calculation of stratigraphic consistency indices is very different from the calculation of debt indices. Nevertheless, if factors other than R affect one type of tree-based metric, then we must be concerned that they will affect the other. Previous simulations indicate that MST inaccuracy increases stratigraphic debt at any particular R (Wagner 2000a). Smith (1994) and others (e.g., Fisher 1991; Benton and Storrs 1994) note that phylogenetically inferred gaps represent minimum possible gaps. The nuisance parameters listed above might exaggerate differences between

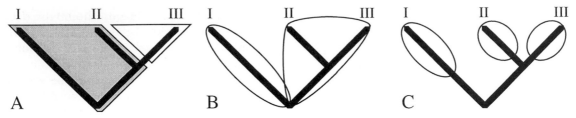

FIGURE 2. Applied taxonomic philosophies. A, Paraphyletic (shaded) vs. monophyletic. B, Stem-based monophyletic taxa, where all species after a basal divergence are placed in taxon I or taxon II-III (see de Queiroz and Gauthier 1990, 1992). C, Apomorphy-based monophyletic taxa, where all species sharing diagnostic synapomorphies are placed in taxa. Because these diagnostic characters can appear any time after divergence, the hypothesized temporal duration of taxon I need not extend to the divergence between I and II-III (Wagner and Sidor 2000). Some taxa simply might not be assignable to any recognized higher taxon. Under both A and C, taxon definitions (memberships) are based on morphologic diagnoses. Under B, taxon definitions are based on inferred phylogenetic relationships. As such, the taxa might not be diagnosable by any one character.

the implied and actual gaps. To explore this possibility, I repeat Wagner and Sidor's simulations here.

All simulations are of 25 sampled taxa. Unless otherwise stated, all simulated clades evolve under bifurcating cladogenesis (where ancestral species disappear at speciation events) with per-stage $\lambda = 0.5$ and per-stage $\mu = 0.4$. Variations on these simulations explore how stratigraphic debt is affected given six different nuisance parameters: (1) variation in R over time, (2) variation in R within clades, (3) different speciation patterns among clades, (4) different λ and μ among clades, (5) different applied taxonomic philosophy among clades, and (6) the assumption that all taxa are mono-

TABLE 1. Effects of nuisance parameters on sampling levels implied by phylogenies and probability-based methods. "+" indicates that the quality of the record is overestimated whereas "−" means that it is underestimated. "R" denotes sampling intensity. "*" indicates that estimates of sampling intensity must simultaneously estimate extinction intensity. This is important when considering specification patterns, because anagenetic elimination of ancestors elevates the apparent extinction intensity of taxa.

Variable	Phylogeny-based implied stratigraphic debt	Probability-based implied sampling intensity
Variable sampling over time	+	+
Variable sampling among taxa	+	+
Low extinction rate	+	*
Budding rather than bifurcating cladogenesis	Low R: + High R: −	*
Use of apomorphic taxa only	High R: −	?
Assuming all taxa monophyletic	High R: −	None
Inaccurate topology	−	None

phyletic when taxa (morphotypes) are sampled without regard to their phylogenetic status. It is known that the first two nuisance parameters lead quantitative methods to overestimate the average overall quality of the fossil record (Foote and Raup 1996). The fourth set of parameters is important because average sampling intensity often cannot be calculated without reference to these parameters (Foote 1997; Solow and Smith 1997). It is not clear how (if at all) the remaining three parameters affect quantitative assessments of the fossil record.

Temporal variation in R decreases stratigraphic debt appreciably when median R is less than 0.35 (Table 1, Fig. 3A), especially when variation in R exceeds the median R. At high median R, very high variation in sampling slightly inflates stratigraphic debt. However, the deviation from expected debt at consistent R is much less when median R is high than when it is low. Variation in R within clades also inflates stratigraphic debt at low median R and inflates stratigraphic debt at high median R (Fig. 3B). Intraclade variation in R inflates debt at high typical R much more than does temporal variation in R. However, the underestimate at low R again exceeds the overestimate at high R. Thus, variation in R leads debt metrics to overestimates median R, just as it leads probabilistic methods to overestimate average R (Foote and Raup 1996).

At any given constant R, if extinction (μ) and cladogenetic (λ) intensities are low relative to R, then clades with low μ and λ have more stratigraphic debt than do clades with

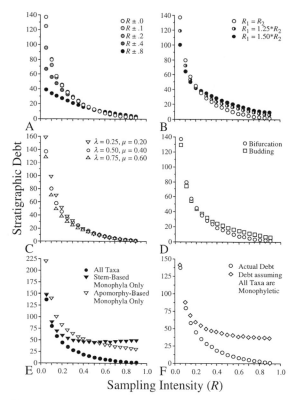

FIGURE 3. Effect of nuisance parameters on stratigraphic debt. Unless otherwise stated, all simulated clades evolved under bifurcation with speciation (λ) and extinction (μ) intensities of 0.5 and 0.4 per stage, with sampling intensity (R) consistent over time and across clades. Sampled taxa represent unique morphotypes and therefore can be paraphyletic or monophyletic. All results represent averages from 1000 simulations per R. A, Variation in R. B, Variation in R across clades. C, Variation in λ and μ. D, Bifurcating speciation (where ancestors become pseudoextinct) versus budding speciation (where ancestors persist). E, Applied taxonomic philosophies (see Fig. 2). F, Assuming that all taxa are monophyletic when sampling is conducted irrespective of paraphyly and monophyly.

high μ and λ (Fig. 3C). Low μ relative to R results in long average durations for individual taxa and also increases the probability that a near relative of a species exists at a much later point in time. Conversely, high μ relative to R results in short typical durations and reduces the probability of a near relative existing at a much later point in time. The probability of a relative existing at that later time increases if the clade happens to be very diverse. However, that high diversity will increase the probability of sampling a temporal and phylogenetic intermediate, which then reduces

the probability of a large amount of debt. Thus, an association between μ and stratigraphic debt should not be surprising. Extinction has a similar effect on probabilistic tests of R. Given a data set, if one estimates R while assuming μ, then one will infer a higher R if one assumes high μ than if one assumes a low μ (Foote 1997). Because of this Foote and others (e.g., Solow and Smith 1997) advocate testing R and μ simultaneously.

Speciation patterns also affect stratigraphic debt. When R is high, clades in which budding cladogenesis predominates have more stratigraphic debt at high R than do clades in which bifurcating cladogenesis predominates. Conversely, when R is low, clades in which budding cladogenesis predominates have less stratigraphic debt than do clades in which bifurcating cladogenesis predominates (Fig. 3D).

Trees including both paraphyletic and monophyletic taxa induce much less debt than do trees including only monophyletic taxa. This is not surprising, because at high sampling rates it is difficult not to sample paraphyletic taxa (Foote 1996). Deliberately excluding paraphyla therefore creates artificial gaps in sampling. Stem-based monophyla (Fig. 2B) exaggerate debt most at high R, whereas apomorphy-based (Fig. 2C) monophyla exaggerate most at low R. However, both types of monophyla make a wide range of sampling intensities predict similar stratigraphic debts.

Nearly all paleontological taxa are morphologically diagnosed rather than phylogenetically defined. Moreover, most character change among fossil taxa induces homoplasy (Wagner 2000b), which means that unique character combinations, not apomorphies, diagnose most fossil taxa. Thus, it is nearly impossible to distinguish paraphyletic and monophyletic taxa a priori. This in turn makes it improbable that debt analyses can deliberately exclude paraphyletic taxa. (The view that taxa and monophyla are synonymous represents arbitrary philosophy and is not considered here.) However, some studies assume that no sampled taxa are paraphyletic (e.g., Hitchin and Benton 1997). The probability of sampling an ancestral taxon increases as R increases (Fortey and Jefferies 1982; Foote 1996), and correctly recognizing paraphyla usually decreases stratigraphic debt (e.g., see Fig. 1). Thus, any bias introduced by deliberately ignoring paraphyletic taxa should be greatest at high R and least at low R. Simulations corroborate these predictions by showing that excluding paraphyletic taxa raises stratigraphic debt, especially at high R (Fig. 3F). These results also corroborate the view that possible paraphyly should be taken into account when tallying stratigraphic debt (e.g., Fisher 1991; Smith 1994; Clyde and Fisher 1997).

Several studies have attempted to compare the qualities of fossil records among different groups using metrics based on stratigraphic debt (e.g., Hitchin and Benton 1997). However, interpreting these results as reflecting sampling alone requires assumptions about numerous nuisance parameters. Thus, multiple explanations are possible for any significant differences among groups.

In summary, simulations indicate that nuisance parameters affecting quantitative assessments of sampling also affect tree-based assessments of sampling. In addition, these simulations and others (e.g., Wagner 2000a) indicate that parameters that do not have obvious effects on quantitative methods—such as taxonomic philosophy, assumptions about monophyly, and phylogenetic accuracy—also affect tree-based assessments of sampling. Quantitative tests can be amended to relax assumptions concerning random sampling (Marshall 1997) or to include tests for extinction rates (Foote 1997; Solow and Smith 1997) and even sampling consistency over time or across a clade (e.g., Wagner 2000c). Because of this, one also can use such tests to examine how these nuisance parameters affect apparent stratigraphic debt within a particular clade. In fact, when comparing debt among numerous clades (e.g., Hitchin and Benton 1997), quantitative tests will become necessary to test possible explanations about why differences among clades exist.

Some Implications of These Results.—Tree-based sampling metrics depict similar sampling levels among Paleozoic and post-Paleozoic clades (Benton et al. 2000). Benton et al. suggest that when we do sample Paleozoic

strata, we sample species as well as we do in post-Paleozoic strata. These simulations suggest a related explanation: Paleozoic sampling rates are more heterogeneous than post-Paleozoic rates, which would improve gap-metrics regardless of the level of sampling within the well-sampled intervals. The simulations also suggest another explanation. Benton et al.'s study assumes that all taxa are monophyletic. However, a wide range of sampling intensities become indistinguishable if one deliberately excludes paraphyletic taxa (which are difficult to recognize without species-level analyses) or if one simply assumes that all sampled taxa are monophyletic. (Few if any of the taxa used in their study can be stem-based, because low-level phylogenetic analyses of the taxa do not exist.) Empirical estimates of sampling intensity (e.g., Foote 1997; Wagner 1997, 1998, 1999) fall well within the blurred range. Thus, initial assumptions might have precluded Benton et al. from finding significant differences between Paleozoic and post-Paleozoic clades.

Modeled versus Empirical Tests of Sampling Hypotheses.—Norell and Novacek (1992a) suggest that tree-based assessments of the fossil record are empirical whereas probabilistic tests represent modeling. A primary theme of this paper is the elimination of modeling in favor of empirical tests whenever possible, so if their view is correct, then we should prefer tree-based metrics. However, probabilistic approaches require making predictions from hypotheses and then examining whether data deviate appreciably from those predictions. This approach epitomizes empirical hypothesis testing. Because unknowns other than the hypothesis in question affect probabilistic predictions, one might conclude that sampling is significantly worse for one taxon than for another when variation in R is what really differed. However, this simply represents a failure of workers to test all available explanations adequately, not the failure of a general research program. Moreover, antidotes for such problems exist. One can use probability analyses to test multiparameter hypotheses that include hypotheses about unknowns such as extinction intensity (Solow 1993; Foote 1997; Solow and Smith 1997) or sampling consistency (Wagner 2000c) as well as hypotheses about sampling intensity. Also, probabilistic tests exist that can relax assumptions about some unknowns (e.g., random sampling [Marshall 1994, 1997]).

The simulations above demonstrate that parameters besides sampling intensity affect tree-based inferences of sampling. Most critically, tree-based sampling inference requires assuming that an unknown (the phylogeny) is true. Using an assumed unknown to make inferences (or to attempt tests) about another unknown (the quality of the fossil record) epitomizes modeling. Because probability can both test nuisance parameters and the hypothesis of interest, and because assumed phylogenies are not required, probability analyses actually are more empirical and less model-dependent than are tree-based analyses.

Testing Historical Hypotheses with Probability and Stratigraphy: A Critical Review of the Criticisms

Workers have devised several quantitative tests of hypothesized phylogenies with stratigraphic data (Fisher 1991, 1994; Cheetham and Jackson 1995; Wagner 1995, 1998; Huelsenbeck and Rannala 1997). Workers also have proposed quantitative methods for testing phylogenetic hypotheses with character data (see Huelsenbeck and Crandall 1997). Such approaches have focused on molecular data, but workers have proposed quantitative tests using morphologic data also (Felsenstein 1988; Wagner 1998).

Criticisms of phylogenetic tests using probability and/or stratigraphic data include the following: (1) probabilistic methods (including those using stratigraphy) seek to verify hypotheses rather than falsify hypotheses; (2) probability cannot be applied to historical hypotheses; (3) only hierarchically distributed data offer information about phylogenetic relationships; and (4) stratigraphic gaps are "negative" evidence. Because of these allegations, it is necessary to review the different arguments before discussing maximum likelihood tests of phylogenetic and character evolution hypotheses using morphologic characters and stratigraphic data.

1. *Parsimony Analyses Are Falsificationist*

Whereas "Stratigraphic"/Likelihood Methods Are Verificationist.—Popper (1959: pp. 76–77) explicitly claims to model falsificationism after modus tollens deduction (i.e., "if p then q"; "not q"; therefore "not p"). However, Popper considered only "universal" statements (e.g., "if p then q") to represent hypotheses (Hull 1983). Thus, falsification really represents predicate logic instead of propositional logic and should be stated: "'if p' and 'if p then q'"; "p but not q"; therefore "not 'if p then q.'"

Popper models verification on inductive logic ("all are p"; "all have q"; therefore "if p then q"). However, other workers use induction to describe a different argument, i.e., "if p then q; q; therefore p," where "p" is the hypothesis/explanation, not "if p then q" (e.g., Newton-Smith 1981; Lipton 1991). This argument is orthogonal to both deduction and induction and is labeled "abduction" (Peirce 1878). The difference between induction and abduction is essentially the difference between suggesting that all elephants are gray (induction) and hypothesizing that an animal in an elephant (abduction). However, both abductive and inductive arguments present conclusions supporting some unknown. Conversely, deductive statements present conclusions that potentially reject (but never truly support) some unknown.

Several workers argue that MSTs falsify phylogenetic hypotheses whereas trees derived from likelihood and "stratigraphic" methods attempt to verify phylogenetic hypotheses (Rieppel 1997; Siddall and Kluge 1997). However, Hull (1983) notes that phylogenies do not meet Popper's criteria for a falsifiable hypothesis because phylogenies represent specific explanations for observations rather than universal explanations. Hull also notes that parsimony is not a Popperian criterion for falsifying hypotheses. O'Keefe and Sander (1999) observe that a strict Popperian would not accept cladistic falsifications because cladistic characters do not meet Popper's criteria for observations (see below).

Additional disagreements exist concerning the logical frameworks of different methods. Sober (1988: p. 50) argues that MSTs are inferences about, not tests of, phylogeny. Because one must assume a general relationship between synapomorphies and homology (i.e., some "if p then q" statement), the inference is abductive. On the other hand, probability and likelihood are derived, in part, from combinatorial logic, which in turn is derived from deductive logic (Grimaldi 1994). Like other deductive arguments, probability tests reject (or fail to reject) hypotheses. Edwards (1992: p. 211) notes that likelihood approaches such as the method of support are "not greatly at variance" with the goals of falsification because likelihood tests examine the amount of evidence against alternative hypotheses. Finally, proponents of methods utilizing stratigraphy, such as Fisher (1994) and Wagner (1995), argue that stratigraphic data provide evidence against hypothesized phylogenies rather than evidence supporting them. Assessing these claims and counter-claims requires contrasting the logical arguments implicit to each approach.

Platnick (1977) suggests that MSTs falsify other trees. In a logical framework, Platnick's argument is as follows:

Premise 1: Implied homoplasies are evidence against a hypothesized phylogeny.

Observation: MST has the fewest implied homoplasies.

∴ MST has the least evidence against it.

Even if we expand falsification to include standard modus *tollens* arguments, this still does not represent a deductive statement. Platnick's "weak falsificationist" (Sober 1988: pp. 125–126) argument lacks an explicit premise relating phylogenies to homoplasies and thus is ad hoc in formal logic. One can formalize Platnick's argument as a likelihood argument (see also Farris 1973; Felsenstein 1981b). However, this will require explicit assumptions about rates of character change (see below).

Phylogenetic hypotheses implicitly include hypotheses about homologies. Patterson (1982) suggests that all homologies are congruent (i.e., share a common origin on an MST). (Patterson also suggests that all homologies are conjunct [i.e., located in similar organismal locations] and similar [i.e., coded identically in some way]; however, as homoplasies also are conjunct and similar, these premises are not unique predictions for testing phylogenetic hypotheses.) Patterson's premise permits the following deductive argument for testing whether shared states are homologies:

Premise 1: Homologous are congruent (= synapomorphies on an MST).

Observation: Shared states are incongruent (= not synapomorphies on any MST).

∴ Shared states are not homologies.

Trees longer than the MST(s) require that some incongruent shared states are homologies. Because we assume that homologies cannot be incongruent, we must conclude that a longer tree cannot be the phylogeny. If there is a single MST, then deductive inference becomes possible via deductive elimination ("disjunctive syllogism"): given finite alternatives, falsification of all but one hypothesis necessarily confirms that final hypothesis.

Patterson's argument is logically valid, as the conclusion necessarily follows given the assumptions. However, O'Keefe and Sander (1999) note that the observable condition of identically coded states (congruent or incongruent) is determined in part by evolutionary patterns among other characters. Thus, congruence is not a property of characters, but is instead a property of character matrices. A more critical problem is that Patterson's premise (i.e., if homologous then congruent) is demonstrably unsound. Numerous simulation (e.g., Rohlf et al. 1990; Huelsenbeck 1991; Mooers et al. 1995) and laboratory (Atchley and Fitch 1992; Hillis and Huelsenbeck 1992) studies indicate that homologies can be incongruent on MSTs, especially when frequencies of homoplasy are high. Simulations indicate that the "if homology then congruent" premise becomes increasingly unsound as character evolution becomes complicated (varying rates of change, nonindependence among characters, etc.) (Kuhner and Felsenstein 1994; Wagner 1998; Huelsenbeck and Nielsen 1999). Finally, studies that reveal inconsistent patterns of congruence in different character sets (e.g., Shaffer et al. 1991; Brochu 1997) empirically demonstrate that the premise is not universally true and thus falsify a hypothesis that all homologies will be congruent. Because a conclusion is valid only if each premise is true, we no longer have a valid deductive argument concerning homologies or phylogenies.

Patterson (1982) himself notes that the "if homology then congruent" premise is not guaranteed to be true, but that homology is the most "probable" explanation for congruence. This concession renders the "test of congruence" a qualitative Bayesian inference in which the prior probability of homology is considered greater than the prior probability of homoplasy. Jefferys and Berger (1992) explore the relationship between parsimony and Bayesian inference in detail.

Deriving a sound and valid deductive test for hypothesized phylogenies is not possible for one simple reason: phylogenies do not by themselves determine character state distributions. As such, they always offer underdetermined (and therefore probabilistic) premises (see discussion of "determined" and "underdetermined" hypotheses below). This observation does not deny that historical patterns are the product of deterministic processes (contra Rieppel 1997). Instead, it denies that phylogeny alone determines that pattern.

Sober (1988) suggests that parsimony analyses are abductive inferences about phylogeny, not deductive tests. Sober's abductive argument is the following:

Premise 1: Homology explains congruence.
Observation: Similarities are congruent.

∴ Hypothesize that similar features are homologies.

Peirce (1878) describes this as the logical framework for hypotheses such as "the animal is an elephant" or "the characters are homologous." Such an argument avoids logical fallacy because the conclusion is a hypothesis, not an assertion. Two points bear noting here. First, the argument that character data better support short trees than they do long trees is opposite of arguing that data contradict a hypothesis. Second, the use of parsimony to generate null hypotheses is how non-systematists use parsimony (see, e.g., Hoffmann et al. 1997): if multiple hypotheses remain untested, then the simplest one is favored as a confession of ignorance. However, the nonparsimonious alternatives are not considered to be rejected or falsified in any way.

"Stratigraphic" and likelihood approaches typically follow logical arguments different from Sober's parsimony argument. Fisher's (1991, 1994) stratocladistics uses an amended form of Platnick's argument:

Premise 1: Implied homoplasies are evidence against a hypothesized phylogeny.
Premise 2: Implied stratigraphic gaps are evidence against a hypothesized phylogeny.

∴ Tree with fewest combined gaps and homoplasies has the least evidence against it.

Like Platnick's argument, stratocladistics attempts to gather evidence against a hypothesis. Also like Platnick's argument, stratocladistics lacks explicit premises relating the data to the hypotheses. However, stratocladistics can be formalized as a simple likelihood test (Wagner 1999). This type of approach will be further developed below.

Alternative phylogenetic methods, including several "stratigraphic" methods, have explicit premises based on probability and likelihood. Among many philosophers of science, induction has been redefined to include all arguments in which the conclusions are not necessary (e.g., Martin 1997). Under such a definition probability and likelihood qualify as inductive (Wiley 1975). However, the criticisms of inductive logic are of the direction of the argument (i.e., seeking support for hypotheses rather than evidence against such hypotheses) and are based on the logicians' definition given above. The direction of probability arguments is the same as that of deduction. Probability can reject a hypothesis as implausible if data deviate from predictions, but probability does not measure how "true" the hypothesis is. Similarly, a high likelihood tells us only that the predictions of a hypothesis are met, not that it is true. Whether one calls probability deductive or inductive is unimportant. However, labeling parsimony inductive does not make probability "verificationist" any more than labeling parsimony "weak falsification" makes parsimony a hypothetico-deductive approach.

Necessary and possible predictions distinguish Boolean (determined) logic from probabilistic (underdetermined) logic, not deduction from induction (or abduction). Probability arguments are deductive arguments in which the "if p then q" statements have values from one (true) and zero (false) instead of either zero or one. ("Fuzzy" logic [Zadeh 1965] takes this one step further by assigning a fraction to observations also.) Boolean deduction actually is a special case of probability where the probabilities of data given a hypothesis are either one or zero. In such cases, the likelihoods of hypotheses given data also are one or zero. Moreover, this special case rarely if ever applies to real world problems. Hume's (1748) observation that no amount of data can prove an "if p then q" statement (which often is offered as an argument against inductive reasoning) means that the assumptions of Boolean deductive arguments cannot be verified. Thus, nearly all deductive statements in science as probabilistic ones.

Several workers (Marshall 1990; Paul 1992; Cheetham and Jackson 1995; Wagner 1995) test phylogenetic hypotheses using confidence intervals on stratigraphic ranges (Strauss and Sadler 1989). Their argument is essentially a fuzzy predicate modus tollens:

Premise 1: if congruent shared states are homologous among taxa, then taxa linked by those states span some duration of time.

Premise 2: Some mathematically estimated subset of all possible stratigraphic data (i.e., distributions and numbers of finds and absences) is consistent with that hypothesized temporal duration.

Observation: Distributions/numbers of finds/misses fall outside that subset.

∴ The hypothesized gap is implausible.
∴ The congruences (in this instance) are unlikely to be homologies.

The argument here is a hypothetico-deductive one: character data are used to infer hypotheses and stratigraphic data are used eliminate those hypotheses from further consideration.

In Boolean logic, one can separate hypotheses that predict observed data from those that forbid the same data. This represents a special case of a likelihood argument in which the probability of data given a hypothesis is

either zero or one. Two important points emerge here. First, multiple hypotheses can have likelihoods of one. This only means that more than one hypothesis exactly predicts the observed data. Second, likelihood best describes the evidence against a hypothesis given data. A likelihood of one means that the data offer no evidence against a hypothesis. This represents corroboration but not confirmation, because other hypotheses also might have high likelihoods. Conversely, a likelihood of zero means that the data are entirely at odds with the hypothesis.

In reality, probabilities and likelihoods will be between zero and one. However, data usually will deviate far more from the predictions of some hypotheses than from the predictions of others. Likelihood tests, using the "method of support," can reject the former class of explanations. Huelsenbeck and Crandall (1997) provide examples where particular phylogenetic hypotheses are significantly worse than others are given the observed character data. Stratigraphic data provide another basis for doing this (Huelsenbeck and Rannala 1997).

One also can use likelihood as an inferential method to find a single best-supported tree (e.g., Felsenstein 1973). However, the single most likely tree usually will not reject numerous alternative trees of nearly equal likelihood. Edwards and Cavalli-Sforza (1964) originally presented minimum evolution (which includes minimum-steps parsimony) as a maximum likelihood estimate given several simplifying (and biologically unrealistic) assumptions (see also Farris 1973; Felsenstein 1981b; Edwards 1996). This offers formal premises for parsimony that are lacking in philosophical justifications. Stratocladistics also is equivalent to a simplified likelihood estimate in which gaps were considered as improbable as morphologic changes (Wagner 1999).

2. *Historical Data Cannot Be Analyzed with Probability.*—Kluge (1997; also Siddall and Kluge 1997) claims that historical hypotheses describe unique events and that probability cannot be applied to unique events. However, phylogenetic studies analyze categorized events, not the "unique" events themselves. Molecular data present the most obvious examples. All adenines (A) are identical to all

other adenines when one is analyzing DNA. The "historian's" argument that two A's derived separately are not truly the same A is irrelevant when estimating the probability of evolving an A. Kluge's argument therefore is irrelevant to phylogenetic analysis when one is evaluating the evidence against a tree presented by an additional posited derivation of A. The same holds for morphologic data. For example, "ornament present" among gastropods might (and usually does) include numerous derivations of that category.

Of course, if one could discern historically unique events in advance, then discussion about probability and likelihood would be irrelevant because such methods would be unnecessary. If each character state reflected a unique event, then every character matrix would show perfect hierarchical structure (i.e., all characters would be compatible with one another and all would have consistency indices of 1.0 on the shortest trees). Unfortunately, such ideal data sets are nonexistent in the real world (Archie 1989; Sanderson and Donoghue 1989).

A second flaw in Kluge's argument is that it apparently confounds probabilities, which measures the soundness of an "if p then q" statement, with frequencies, which measure observed q's given p. As discussed above, the correct criterion for applying probability is whether the premises are underdetermined. As also noted above, phylogeny alone does note determine character state distributions. Thus, probability is necessary to predict character state distributions.

Kluge (1997) also claims that historical hypotheses require "retrodictions" whereas probability deals only with predictions. However, prediction and retrodiction are synonymous in logic. In both cases, they reflect a premise stating what data should be if a hypothesis is true. Such statements are agnostic with respect to time. Thus, Kluge's distinction is entirely artificial.

3. *Hierarchical versus Linear Information.*— Smith (1994; also, Rieppel and Grande 1994) argues that because phylogenies are hierarchical structures, information about phylogeny exists only in hierarchically distributed data. Character data represent such data

whereas stratigraphic data do not. Instead, stratigraphic data are "linearly" distributed across a phylogeny—ancestors precede descendants, but the fact that two species are contemporaneous is not evidence that they are more closely related to each other than either is to an older or younger species. Also, whereas character data are inherited from common ancestors, stratigraphic ranges are not (C. Brochu personal communication 1999).

These arguments are appropriate if we restrict phylogenetic methods to those making inferences (i.e., induction/abduction). In such cases, we are interested only in evidence supporting a hypothesis. However, if we include methods attempting tests, then we are concerned with evidence against a hypothesis. Whenever considering whether data offer evidence against a hypothesis, there is only one criterion: whether that hypothesis makes predictions about that data. Thus, if stratigraphic data should be distributed linearly across phylogeny, then nonlinear distributions (i.e., stratigraphic gaps) are evidence against any hypothesized phylogeny (Fisher 1982). Whether one also can infer hypotheses from those data is not relevant.

Huelsenbeck and Rannala (1997) note that one can infer phylogeny from stratigraphic data if all ancestral morphologies become pseudoextinct at speciation and if speciation/extinction events never coincide with one another. However, if speciation follows budding patterns or if multiple speciation/extinction events happen within the limits of temporal resolution, then phylogeny cannot be inferred from stratigraphy alone even with a perfect fossil record.

4. *Stratigraphic Gaps Are Negative Evidence.*— The conclusion offered above might be negated by arguing that stratigraphic gaps are "negative evidence" and thus not information of any sort (e.g., Rieppel and Grande 1994; Heyning and Thacker 1999). However, the definition of negative evidence used by these authors confounds modus tollens with negative evidence. A modus tollens argument makes conclusions when predicted data are not observed when there is opportunity to observe such data. A true negative-evidence argument makes conclusions in the absence of oppor-

tunities in which one could observe corroborating results. In laboratory sciences, opportunities to observe are particular experiments. In paleontology, the opportunities to observe are fossiliferous beds, not occurrences of the species in question.

Taphonomic controls (Bottjer and Jablonski 1988) allow us to separate sampling opportunities from situations where there is no opportunity to sample that taxon. Consider a hypothesis that a gastropod species' duration exceeded its stratigraphic range by a particular amount. This hypothesis predicts that the species should be present in some proportion of gastropod-bearing beds within that duration but absent in gastropod-bearing beds outside that duration. If the species has an aragonitic shell and is found with other aragonitic mollusks within a particular biogeographic unit, then the unit of evidence is not the presence of the particular species. Instead, the units of evidence are beds from the same paleoenvironments within that biogeographic unit and containing those aragonitic mollusks (J. Alroy personal communication in Wagner 1995; Marshall and Ward 1996). Beds do not represent pertinent observations (and thus do not represent absences) if those beds (1) are from different facies, (2) are taphonomically/diagenetically altered, (3) include no aragonitic mollusks, or (4) are from different biogeographic units (see Wagner 1995). Fisher (1994) also recognizes the importance of taphonomic controls when stating that only intervals known to bear members of a clade should be used when tallying stratigraphic debt.

Simultaneously Testing Hypothesized Phylogenies and Hypothesized Character Evolution with Characters and Stratigraphy: A Likelihood Approach

The Utility of Likelihood for Testing Hypothesized Phylogenies and Associated Nuisance Parameters.—Empirical hypothesis testing requires that a hypothesis make explicit predictions about data. As noted above, phylogenetic hypotheses do not make predictions about character state distributions without invoking nuisance parameters. Felsenstein (1973: Table 1) identifies 19 factors besides phylogeny and 768 combinations of those factors that affect

predictions about character state distributions. Examples include relative rates of change, consistency of those rates over time and across phylogeny, reversibility of character states, and the order of character state evolution. Felsenstein further notes that his list was by no means exhaustive. One option for dealing with such parameters is to make explicit assumptions about them and thus treat character evolution and other unknowns as models (Edwards and Cavalli-Sforza 1964; Felsenstein 1973). Another option is to integrate likelihoods over all possible values for the nuisance parameters and thus consider all possibilities (Barnard et al. 1962). A third option is to test hypotheses about phylogeny and other parameters simultaneously (e.g., Yang 1994). The second and third approaches are explored here.

Some workers argue that parsimony analyses with equally weighted characters and unordered states assume only that phylogeny explains character congruence (e.g., Brady 1985; Kitching et al. 1998), which would render Felsenstein's other factors unimportant. However, equating congruence with homology when congruence is based on equally weighted, unordered characters implicitly assumes that all characters and all character states offer equal evidence for or against a tree. One can derive this assumption validly only if one assumes (among other things) that all character and character state changes are equally plausible, and thus that all characters and character states evolve at the same rates (Felsenstein 1981b; Sober 1988; Swofford and Olsen 1990; Wheeler 1990; Maddison and Maddison 1992). If, as Felsenstein and others maintain, weighting and ordering make assumptions about how characters evolved, and if it is not possible for a method to test alternative weighting and ordering schemes, then character weights and orderings are models in cladistic analyses.

The Likelihood of a Hypothesized Phylogeny Given the Most Likely Hypothesis of Character Evolution for That Phylogeny.—Likelihood tests of character evolution hypotheses require determining the probability of deriving an observed distribution of character states. Fundamentally, this is done on a branch-by-branch basis. Although likelihood analyses assume that stochastic processes generate character state distributions, they do not assume a random distribution of data. Instead, likelihood assumes that characters and character states are distributed according to a Markov process. In other words, states at any point in the tree are determined partly (but never entirely) by states in immediate lower (ancestral) regions. Given a pair of sister taxa, there are some reconstructed ancestral conditions and some rate(s) of character change that maximize the probability of the observed states for those two taxa. It follows that there is a set of reconstructed ancestral conditions and mode(s) of character change across any tree that maximize the probability of the observed states for all taxa.

If character evolution occurs continuously, then the likelihood of a branch (br) and a rate of change (π) given inferred changes for any character i is given by a Poisson distribution (Felsenstein 1973):

$$L[\text{branch}, \pi \mid \partial_i] = \frac{\pi t^{\partial_i} e^{-\pi t_{br}}}{\partial_i!} \qquad (1)$$

where π gives the probability of change per unit time, t, t_{br} is the duration spanned by branch br, and ∂_i, is the net number of changes. This simply gives the Poisson probability of ∂_i changes given πt expected changes. Note that this tests a hypothesis concerning a constant rate of change and thus might be rejected if rates vary over time.

Several workers suggest that character evolution is concentrated into particular intervals (e.g., Wright 1931, 1932; Mayr 1963; Eldredge 1971; Eldredge and Gould 1972). If so, then the likelihood of a branch (br) and a rate of change (π) given inferred changes for any character i is given by a binomial distribution:

$$L[\text{branch}, \pi \mid \partial_i]$$
$$= \binom{t_{br}}{\partial_i} \times \pi^{\partial_i} \times (1 - \pi)^{t_{br} - \partial_i} \qquad (2)$$

where t_{br} now gives the number of speciation events. Note that ∂_i now cannot exceed t_{br}, as is appropriate if change occurs only during speciation. Equations 1 and 2 will converge when t is great and/or π is small, but will differ appreciably when t_{br} is low and π is great. Note also that recognizing punctuated or gradual patterns makes few (if any) assumptions about speciation processes, as processes that can yield phyletic gradualism also can yield punctuated patterns (Wright 1931; Lande 1975; Newman et al. 1985).

For multiple independently evolving characters, the likelihood of the branch is simply the product over all characters:

$$L[\text{branch}, \pi \mid \Delta_{br}]$$
$$= \prod_{i=1}^{\text{characters}} L[\text{branch}, \pi \mid \partial_i] \qquad (3)$$

where $L[\text{branch}, \pi \mid \partial_i]$ is either equation 1 or 2, Δ_{br} is an array of character changes along the branch (br), and X is the total number of characters that might change. X is unknown, but it can be estimated for a given topology using the per-step frequency of new character generation. A best-fit number can be derived from rarefaction equations (Hurlbert 1971; see Wagner 2000b).

The likelihood of π and the hypothesized path of character evolution, τ, is

$$L[\tau, \pi \mid \Delta_1 \ldots \Delta_Y]$$
$$= \prod_{br=1}^{\text{branches}} \prod_{i=1}^{\text{characters}} L[\text{branch}, \pi \mid \partial_i] \qquad (4)$$

where $L[\text{branch}, \pi \mid \partial_i]$ again is either equation (1) or (2) and Y is the number of branches. In other words, the likelihood of π and τ is simply the product of the probability of each character changing or remaining static along each branch of the tree.

Testing for Heterogeneous Rates of Change among Characters.—The simplest (and therefore null) hypothesis of character evolution posits a single π for all characters across the phylogeny (Jukes and Cantor 1969). One can test this relative to a hypothesis that π's vary among characters (i.e., a multiparameter hypothesis M). The likelihood of τ and M is given by

$$L[\tau, M_Z | \Delta_1 \ldots \Delta_Y]$$

$$= \prod_{br=1}^{branches} \prod_{j=1}^{rate\ sets} \prod_{i=1}^{chars\ in\ rate\ set\ j} L[branch,\ \pi_j | \partial_i]$$

$$(5)$$

where $L[branch,\ \pi_j | \partial_i]$ again is either equation 1 or 2, M_Z is a hypothesis positing Z different π's and X_j is the number of characters evolving under rate π_j (Hasagawa et al. 1985). Note that j now gives the number of a character within a set of characters sharing π_j. Thus, equation (4) is simply a special case of equation (5) in which $Z = 1$. Thus, the likelihood of a phylogeny (i.e., branches 1 through Y) and a collection of rates (i.e., π_1 through π_Z) is simply the probability of each character changing or remaining static along each branch. For any given number of rates (Z), this will be maximized for a particular set of rates and a particular assignment of characters to those rates.

If characters necessarily change different numbers of times, then the likelihood of τ and M almost always increases as Z increases (i.e., π's are added). There can be as many π's as characters (i.e., $Z = X$) or even more if shifts in rates across individual branches are considered. However, we reject one rate in favor of 2+ rates only if it yields a significant increase in likelihood. One assesses this using a log-likelihood ratio statistic:

$$\Lambda = 2 \times (\ln L[\tau, M_Z | \Delta_1 \ldots \Delta_Y]$$
$$- \ln L[\tau, M_{Z-k} | \Delta_1 \ldots \Delta_Y]) \quad (6)$$

where Z and $Z - k$ represent the number of different π's posited by M_Z and M_{Z-k}. The test statistic (often labeled G or δ) is evaluated by a chi-square distribution with degrees of freedom determined by the difference in parameters (i.e., k) (Sokal and Rohlf 1981: p. 695) *if* the simpler hypothesis represents a special case of the more complicated hypothesis. The hypothesis that characters A and B have the same rate of change is a special case of a set of hypotheses where A and B have their own rates, which might or might not be equal. However, if the phylogenetic topologies differ, then the two hypotheses do not represent spe-

cial cases of one another (Goldman 1993). This is discussed in greater detail below.

The test treats hypotheses with fewer varying parameters as null hypotheses relative to hypotheses with more varying parameters. Thus, even though more complicated hypotheses always provide better explanations than simpler ones, we prefer the simpler hypotheses until data allow us to reject the simpler hypotheses. As noted above, this is the usual manner in which hypothesis-testing uses parsimony.

Accommodating Uncertainty about Unsampled Ancestral Morphotypes.—One of two major differences between parsimony and likelihood concerns character states for unsampled (hypothetical) ancestors. Whereas parsimony attempts to assign particular states to hypothetical ancestors, likelihood need not do so (Felsenstein 1973, 1981a; see also Pagel 1994, 1999; Schluter et al. 1997; Mooers and Schluter 1999). Consider a simple hypothetical example with one binary character and three taxa, none of which are considered ancestral to another (Fig. 4a). Given the two hypothetical ancestors, there are four possible histories, each of which has some probability of producing the observed distribution of character states. The likelihood of the tree and/or character evolution hypothesis is the sum of the probabilities, not the likelihood of any one path.

The number of possible histories is states[h], where h is the number of hypothetical ancestors. A three-state character in Figure 4 would have nine possible histories. Alternatively, a four-taxon tree (Fig. 4b) with no sampled ancestors would have eight possible histories for a binary character. However, if a phylogeny posits that a sampled species is ancestral to another sampled species, then the "nodal" states obviously are identical to the states of that species and the number of possible histories is decreased.

Uncertainty about ancestral character states has strong and somewhat counterintuitive implications for the likelihood of a tree (see also Pagel 1999: p. 620). Consider an extremely simple example, where each branch in Figure 4a represents a single speciation event and morphologic change occurs only during speciation. Two variables become important, the

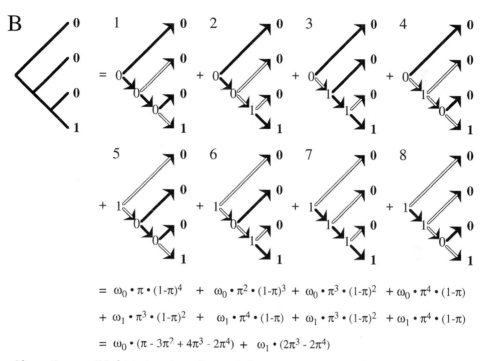

$$= \omega_0 \cdot \pi \cdot (1\text{-}\pi)^3 + \omega_0 \cdot \pi^2 \cdot (1\text{-}\pi)^2 + \omega_1 \cdot \pi^2 \cdot (1\text{-}\pi)^2 + \omega_1 \cdot \pi^3 \cdot (1\text{-}\pi)$$

$$= \omega_0 \cdot \pi \cdot (1\text{-}\pi)^2 + \omega_1 \cdot \pi^2 \cdot (1\text{-}\pi)$$

$$= \omega_0 \cdot \pi \cdot (1\text{-}\pi)^4 \quad + \quad \omega_0 \cdot \pi^2 \cdot (1\text{-}\pi)^3 + \omega_0 \cdot \pi^3 \cdot (1\text{-}\pi)^2 + \omega_0 \cdot \pi^4 \cdot (1\text{-}\pi)$$

$$+ \quad \omega_1 \cdot \pi^3 \cdot (1\text{-}\pi)^2 \quad + \quad \omega_1 \cdot \pi^4 \cdot (1\text{-}\pi) \quad + \omega_1 \cdot \pi^3 \cdot (1\text{-}\pi)^2 \quad + \omega_1 \cdot \pi^4 \cdot (1\text{-}\pi)$$

$$= \omega_0 \cdot (\pi - 3\pi^2 + 4\pi^3 - 2\pi^4) + \omega_1 \cdot (2\pi^3 - 2\pi^4)$$

FIGURE 4. Alternative possible histories given observed character states among three (A) and four (B) taxa. The probability of the observed data given the cladogram is not simply the probability of one change distributed over the five branches, but the sum of probabilities for all possible histories. The equations are tree likelihoods given some probability of change, π, and some probability for the initial condition, ω, while assuming that each branch represents one speciation event and character change is punctuated. The most likely π is given when the derivative of these equations (i.e., $[1 - 2\pi]$ for A and $[1 - 6\pi + 18\pi^2 - 16\pi^3]$ for B) equal zero. For A, this is when $\pi = 0.5$. For B, this is when $\pi = 0.727$. Assuming continuous change along branches and variable branch lengths increases the most likely rates (see text).

probability of change per branch (π) and the probability that the initial character state is either 0 (ω_0) or 1 (ω_1). If we considered only the most parsimonious alternative (Fig. 4a.1), then $\omega_0 = 1$ and the likelihood of the tree is $\pi \cdot (1-\pi)^3$. The likelihood is maximized when the derivative of the likelihood equation equals zero (Edwards 1992). In this case, that is when $\pi = 0.25$. However, the ancestral conditions are unknowns, not data. Even if we

continue to assume that $\omega_0 = 1$, the likelihood of the tree is the probability of tree 4a.1 and tree 4a.2. The likelihood now is $\pi \cdot (1 - \pi)^2$, which is maximized when $\pi = 0.33$. Finally, we must consider that ω_1 might be greater than 0. If $\omega_0 = \omega_1 = 0.5$, then the likelihood of the tree is $\pi \cdot (1 - \pi)$, which is maximized when $\pi = 0.5$.

The situation becomes worse as the number of taxa increases. If we assume $\omega_0 = 0$ with four taxa (tree 4b), then the likelihood of the tree is sum of trees 4b.1 – 4b.4. This is maximized when $\pi = 0.5$. If $\omega_0 = \omega_1 = 0.5$, then the likelihood is maximized when $\pi = 0.72$. In other words, the tree's likelihood is maximized when change is nearly three times more probable than stasis!

If we assume that $\omega_0 = 0$, then a multistate character yields the same likelihood, because the probabilities depend on whether hypothesized ancestral states match the observed states. However, if we assume that $\omega_0 = 1/$ states, then the tree's likelihood is maximized at some $\pi > 0.5$. This is because we essentially grant higher prior probability to evolutionary scenarios requiring greater numbers of changes whose likelihoods increase as rates of change increase.

Assuming continuous change also worsens the situation, because we now must consider the possibility of multiple changes per branch yielding zero or one net changes (see below). Although the individual likelihoods of these scenarios are low even under high rates, they greatly outnumber the scenarios with few changes. This results in high rates receiving higher likelihoods.

Continuous change reveals why high rates have high likelihoods. There are more scenarios requiring many changes than there are scenarios requiring few changes, and this proportion increases as the number of taxa increase. The likelihoods of the many "multichange" hypotheses are greatest when hypothesizing high rates of change. However, such rates are quite implausible when one considers additional data. Under such high rates of change, there should be very little hierarchical structure in character data. Metrics such as compatibility (Le Quesne 1969; Estabrook et al. 1975) can assess hierarchical signal (Meacham 1984; Alroy 1994). Simulated evolution of matrices with rates of change set at 0.25 per branch produces characters with much less compatibility than real characters from most real matrices of the same size (Wagner unpublished results). Such information must be used either to expand the data used to calculate likelihoods or to produce prior probabilities for Bayesian analyses.

Given these difficulties, these analyses will use not all possible ancestral conditions, but the most likely collection of ancestral states given an assumption that probabilities of change are less than 0.5 per speciation. For the examples in Figure 4, this will match the parsimony optimization (e.g., trees 4a.1 and 4b.1), especially if we root the character states using outgroups. However, when one accounts for variable branch lengths (see next section), then nonparsimonious ancestral conditions might be more likely than parsimonious ones. This is most plausible when inferred ancestral morphologies are sampled long after the appearances of inferred descendant morphologies. Unfortunately, failing to integrate the likelihood over all possible ancestral character states means that likelihood ceases to be statistically consistent (D. Swofford personal communication 2000). Rectifying this problem will be left for future work.

A second reason for using the most likely set of ancestral conditions is somewhat peculiar to paleontological data. With many fossil records, we probably sample ancestral morphotypes. Observed ancestors reduce the nodes on which multiple ancestral conditions might be considered, which reduces the likelihood of a tree. Part of the purpose of the tests proposed herein is to contrast hypotheses of ancestor-descendant relationships with simple sister-species relationships. Thus, explicit hypotheses of character state evolution need to be tested.

Accommodating Variable (Temporal) Branch Lengths.—Whereas parsimony assumes that character change probability is equal on all branches, likelihood assumes that character change probability is greater on branches spanning long periods of time than on branches spanning short periods of time (Felsenstein 1981b). Several studies support the assumption that the durations of branches (t_{br}) predict amounts of morphologic change along those branches (Smith et al. 1992; Jackson and Cheetham 1994; Sidor and Hopson 1998). Accounting for variation in branch durations across any hypothesized tree therefore will affect the likelihood of that tree and any hypothesis of character evolution. If we assume that each branch encompasses the same amount of evolution (i.e., a single t_{br}) and that all characters change independently, then the most likely tree will be the one with the fewest changes. This will be true even if a single parameter M is rejected. However, unless sampling is 100% complete, the amount of evolution (in time or speciation events) probably will vary along each branch. The amount of evolution along a branch affects the predicted amount of character change along that branch (Felsenstein 1978). Inferring t_{br} is a particularly vexing problem for fossil taxa. Given continuous change and an (A,(B,C)) relationship among contemporaneous taxa, evolution along the branch leading to A should approximately equal the amount along the branches leading from the base of the tree to either B or C. However, sampling taxa from different points in time invalidates this conclusion. For example, if A is ancestral to B and B is ancestral to C, then there should be no changes on the branches leading directly to A and B. Conversely, if A is first sampled after B and C, then we should expect more change along that branch than along the others.

Patterns of character change further complicate inferring t_{br}. If character change is continuous, then one can use the temporal duration of an implied stratigraphic gap for t_{br}. However, if character change is punctuated ("speciational"), the t_{br} is the number of speciation events along a branch. Whether those truly constitute speciation events actually is irrelevant—phylogenetic analyses (as well as tests concerning sampling, "speciation," and extinction intensities) of fossil taxa almost inevitably use novel combinations of characters rather than other species concepts (e.g., a biological species concept [Mayr 1942]). Thus, the evolution and subsequent sampling of a new character combination is tantamount to the evolution and sampling of a new fossil species. Also, the use of discrete characters in phylogenetic analyses essentially enforces a punctuated view of speciation and character evolution (Gayon 1990). This certainly is true if single differences are considered sufficient to separate taxa.

Determining the number of unsampled species along any given branch is not straightforward. One alternative approach is to treat the number of speciation events as an additional

hypothesis. Given G, the time elapsed along a branch implied by stratigraphy, average extinction rates can be used to estimated the probability of t_{br} species spanning G along branch br:

$$P[t_{br}|G, \mu) = \frac{(\mu G)^{t_{br}} \times e^{-\mu G}}{t_{br}!}. \tag{7}$$

Equation (7) introduces yet another nuisance parameter, μ, the extinction intensity. Foote (1997) presents methods for testing hypotheses about μ and sampling intensity, R, given stratigraphic range data (SR). The probability of two independent observations given a single hypothesis is simply the product of those probabilities. Thus,

$$P[t_{br}, SR \mid G, \mu, R) = P''t_{br} \mid G, \mu)$$
$$\times P[SR \mid R, \mu). \tag{8}$$

This value will be maximized at a different μ for each possible t_{br} given a particular G.

Multiple speciation events along a branch (or extended continuous change along a branch) means that single characters might change multiple times along a branch. Given any number of steps, the proportion of possible paths that lead from one state to one or more other states and back to the original state is

$$\frac{\sum_{i=1}^{steps-1} -1^i \times (states - 1)^i}{(states - 1)^{steps}} \tag{9}$$

where the denominator is the number of possible outcomes and the numerator is the number of outcomes that return to the initial state (see Appendices 1–3). The proportion of paths that lead from one state to another is

$$\frac{\sum_{i=1}^{steps} -1^{i-1} \times (states - 1)^i}{(states - 1)^{steps+1}}. \tag{10}$$

Here the numerator gives the number of possible outcomes that yield some state other than the initial one. However, we are interested in only the $1/(states-1)$th of those outcomes that yield a particular state, so the denominator gives the total number of possible outcomes (as in eq. 9) times (states − 1). When there are 3+ states, this converges on 1/states as steps increase. However, the proportion of paths will always equal one or zero when characters are binary. The final step for calculating the probability of stasis or change given duration G and other data is to account for the uncertainty in t_{br}. This gives

$$P[\partial = 0 \mid \pi, G]$$

$$= \sum_{t_{br}=0}^{t_{br}=\infty} \frac{(\mu G)^{t_{br}} \times e^{-\mu G}}{t_{br}!}$$

$$\times \sum_{steps=1}^{steps=t_{br}} \frac{\displaystyle\sum_{i=1}^{steps-1} -1^i \times (states - 1)^i}{(states - 1)^{steps}}$$

$$\times L[branch, \pi_j \mid steps] \quad \text{and} \quad (11)$$

$$P[\partial = 1 \mid \pi, G]$$

$$= \sum_{t_{br}=0}^{t_{br}=\infty} \frac{(\mu G)^{t_{br}} \times e^{-\mu G}}{t_{br}!}$$

$$\times \sum_{steps=1}^{steps=t_{br}} \frac{\displaystyle\sum_{i=1}^{steps} -1^{i-1} \times (states - 1)^i}{(states - 1)^{steps+1}}$$

$$\times L[branch, \pi_j \mid steps] \quad (12)$$

where $L[branch, \pi_j \mid steps]$ is either equation (1) or (2), This leaves the likelihood of a topology and hypothesis of character evolution as

$$L[\tau, M_Z \mid \Delta 1 \ldots \Delta Y]$$

$$= \prod_{br=1}^{branches} \prod_{j=1}^{rate\ sets} \prod_{i=1}^{chars\ in\ rate\ set\ j} P[\partial_i = x \mid \pi, G]$$

$$(13)$$

where $P[\partial = x \mid \pi, G]$ is equation (11) when there is no net change along a branch and equation (12) when there is change along a branch.

Because likelihood accounts for the possibility of multiple changes along cladistic branches, trees no longer determine the number of steps. Instead, I will report the minimum possible steps for each tree or for particular states.

Testing Hypotheses of Independent Character Evolution.—The calculations above assume that characters change independently. However, if changes in character i sometimes or always induce changes in character j, then $P[\partial_i, \partial_j \mid G] > P[\partial_i \mid G] \times P[\partial_j \mid G]$. Two types of dependence exist: change in character j might be more (or less) probable when character i changes or change in character j might have a different probability when character i has one state than when character i has another state. Thus, hypotheses of character correlation affect predictions about character state distributions (Felsenstein 1973). Correlated character evolution represents a two- (or more) parameter hypothesis whereas independent evolution represents a single-parameter hypothesis. If one hypothesizes that change in character j is more probable when character i changes, then one π_j is used when $\partial_i = 1$ and another π_j is used when $\partial_i = 0$. If an extreme case (where i and j are completely dependent), the $\pi_j = 1$ when $\partial_i = 1$ and $\pi_j = 0$ when $\partial_i = 0$. Alternatively, one might hypothesize that the condition (state) of character i influences the evolution of character j. For example, this could occur when multielement structures that are functional complexes (e.g., skulls or shells) are subject to trends. In this case, one might hypothesize a different π_j for different combined states of i and j. These are π_j^{00} when characters i and j both have state 0, π_j^{10} when character i has state 1 and character j has state 0, π_j^{01} when character i has state 0 and character j has state 1, and π_j^{11} when characters i and j both have state 1. Note this tests a hypothesis of biased gains/loses for one character that is determined by another character (see Sanderson 1993). Substituting the conditional probabilities of change into equation (13) gives the likelihood of a hypothesized topology, general character evolution, and correlated character evolution. Correlated character evolution represents an additional set of parameters that will improve the likelihood of the package of hypotheses. Each hypothesized set of correlated characters adds $N - 1$ parameters, with N being the number of characters in the set (Pagel 1994). A multiparameter hypothesis including correlated characters and an otherwise identical multiparameter hypothesis lacking correlated character hypotheses now can be evaluated by a log-likelihood ratio test with degrees of freedom equal to the difference in parameters. O'Keefe and Wagner (1999) test for nonindependence of character pairs by examining similarity coefficients of pairwise compatibility. This measures the proportion of characters with which two characters are both compatible. Multivariate and Monte Carlo tests sometimes indicate that character pairs have significantly more shared compatibilities than expected if they had been evolving independently. Such sets then can be examined to determine whether a multiparameter hypothesis with correlated change allows us to reject a hypothesis without that correlated change.

Testing "Clock" Hypotheses.—The equations presented above assume that rates remain constant through clade history. The hypothesis that characters have different π's among different taxa (Huelsenbeck and Crandall 1997) represents a more complicated hypothesis. The test here is very similar to tests for different π's among taxa. If character changes are concentrated in particular regions of the phylogeny, then one can calculate the likelihood of the multiparameter hypothesis using two π's for one or more characters. Each change invokes an additional parameter. Because hypotheses changing π for one or few branches rarely will increase likelihoods greatly, such hypotheses rarely will permit the rejection of simpler hypotheses.

Testing Hypotheses about State Evolution within Characters.—Likelihood analyses of nucleotide sequences typically test hypotheses about differential state evolution within characters (e.g., transition:transversion biases) before testing hypotheses about differential evolution among characters (sites) (Felsenstein 1981a; Huelsenbeck and Crandall 1997). This is possible because molecular characters share the same states. Because states among morphologic characters usually have no analogue among other morphological characters, there is no reason to expect analogues of transition: transversion biases among morphologic characters (D. Swofford personal communication 2000). Nevertheless, workers have proposed general models of state evolution such as irreversibility (Camin and Sokal 1965) and "Dollo's Law" (i.e., single gains and multiple losses [Rensch 1960; Farris 1977]). Step-matrices (Sankoff and Rousseau 1975; Maddison

and Maddison 1992) allow less strict variants of these models, where losses/decreases might be easier than gains/increases or viceversa. Driven trend hypotheses (McShea 1994) posit such patterns in morphologic evolution. Sanderson (1993) gives a test that is similar to a test for correlated character evolution and which can reject the simpler hypothesis of unbiased gains/losses. For character i, one contrasts the hypothesis of a single π_i with a hypothesis of $\pi_i^0 \neq \pi_i^1$ (i.e., the probability of change if the ancestral condition is zero does not equal the probability of change if the ancestral condition is one). Unless a tree posits as many gains as loses, the likelihood of a tree and $\pi_i^0 \neq \pi_i^1$ will be greater than the likelihood of a tree and $\pi_i^0 = \pi_i^1$. Similarly, if one posits biased increases and decreases for an ordered multistate character i, then one tests a null hypothesis that $\pi_i^- = \pi_i^+ = \pi_i/2$ (i.e., there is a single probability for increase ["+"] and decrease ["−"], both of which are one half the probability of character i changing). This can be expressed as an array:

Transition	Probability
Decrease	π^-
Stasis	$1 - (\pi^- + \pi^+)$
Increase	$\pi^+.$

They hypothesis of a driven trend (i.e., biased change in one direction) posits that $\pi_i^- \neq \pi_i^+$ and thus adds a parameter. Thus, the increase in likelihood must be appreciable in order to reject the null hypothesis of no bias.

Testing Hypothesized Gaps and Sampling Intensities.—A basic tenet of this paper is that if a hypothesis makes a prediction about data, then that hypothesis is testable by those data. Because hypothesized phylogenies imply gaps in sampling, hypothesized phylogenies make explicit predictions about stratigraphic sampling. One prediction concerns sampling intensity, R. Wagner (1998) argues that the fossil record of hyaenids presents substantial evidence against the very low R implicit to the hyaenid MST.

In Wagner's analysis, Monte Carlo simulations determine the most likely R to a given amount of stratigraphic debt. However, given R and extinction (μ), a range of stratigraphic debts are plausible in addition to the most probable amount. This leaves two items, stratigraphic ranges (SR) and stratigraphic debt (SD), over which R makes probabilistic predictions. The likelihood of any given R therefore is simply

$$L[R, \mu \mid SD, SR] = L[R \mid SD] \times L[R, \mu \mid SR] \tag{14}$$

The extinction parameter represents a nuisance parameter here. Wagner (1998) used a value of μ that maximized the likelihood of each R when determining $L[R \mid SR]$. Those values of μ were used when estimating $L[R \mid SD]$ also.

The test presented here does not consider temporal variation in R. However, Foote (personal communication 1999) developed methods for testing multiparameter hypotheses about sampling, origination, and extinction over time given range data. The method presented here should be modified easily to accommodate this more powerful test. Thus, reduced amounts of debt due to variable sampling will be distinguishable from reduced amounts of debt due to high average sampling.

Example: The Likelihoods of Alternative Hypotheses of Hyaenid Evolution.—Werdelin and Solounias (1991) present a phylogenetic analysis of 18 hyaenid species using 20 discretely coded morphological characters. Although Werdelin and Solounias evaluate the characters and states in careful detail, the resulting character matrix possesses very little hierarchical structure. The MSTs (assuming unordered multistate characters) posit at least 48 steps distributed among 19 characters and 26 states (the final character is invariant) (Fig. 5). Pairwise character compatibility rapidly decays with the addition of new taxa and the character space rapidly exhausts (Wagner 2000b). Monte Carlo analyses (Wagner 1998) indicate that, given the number of steps, taxa, and characters, the most likely tree length is at least 57 steps even if all characters evolved at the same rate (i.e., the simplest possible hypothesis). More complicated hypotheses of character evolution require even more steps before the observed congruence becomes probable. These findings suggest that the

MSTs are not reliable inferences of hyaenid phylogeny. The likelihood approach outlined above will be used to test hypotheses of hyaenid relationships and character evolution.

Hyaenid MSTs suggest that the 19 variable characters appear over the first 28 steps and that the final 21 steps simply reshuffle existing characters (Fig. 6). Given the rate at which new characters are added, the single best-fit rarefaction curves is for 24 potentially variable characters. Therefore, four additional characters are considered present but unchanged on each branch. Werdelin and Solounias (1991) note that there is no evidence of continuous change among the coded characters within hyaenid lineages. This suggests a punctuated pattern of character change and the analyses assume a binomial distribution of changes along branches. One derives the minimum time spanned by each branch by plotting the MST against observed stratigraphic ranges (Fig. 7) (see Smith 1994). The probability of t_{br} speciation events along each branch then is evaluated using equation (8), with the likelihood of μ determined from observed stratigraphic ranges (illustrated in Wagner 1998). In each case, a μ maximizing the probability of each t_{br} is used. However, $P[t_{br} \mid G]$ is nearly identical for all t_{br} and G when using the most likely μ as a model (Fig. 8).

The single best one-rate hypothesis posits π = 0.037 changes per speciation (Table 2). The likelihood of the best is two-π hypothesis is substantially better (log-likelihood or support, S, = −178.6 vs. −165.6). Improvement continues with the addition of rate parameters. The most likely hypothesis given the MST has 19 rate parameters (S = −158.08). Log-likelihood ratio tests show that the best two- and three-parameter hypotheses strongly reject simpler hypotheses (Table 3). However, more complicated hypotheses fail to improve support significantly (Tables 2, 3).

Multivariate analyses of character compatibility (O'Keefe and Wagner 1999) suggest that characters 16 (basioccipital shape), 17 (premaxillary-maxillary suture position), and 19 (position of the major palatine foramen) do not change independently. The MST posits that all three characters change once along the same branch. An alternative hypothesis that

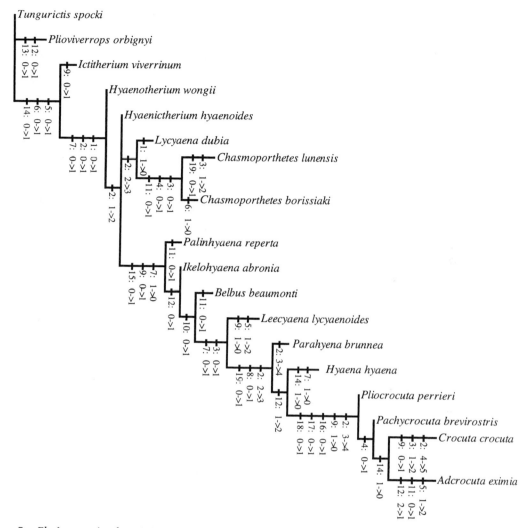

FIGURE 5. Phylogram for the minimum-steps parsimony tree for 18 hyaenid species, assuming an unordered and equal-weights model of evolution. The character data are from Werdelin and Solounias 1991. The tree requires a minimum of 48 steps (Maximum Consistency Index = 0.552, or at least 1.85 steps per derived state). Note that likelihood posits a slightly different character optimization, which gives steps to some zero-length branches.

the three characters are completely noninde-pendent has two more parameters than does the independent-character hypothesis. How-ever, the correlated-character hypothesis yields an increase in support (8.8) that justifies the added parameters (Table 3).

Many MST branches span long intervals but invoke little or no necessary change. There-fore, hypotheses with variable π or μ across phylogeny therefore have greater support than hypotheses with consistent π or μ. How-ever, the increase in support is insufficient to justify these added parameters. In other words, the null hypothesis of consistent π's cannot be rejected.

Hypotheses about biased gains or acquisi-tions also fail to increase support appreciably. This is not surprising, because only one char-acter has a minimum number of changes ex ceeding four. That one character (character 2, ratio of upper fourth premolar length to upper first molar width) shows at least four decreas-es and at least one increase in the multiparam-eter hypothesis. Given an overall rate of change of $\pi_2 = 0.082$, the most likely unbiased change hypothesis is

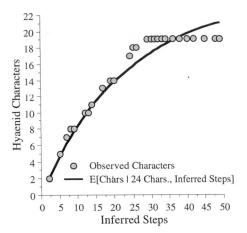

FIGURE 6. Maximum per-step rate of new-character accumulation for hyaenids. Based on an MST derived from Werdelin and Solounias 1991. The hypothesis of continuous accumulation is strongly rejected in favor of a finite-space hypothesis of 24 total characters.

Transition	Probability
Decrease	$\pi_2^- = 0.041$
Stasis	$1 - (\pi^- + \pi^+) = 0.918$
Increase	$\pi_2^+ = 0.041$

The most likely biased change hypothesis is

Transition	Probability
Decrease	$\pi_2^- = 0.067$
Stasis	$1 - (\pi^- + \pi^+) = 0.918$
Increase	$\pi_2^+ = 0.015$

The log-likelihood of the tree and three-rate character hypotheses with character 2 showing no bias now is -165.31. The apparent "decrease" in support compared to the three-rate hypothesis in which biased change is not examined is an artifact of testing a more explicit

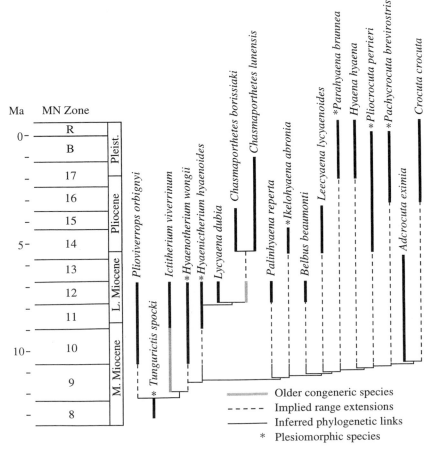

FIGURE 7. Inferred phylogeny consistent with a minimum-steps parsimony tree for hyaenids. Modified from figures presented in Flynn 1996 and Wagner 1998. The tree posits 48 units of stratigraphic debt (in Mein Neogene [MN] zones).

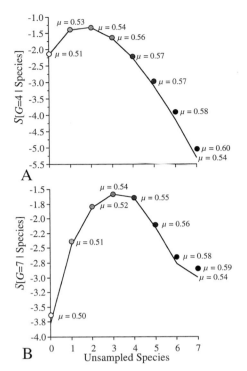

FIGURE 8. The likelihood of t_{br} species spanning a temporal duration, G. Solid curve gives the log-likelihood (support) assuming the most likely overall extinction intensity ($\mu = 0.54$). Open circles give the maximum support (based on the likelihood of t_{br} species given μ times the likelihood of μ given stratigraphic ranges) and the required extinction intensity. Note that μ can be used as a model with little effect on the net likelihood. A, Duration of four MN zones. B, Duration of seven MN zones.

hypothesis. The biased-change hypothesis increases the likelihood of the entire hypothesis to -164.04. The increase in support (1.28) is insufficient to reject the null hypothesis of unbiased change in molar size.

TABLE 3. Results of log-likelihood ratio tests given τ_1. M_5, with three rates of change and complete correlation between characters 16, 17, and 19 rejects all other hypotheses. M_6, not shown here, posits four rates and complete correlation but does not yield significantly greater support.

M_1	●			
M_2	**3.5×10^{-7}**	●		
M_3	1.4×10^{-7}	**0.018**	●	
M_4	3.7×10^{-7}	0.034	0.265	●
M_5	3.2×10^{-10}	2.1×10^{-5}	**8.6×10^{-5}**	2.8×10^{-5}

One can test the MST (hereafter, τ_1) and accompanying character-evolution hypothesis by determining the likelihoods of alternative hypotheses of relationships and character evolution. Numerous alternative phylogenies were examined, but only one is discussed here. That alternative, τ_2, posits a minimum of 56 steps (Fig. 9). τ_2 is substantially less pectinate than τ_1 and differs in two appreciable topologic features. First, it removes the base of a particular pectinate portion of τ_1 into a separate clade. Second, it rearranges the relationships among the extant taxa from (*Hyaena hyaena*, (*Parahyaena brunnea*, *Crocuta crocuta*)) to (*C. crocuta*, (*P. brunnea*, *H. hyaena*)). Another major difference here is that the Miocene *Adcrocuta eximia* is no longer considered the immediate sister taxon of the extant *C. crocuta*. Plotting τ_2 against stratigraphy (Fig. 10) shows that the topologic changes substantial reduce stratigraphic debt (15 vs. 48 Mein Neogene [MN] zones).

Likelihood tests of character-evolution hypotheses for τ_2 reach similar conclusions as

TABLE 2. Support (S = log-likelihood) of topology (τ) and character evolution (M) hypotheses given observed characters (C). Hyaenid MST is used here. "Pa" gives the number of parameters. "NI" indicates hypothesized nonindependence. Two times the difference in log-likelihoods is assessed by a chi-square distribution with degrees of freedom equal to the difference in parameters.

$S[\tau,M\mid C]$	Pa	M	Characters
-178.6	1	$\pi_1 = 0.037$	
-165.6	2	$\pi_1 = 0.016$	1, 4, 6, 8, 10, 14, 15, 16, 17, 18, 19, 20, 21, 22, 23, 24
		$\pi_2 = 0.082$	2, 3, 5, 7, 9, 11, 12, 13
-162.8	3	$\pi_1 = 0.010$	8, 10, 14, 15, 17, 18, ,19, 20, 21, 22, 23, 24
		$\pi_2 = 0.037$	1, 4, 6, 16
		$\pi_3 = 0.082$	2, 3, 5, 7, 9, 11, 12, 13
-162.2	4	$\pi_1 = 0.010$	8, 10, 14, 15, 17, 18, 19, 20, 21, 22, 23, 24
		$\pi_2 = 0.037$	1, 4, 6, 16
		$\pi_3 = 0.072$	3, 5, 7, 11, 13
		$\pi_4 = 0.117$	2, 9, 12
-153.4	5	M_3, NI	16, 17, 19 completely codependent

those for τ_1. Additional rate parameters rapidly increase support (Table 4), with a three-rate hypothesis positing complete correlation between characters 16, 17, and 19 rejecting all simpler hypotheses (Table 5).

Under τ_2, character 2 shows a minimum of eight changes, with potentially all being decreases. The most likely unbiased change hypothesis is

Transition	Probability
Decrease	$\pi_2^- = 0.082$
Stasis	$1 - (\pi^- + \pi^+) = 0.837$
Increase	$\pi_2^+ = 0.082$

The most likely biased change hypothesis is

Transition	Probability
Decrease	$\pi_2^- = 0.163$
Stasis	$1 - (\pi^- + \pi^+) = 0.837$
Increase	$\pi_2^+ = 0.000$

The support for τ_2 + three rates + nonindependence + unbiased change for character 2 given data is 150.40. (Again, the likelihoods are higher here than in Table 5 because the analyses test more specific hypotheses.) The support of τ_2 + three rates + nonindependence + unbiased change for character 2 given data is 146.04. The log-likelihood ratio statistic ($\Lambda = 8.71$) allows us to reject the unbiased change hypothesis in this case ($p = 0.003$).

The support for τ_2's multiparameter hypothesis is appreciably greater than that for τ_1 multiparameter hypothesis (-146.04 vs. -153.44). The main reason for this is that τ_1 includes many branches with little or no change that span several MN zones. For example, the parsimony optimization predicts that an ancestral species identical to *Parahyaena brunnea* existed for over seven MN zones. The probability of little or no change over so much time (or of all changes being reversed subsequently) is very low given how plastic hyaenid characters are over the rest of the phylogeny. Making the ancestral states differ from those of *P. brunnea* increases the probability of observing the realized distribution of states, but not appreciably. Adding parameters by decreasing π's does not increase sup-

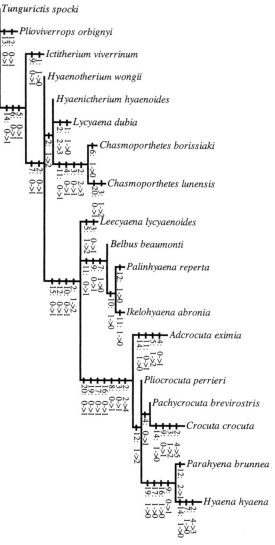

FIGURE 9. Phylogram for an alternative phylogenetic hypothesis. See Figure 5. The most likely character optimization differs only slightly from that illustrated.

port sufficiently to justify those added parameters.

We now wish to test the H_1 (τ_1 + three rates + correlated change) against H_2 (τ_2 + three rates + biased molar reduction + correlated change). However, Goldman (1993) notes that phylogenies do not represent special cases of one another. Thus, it is difficult to determine the difference in parameter numbers between two phylogenies. Goldman and others (e.g., Huelsenbeck et al. 1996) recommend Monte Carlo approaches for generating test distributions. This involves using the null hypoth-

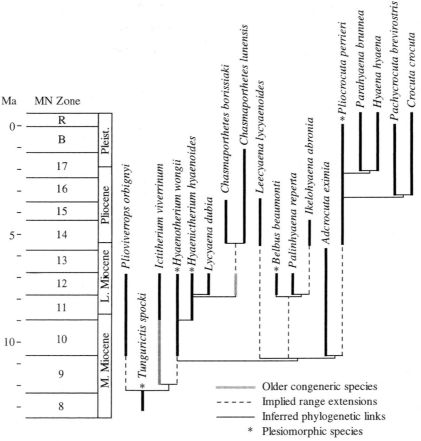

FIGURE 10. Inferred phylogeny consistent with Figure 9. The tree posits 15 units of stratigraphic debt (in MN zones).

esis (in this case, H_1) as a model in the simulation, then determining the likelihoods of both the actual phylogeny and the most likely phylogeny given a character-evolution hypothesis as complicated as that of H_2. The differences in log-likelihoods provide a distri-

bution to evaluate whether the real difference in likelihoods surpasses that expected if the null hypothesis actually is correct. Unfortunately, such simulations are time consuming and could not be performed here.

When two hypotheses are equally complex

TABLE 4. Support of τ_2 and character evolution (M) hypotheses given observed characters (C). See Table 3 for additional information.

$s[\tau,M\mid C]$	Pa	M	Characters
−177.3	1	$\pi_1 = 0.068$	
−162.6	2	$\pi_1 = 0.010$	8, 13, 15, 18, 21, 22, 23, 24
		$\pi_2 = 0.103$	1, 2, 3, 4, 5, 6, 7, 9, 10, 11, 12, 14, 16, 17, 19, 20
−158.8	3	$\pi_1 = 0.010$	8, 13, 15, 18, 21, 22, 23, 24
		$\pi_2 = 0.069$	1, 4, 5, 6, 7, 10, 16, 17, 19, 20
		$\pi_3 = 0.163$	2, 3, 9, 11, 12, 14
−157.7	4	$\pi_1 = 0.010$	8, 13, 15, 18, 21, 22, 23, 24
		$\pi_2 = 0.069$	1, 4, 5, 6, 7, 10, 16, 17, 19, 20
		$\pi_3 = 0.139$	3, 9, 11, 12, 14
		$\pi_4 = 0.289$	2
−146.0	5	M_3, NI	16, 17, 19 completely codependent

TABLE 5. Results of log-likelihood ratio tests given τ_2. Again, M_5, with three rates of change and complete correlation between characters 16, 17, and 19, rejects all other hypotheses.

M_1	•			
M_2	**6.0×10^{-8}**	•		
M_3	8.8×10^{-9}	**5.5×10^{-3}**	•	
M_4	1.6×10^{-8}	7.5×10^{-3}	0.151	•
M_5	8.5×10^{-13}	3.0×10^{-7}	**3.0×10^{-6}**	1.3×10^{-6}

but nonhierarchical, Edwards (1992) suggests using a simple rejection criterion of a difference in support of 2 ($\Lambda = 4$). Because H_2 includes biased reduction of molars, it is somewhat more complicated than H_1. However, we can compare H_1 with a slightly simpler version of H_2 with unbiased molar change. $H_{2\text{-biased change}}$ and H_1 now share approximately equal complexity. The difference in support is 7.4 ($\Lambda = 14.8$), which allow us to reject H_1 in favor of $H_{2\text{-biased change}}$. H_2 in turn allows us to reject $H_{2\text{-biased change}}$, which allows us to state that H_2 rejects H_1 as a syllogism.

It bears stressing that these results are not based on a presupposed model of evolution that favors τ_2 over τ_1. Instead, the tests reject one multiparameter hypothesis of character evolution (with τ representing the historical distribution of characters) in favor of another.

The Likelihood of a Phylogenetic Hypothesis Given Stratigraphic Data.—Treating sampling intensity as a hypothesis given stratigraphic debt has a small effect on the hyaenid example. Here, τ_1 posits SD = 48 MN Zones (Fig. 7) whereas τ_2 posits SD = 15 MN zones (Fig. 10). The probability of SD = 48 is maximized at sampling intensity = 0.18. $P[SD=15 \mid R]$ is maximized at $R = 0.37$ (Fig. 11). If we use those R's to typify τ_1 and τ_2, then the log-likelihood of τ_1 and the sampling intensity given stratigraphic ranges ($S[\tau_1, R \mid SR]$) = -5.78 whereas $S[\tau_2, R \mid SR]$ = -1.02. (Again, both values assume the most likely μ given the hypothesized R.) However, if we use the log-likelihood of a tree and sampling intensity given stratigraphic data and implied stratigraphic debt, then $S[\tau_1, R \mid SD, SR]$ = -5.57 whereas $S[\tau_2, R \mid SD, SR]$ = -0.57. This results in a slightly greater increase of Λ (10.00 vs. 9.52) when testing τ_1 against τ_2, and thus a slightly stronger rejection of τ_1. Whether one treats the

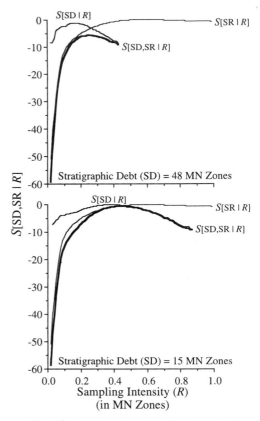

FIGURE 11. The effects of treating the relationship between topology (τ) and sampling intensity (R) as an assumption and as a hypothesis, based on implicit stratigraphic debt (SD) and observed stratigraphic ranges (SR). A, $S[R \mid SD]$ gives the log-likelihood (or support, S) of SD = 48 distributed among 18 taxa given alternative R's. Support is maximized at $R = 0.18$. However, support for that hypothesized sampling intensity given stratigraphic ranges is quite low (based on methods presented in Foote 1997; see analyses in Wagner 1998). Increasing hypothesized R decreases $S[R \mid SD]$ but increases $S[R \mid SR]$ so that $S[R \mid SD, SR]$ is greatest at $R = 0.23$. B, The same approach given SD = 15. Whereas $S[R \mid SD]$ is greatest at $R = 0.37$, $S[R \mid SD, SR]$ is greatest at $R = 0.45$.

relationship between R and τ as an assumption or a hypothesis, stratigraphic data contradict τ_1 far more strongly than they do τ_2. Thus, both stratigraphic and character data firmly reject the hyaenid MST in favor of an alternative phylogenetic hypothesis.

Summary.—The likelihood test outlined above might appear to require many more assumptions than does minimum-steps parsimony. This is only because parsimony makes no reference to some important factors, such as the expected amount of change along branches given evolutionary time. It therefore is incorrect to state that minimum-steps parsimony makes no assumptions about branch lengths. Instead, minimum-steps parsimony assumes that branch lengths do not affect predictions about character state distributions. Similarly, it is incorrect to state that minimum-steps parsimony makes no assumptions about preservation rates in the fossil record. Instead, minimum-steps parsimony assumes that phylogenies make no predictions about preservation rates. Assuming that a factor is irrelevant is very different from making no assumption about that factor.

In other respects, likelihood can reduce assumptions required by minimum-steps parsimony. Character weighting, character state transitions, and character independence all affect the outcomes of parsimony analyses. Because these parameters cannot be varied and tested during analysis, they must remain models in minimums steps parsimony analyses. Conversely, likelihood permits testing of these variables in multiparameter hypotheses.

Conclusions

Recent years have witnessed a growing awareness of the importance of phylogeny for explaining historical patterns. Perceptions of phylogenetic relationships clearly affect perceptions of character evolution, as well as other macroevolutionary patterns (e.g., diversification) not discussed here. Perceptions of character evolution affect perceptions of phylogenetic relationships, especially when discussing driven trends, correlated character change, and shifts in rates of character evolution. Such hypotheses make different predictions about the distributions of characters given a phylogenetic topology than do hypotheses of unbiased and independent changes occurring at consistent rates. This makes testing hypotheses about phylogenetic relationships inseparable from testing hypotheses about character evolution. This is not to say that tree-based approaches are without merit. Instead, it is to say that tree-based approaches should treat phylogeny as one parameter among a set of hypotheses, not as a model in tests of more particular hypotheses. Such an approach shifts phylogenetic analyses from inferences to actual tests. Such an approach also reduces

the number of necessary assumptions by identifying factors affecting the predictions of phylogenetic hypotheses, and then testing alternative hypotheses for those parameters. Ultimately, tests of these evolutionary parameters will teach us more about macroevolution in general and phylogeny in particular than modeled phylogenies will teach us about those evolutionary parameters.

Acknowledgments

Some of the ideas described herein originally were developed for the Nature Online Debate "Is the fossil record adequate?" However, the paper's format was modeled on Paul McCartney's "Uncle Albert / Admiral Halsey." For discussion and comments, I thank C. Brochu, M. Foote, J. Huelsenbeck, R. Lupia, J. Marcot, and F. Thesis. D. Swofford provided an especially helpful review and discussion. Discussions with T. Grantham helped clarify the different definitions assigned by logicians and philosophers to formerly homologous words. Finally, I thank O. Rieppel for discussions concerning the pattern cladistic viewpoint. (But I still don't buy it!) This work was funded in part by National Science Foundation research grant EAR-99-03238.

Literature Cited

Alroy, J. 1994. Four permutation tests for the presence of phylogenetic structure. Systematic Biology 43:430–437.

Archie, J. W. 1989. A randomization test for phylogenetic information in systematic data. Systematic Zoology 38:239–252.

Atchley, W. R., and W. M. Fitch. 1992. Gene trees and the origins of inbred strains of mice. Science 254:554–558.

Barnard, G. A., G. M. Jenkins, and C. B. Winston. 1962. Likelihood inference and time series. Journal of the Royal Statistics Society A 125:321–372.

Benton, M. J. 1995. Testing the time axis of phylogenies. Philosophical Transactions of the Royal Society of London B 349:5–10.

———. 1998. Molecular and morphological phylogenies of mammals: congruence with stratigraphic data. Molecular Phylogenetics and Evolution 9:398–407.

Benton, M. J., and R. Hitchin. 1997. Congruence between phylogenetic and stratigraphic data on the history of life. Proceedings of the Royal Society of London B 264:885–890.

Benton, M. J., and G. W. Storrs. 1994. Testing the quality of the fossil record: paleontological knowledge is improving. Geology 22:111–114.

Benton, M. J., M. A. Wills, and R. Hitchin. 2000. Quality of the fossil record through time. Nature 403:534–537.

Bottjer, D. J., and D. Jablonski. 1988. Paleoenvironmental patterns in the evolution of post-Paleozoic benthic marine invertebrates. Palaios 3:540–560.

Brady, R. H. 1985. On the independence of systematics. Cladistics 1:113–126.

Brochu, C. 1997. Morphology, fossils, divergence timing, and the phylogenetic relationships of *Gavialis*. Systematic Biology 46:479–522.

Camin, J. H., and R. R. Sokal. 1965. A method for deducing branching sequences in phylogeny. Evolution 19:311–326.

Cheetham, A. H., and J. B. C. Jackson. 1995. Process from pattern: tests for selection versus random change in punctuated bryozoan speciation. Pp. 184–207 in D. H. Erwin and R. L. Anstey, eds. New approaches for studying speciation in the fossil record. Columbia University Press, New York.

Cloutier, R. 1991. Patterns, trends, and rates of evolution with the Actinistia. Environmental Biology of Fishes 32:23–58.

Clyde, W. C., and D. C. Fisher. 1997. Comparing the fit of stratigraphic and morphologic data in phylogenetic analysis. Paleobiology 23:1–19.

de Queiroz, K., and J. Gauthier. 1990. Phylogeny as a central principle in taxonomy: phylogenetic definitions of taxon names. Systematic Zoology 39:307–332.

———. 1992. Phylogenetic taxonomy. Annual Review of Ecology and Systematics 23:449–480.

Edwards, A. W. F. 1992. Likelihood—expanded edition. Johns Hopkins University Press, Baltimore.

———. 1996. The origin and early development of the method of minimum evolution for the reconstruction of phylogenetic trees. Systematic Biology 45:79–91.

Edwards, A. W. F., and L. L. Cavalli-Sforza. 1964. Reconstruction of evolutionary trees. Pp. 67–76 in J. H. Heywood and J. McNeil, eds. Phenetic and phylogenetic classification. Systematic Association, London.

Eldredge, N. 1971. The allopatric model and phylogeny in Paleozoic invertebrates. Evolution 25:156–167.

Eldredge, N., and S. J. Gould. 1972. Punctuated equilibria: an alternative to phyletic gradualism. Pp. 82–115 in T. J. M. Schopf, ed. Models in paleobiology. Freeman, Cooper, San Francisco.

Estabrook, G. F., C. S. Johnson Jr., and F. R. McMorris. 1975. An idealized concept of the true cladistic character. Mathematical Biosciences 23:263–272.

Farris, J. S. 1973. A probability model for inferring evolutionary trees. Systematic Zoology 22:250–256.

———. 1977. Phylogenetic analysis under Dollo's Law. Systematic Biology 26:77–88.

Felsenstein, J. 1973. Maximum-likelihood and minimum-steps methods for estimating evolutionary trees from data on discrete characters. Systematic Zoology 22:240–249.

———. 1978. Cases in which parsimony or compatibility methods will be positively misleading. Systematic Zoology 27:401–410.

———. 1981a. Evolutionary trees from DNA sequences: a maximum likelihood approach. Journal of Molecular Evolution 17:368–376.

———. 1981b. A likelihood approach to character weighting and what it tells us about parsimony and compatibility. Biological Journal of the Linnean Society 16:183–196.

———. 1985. Phylogenies and the comparative method. American Naturalist 125:1–15.

———. 1988. Phylogenies and quantitative characters. Annual Review of Ecology and Systematics 19:445–471.

Fisher, D. C. 1982. Phylogenetic and macroevolutionary patterns within the Xiphosurida. Pp. 175–180 in B. Mamet and M. J. Copeland, eds. Proceedings of the third North American paleontological convention. Geological Survey of Canada, Montreal.

———. 1988. Stratocladistics: integrating stratigraphic and morphologic data in phylogenetic inference. Geological Society of America Abstracts with Programs 20:A186.

———. 1991. Phylogenetic analysis and its implication in evolutionary paleobiology. In N. L. Gilinsky and P. W. Signor, eds.

Analytical paleobiology. Short Course in Paleontology 4:103–122. Paleontological Society, Knoxville, Tenn.

———. 1994. Stratocladistics: morphological and temporal patterns and their relation to phylogenetic process. Pp. 133–171 in L. Grande and O. Rieppel, eds. Interpreting the hierarchy of nature—from systematic patterns to evolutionary process theories. Academic Press, Orlando.

Flynn, J. J. 1996. Carnivoran phylogeny and rates of evolution: morphological, taxic, and molecular. Pp. 542–581 in J. Gittleman, ed. Carnivoran behavior, ecology and evolution, Vol. 2. Comstock, Ithaca, N.Y.

Foote, M. 1996. On the probability of ancestors in the fossil record. Paleobiology 22:141–151.

———. 1997. Estimating taxonomic durations and preservation probability. Paleobiology 23:278–300.

Foote, M., J. P. Hunter, C. M. Janis, and J. J. Sepkoski Jr. 1999. Evolutionary and preservational constraints on the origins of major biologic groups: limiting divergence times of eutherian mammals. Science 283:1310–1314.

Foote, M., and D. M. Raup. 1996. Fossil preservation and the stratigraphic ranges of taxa. Paleobiology 22:121–140.

Foote, M., and J. J. Sepkoski Jr. 1999. Absolute measures of the completeness of the fossil record. Nature 398:415–417.

Fortey, R. A., and R. P. S. Jefferies. 1982. Fossils and phylogeny—a compromise approach. Pp. 197–234 in Joysey and Friday 1982.

Gauthier, J., A. G. Kluge, and T. Rowe. 1988. Amniote phylogeny and the importance of fossils. Cladistics 4:105–209.

Gayon, J. 1990. Critics and criticisms of the modern synthesis—the viewpoint of a philosopher. Evolutionary Biology 24:1–49.

Goldman, N. 1993. Statistical tests of models of DNA substitution. Journal of Molecular Evolution 36:182–198.

Gould, S. J., and R. C. Lewontin. 1979. The spandrels of San Marco and the Panglossian Paradigm: a critique of the adaptationist programme. Proceedings of the Royal Society of London B 205:581–598.

Graybeal, A. 1998. Is it better to add taxa or characters to a difficult phylogenetic problem? A simulation study. Systematic Biology 47:9–17.

Grimaldi, R. P. 1994. Discrete and combinatorial mathematics, 3d ed. Addison-Wesley, New York.

Harvey, P. H., and M. D. Pagel. 1991. The comparative method in evolutionary biology. Oxford University Press, Oxford.

Hasagawa, M., H. Kishino, and T. Yano. 1985. Dating of the human-ape splitting by a molecular clock of mitochondrial DNA. Journal of Molecular Evolution 22:160–174.

Heyning, J. E., and C. Thacker. 1999. Phylogenies, temporal data, and negative evidence. Science 285:1179.

Hillis, D. M., and J. P. Huelsenbeck. 1992. Signal, noise, and reliability in molecular phylogenetic analysis. Journal of Heredity 83:189–195.

Hitchin, R., and M. J. Benton. 1997. Stratigraphic indices and tree balance. Systematic Biology 46:563–569.

Hoffmann, R., V. I. Minkin, and B. K. Carpenter. 1997. Ockham's Razor and chemistry. HYLE—An International Journal for the Philosophy of Chemistry 3:3–28.

Holland, S. M. 2000. The quality of the fossil record: a sequence stratigraphic perspective. In D. H. Erwin and S. L. Wing, eds. Deep time: *Paleobiology*'s perspective. Paleobiology 26(Suppl. to No.4):148–168.

Huelsenbeck, J. P. 1991. When are fossils better than extant taxa in phylogenetic analysis? Systematic Zoology 40:458–469.

———. 1994. Comparing the stratigraphic record to estimates of phylogeny. Paleobiology 20:470–483.

Huelsenbeck, J. P., and K. A. Crandall. 1997. Phylogeny estimation and hypothesis testing using maximum likelihood. Annual Review of Ecology and Systematics 28:437–466.

Huelsenbeck, J. P., and M. Kirkpatrick. 1996. Do phylogenetic methods produce trees with biased shapes? Evolution 50: 1418–1424.

Huelsenbeck, J. P., and R. Nielsen. 1999. Effects of nonindependent substitution on phylogenetic accuracy. Systematic Biology 48:317–328.

Huelsenbeck, J. P., and B. Rannala. 1997. Maximum likelihood estimation of topology and node times using stratigraphic data. Paleobiology 23:174–180.

Huelsenbeck, J. P., D. M. Hillis, and R. Jones. 1996. Parametric bootstrapping in molecular phylogenetics: applications and performance. Pp. 19–45 in J. D. Ferraris and S. R. Palumbi, eds. Molecular zoology: advances, strategies and protocols. Wiley-Liss, New York.

Hull, D. L. 1983. Karl Popper and Plato's metaphor. Pp. 177–189 in N. L. Platnick and V. A. Funk, eds. Advances in cladistics. Columbia University Press, New York.

Hume, D. 1748. An inquiry concerning human understanding. Bobbs Merrill, Indianapolis.

Hurlbert, S. H. 1971. The nonconcept of species diversity: a critique and alternative parameters. Ecology 52:577–586.

Jackson, J. B. C., and A. H. Cheetham. 1994. Phylogeny reconstruction and the tempo of speciation in cheilostome Bryozoa. Paleobiology 20:407–423.

Jefferys, W. H., and J. O. Berger. 1992. Ockham's Razor and Bayesian analysis. American Scientist 80:64–72.

Joysey, K. A., and A. E. Friday, eds. 1982. Problems of phylogenetic reconstruction. Academic Press, London.

Jukes, T. H., and C. R. Cantor. 1969. Evolution of protein molecules. Pp. 21–132 in H. M. Munro, ed. Mammalian protein metabolism. Academic, New York.

Kitching, I. J., P. L. Forey, C. Humphries, and D. M. Williams. 1998. Cladistics, 2d ed. The theory and practice of parsimony analysis. Oxford University Press, Oxford.

Kluge, A. G. 1997. Testability and the refutation and corroboration of cladistic hypotheses. Cladistics 13:81–96.

Kluge, A. G., and J. S. Farris. 1969. Quantitative phyletics and the evolution of anurans. Systematic Zoology 18:1–32.

Kuhner, M. K., and J. Felsenstein. 1994. A simulation comparison of phylogeny algorithms under equal and unequal evolutionary rates. Molecular Biology and Evolution 11:459–468.

Lande, R. 1975. Natural selection and random genetic drift in phenotypic evolution. Evolution 30:314–334.

Le Quesne, W. J. 1969. A method of selection of characters in numerical taxonomy. Systematic Zoology 18:201–205.

Lipton, P. 1991. Inference to the best explanation. Routledge, London.

Maddison, W. P., and D. R. Maddison. 1992. MacClade, analysis of phylogeny and character evolution. Sinauer, Sunderland, Mass.

Marshall, C. R. 1990. Confidence intervals on stratigraphic ranges. Paleobiology 16:1–10.

———. 1994. Confidence intervals on stratigraphic ranges: partial relaxation of the assumption of randomly distributed fossil horizons. Paleobiology 20:459–469.

———. 1997. Confidence intervals on stratigraphic ranges with nonrandom distributions of fossil horizons. Paleobiology 23:165–173.

Marshall, C. R., and P. D. Ward. 1996. Sudden and gradual molluscan extinctions in the latest Cretaceous of the western European Tethys. Science 274:1360–1363.

Martin, R. M. 1997. Scientific thinking. Broadview, Toronto.

Mayr, E. 1942. Systematics and the origin of species. Columbia University Press, New York.

———. 1963. Animal species and evolution. Harvard University Press, Cambridge.

McShea, D. W. 1994. Mechanisms of large-scale evolutionary trends. Evolution 48:1747–1763.

Meacham, C. A. 1984. Evaluating characters by character compatibility analysis. Pp. 152–165 in T. Duncan and T. F. Stuessy, eds. Cladistics: perspectives on the reconstruction of evolutionary history. Columbia University Press, New York.

Mooers, A. Ø., and D. Schluter. 1999. Reconstructing ancestor states with maximum likelihood: support for one- and two-rate models. Systematic Biology 48:623–633.

Mooers, A. Ø., R. D. M. Page, A. Purvis, and P. H. Harvey. 1995. Phylogenetic noise leads to unbalanced cladistic tree reconstructions. Systematic Biology 44:332–342.

Newman, C. M., J. E. Cohen, and C. Kipnis. 1985. Neo-darwinian evolution implies punctuated equilibria. Nature 315:400–401.

Newton-Smith, W. H. 1981. The rationality of science. Routledge, London.

Norell, M. A. 1992. Taxic origin and temporal diversity: the effect of phylogeny. Pp. 89–118 in M. J. Novacek and Q. D. Wheeler, eds. Extinction and phylogeny. Columbia University Press, New York.

———. 1993. Tree-based approaches to understanding history: comments on ranks, rules, and the quality of the fossil record. American Journal of Science 293-A:407–417.

———. 1996. Ghost taxa, ancestors and assumptions—a comment on Wagner. Paleobiology 22:454–455.

Norell, M. A., and M. J. Novacek. 1992a. Congruence between superpositional and phylogenetic patterns: comparing cladistic patterns with fossil records. Cladistics 8:319–337.

———. 1992b. The fossil record and evolution: comparing cladistic and paleontologic evidence for vertebrate history. Science 255:1690–1693.

O'Keefe, F. R., and P. M. Sander. 1999. Paleontological paradigms and inferences of phylogenetic pattern: a case study. Paleobiology 25:518–533.

O'Keefe, F. R., and P. J. Wagner. 1999. A compatibility-based method for assess the independence of cladistic characters. Geological Society of America Abstracts with Programs 32: A30.

Pagel, M. 1994. Detecting correlated evolution on phylogenies: a general method for the comparative analysis of discrete characters. Proceedings of the Royal Society of London B 255: 37–45.

———. 1999. The maximum likelihood approach to reconstructing ancestral character states of discrete characters on phylogenies. Systematic Biology 48:612–622.

Patterson, C. 1982. Morphological characters and homology. Pp. 21–74 in Joysey and Friday 1982.

Paul, C. R. C. 1982. The adequacy of the fossil record. Pp. 75–117 in Joysey and Friday 1982.

———. 1985. The adequacy of the fossil record revisited. Pp. 1–16 in J. C. W. Cope and P. W. Skelton, eds. Evolutionary case histories from the fossil record. Palaeontological Society, London.

———. 1992. The recognition of ancestors. Historical Biology 6: 239–250.

Peirce, C. S. 1878. Deduction, induction, and hypothesis. Popular Science Monthly 13:470–482.

Platnick, N. I. 1977. Cladograms, phylogenetic trees, and hypothesis testing. Systematic Zoology 26:438–442.

Popper, K. R. 1959. The logic of scientific discovery. Routledge, London.

Raup, D. M., and S. J. Gould. 1974. Stochastic simulation and evolution of morphology—towards a nomothetic paleontology. Systematic Zoology 23:305–322.

Raup, D. M., S. J. Gould, T. J. M. Schopf, and D. S. Simberloff. 1973. Stochastic models of phylogeny and the evolution of diversity. Journal of Geology 81:525–542.

Rensch, B. 1960. Evolution above the species level. Columbia University Press, New York.

Rieppel, O. 1997. Falsificationist versus verificationist approaches to history. Journal of Vertebrate Paleontology 17A:71A.

Rieppel, O., and L. Grande. 1994. Summary and comments on systematic pattern and evolutionary process. Pp. 133–171 in L. Grande and O. Rieppel, eds. Interpreting the hierarchy of nature—from systematic patterns to evolutionary process theories. Academic Press, Orlando.

Rohlf, F. J., W. S. Chang, R. R. Sokal, and J. Kim. 1990. Accuracy of estimated phylogenies: effects of tree topology and evolutionary model. Evolution 44:1671–1684.

Sanderson, M. J. 1993. Reversibility in evolution: a maximum likelihood approach to character gain/loss bias in phylogenies. Evolution 47:236–252.

Sanderson, M. J., and M. J. Donoghue. 1989. Patterns of variation in levels of homoplasy. Evolution 43:1781–1795.

———. 1994. Shifts in diversification rate with the origin of angiosperms. Science 264:1590–1593.

Sankoff, D., and P. Rousseau. 1975. Locating the vertices of a Steiner tree in arbitrary space. Mathematical Programming 9: 240–246.

Schluter, D., T. Price, A. Ø. Mooers, and D. Ludwig. 1997. Likelihood of ancestor states in adaptive radiation. Evolution 51: 1699–1711.

Shaffer, H. B., J. M. Clark, and F. Kraus. 1991. When molecules and morphology clash: a phylogenetic analysis of the North American ambystomatid salamanders (Caudata: Ambystomatidae). Systematic Zoology 40:284–303.

Siddall, M. E., and A. G. Kluge. 1997. Probabilism and phylogenetic inference. Cladistics 13:313–336.

Sidor, C. A., and J. A. Hopson. 1998. Ghost lineages and "mammalness": assessing the temporal pattern of character acquisition in the Synapsida. Paleobiology 24:254–273.

Smith, A. B. 1988. Patterns of diversification and extinction in early Palaeozoic echinoderms. Palaeontology 31:799–828.

———. 1994. Systematics and the fossil record: documenting evolutionary patterns. Blackwell Scientific, Oxford.

Smith, A. B., and D. T. J. Littlewood. 1994. Paleontological data and molecular phylogenetic analysis. Paleobiology 20:259– 273.

Smith, A. B., B. Lafay, and R. Christen. 1992. Comparative variation of morphological and molecular evolution through geologic time: 28S ribosomal RNA versus morphology in echinoids. Philosophical Transactions of the Royal Society of London B 338:365–382.

Sneath, P. H. A. 1995. Thirty years of numerical taxonomy. Systematic Biology 44:281–298.

Sober, E. 1988. The nature of selection: evolutionary theory in philosophical focus. MIT Press, Cambridge.

Sokal, R. R., and F. J. Rohlf. 1981. Biometry, 2d ed. W. H. Freeman, New York.

Solow, A. R. 1993. Inferring extinction in a declining population. Ecology 74:962–963.

———. 1996. Tests and confidence intervals for a common upper endpoint in fossil taxa. Paleobiology 22:406–410.

Solow, A. R., and W. Smith. 1997. On fossil preservation and the stratigraphic ranges of taxa. Paleobiology 23:271–277.

Springer, M. S., and A. Lilje. 1988. Biostratigraphy and gap analysis: the expected sequence of biostratigraphic events. Journal of Geology 96:228–236.

Strauss, D., and P. M. Sadler. 1989. Classical confidence intervals and Bayesian probability estimates for ends of local taxon ranges. Mathematical Geology 21:411–427.

Swofford, D. L., and G. J. Olsen. 1990. Phylogeny reconstruction. Pp. 411–501 in D. M. Hillis and G. Moritz, eds. Molecular Systematics. Sinauer, Sunderland, Mass.

Wagner, P. J. 1995. Stratigraphic tests of cladistic hypotheses. Paleobiology 21:153–178.

———. 1997. Patterns of morphologic diversification among the Rostroconchia. Paleobiology 23:115–150.

———. 1998. A likelihood approach for estimating phylogenetic relationships among fossil taxa. Paleobiology 24:430–449.

———. 1999. Phylogenetics of Ordovician–Silurian Lophospiridae (Gastropoda: Murchisoniina): the importance of stratigraphic data. American Malacological Bulletin 15:1–31.

———. 2000a. The quality of the fossil record and the accuracy of phylogenetic inferences about sampling and diversity. Systematic Biology 49:65–86.

———. 2000b. Exhaustion of morphologic character states among fossil taxa. Evolution 54:365–386.

———. 2000c. Likelihood tests of hypothesized durations: determining and accommodating biasing factors. Paleobiology 26:431–449.

Wagner, P. J., and C. A. Sidor. 2000. Age rank:clade rank metrics—sampling, taxonomy, and the meaning of "Stratigraphic Consistency." Systematic Biology 49:463–479.

Weiss, R. E., and C. R. Marshall. 1999. The uncertainty in the true end point of a fossil's stratigraphic ranges when stratigraphic sections are sampled discretely. Mathematical Geology 31:435–453.

Werdelin, L., and N. Solounias. 1991. The Hyaenidae: taxonomic systematics and evolution. Fossils and Strata 30:1–104.

Wheeler, W. C. 1990. Combinatorial weights in phylogenetic analysis: a statistical parsimony procedure. Cladistics 6:269– 275.

Wiley, E. O. 1975. Karl R. Popper, systematics, and classification: a reply to Walter Bock and other evolutionary taxonomists. Systematic Zoology 24:233–243.

Wills, M. A. 1998. Crustacean disparity through the Phanerozoic: comparing morphological and stratigraphic data. Biological Journal of the Linnean Society 65:455–500.

———. 1999. The congruence between phylogeny and stratigraphy: randomization tests and the Gap Excess Ratio. Systematic Biology 48:559–580.

Wright, S. 1931. Evolution in Mendelian populations. Genetics 16:97–159.

———. 1932. The roles of mutation, inbreeding, crossbreeding and selection in evolution. Proceedings of the Sixth International Congress of Genetics 1:356–366.

Yang, Z. 1994. Maximum likelihood phylogenetic estimation from DNA sequences with variable rates over sites: approximate methods. Journal of Molecular Evolution 39:306–314.

Zadeh, L. A. 1965. Fuzzy sets. Information Control 8:338–353.

Appendix 1

Derivation of Numbers of Evolutionary Paths Leading to Particular States

Given n states and any initial state, there are $(n - 1)^x$ possible paths (histories) after x steps. If we do not include "silent" changes, then there is no way to return to the initial state in one step. However, the initial state can be rederived in the second step from any of the derived states, leaving $(n - 1)$ paths of the $(n - 1)^2$ possible paths. For example, if there are 4 states, then the three of the possible 9 histories are a → (b,c,d) → a. Thus, the number of ways to revert to the initial state after x steps is always the number of ways to produce any derived state at x − 1 steps.

Given n states, there are $(n - 1)$ possible ways to evolve any derived state after one step. At the second step, each of these might revert to the initial state, leaving $(n - 2)$ paths from each derived character that lead to another derived character. This can be written either as $n^2 - 3n + 2$ or as $(n - 1)^2 - (n - 1)$. This is simply the total number of paths at two steps minus the number that yield the initial state at two steps. Because the number of paths yielding the initial state at x steps equals the number of paths yielding any of the derived at $(x - 1)$ steps, the number of ways to evolve any derived state after x steps is simply the total number of paths minus the number of paths that yielded derived states after $(x - 1)$ steps.

At three steps, the initial states can evolve from any of the derived states, for which there were $(n - 1)^2 - (n - 1)$ possible paths. Any derived state can evolve in any of the remaining $(n - 1)^3 - ([n - 1]^2 - [n - 1])$ paths. This can be written as

$$(n - 1)^3 - (n - 1)^2 + (n - 1) \quad \text{or}$$

$$(-1^2 \times [n - 1]^3) + (-1^1 \times [n - 1]^2) + (-1^0 \times [n - 1]).$$

This now can be written as a summation,

$$\sum_{i=1}^{\text{steps}} -1^{i-1} \times (\text{states} - 1)^i.$$

To determine the number of paths that lead to a particular derived state, one simply divides the summation by $(\text{states} - 1)$, giving

$$\frac{\sum_{i=1}^{\text{steps}} -1^{i-1} \times (\text{states} - 1)^i}{\text{states} - 1}.$$

A similar summation can be used to calculate the number of paths leading from the initial state back to the initial state. At three steps, this is $(n - 1)^2 - (n - 1)$ or

$$(-1^2 \times [n - 1]^2) + (-1^1 \times [n - 1]).$$

This can be written as

$$\sum_{i=1}^{\text{steps}-1} -1^i \times (\text{states} - 1)^i.$$

To determine the proportion of total paths leading to any given state, the summations are simply divided by the total possible paths at x steps. For the proportion returning to the initial condition, this gives

$$\frac{\sum_{i=1}^{\text{steps}-1} -1^i \times (\text{states} - 1)^i}{(\text{states} - 1)^{\text{steps}}}.$$

For the proportion of paths producing a particular derived state, this is

$$\frac{\sum_{i=1}^{\text{steps}} -1^{i-1} \times (\text{states} - 1)^i}{(\text{states} - 1)^{\text{steps}+1}}.$$

Appendix 2

Numbers of paths leading to any derived state.

Steps	States					
	2	3	4	5	6	7
2	0	2	6	12	20	30
3	1	6	21	52	105	186
4	0	10	60	204	520	1110
5	1	22	183	820	2605	6666
6	0	42	546	3276	13,020	39,990
7	1	86	1641	1310	65,105	239,946

Appendix 3

Numbers of paths leading back to the initial state.

Steps	States					
	2	3	4	5	6	7
2	1	2	3	4	5	6
3	0	2	6	12	20	30
4	1	6	21	52	105	186
5	0	10	60	204	520	1110
6	1	22	183	820	2605	6666
7	0	42	546	3276	13,020	39,990

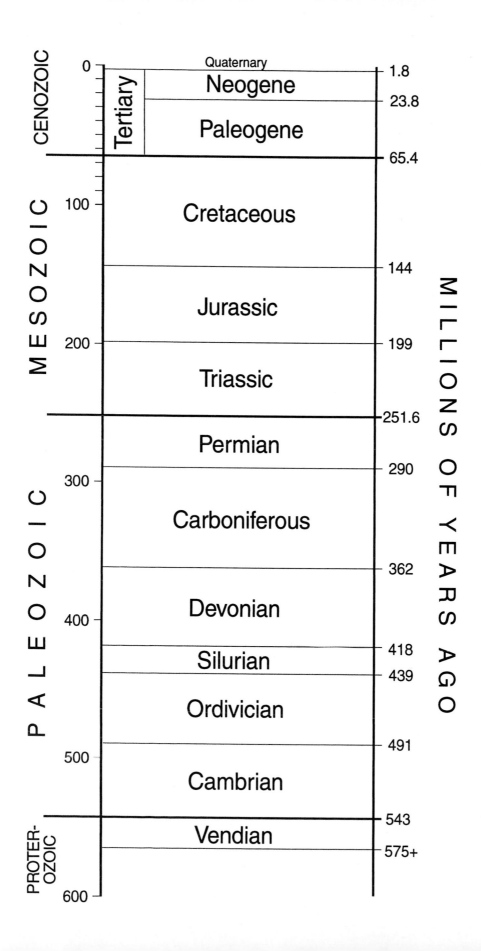